International Conference on Naval Architecture and Ocean Engineering (ICNAOE 2024)

E3S Web of Conferences Volume 606 (2025)

Beijing, China
22-24 November 2024

ISBN: 979-8-3313-1324-1

Printed from e-media with permission by:

Curran Associates, Inc.
57 Morehouse Lane
Red Hook, NY 12571

Some format issues inherent in the e-media version may also appear in this print version.

For additional information, please contact EDP Sciences – Web of Conferences
at the address below.

EDP Sciences – Web of Conferences
17, Avenue du Hoggar
Parc d'Activité de Courtabœuf
BP 112
F-91944 Les Ulis Cedex A
France

Phone: +33 (0) 1 69 18 75 75
Fax: +33 (0) 1 69 28 84 91

contact-edps@webofconferences.org

Additional copies of this publication are available from:

Curran Associates, Inc.
57 Morehouse Lane
Red Hook, NY 12571 USA
Phone: 845-758-0400
Fax: 845-758-2633
Email: curran@proceedings.com
Web: www.proceedings.com

TABLE OF CONTENTS

CARBON MANAGEMENT AND ENVIRONMENTAL PROTECTION TECHNOLOGIES

GREEN CONSTRUCTION AND INFRASTRUCTURE TECHNOLOGIES

RENEWABLE ENERGY APPLICATIONS AND EFFICIENCY ENHANCEMENTS

https://doi.org/10.1051/e3sconf/202560600001

Preface

It is with great pleasure that we present the proceedings of the 2024 International Conference on Naval Architecture and Ocean Engineering (ICNAOE 2024). This conference, held in Beijing, China, brought together experts, researchers, and practitioners from around the globe to exchange ideas, share innovations, and discuss challenges in the fields of naval architecture, ocean engineering, and beyond.

Naval architecture and ocean engineering play an essential role in addressing the complex challenges of modern maritime industries and the sustainable development of our oceans. However, as the global demand for clean and renewable energy continues to grow, the integration of new energy systems into maritime and oceanic applications has become a crucial area of focus. To reflect this evolving interdisciplinary collaboration, the ICNAOE 2024 proceedings encompass a diverse range of topics, including naval architecture, ocean engineering technologies, and advancements in renewable energy systems and sustainable solutions.

The proceedings include contributions that the development of advanced marine structures, offshore renewable energy platforms, wave and tidal energy systems, and environmentally friendly vessel designs. In addition, several papers address the application of renewable energy technologies, energy storage solutions, and their implications for sustainable ocean engineering practices. These papers highlight the intersection of traditional maritime expertise with emerging energy technologies, showcasing the transformative potential of interdisciplinary collaboration.

We extend our deepest gratitude to all the authors, reviewers, and organizing committee members who contributed to the success of this conference. Your hard work and dedication have made this event and its proceedings an invaluable resource for advancing knowledge and innovation.

We hope that the research presented in this collection will inspire further collaboration and progress in naval architecture, ocean engineering, and renewable energy development, fostering a sustainable future for the maritime and energy industries.

Statement of Peer review

In submitting conference proceedings to *Web of Conferences,* the editors of the proceedings certify to the Publisher that

1. They adhere to its **Policy on Publishing Integrity** in order to safeguard good scientific practice in publishing.
2. All articles have been subjected to peer review administered by the proceedings editors.
3. Reviews have been conducted by expert referees, who have been requested to provide unbiased and constructive comments aimed, whenever possible, at improving the work.
4. Proceedings editors have taken all reasonable steps to ensure the quality of the materials they publish and their decision to accept or reject a paper for publication has been based only on the merits of the work and the relevance to the journal.

2024 International Conference on Naval Architecture and Ocean Engineering (ICNAOE 2024), November 22-24, 2024, Beijing, China

Proceedings editor(s): Eugen Rusu

Date and editor's signature Eugen Rusu, November 22, 2024

17, avenue du Hoggar - PA de Courtabœuf – BP 112 - 91944 Les Ulis Cedex A (France)
Tél. : 33 (0)1 69 18 75 75 - Fax : 33(0)1 69 07 45 17 – www.edpsciences.org

Comparison of Ignition Control Techniques in HCCI Engines from Theoretical and Practical Perspectives

Jiaqi Yang[1*]

[1]Sendelta International Academy, 518108 Shenzhen, China

Abstract. Homogeneous Charge Compression Ignition (HCCI) is a cutting-edge combustion engine technology that merges the benefits of both diesel and gasoline engines. In HCCI engines, a uniform mixture of fuel and air enters the combustion chamber during the intake phase, which significantly lowers the average temperature within the chamber. This temperature reduction is crucial as it minimizes the formation of nitrogen oxides (NOx), pollutants that typically form at high temperatures, thus making HCCI engines more environmentally friendly. HCCI engines employ compression ignition, offering greater thermal efficiency than spark ignition engines, which are limited by the potential for knocking. These features make HCCI a promising technology for sustainable development. However, one of the main challenges preventing widespread adoption is the difficulty in precisely controlling the ignition timing. This paper aims to delve into the theoretical and practical aspects of ignition control methods in HCCI engines, highlighting the differences and challenges between theoretical models and real-world applications.

1 Introduction

Homogeneous Charge Compression Ignition (HCCI) engines were initially researched as a replacement for two-stroke engines. The initial research on this combustion process was carried out in 1979. In 1992, Stockinger introduced the first four-cylinder gasoline engine utilizing the HCCI mode [1]. HCCI engines are notable for their negligible emissions of NOx and soot because the auto-ignition technique in HCCI gives them a significantly lower average temperature in combustion. This offers significant potential for sustainable development [2]. Currently, numerous companies and individuals, including Toyota, Nissan, and Mercedes-Benz, are dedicating considerable effort to researching HCCI engines. However, controlling the timing of auto-ignition in HCCI engines is challenging., which is an obstacle to HCCI's wide application. The reason is that HCCI engines have neither spark plugs nor a fuel injector, which means HCCI does not have a direct control unit for auto-ignition timing [3]. Furthermore, there are theoretical and practical techniques for controlling auto-ignition. This essay will make a comparison between auto-ignition control techniques

* Corresponding author: jy3422@columbia.edu

in theory and reality based on research from reputable scientific sources of theoretical and practical auto-ignition control techniques in HCCI engines.

2 Basic working principle and unique superiorities of auto-ignition

Auto-ignition in HCCI engines occurs during the compression stroke. In this phase, the engine compresses a uniform mixture of fuel and air, which has been drawn in during the intake stroke. This compression process continues until the mixture reaches the point where combustion spontaneously occurs, known as the auto-ignition point. The auto-ignition of a homogeneous air-fuel mixture results in a relatively low average combustion temperature due to its inherently low ignition threshold. Furthermore, since NOx is produced in high-temperature combustion, auto-ignition generates negligible NOx and soot. Besides, low combustion temperatures can reduce heat transfer losses to the cylinder walls, which can improve thermal efficiency [4]. In general, auto-ignition has two unique advantages against conventional engines, which are low nitrogen oxide emission and high thermal efficiency (Fig. 1).

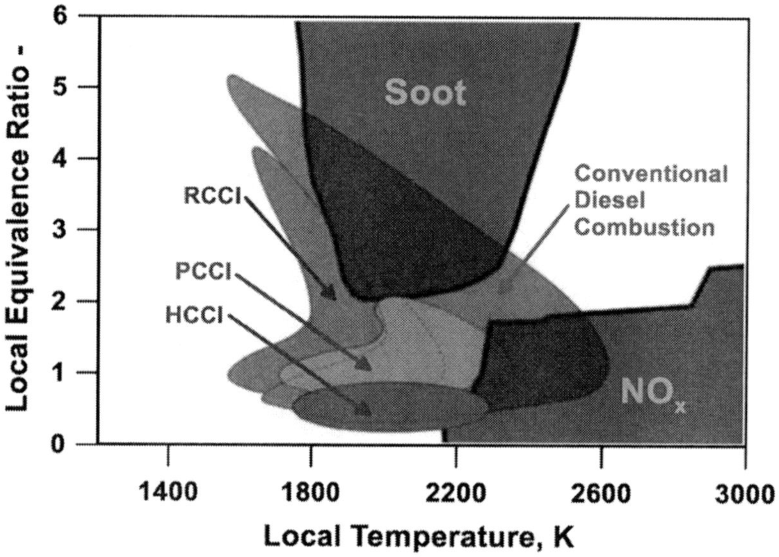

Fig. 1. Comparison of NOx emissions from HCCI engines and conventional engines [5].

3 Key challenges to control auto-ignition

However, there is difficulty in controlling the timing of ignition in HCCI engines. A key challenge is the lack of direct ignition control, which is typically managed by a spark plug in conventional SI engines. Moreover, diesel engines use fuel injectors to control ignition. However, there is a direct ignition control mechanism within HCCI engines that relies on auto-ignition., as shown in Fig. 2. Besides, auto-ignition timing in HCCI is sensitive to operating conditions, which are temperature, pressure, and mixture composition. Even tiny variations in these conditions can cause significant changes in timing control, making it hard to achieve consistent and precise control.

Further, in HCCI, the mixture of fuel and air ignites almost simultaneously, and this will lead to a rapid change in temperature and pressure, even knocking. Also, different fuel has various chemical properties, so control ignition strategies are various as changes in fuel [3].

Thus, the lack of a direct ignition mechanism and high sensitivity to in-cylinder conditions, variability of fuel, and rapid and simultaneous combustion are the main challenges to controlling auto-ignition in HCCI engines.

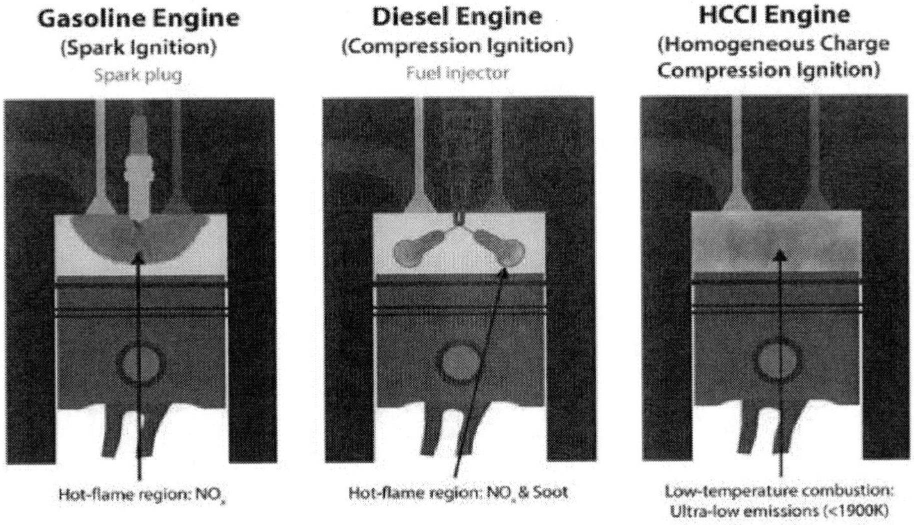

Fig. 2. Comparison of ignition mechanism of SI, CI, and HCCI engines [6].

4 Key challenges to control auto-ignition

4.1 Fuel management

Fuel management is a primary method for theoretical auto-ignition control, encompassing both dual fuel systems and staged fuel injection [7]. This approach involves the strategic regulation and delivery of fuel to optimize the ignition process. Dual fuel systems combine different types of fuel to enhance combustion efficiency and performance, while staged fuel injection involves injecting fuel at multiple stages to better control the combustion process and reduce emissions. Together, these techniques offer a comprehensive strategy for managing auto-ignition and improving overall engine performance.

4.1.1 Dual fuel systems

In a dual fuel system, two distinct types of fuel with varying chemical characteristics are utilized concurrently. This method involves adjusting the fuel's reactivity to simplify the control of auto-ignition [8]. To be specific, in a dual fuel system that uses gasoline and diesel, diesel's high cetane number causes early ignition, while gasoline's high octane number delays ignition. This can simplify the control of the ignition timing. However, the dual fuel system has a complex structure (Fig. 3), making it hard to be widely applied.

4.1.2 Staged fuel injection

Staged Fuel Injection is a technique where fuel is introduced into the engine cylinder at multiple intervals rather than all at once. This method aims to enhance the control over the temperature and uniformity of the air-fuel mixture inside the cylinder. By doing so, it allows for more precise management of the ignition timing and improves the stability of combustion, particularly in HCCI engines. This controlled approach helps in achieving better combustion efficiency and reduces the likelihood of knocking, thereby optimizing the engine's overall performance. In an experiment that compares HCCI engines with direct fuel injection and secondary injection, the HCCI engines that have a secondary injection have the lowest probability of experiencing the knock [9].

4.1.3 Limitations of fuel management

Although dual fuel systems and staged fuel injection have the potential to simplify auto-ignition control and reduce the probability of knocking, there are still obstacles to their application. For both approaches, the demand for complicated hardware and control strategies significantly increases their cost and system complexity, making them hard to be widely used.

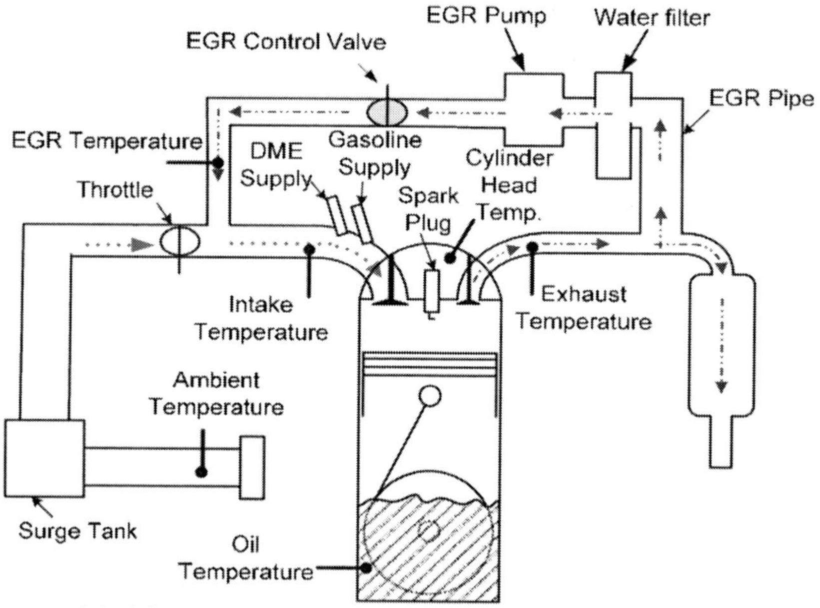

Fig. 3. Structure of dual fuel system in HCCI engines [10].

4.2 Controlling the temperature evolution

Temperature evolution refers to managing the in-cylinder temperature evolution, this allows a more accurate control of the timing of the auto-ignition process. A well-controlled temperature change can enhance the uniformity of the mixture within the cylinder, which can adjust the delay of ignition timing and reduce the likelihood of knock. However, this method has a narrow operating range due to the high sensitivity to temperature change in auto-ignition. In addition, accurate temperature control requires complex control systems with advanced sensors and actuators, which significantly increases the cost. Besides, it is very difficult to control the in-cylinder temperature at high loads, so a temperature evolution system will cause a limited cylinder capacity. Overall, these difficulties make it hard to apply this theory broadly.

5 Practical methods to control auto-ignition

5.1 Modifying the gas temperature

In HCCI engines, the control of gas temperature can be effectively managed through the use of Variable Compression Ratio (VCR) and Variable Valve Timing (VVT) systems. These technologies play a crucial role in regulating the timing and characteristics of the combustion process. By adjusting the compression ratio and valve timing, VCR and VVT can optimize the combustion phase, ensuring efficient fuel combustion and reducing emissions. This precise control helps maintain the desired temperature range within the engine, improving overall performance and efficiency. Additionally, VCR allows for adjusting the compression ratio to address issues related to premature auto-ignition. However, the efficiency will decrease as the compression ratio decreases. Nonetheless, generally speaking, Variable Compression Ratio can undoubtedly expand the operation ratio of auto-ignition [9].

5.2 Controlling the intake charge temperature

The temperature of the intake mixture temperature in one of the main factors in controlling HCCI engines. One major way of changing the Intake charge temperature to adjust ignition timing is to heat the intake of air. Higher intake charge temperature can also reduce the duration of combustion and adjust the delay of auto-ignition. An experiment was carried out to assess how intake air temperature affects engine performance and emissions. The results highlighted a crucial discovery: when the intake air temperature surpasses 90 degrees Celsius, significant changes were observed. Specifically, emissions of CO and HC began to decrease. However, this method has the potential to increase the probability of knocking.

5.3 Altering the intake charge composition

One common method for altering the intake charge composition in HCCI engines is through the use of exhaust gas recirculation (EGR). While EGR is typically employed in conventional diesel engines to lower nitrogen oxide emissions, in HCCI engines, it serves a dual purpose. EGR plays a crucial role in controlling combustion timing and mitigating auto-ignition issues by reintroducing exhaust gases into the intake manifold. This recirculation can slow down the rate of heat release, thereby reducing the chances of engine knocking [9]. Moreover, EGR is known for its ability to lower emissions, particularly NOx and PM. It is worth noting that HCCI engines naturally emit lower levels of NOx and particulate matter compared to traditional diesel engines, even in the absence of EGR systems.

6 Conclusion

HCCI engines are an emerging technology with the advantage of significantly lower emissions of NOx and particulate matter compared to traditional internal combustion engines. Further, they offer higher thermal efficiency. However, one of the main challenges with HCCI engines is controlling the timing of the auto-ignition process. This difficulty arises due to the lack of a direct ignition system, the engine's high sensitivity to in-cylinder conditions, variations in fuel properties, and the rapid, simultaneous combustion process. Additionally, the difficulty in managing the timing of auto-ignition presents a significant barrier to the widespread adoption of HCCI engines. To deal with this issue, scientists have developed numerous theoretical and practical approaches. Strategies, in theory, include fuel management and control of temperature evolution.

Nevertheless, both strategies have the disadvantages of high demand for complex equipment and narrow operating range, so it is hard to apply these techniques broadly. However, practical strategies of auto-ignition controlling techniques in HCCI can solve the problems that theoretic methods face. They can adjust the timing of auto-ignition does not have a high demand for complex mechanisms. Moreover, does not have narrow operating conditions, especially VCR, which can extend the operation ratio of HCCI engines. However, there is still space for improvement in practical methods. VCR technology can lead to a reduction in thermal efficiency when the compression ratio is decreased. HCCI engines, managing the intake charge temperature is crucial because these engines are highly sensitive to combustion conditions, which can increase the risk of knock. Future advancements may focus on developing precise and cost-effective sensors to regulate auto-ignition in HCCI engines. Additionally, artificial intelligence could be implemented to monitor in-cylinder conditions and optimize the timing of combustion in these engines.

References

1. Telli, G., Altafini, C., Costa, C., et al. A comprehensive review of homogeneous charge compression ignition (HCCI) engines: Advantages, challenges, and evolution. SAE Technical Papers (2021)
2. He, Z., Yang, X., & Zhang, P., HCCI Engine: Advancements in combustion control and potential innovation in power configuration for vehicles. Highlights in Science, Engineering and Technology, **88**, 821 (2024)
3. Duan, X., Lai, M., Jansons, M., et al. A review of controlling strategies of the ignition)timing and combustion phase in homogeneous charge compression ignition (HCCI) engine. Fuel, **285**, 119142 (2021)
4. Iida, N., Yamasaki, Y., Sato, S., et al. Study on auto-ignition and combustion mechanism of HCCI engine. SAE Transactions, **113**, 1977 (2004)
5. De Bellis, V., Malfi, E., Lanotte, et al. A tabulated chemistry multi-zone combustion model of HCCI engines supplied with pure fuel and fuel blends. Energies, **16**, 265 (2022)
6. Minh, T., Minh, H., & Duy, V., A review of internal combustion engines powered by renewable energy based on ethanol fuel and HCCI technology. AIMS Energy, **10**, 1005 (2022)
7. Veza, I., Said, M., Latiff, Z., et al. Strategies to form homogeneous mixture and methods to control auto-ignition of HCCI engine. International Journal of Automotive and Mechanical Engineering, **18**, 9253 (2021)
8. Aldawood A, Mosbach S, Kraft M. HCCI combustion control using the dual-fuel approach: Experimental and modeling investigations. SAE Technical Paper (2012)
9. Wang, Z., Shuai, S., Wang, J., et al. A computational study of direct injection gasoline HCCI engine with secondary injection. Fuel, **85**, 1831 (2006)
10. Wu, Y., Chen, B., & Wang, J., Application of HCCI engine in motorcycle for emission reduction and energy saving. Aerosol and Air Quality Research, **15**, 2140 (2015)

E3S Web of Conferences **606**, 01002 (2025) https://doi.org/10.1051/e3sconf/202560601002
ICNAOE 2024

Critical Development of Electric Fixed-Wing Manned Civil Aircraft

Yifei Yu[1]*

[1]Western Canada High School, T2S 0B5 Calgary, Canada

Abstract. This paper briefly introduces the current role of civil aviation in the emission of greenhouse gases and the history of electric aviation, pointing out the necessity of developing electric civil aircraft to reach strict environmental policies such as the EU flightpath 2050. The four common architectures for electric airplanes, which are the all-electric, series hybrid electric, parallel hybrid electric and turboelectric, are briefly explained. This paper explains their respective construction, advantages and disadvantages, and technical feasibility and gives some examples of developing projects of all-electric and turboelectric architecture. The major problem for large commercial e-aircraft, energy density, is also discussed and the energy density required for various structures to achieve the highest efficiency is given. For several small business and regional aircraft that have been put into use or completed test flights, this paper explores their economics and whether their current cost is competitive with similar conventional gas turbine aircraft.

1 Introduction

In 2019, commercial aviation contributed 920 Mt of CO_2 to global emissions, 85% of which came from passenger aircraft [1]. Considering the increase in population income and the decrease in flying costs, the number is likely to grow in the future. However, emission reduction is a very hot topic now due to environmental and political factors, which is also the case in aviation. European Union, for example, has already proposed its own aviation development goal, aiming to reduce 75% of CO_2 emissions and 90% of NO_x emissions of aircraft by 2050, with a midway goal to reduce 55% of greenhouse gas emissions by 2030 [2]. In terms of current technology, within a reasonable budget, electrification is the only approach to achieving the goal. Though there is no electric passenger aviation in use now, it is a very promising area attracting investments and discussions from large companies and organizations worldwide, including NASA, Boeing, Airbus, Rolls-Royce, etc. [3]. There are 3 ongoing large commercial electric aircraft projects, with more business/regional aircraft projects [4].

The electrified plane is not totally a new thing. The first electric flying model was introduced in 1957 [5]. The first application of electric fixed-wing manned aircraft was produced in 1971, yet it does not have the self-launching ability, which means that it cannot

* Corresponding author: alanyu060929@berkeley.edu

take off with its power. Most of the commercial electric fixed-wing aviation in use now are also motor gliders. Another kind of electric aircraft is called solar-powered aircraft, usually carrying lots of solar panels and a small amount of batteries as a supplement.

An example is the Solar Impulse produced by Swiss [6]. However, due to the limitation of battery, namely, the limitation of energy density, none of the existing electric planes can carry multiple passengers at a reasonable cost. To achieve the goal of 75% CO_2 emission reduction, there must be a new architecture for electric aviation.

2 Operation architecture

Researchers have proposed four technical ideas for electric aircraft. This paper will briefly introduce each architecture and then focus on all-electric and turboelectric architecture, which is the most common architecture among electric projects.

2.1 All-electric

In all-electric architecture, the entire aircraft propulsion comes from electric fans, and the battery is the only source of power. As the all-electric architecture does not consume any fuel, it is the only option capable of achieving 0 emissions. It is highly viable and cost-efficient in terms of propulsion systems. Most of the VTOL aircraft projects choose all-electric as their approach. Nonetheless, the feasibility is highly dependent on the energy density of the battery [6,7].

As a reference, common mobile lithium-ion batteries have a specific energy range from 150-250 Wh/kg, and the advanced LLZTO solid electrolyte batteries can reach a specific energy of up to 600 Wh/kg [8]. Therefore, as shown in Table 1, it is not surprising that most of the ongoing all-electric projects are small regional/business aircraft that are not highly dependent on energy density. On the other hand, to compete with conventional aircraft, in terms of construction cost, the price for batteries should be below USD 100 per kWh [6].

Table 1. Current conceptual design/ongoing project for all-electric aircraft [9]

Project	Aviation type	Technical details
Ampaire Tailwind	Regional aircraft	93 kt cruise speed 10000 FT cruise alt 5000 kg MTOW 9 passenger capacity 161 km range
Eviation Alice	Business aircraft	220 kt cruise speed 32000 FT cruise alt 6350 kg MTOW 9 passenger capacity 814 km range A prototype has finished test flight
Tecnam P-Volt	Regional aircraft	9 passenger capacity
Heart Aerospace ES19	Regional aircraft	180 kt cruise speed 8600 kg MTOW 19 passenger capacity 400 km range
Wright Electric/Easy Jet	Commercial aircraft	186 passenger capacity 540 km range

2.2 Series hybrid electric architecture

In series hybrid electric architecture, the motor-driven fans are still responsible for all of the propulsive power and will also carry a battery as the source of power. The difference between series hybrid and all-electric is that the former will be equipped with a gas turbine, which is designed to be capable of operating independently from the electric system yet will not provide thrust to the aircraft. Instead, the gas turbine will provide electricity for the battery through a generator and converter. Series hybrid electric architecture is a transitional design for all-electric architecture. As the specific energy of batteries is still not high enough to support the aircraft with a reasonable range alone, traditional fuel is needed.

2.3 Parallel hybrid electric architecture

In parallel hybrid electric architecture, gas turbines become the major provider of propulsive power. Batteries will still be equipped with generators and converters to charge the battery, but the electric propulsive system will only play an auxiliary role, which will help reduce fuel consumption during the cruise phase and improve propulsion in low-speed situations. In some designs, the electric motor and the gas turbine are integrated into the same engine module, saving more space.

2.4 Turboelectric architecture

The turboelectric architecture does not have a battery. Instead, a gas turbine will provide all electricity for electric fans, which in turn will provide propulsive power for the plane. This design has a very low technical requirement, and due to the absence of a battery, it can be very light compared to other architectures. Therefore, it is the most common design among large commercial aircraft, as shown in Table 2.

Compared to traditional turbines, the turboelectric architecture is allowed to exploit the benefit of geared turbofan engines fully. For example, in the traditional engine, the gas turbine and fan have to work at the same rate. While high rotation speed allows the gas turbine to output energy fully, it may cause some problems for the fan. The excessive rotation speed will cause lower efficiency, especially when the speed is close to the speed of sound, where the shockwave begins to form which will significantly affect the aerodynamic performance. However, for turboelectric architecture, the rotation speed for the turbine and fan is decoupled, which allows both to work at their optimal rate.

Also, as the number of turbines and fans is decoupled, the turboelectric architecture can achieve a much higher bypass ratio. A high bypass ratio engine offers several advantages in aviation like higher fuel efficiency and reduced noise [10]. According to NASA, with advanced aerodynamic design, turboelectric architecture can achieve a 70% reduction in fuel burn compared to traditional turbines [11].

Another derivative of turboelectric architecture is partial turboelectric architecture. The turbine will not only provide electricity but also directly provide propulsive power for the aircraft.

Table 2. Current conceptual design/ongoing project for turboelectric aircraft [9,11]

Project	Type	Technical details
EAG Hybrid Electric	Commercial aircraft	70 passenger capacity 1482 km range
NASA STARC-ABL	Large commercial aircraft	0.8 Mach cruise speed 12 percent fuel burn reduction Partial turboelectric

Boeing SUGAR Freeze	Large commercial aircraft	56% fuel burn reduction 1450 km range Partial turboelectric-fuel cell
NASA N3-X	Large commercial aircraft	70% fuel burn reduction Blended wing body design Full turboelectric based on superconducting electric machines

3 Energy density

One of the greatest challenges of electric aviation is the energy density. As far as the existing technology is concerned, even though the thermal efficiency of motors is way higher than gas turbines, the overall energy density for batteries is still much lower than traditional engines. As the take-off mass is limited, which means that the amount of battery is restricted, the energy density is critical to the range of the aircraft, a vital factor when competing with traditional aviation. Another potential weight problem is that the weight of the battery, unlike fuels, will not change during the flight, resulting in additional weight compared to traditional aviation. Take the Boeing 777-200 as an example. The maximum take-off weight (MTOW) is 545,000 lbs., while the maximum fuel capacity is 31,000 gallons (about 207,700 lbs.). The fuel weight accounts for 38% of the total take-off weight. 'Fuel that does not lighten' will burden the electric aircraft heavily. Thus, it is not surprising that, based on existing gas turbine and motor technology, for large commercial aircraft similar in size to the Boeing 777-200, the turboelectric architecture (the architecture without a battery) is the most weight-efficient. Low electric thrust ratio architecture, for instance, parallel hybrid, will become competitive from 3 to 3.5 kWh/kg [12]. The requirement for smaller aircraft, namely regional aircraft and business aircraft, will be relatively low, as they do not have long-range requirements, and the battery only accounts for a small portion of aircraft weight, alleviating the problem of unchanged weight to some extent.

4 Cost and competitiveness

The general cost of electric aircraft is a significant factor for its market. Considering that there is no large commercial electric aircraft that has completed a test flight, the comparison between large traditional/electric aircraft will be to theoretical that there is no practical significance. However, comparison for regional aircraft is possible now, which also plays a notable role in civil aviation, especially in low population density areas such as Northern Europe. Electric aircraft will have a huge advantage in terms of fuel and maintenance cost, yet due to the lack of mass production, first-generation electric aircraft usually have higher built costs compared to similar-size traditional aircraft. Also, a recognizable disadvantage is that electric planes have to change batteries in their service life, which will cause extra spending. Based on the regional electric aircraft project that has completed test flights, there is an essential difference between traditional aviation and electric aircraft: for traditional aircraft, the cost per available seat km (ASK, an indicator often used to determine the cost-efficiency of aircraft) will become lower when the seating capacity becomes greater; for the electric aircraft, the ASK will reach the bottom point when the seating capacity is about 20, and raise again when the capacity is greater than 20. This is due to the immature technology in larger electric aviation, and with the development of technology, it is likely to be reduced in the future. Right now, the ASK for first-generation electric aircraft (0.72-0.24 USD per ASK) is still higher than traditional regional aircraft (0.45-0.20 USD per ASK). If a price reduction of 25% can be achieved, electric aircraft will be able to compete with traditional regional aircraft in some markets [1].

5 Conclusion

Although most concepts are still in the experimental stage, a considerable number of civilian electric aircraft have completed test flights, with many more ongoing projects covering areas from eVTOL to regional aircraft to large commercial aircraft. NASA claimed that their advanced design will enter service before 2050, while EAG said that their 100-seat commercial aircraft will finish text flight by 2025. However, in the foreseeable future, limited by battery technology, the all-electric architecture is only applicable for small aircraft with less than 50 seats, the battery of which accounts for a relatively small amount of mass so that the specific energy is not that critical compared to large commercial aircraft. Achieving 0 emissions in the field of civil aviation is still a distant goal without substantial breakthroughs in the specific energy density of batteries and superconductor technology. Nonetheless, even with the turboelectric architecture, the design closest to traditional gas turbines, it is possible to reduce CO_2 and NO_x emissions to a great extent.

On the other hand, reducing cost is also a critical aspect for electric aircraft that are already in use and that will be in use when competing with traditional aviation. This will be reflected in two aspects: cost per unit mass for batteries and maintenance cost. The development of battery materials will also have a great impact on electric aircraft. As nations are setting strict requirements on future emission reduction, electric aviation is attracting more and more attention and investment. The author is also looking forward to further development of e-aircrafts.

References

1. Avogadro, N., & Redondi, R., Demystifying electric aircraft's role in aviation decarbonization: Are first-generation electric aircraft cost-effective? Transportation Research Part D: Transport and Environment, **130**, 104191 (2024)
2. European Commission. *Reducing Emissions from Aviation*; on-line documentation; European Commission: Brussels, Belgium, 2021;
3. Wheeler, P., Sirimanna, T. S., Bozhko, S., & Haran, K. S., Electric/hybrid-electric aircraft propulsion systems. Proceedings of the IEEE, **109**, 1115 (2021)
4. ICAO Environment, available online: Electric and Hybrid Aircraft Platform for Innovation (E-HAPI) (ICAO.int)
5. Cardwell, Diane (1 May 2013). "Cross-Country Solar Plane Expedition Set for Take-off". The New York Times. Retrieved 2 May 2013.
6. Viswanathan, V., & Knapp, B. M., Potential for electric aircraft. Nature Sustainability, **2**, 88 (2019)
7. Gnadt, A. R., Speth, R. L., Sabnis, J. S., & Barrett, S. R., Technical and environmental assessment of all-electric 180-passenger commercial aircraft. Progress in Aerospace Sciences, **105**, 1 (2019)
8. Yang, M., Fu, L., Luo, Z., et al. High specific energy $Li_7La_3Zr_2O_{12}$ solid electrolyte based thermal battery. Journal of The Electrochemical Society, **168**, 120551 (2021)
9. ICAO environment. https://www.icao.int/environmental-protection/Pages/electric-aircraft.aspx

10. Brelje, B. J., & Martins, J. R., Electric, hybrid, and turboelectric fixed-wing aircraft: A review of concepts, models, and design approaches. Progress in Aerospace Sciences, **104**, 1 (2019)
11. Jansen, R., Bowman, C., Jankovsky, A., et al. Overview of NASA electrified aircraft propulsion (EAP) research for large subsonic transports. In 53rd AIAA/SAE/ASEE joint propulsion conference, **4701** (2017)
12. Gimenez, F. R., Mady, C. E. K., & Henriques, I. B., Assessment of different more-electric and hybrid-electric configurations for long-range multi-engine aircraft. Journal of Cleaner Production, **392**, 136171 (2023)

E3S Web of Conferences 606, 01003 (2025) https://doi.org/10.1051/e3sconf/202560601003
ICNAOE 2024

Electric Aviation Development and Future Prospect

Zhoujinpeng Li[1]*

[1]North America International School, 201102 Shanghai, China

Abstract. Electric vehicles have gained significant traction, with hybrid and fully electric cars and trucks increasingly replacing traditional fuel-powered vehicles. However, the development of electrified aircraft has received comparatively less public attention. This research study aims to explore the evolution of electric aviation and its future potential. It delves into the history of electric aviation and examines various types of electric propulsion systems, including all-electric, series hybrid, parallel hybrid, and turboelectric configurations. The study also analyzes the advantages and disadvantages of current electrified aircraft technologies. Additionally, it considers the policies and investments necessary to stimulate the advancement of electric aviation. As concerns over environmental issues and energy shortages grow, electrified aircraft could emerge as a viable alternative to gas turbine aircraft, offering reduced carbon emissions and lower operational costs. The purpose of this paper is to provide a comprehensive overview of the current state and future prospects of electric aviation, with a particular emphasis on its potential to revolutionize the aviation industry.

1 Introduction

The electrification of aircraft began much earlier than most people expected on the 12th of Oct 1973. The inaugural flight of an electric-powered aircraft occurred in Wels, Austria, the Militky-Brditschka Elektroflieger No.1. The flight lasted approximately nine minutes and reached an altitude of approximately 300 meters [1] (Fig. 1). However, after MB-E1, the development of electrified aircraft stagnated. Thirty years later, some companies realized the value of the electrification of aircraft, and development started again. The electrification of vehicles is more environmentally friendly and efficient, and carbon emissions and toxic gas emissions can also be reduced significantly. At the same time, electric vehicles follow the trend of environmental protection and the public's perception of some inevitable defects of traditional fuel carriers. Thus, using electrified vehicles to replace traditional fuel vehicles has become one of the most viable options. As the battery and charging technology improved enormously, lots of electric vehicles entered the market to replace traditional fuel vehicles [1].

Similar to electrified vehicles on the ground, electric machines for aircraft have different types of architectures, including all-electric, hybrid-electric, and turboelectric [2]. All of

* Corresponding author: gjiud@ldy.edu.rs

these structures have their advantages and disadvantages in different aspects of aircraft. At the same time, some barriers are still hard to overcome, like the weight of electric components and battery packs. This research provides insight into the architecture of the electrification of aircraft and the prospect of electrified aircraft.

Fig. 1. MB-E1 (Militky-Brditschka Elektroflieger No.1) [1].

2 Different architectures of electric machines

In turbine engines, the turbine's speed and the propeller's rotational speed are linked. [3]. This coupling is a barrier to thermal efficiency, so electrified vehicles are used to overcome this defect and improve overall thermal efficiency. The fan and turbine speeds in direct current systems are disconnected, allowing each to be run at its optimal operating section. [3]. Similar to internal combustion engine vehicles and their electrification, electric machines for aircraft have full-electric and hybrid architecture. This is because the energy contained in the battery is much smaller than energy in the same volume of fuel. Thus, the hybrid electric machine is designed to reduce weight and have a higher thermal efficiency compared to gas turbine and all-electric architecture. All these electric machine architectures can be classified into four types: all-electric, series hybrid, parallel hybrid, and turboelectric [2]. In this research study, all these architectures will be introduced from the perspective of structures and advantages.

2.1 All-electric architecture

Fig. 2 shows the full-electric architecture. Full-electric architecture is expected to be the most economical and viable choice to reduce carbon emission and toxic gas release for both fixed-wing aircraft and vertical take-off and landing aircraft [2]. Aircraft (plural) have to carry a battery, and the battery's weight is the main problem all-electric aircraft have. Batteries store less energy in the same volume compared with fuel, so aircraft using this architecture always have a relatively short range. After the battery generates current, the current goes through the converter to the motor, and then one or more propellers will rotate to produce force for taking off and flying. For flight, mission ranges that are less than 600 nautical miles, NOx emissions at the take-off area can be remarkably reduced by 40%, and direct CO_2 emissions can be reduced by 15% [2]. In short, all-electric architecture is the most environmentally friendly structure, as the battery-specific energy increases can be used in a wider working environment.

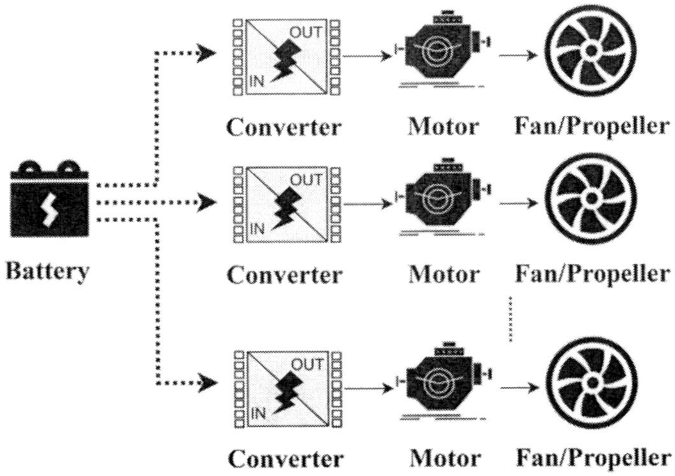

Fig. 2. All-electric architecture [2].

2.2 Serious hybrid electric architecture

The configuration of the series hybrid electric architecture is illustrated in Fig. 3, and it uses a single gas turbine to generate electricity, and electricity is stored in the battery on the aircraft [2]. Then, it is the same as all-electric architecture. The motors drive some small propellers. Either an energy storage device or a generator powered by a gas turbine provides the electrical power. With this configuration, the transmission system and electrical conversion enable the gas turbine system to separate from the propulsors [4]. The ability to run the gas turbine independently of propeller speed is a built-in advantage of the serious hybrid electric architecture [4]. The thermal efficiency of the gas turbine can reach its maximum efficiency because the power transmission system and the electric power generation system are separated [2]. Hybrid electric architecture can store the electric energy in a battery to supplement the propulsive power in different operational segments [4]. Compared with all-electric aircraft, hybrid architecture uses fuel as a source of energy. Because of the higher fuel energy capacity, aircraft can store a small battery. Thus, serious hybrid aircraft can be lighter and have a longer flight range.

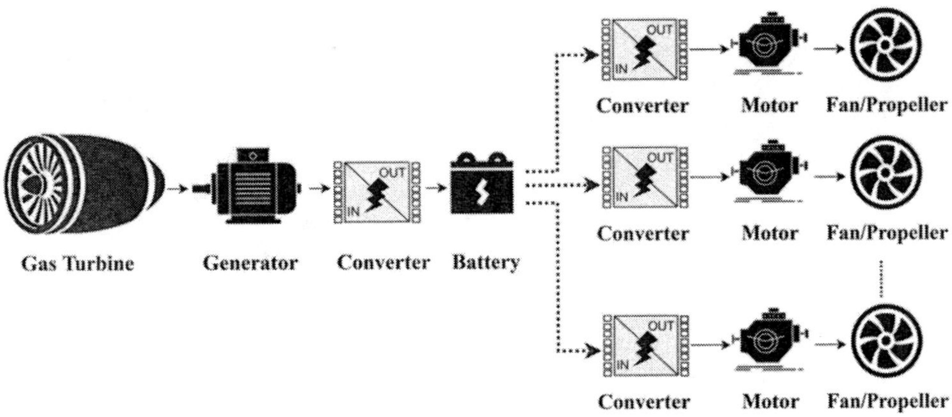

Fig. 3. Serious hybrid electric architecture [2].

2.3 Parallel hybrid electric architecture

As shown in Fig. 4, in the parallel hybrid electric architecture, gas turbines are connected parallelly to produce electricity and store it in the battery. Both motors and gas turbines are stored on a shaft to drive the propeller [2]. Low-pressure and high-pressure shafts can the electric power boost both on-take [2]. Compared with serious hybrid electric architecture, the parallel structure has fewer electric components, so it will save weight and space for the aircraft [2]. This is a significant advantage for the aircraft, and it will make the flight more economically competitive.

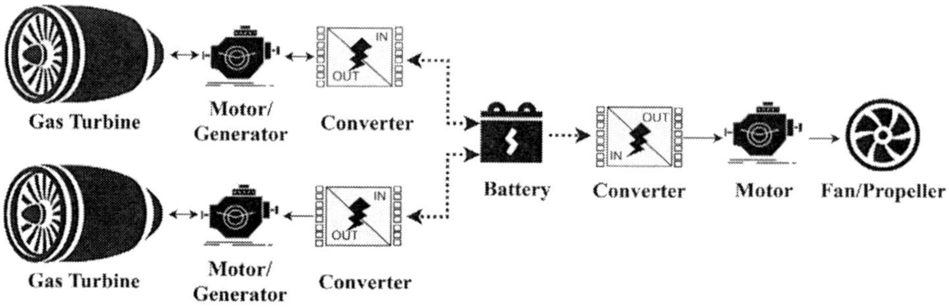

Fig. 4. Parallel hybrid electric architecture [2].

2.4 Turboelectric architecture

As illustrated in Fig. 5, the turboelectric configuration comprises a gas turbine that drives a generator, thereby producing electricity for each motor. The electricity is transmitted to the individual motors and subsequently drives the propellers [2]. An electricity storage package is not included in this configuration, and turboelectric aircraft use fuel as their only energy resource. This structure is the most promising architecture because it does not relate to the battery energy storage technology [2]. In addition, the volume of electric components it needs is smaller due to the movement of the battery pack. Thus, it can also reduce weight to have better flight economy efficiency.

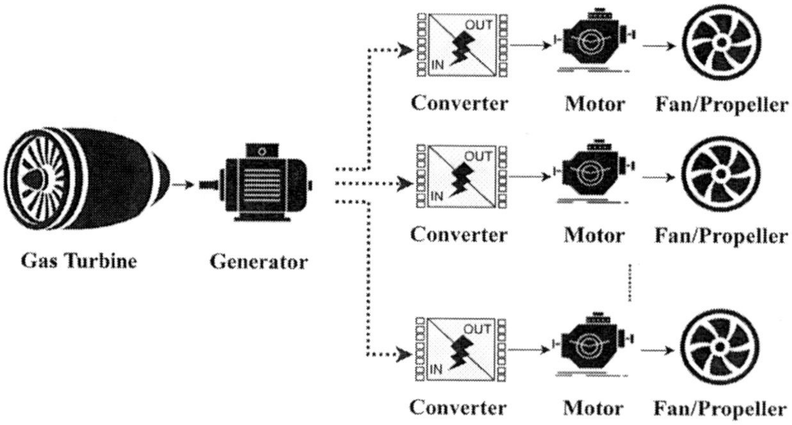

Fig. 5. Turboelectric architecture [2].

3 Merits and demerits of electrified aircraft compared with gas turbine aircraft

3.1 Advantages

3.1.1 Thermal efficiency

Electrified aircraft have a higher thermal efficiency compared with gas turbine aircraft. Electric machines are necessary for all electrified aircraft, and researchers are trying to build up a lightweight, high-efficiency electric machine [5]. The large airliner needs megawatt-level components because they can be synchronously stored in different architectures like turboelectric and hybrid electric [5]. Smaller aircraft can use hybrid electric and turboelectric propulsion systems, whereas single-aisle airplanes can use a combination of both forms of propulsion. [5].

NASA is sponsoring Ohio State University and the University of Illinois to develop megawatt-class motors, and their efficiency goal of the electric machines is all over 95% [5]. However, the production process of electricity has some energy loss, so the thermal efficiency of electricity production is about 65%. This needs to be considered when this research is calculating the electric machine's total thermal efficiency. In comparison, research from the National Academies Press states that turbine thermodynamic efficiencies of no more than 55% and propulsive efficiencies of approximately 70% define the most efficient airplane gas turbines in use or under development in the past ten years [6]. From the product of two efficiencies, these gas turbine aircraft have an overall efficiency of about 40% [6]. Gas turbine aircraft cannot always work in a zone with the highest thermal efficiency, but electric machines can. The aggregate efficiency of the electric machine, defined as the product of the electric machine's thermal efficiency and the efficiency of electricity production, is markedly superior to the thermodynamic efficiency of the gas turbine.

3.1.2 Pollution

Reducing carbon emissions and meeting the trend of environmental protection are the main reasons why electrified aircraft have developed. The two principal forms of air pollution emitted by gas turbines are smoke and nitrogen oxides at maximum speed and unburned fuel and carbon monoxide at idle working conditions [7]. Traces of smoke are a visual nuisance, and the other pollutants are either toxic, irritating, or both [7].

Electric machines have almost no pollution when they are working compared with gas turbines; the only pollution electric machines have is from the carbon emission when electricity is produced and the used battery package. However, more and more companies have launched their battery recycling plan because some materials in the battery are still valuable–for example, lithium, cobalt, manganese, and nickel cells [8]. In this way, recycling batteries will greatly reduce the impact on the plant-growing environment due to toxic chemicals in the batteries. At the same time, firms can reduce the cost of materials to increase their profit and develop sustainably.

3.2 Disadvantages

3.2.1 Convenience

Lots of private gas turbine aircraft use different designs to reduce drug force, and the goal of these designs is to have better flight performance. To be more specific, some private aircraft can even fly cross-ocean airlines, and they have an outstanding maximum cruise speed and maximum altitude. As shown in Fig. 6, G650 from Gulfstream has a maximum flight range of 7000 nautical miles, about 13000 km. This range is enough to have non-stop flights from

Europe to North America, Europe to Africa, or Europe to much of Asia. At the same time, the G650 has a long-range cruise speed of 488 knots and a maximum cruise speed of 516 knots. It also has a maximum cruising altitude of 51,000 feet, which is placed at the joint top of the list for the highest private jet cruising altitudes. All these satisfying flight performances are because fuel has a high energy capacity.

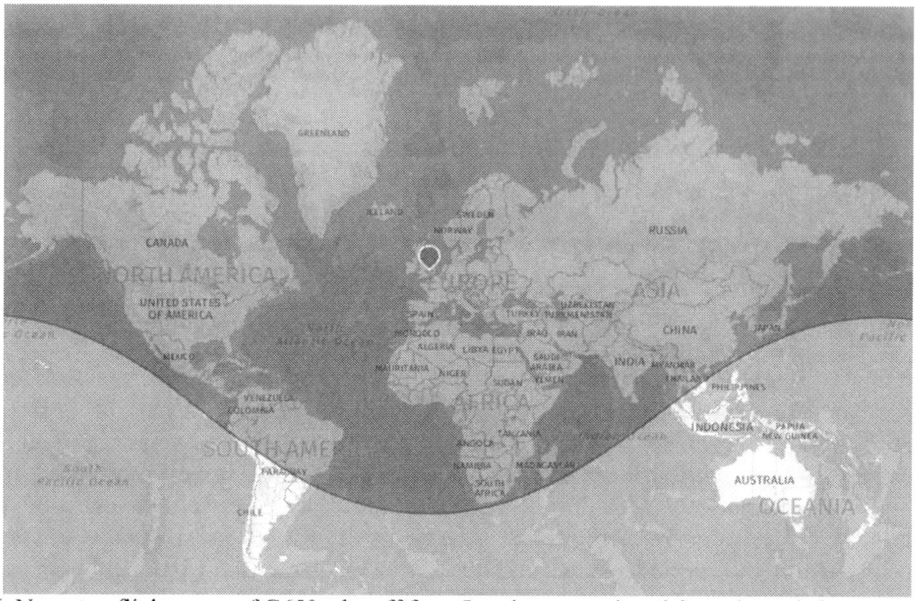

Fig. 6. Non-stop flight range of G650 take off from London, reproduced from the website: https://compareprivateplanes.com/listing/large/g650#performance.

Electrified aircraft have heavy battery packs and electrical components, so their weight is heavier than gas turbine aircraft with similar carrying capacity. Electrified aircraft Eviation Alice was first developed in Israel and is powered by two electric motor machines. It can carry 9 passengers and 2 pilots, similar to the passenger carrying capacity of G650. Nevertheless, it has a relatively small non-stop flight range and low cruising velocity compared to the G650. The flight range of Eviation Alice is only 460 km, about 4% of the maximum range of G650, and a maximum velocity of 260 knots.

Another problem is that electrified aircraft need to be changed, and the energy charging efficiency is very slow. For Eviation Alice, 30 minutes of charging time is required for each hour flight. This problem is another problem passengers will consider when they are planning for their flight.

3.2.2 Safety

Safety is always put in the first place for both passengers and an aircraft company. Compared with mature fuel gas turbine technology and structure. Therefore, if electrified aircraft want to be widely accepted by the public, electrified aircraft companies still need to solve potential safety hazards. One of the potential risks is the battery pack, which contains a huge amount of energy and is very unstable. Concurrently, the rechargeable Li-ion battery technology has reached a sufficient level of maturity, and the charge and discharge rates of these batteries have also been enhanced [9].

On the other hand, the incidence of safety related issues has also increased as battery performance improved and adoption enhanced [9]. The onset of overheating, heat accumulation and gas release process, and combustion and explosion are the thermal runaway has three stages [9]. The best solution to control the accident rate is to minimize the incidence of manufacturing defects [9].

4 Future prospect

Electrified vehicles are the main method of transportation in the foreseeable future. Many global organizations and governments are promoting the electrification industry, including the electrified aircraft manufacturing industry.

4.1 Global policies about electrified aircraft

The International Civil Aviation Organization, ICAO, officially launched the "ICAO Assistance, Capacity-building and Training for Sustainable Aviation Fuels (ACT-SAF) program" on 1st June 2022 [10]. ICAO aims to showcase what ICAO has achieved and continues to strive for in decarbonizing aviation [10]. Governments also expect to boost the employment rate and develop the economy through the development of new electrified aircraft industries, and they hope to stimulate technological development through release conditions for electric machine engines. On 27 September 2021, the Federal Aviation Administration (FAA) published Special Conditions for Electric Engine Airworthiness [11]. This was a special condition that was imposed on the first project for the certification of an electric motor. The condition represents a huge step forward in the commercial process [11].

4.2 Commercial value

The considerable quantity of energy necessary to transport hundreds of passengers over distances of hundreds or thousands of miles represents a significant challenge for the use of battery energy storage in commercial transportation. The disparity in energy density between batteries and hydrocarbon fuels implies that the range of an electrified aircraft will be considerably diminished in comparison to an equivalent aircraft powered by a gas turbine. [12]. Therefore, short-range commercial flights for Urban Air Mobility make it easier to achieve results in stages [12]. A number of companies have an outstanding performance in the commercial development of electrified aircraft. They have secured significant investments, built factories and developed new technologies in line with future market conditions. Archer is a company located in the U.S. It will provide short-range flights, which are called electric air taxis [13]. Archer received a supplementary investment of $55 million from Stellantis in accordance with the terms of their strategic funding agreement. This was subsequent to the attainment of the transition flight test milestone by Archer in the preceding month. [13]. Through this transaction, we can see the market's optimistic judgment on the development of electrified flying machines.

5 Conclusion

To sum up, the research concludes the different architectures of electric machines and summarizes the pros and cons of electrified aircraft. The existing advantages and disadvantages of electrified vehicles reveal that there are many barriers if electrified vehicles are going to be commercially available and have flying performance that completely exceeds that of conventional gas turbine aircraft. This is mainly due to the weight and size of the battery packs, which can severely reduce the range, cruise speed and carrying ability of the aircraft.

Fortunately, growing environmental concerns from the public are stimulating development and investment in the field of high-performance battery packs. As a result, technology in this area is developing at an extremely fast pace. In the future, there is a high probability that the battery energy capacity limit can be broken with the development in the field of materials science. Driven by Governments and the market, electrified aircraft will complete their commercialization and eventually replace gas-turbine aircraft in most situations. At the same time, electrified vehicles will also create new markets, such as cheap short-range flights, and continue to attract more investment.

References

1. E. Greenwood, K. Brentner, R. Rau, et al., Challenges and opportunities for low noise electric aircraft. Int. J. Aeroacoust. **21**, 1 (2022)
2. J. Zhang, I. Roumeliotis, A. Zolotas, Sustainable aviation electrification: A comprehensive review of electric propulsion system architectures. Energy Management, and Control. Sustainability **14**, 5880 (2022)
3. S. Sahoo, X. Zhao, K. Kyprianidis, A review of concepts, benefits, and challenges for future Electrical Propulsion-Based Aircraft. Aerospace. **7**, 44 (2020)
4. B. Brelje, J. Martins, Electric, hybrid, and turboelectric fixed-wing aircraft: A review of concepts, models, and design approaches, Prog. Aerosp. Sci. **104**, 1 (2019)
5. NASA Glenn Research Center. Electric Machines. November 16, 2022. Retrieved on July 15, 2024. Retrieved from: https://www1.grc.nasa.gov/aeronautics/eap/technology/electric-machines/
6. National Academies of Sciences, Engineering, and Medicine, Commercial Aircraft Propulsion and Energy Systems Research: Reducing Global Carbon Emissions (The National Academies Press, Washington, DC, 2016) https://doi.org/10.17226/23490
7. T. Durrant, The control of atmospheric pollution from gas turbine engines, SAE Tech. Pap. 680347 (1968)
8. Car and Driver. Everything You Need to Know about EV Battery Disposal. Jun 10, 2023. Retrieved on July 17, 2024. Retrieved from: https://www.caranddriver.com/features/a44022888/electric-car-battery-recycling/
9. S. Sripad, A. Bills, V. Viswanathan, A review of safety considerations for batteries in aircraft with electric propulsion. MRS Bull. **46**, 435 (2021)
10. International Civil Aviation Organization. ICAO Assistance, Capacity-building and Training for Sustainable Aviation Fuels (ICAO ACT-SAF). Retrieved on July 17, 2024. Retrieved from: https://www.icao.int/environmental-protection/Pages/act-saf.aspx
11. Federal Aviation Administration. Special Conditions: magniX USA, Inc., magni250 and magni500 Model Engines. November 19, 2020. Retrieved on July 17, 2024. Retrieved from: https://www.federalregister.gov/documents/2020/11/19/2020-23434/special-conditions-magnix-usa-inc-magni250-and-magni500-model-engines
12. J. Langford, D. Hall. Electrified aircraft propulsion, The Bridge, **50,** 21 (2020)
13. Archer Aviation. Stellantis Invests Additional $55 Million Into Archer Following Recent Flight Test Milestone. July 2, 2024. Retrieved on July 16, 2024. Retrieved from: https://news.archer.com/stellantis-invests-additional-55-million-into-archer-following-recent-flight-test-milestone

Performance Analysis and Future Trends of the Mazda Wankel Rotary Engine

Jianhua Wu[1]*

[1]Suzhou Industrial Park Foreign Language School, 215000 Suzhou, China

Abstract. As the global focus shifts towards environmental protection and sustainable energy development, electric vehicles are increasingly seen as the future of transportation. However, internal combustion engines (ICEs) continue to offer various irreplaceable advantages, particularly in terms of cost-effectiveness and reliability. The Mazda Wankel rotary engine, in spite of a lower compression ratio compared to traditional reciprocating engines, stands out for its compact size, lightweight design, and ability to deliver high horsepower with rapid operation. This unique combination of features makes the Wankel engine a noteworthy alternative to conventional ICEs. This paper will delve into the performance characteristics of the Mazda Wankel rotary engine, highlighting its strengths and limitations. Additionally, the study will explore the potential future trends for this engine technology, considering ongoing advancements and market demands. The analysis aims to provide a comprehensive understanding of the Wankel engine's role in the automotive industry amidst the growing prominence of electric vehicles.

1 Introduction

Mazda Wankel Engines are renowned in the automotive industry for their distinctive design and exceptional performance. Unlike conventional piston engines, rotor engines harness power through rotational motion, granting them significant advantages in terms of compactness, lightweight construction, and high revolution per minute (RPM) capabilities. Since the 1960s, Mazda has consistently applied this innovative engine technology to numerous iconic models, such as the acclaimed RX-7 and RX-8 series, continuously advancing and refining its technical principles while tracing a rich historical development.

Furthermore, the Wankel Rotary Engine needs some technique improvements to eliminate its inevitable shortcomings. Since it has been founded for decades, however, in the global background of emission reduction, the problems of the Wankel Rotary Engine are relatively low combustion efficiency, high fuel consumption, and unburnt hydrocarbon emissions, attributed to constraints on flame speed, elongated combustion chambers, and low compression ratios [1]. Nevertheless, new technology like the hydrogen-fueled Wankel rotary engine can achieve zero emission [2]. After solving the main problem, the improved

* Corresponding author: tonywu0719@ldy.edu.rs

Wankel engine can be used in multiple fields related to transportation. For example, unmanned aerial vehicles can settle the hydrogen-fueled Wankel Rotary Engine.

This paper will conduct an in-depth exploration of the technical principles, historical evolution, and performance characteristics of the Mazda Wankel Engine, as well as its current position and prospects within the modern automobile industry.

2 Technical principles

2.1 Working principles

Felix Heinrich Wankel, a German mechanical engineer, designed the Wankel rotary engine. This engine is a type of pistonless four-stroke internal combustion engine, which is the four cycles of the Otto engine without a piston. Instead of a piston, Wankel uses an almost-triangular(trochoidal) rotor moving around in an almost-elliptical(epitrochoidal) housing [3].

The three tips of the rotor are slid along the walls of the chamber so that during a single complete revolution of the rotor, the air inside is divided into two segments. The larger gear is on one side of the rotor, and it is attached to the tips. They are meant to keep the rotor centered. The rotor has, on one side, toothed gears for hula-hoop rotation to keep it from sliding down; on the other hand, the teeth are not engaged in the act of wobbling or dynamic repositioning but are there solely to keep the rotor in phase with the air. In addition, specifically, a plain Babbitt-style bearing allows for the normal rotor rotation caused by the eccentric shaft, which goes through the small gear and functions as an output shaft. The eccentric shaft takes three cycles along its trajectory in the casing for a single rotor revolution, which happens once.

The combined use of the eccentric shaft and the constancy or "timing" in the rotation of the rotor freezes the relationship between the two tips; thus, the tips form a path known as an epitrochoid, which has a complex equation but looks like an oval with a pinched hole. The rotor tips revolve in such a way as to form the mentioned chambers, each having a different volume for every single rotation, making this rotor an excellent design for propulsion and fluid movement in compressors, corkscrew pumps, etc. The three chambers will change from the intake, then compression, later expansion, to the last exhaust phase. The first representation shows, inside A, the incoming air through the open intake port while the engine is overcoming the intake cycle; the entire system is inhaling a new mixture. At the tightest situation, which is the maximum compression, chamber B is burning the mixture to generate power. Being at about 30° of the intake phase, chamber C would be on the exhaust phase. The Fig.1 shows the entire working process of the Wankel Rotary Engine.

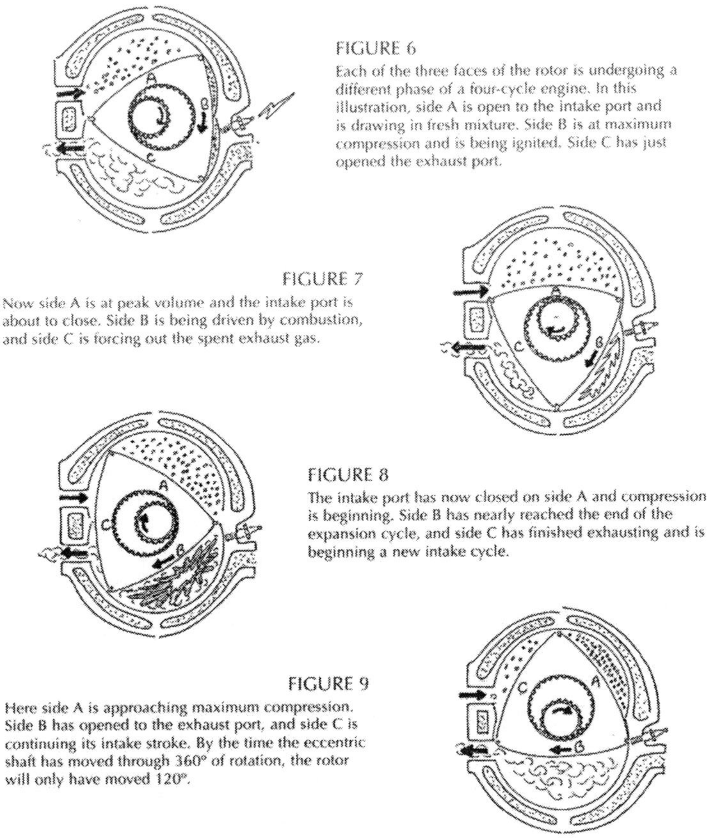

Fig. 1. Working process of Wankel rotary engines [3]

2.2 Power comparison

The working process of Wankel rotary engines is a four-stroke cycle, which is the same as the reciprocating engines. As Fig. 2 shows, it needs two crankshaft revolutions (720 ◦CA-Crank Angle) for a reciprocating engine to fulfill a four-stroke cycle. However, every three revolutions at each rotor revolution (1080° EA-Eccentric Angle) completed by an eccentric shaft of the Wankel rotary engine will finish three four-stroke cycles in total since each side of the rotor revolution fulfills each four-stroke cycle [4]. Since the number of the moving parts of the chamber which used to combust is decreased, it can achieve a higher RPM than traditional piston motor. Thus, each power output created by a Wankel rotary engine can be seen as three times that of a reciprocating engine.

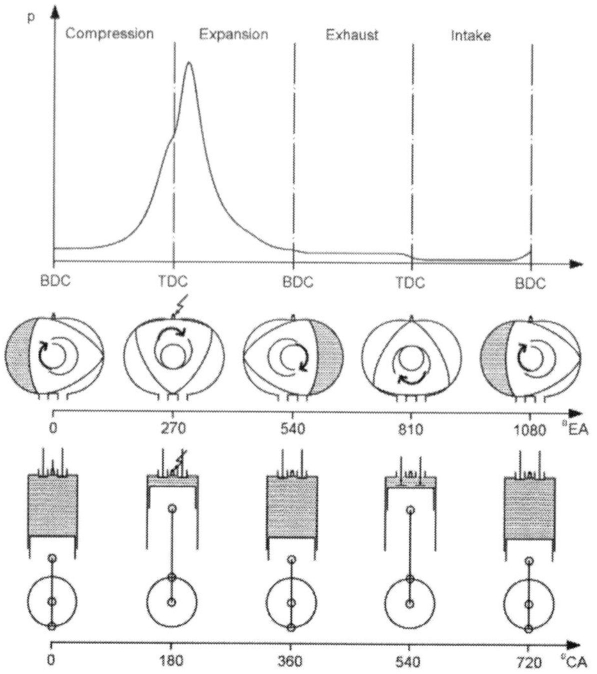

Fig. 2. Four-stroke cycle for reciprocating and Wankel engines [4].

One of the main limitations is related to the design of the rotor sealing system, specifically regarding the apex seal [5]. In this scenario, a single slat is responsible for fulfilling the role typically carried out by piston rings in a conventional engine. Moreover, unfavorable variations in the angle at which the apex seal operates hinder achieving an optimal shape for its sliding surface concerning oil film parameters. As a result, this leads to a decline in mating conditions between elements of the apex seal, ultimately resulting in reduced efficiency and longevity.

2.3 Hydrogen power source

Research on the application of both quantitative and qualitative control of a hydrogen-fueled Wankel rotary engine (WRE) and look for a proper load control strategy that is expected to improve upon current drawbacks [6]. If the operation is carried out at 1500 revolutions per minute, then hydrogen-fueled WREs have a greater efficiency. Hydrogen-fueled WREs have their own shortcomings in comparison to hydrogen-fueled reciprocating engines; these include low brake thermal efficiency, uneven thermal load, high knock numbers, and high NOx emissions as well. The hybrid control (qualitative control combined with some elements of quantitative control) can noticeably increase brake thermal efficiency, reduce the thermal imbalance around the combustion side, reduce NO emission, and reduce the knocking tendency to some extent, with a moderate penalty on power stability. This hybrid load control is also capable of selecting appropriate matching logic depending on the priority order, including power, efficiency, emission, and stability. Brake thermal efficiency of the load control method of 38.4% (on the relative basis) and 4.75% (on the absolute basis) in the operation area. Moreover, 1.4 times the volumetric efficiency and a maximum 80% reduction of maximum pressure rising rate are obtained. On top of that, the emissions in most of the operating range of the torque can be less than 100 parts per million and can be eliminated by post-treatment under stoichiometric combustion. The increase in cyclic variation is also small, less than 4.5%.

Although there are similarities between the knocking properties observed in hydrogen-powered WREs and reciprocating piston engines, it is important to note that hydrogen-fueled WREs exhibit distinct knock characteristics [7]. Consequently, specific measures must be implemented to mitigate knocking in these systems.

In addition, a study examines the efficiency of the Wankel rotary engine (WRE) fueled by hydrogen enriched with ammonia at varying levels of ammonia enrichment (NH3%) and excess air ratios (λ) [8]. Ammonia enrichment has been proven as a highly effective strategy to enhance the dynamic and economic performances of hydrogen-powered WRE. We observed that under specific operating conditions, such as 1200 r/min, a manifold absolute pressure (MAP) of 70 kPa, stoichiometric ratio, and maximum brake torque (MBT) ignition timing conditions, brake torque reached its peak at approximately 30% NH3%, while brake thermal efficiency peaked at around 40% NH3%. These findings indicate a significant improvement of 21% and 29%, respectively, compared to the use of pure hydrogen. However, it is important to note that ammonia enrichment does have drawbacks in terms of increased NO emissions. Notably, the optimal NH3% for hydrogen-fueled WRE differs significantly from what is required by hydrogen-fueled reciprocating piston engines

3 Historical evolution

To begin with, the German NSU vehicle manufacturer was the one who initially possessed the German Wankel engine proposed by Wankel [9]. On January 19, 1960, at a meeting of the German Engineers Association (VDI), which is prevalent today, it was Wankel who made it public that this invention of a new cutting-edge engine was imminent. This news soon reached the car manufacturing powerhouses, such as GM of the US, Daimler Benz of Germany, and Toyota of Japan. Nevertheless, it became apparent from the trial tests that the complexity of production would be more of a bomb than they had anticipated, and they all pulled out. To sign the patent license, Mazda agreed to vivid terms by focusing on getting the contract in February 1961.

In July of the same year, the organization dispatched an eight-member team of engineers, including Kenichi Yamamoto, to undertake a training course at the NSU factory in West Germany, and later, he was named the head of the RE Research Department. Then, for the demonstration, the engineers took a copper board off the Wankel engine while it was on, and the copper plate stood up as proof of the Wankel engine's quietness of operation. Still, after some time, they spotted that the lobes of the rotor chamber would have sharp and wavy wear marks on the cylinder wall due to the rotor vertex seal. That was the obstacle to mass production; after solving the problem, it would now go to the official phase. It was in 1963 (Showa 38) that Mazda set up the "RE Research Department" and appointed Kenichi Yamamoto to head 46 people to work on the Wankel engine for improvement. They would call them “Rotor 47 Warriors” like the “Ako 47 Ronin” if they did not succeed.

On the other hand, drumming the car industry as a result of passenger car liberalization, the Ministry of International Trade and Industry (now the Ministry of Economy, Trade and Industry) of Japan did not lag. It came up with the idea of merging the domestic automobile industry in 1965. Confronted with the current crisis and the possibility of such a merger, President Matsuda put forward the slogan “Technology is always innovative,” which additionally spurred the RE Research Department.

Now, there are some engine types of peculiar to this one - for example, to say 'rotor reduces rotation' when the rotor (being a key component of the Wankel engine) moves in irregular oval orbits. It is important to construct the three faces employed at the three outlets of the cylinder so as to avoid diminishing airtightness between these faces and the cylinder wall by use of a three-dimensional linkage structure involving an apex seal and a corner seal. The apex seal function will follow that of the piston ring rings, and the metal spring sheet will also expand and contract reciprocally between the wall and the apex seal gap spacing. When the sides of the top lose their geometry, they may be referred to as 'devil's fingernails', which are the result of the cylinder wall being scratched by the apex seal. Put another way, the formula for enhancing the apex seal material performance is the development history of Wankel engines.

Once the tests were increasingly repeated, the researchers perceived a great similarity between the wear marks and the vibrations, which rise naturally above a certain point and become static, and so they changed the conformation of a so-called diamond seal. They created a diamond seal, which had a particular shape of a hole at its top and two spread-like holes going across, that were named cross-hollow diamond seals instead. The field trials confirmed that the pairs (diamond-cylinder) had experienced the frequency of resonance variation, thus not leaving any wavy wear marks within 300 hours and thus finalizing the initial problem. Then, by the summer of 1964, Japan Carbon Company (Japan Carbon Co., Ltd.) developed an alternative zinc to the first diamond seal fabricated by Mazda with a new aluminum alloy material mixed with carbon. The Cosmo Sport was unveiled in May of 1967 and assembled with a "carbon aluminum" material as it was to host a rotary power plant in a passenger production vehicle. The Fig. 3 shows the car with the first Mazda Rotary Engine [10].

Fig. 3. The car and the First Mazda Rotary Engine [10]

4 Conclusions

The usage of the Wankel Rotary Engine is in an embarrassing position nowadays since it consumes more fuel than other types of engines. Although it can do more work, it still needs to face the barrier of relatively low combustion efficiency. As technology advances and better fuels become available for the engine, the use of such engines could then be greatly expanded to more civilian vehicles.

The result of the experiment shows that a Wankel engine controls a lesser number of movable parts than the IC engine through the technology of fuel cells because many operational components are taken out but still obtain enough electrical yield. It is vital to have an engine like this, as when a conventional engine working in a car is taken as a case example, a part of the mechanical energy transferred to electrical power, an outcome of so many losses, is wasted. As a result, the Wankel rotary engine must eliminate its barriers of sealing, efficiency, and emissions mentioned in this paper to progress to the extent of becoming a widely used engine rather than an engine that still occupies a mere portion of the engine industry. The improvement in these three key areas, machine design, seal technology, and coat materials, as well as the advancing of novel alternative fuels, have addressed the challenges the Wankel technology had been facing, leading to the superior performance of the engine. The data lead to the prediction of a substantial lowering of oil consumption while the improved Wankel engine design is used, as opposed to the piston engines. Nevertheless, it should be noted that these optimized engines have a significant way to go in controlling hydrocarbon emissions.

References

1. Shen, X., Costall, A., Enhancing the efficiency of rotary thermal propulsion systems. Energies, **17**, 2112 (2024)
2. Jandhyala, A. S., Varun, G. S., & Paliwal, R., An insight into the methodology of the hydro Wankel cell. Heat Transfer Summer Conference, **46938**, 1 (2004)
3. Hege, J. B., The Wankel rotary engine: a history. McFarland (2006)
4. Kutlar, O. A., & Malkaz, F., Two-stroke Wankel type rotary engine: a new approach for higher power density. Energies, **12**, 4096 (2019)
5. Iskra A, Babiak M, Kałużny J, et al. The oil film parameters of the Wankel engine apex seal in aspects of durability of mating elements. Journal of KONES, **17**, 199 (2010)
6. Meng, H., Ji, C., Wang, D., et al., Research on the load control of hydrogen-fueled Wankel rotary engine. International Journal of Hydrogen Energy, **47**, 16665 (2022)
7. Meng, H., Ji, C., Su, T., et al., Analyzing characteristics of knock in a hydrogen-fueled Wankel rotary engine. Energy, **250**, 123828 (2022)
8. Meng, H., Ji, C., Yang, J., et al., Performance analysis of the ammonia-enriched hydrogen-fueled Wankel rotary engine. International Journal of Hydrogen Energy, **49**, 462 (2024)
9. Hopkins, W., Designing and manufacturing a rotary Wankel engine (Doctoral dissertation, Worcester Polytechnic Institute (2023)
10. Mazda, History of Rotary, Retrieved on July 30, 2024. Retrieved from: https://www.mazda.com/en/innovation/stories/rotary/newfrontier/

Feasibility of Hydrogen Energy for High-Power Aerospace Applications: A Comprehensive Assessment

Zhenglin Cai[1*]

[1]Beijing No. 80 High School, 100102 Beijing, China

Abstract. Hydrogen energy offers a clean, versatile, and renewable alternative to traditional fossil fuels, playing a crucial role in global decarbonization efforts. This paper assesses the feasibility of hydrogen energy for high-power applications in aerospace, highlighting its potential to reduce environmental pollution and enhance the energy supply. The study evaluates hydrogen turbofan engines, hydrogen fuel cell powertrains, and hydrogen fuel cell electric fan engines, demonstrating improvements in range and load capacity compared to conventional fuels. Additionally, the technical viability of hydrogen storage and distribution infrastructure, such as cryogenic storage, lightweight tank materials, and hydrogen fuel cell technology, is explored. The analysis focuses on fuel efficiency gains, emission reductions, and the integration of hydrogen-powered aircraft into current systems. Considerations include structural design, aerodynamic characteristics, weight distribution, refueling infrastructure, and regulatory compliance. This comprehensive evaluation underscores the significant potential of hydrogen energy in advancing the aerospace industry and contributing to global Sustainable Development Goals.

1 Introduction

Hydrogen energy holds considerable advantages in the aerospace sector in terms of performance and sustainability [1]. Hydrogen energy is proving to be a game-changer in the aerospace sector, offering substantial advantages in both performance and sustainability. Its high energy density enables longer ranges and greater thrust, while hydrogen fuel cells provide high efficiency with zero emissions, producing only water vapor. This makes hydrogen a clean alternative to traditional jet fuels, reducing greenhouse gases and pollutants [2]. Additionally, hydrogen is capable of being generated sustainably sourced from renewable materials, supporting the aerospace industry's efforts to reduce its carbon footprint [1]. As technology and facilities advance, A crucial role is expected to be played by hydrogen energy in enhancing performance and achieving environmental goals in aviation.

Globally, decarbonization has become a crucial objective in addressing climate change. Decarbonization has become a key strategy for addressing climate change on a global scale [3]. It is essential that greenhouse gas emissions, with a particular focus on carbon dioxide,

* Corresponding author: zhenglincai@hhu.edu.cn

be reduced to meet climate targets. The aerospace sector, known for its high energy consumption and emissions, faces significant pressure to reduce its carbon footprint. The application of hydrogen energy technology in this sector can significantly lower carbon dioxide (CO_2) emissions while enhancing the energy efficiency of aircraft, helping the aerospace industry achieve its decarbonization goals [2]. The advancement of hydrogen energy technology offers a practical solution for the aerospace sector. By adopting hydrogen energy, aircraft can achieve high power output while reducing environmental impacts. This not only meets the global demand for low-carbon development but also drives innovation and progress in aerospace technology [3].

To achieve this goal, various industries are actively exploring more environmentally friendly technologies and solutions. In this context, hydrogen energy, with its significant potential as a clean energy technology, is gradually becoming a focal point in the aerospace sector [4]. Particularly in high-power flight applications, the advantages of hydrogen energy not only offer new pathways for environmental protection but also bring new opportunities for the future development of aerospace technology.

2 Technical feasibility of hydrogen energy

2.1 Performance benefits

2.1.1 High energy density

Hydrogen offers the highest energy density per kilogram of any fuel, providing 120 MJ/kg based on its lower heating value [5]. This makes it an exceptionally potent fuel for aerospace applications. While hydrogen has a lower volumetric energy density of 8 MJ/L for its liquid form, gasoline boasts a higher volumetric energy density of 32 MJ/L [5]. It means hydrogen occupies more space for the same amount of energy. Technological advances in storage—such as metal hydrides, carbon-based materials, metal-organic frameworks (MOFs), and nanomaterials—allow efficient storage and use [6]. These innovations enable aircraft to store substantial amounts of energy in a manageable volume, supporting longer flight ranges and greater thrust [7]. The high energy density of hydrogen thus directly contributes to extended operational capabilities and enhanced performance for aerospace vehicles.

2.1.2 Efficient propulsion systems

Hydrogen's potential in propulsion systems is significant, whether used in combustion engines or fuel cells. In hydrogen combustion engines, hydrogen burns at a high flame temperature, which can be optimized to produce efficient thrust, though modifications to traditional engines are necessary to accommodate hydrogen's unique combustion characteristics [8]. Alternatively, Hydrogen fuel cells utilize an electrochemical reaction to convert chemical energy straight into electrical energy, often achieving efficiency rates above 50% [9]. This high efficiency in fuel cells minimizes energy losses and improves overall performance. Both types of propulsion systems, therefore, offer substantial performance enhancements compared to conventional internal combustion engines, delivering significant thrust and operational efficiency [8].

2.1.3 Reduced weight

The use of hydrogen energy can lead to a reduction in the weight of fuel systems and associated components, thanks to hydrogen's light molecular weight [10]. This weight reduction is crucial in aerospace applications, where minimizing mass is essential for optimizing fuel efficiency and overall performance. Hydrogen fuel cells, which are relatively compact and lightweight compared to traditional internal combustion engines, further contribute to this benefit. By integrating lighter fuel systems and components, aircraft can achieve better fuel efficiency, which improves overall performance and allows for more innovative and aerodynamic aircraft designs [11]. Reduced weight translates into less energy required for lift and propulsion, enhancing overall operational efficiency.

2.1.4 Rapid refueling

Hydrogen refueling systems are being developed to match the speed and convenience of conventional jet fuel refueling, aiming to minimize downtime for aircraft. Faster refueling enables more frequent flights and supports quicker service in both commercial and military contexts. The ability to rapidly replenish hydrogen fuel can provide significant operational advantages and streamline aircraft operations [12].

2.2 Cryogenic storage

The procedure entails cooling hydrogen to temperatures below -253°C to transform it into a liquid form. It significantly enhances the density of hydrogen, allowing for storage with higher efficiency plus transport in comparison with its gaseous form. Despite its advantages, cryogenic storage is energy-intensive, requiring substantial energy input for liquefaction and maintaining cryogenic conditions [13]. Additionally, advanced insulation technologies are necessary to manage the boil-off, or evaporation, of liquid hydrogen, which contributes to high infrastructure costs [14].

2.3 Lightweight tank materials

The development of lightweight tank materials is crucial for the efficient storage of hydrogen, especially in applications such as automotive hydrogen tanks. Advanced materials, including carbon fiber composites, offer significant weight reduction and enhanced structural integrity compared to traditional metals. This reduction in weight is particularly beneficial for vehicle performance and efficiency [15].

2.4 Hydrogen fuel-based engines

In the aerospace sector, hydrogen energy can be applied through various technologies, including hydrogen turbofan engines and hydrogen fuel cell power systems. Each of these technologies offers distinct advantages and disadvantages [9].

2.4.1 Hydrogen turbofan engines

They combine the high energy density of hydrogen in relation to the efficiency of turbofan engines, making high power easily occur on this type of engine. They are suitable for long-range and high-thrust applications. When using hydrogen as fuel, the efficiency can be further enhanced due to hydrogen's high energy density, allowing the engine to provide significant thrust with lower fuel consumption. Hydrogen turbofan engines can also leverage existing turbofan engine technologies with established design principles and manufacturing processes. Nevertheless, since the energy utilization efficiency is slightly lower, engineers should Improve the structural design of existing aircraft to accommodate liquid hydrogen fuel storage and try to eliminate the emissions of water vapor and nitrogen oxides [9].

2.4.2 Hydrogen fuel cell power systems

They offer high energy conversion efficiency, typically around 45-50%, which is significantly higher than traditional internal combustion engines. This high efficiency results from the fuel cells' capability to shift chemical energy into electrical energy with minimal energy loss. Furthermore, hydrogen fuel cells emit only water, with no carbon dioxide or other pollutants, minimizing environmental impact. The modular design of fuel cell systems provides flexibility, making them adaptable for various power needs and configurations. However, people should redesign the aircraft structure to accommodate liquid hydrogen fuel storage and a new hydrogen fuel cell power system. They also hold a low power density, which means they currently cannot achieve high power [9]. (See Table 1)

Table 1. Comparison of hydrogen turbofan engine and hydrogen fuel cell power system [9]

	Hydrogen turbofan engine	Hydrogen fuel cell power system
Efficiency	40%	40%-50%
Environmental Influence	No carbon monoxide, carbon dioxide, sulfide and hydrocarbon emissions; There are water vapor and nitrogen oxide emissions	No carbon monoxide, carbon dioxide, sulfide, hydrocarbons and nitrogen oxide emissions; Steam discharge
Technological Difficulty	Existing aviation turbines need to be modified or redesigned to accommodate hydrogen fuel; Raised hitch requirement; Improving aircraft structure and safety requirements; Meet liquid hydrogen storage requirements	Development of efficient, high power density, fast start fuel cells; Develop motor, electronic power equipment cables, and other electrical components that meet airworthiness standards; Improve aircraft structure and safety requirements to meet liquid hydrogen storage needs
The Ways of Working	Turbofan engines that use hydrogen fuel provide thrust through a fan	Hydrogen fuel cells are employed to generate electricity, while thrust is delivered by electric or ducted fans.
Advantages	Zero carbon emissions, a small amount of pollutant	Zero carbon emissions, zero pollutant emissions;

	emissions; The applicability of hydrogen fuel can be improved on the basis of traditional aviation turbofan engines to form hydrogen fuel turbofan engines. Easy to achieve high power	The power system is easier to combine with distributed electric propulsion technology and the energy utilization efficiency is high
Disadvantages	Improving the structural design of existing aircraft to accommodate liquid hydrogen fuel storage; Emissions of water vapor and nitrogen oxides; The energy utilization efficiency is slightly lower	Redesign the aircraft structure to accommodate liquid hydrogen fuel storage and a new hydrogen fuel cell power system; Discharge steam; Low power density, currently cannot achieve high power

3 Performance capabilities

3.1 Enhancing fuel efficiency

Hydrogen energy technology has the potential to revolutionize fuel efficiency in aviation, offering significant improvements over traditional aviation fuels. One of the standout advantages is the high energy density of hydrogen by weight, which is considerably greater than that of conventional jet fuels. Hydrogen's energy density—around 120-142 MJ/kg—far exceeds that of aviation gasoline (46-47 MJ/kg) or kerosene-based fuels [16]. This means that hydrogen can deliver more energy per unit of weight, which can translate into increased range and operational efficiency for aircraft.

Hydrogen fuel cells, which convert hydrogen into electricity, provide a particularly efficient energy conversion method. They typically achieve efficiency rates of around 55%, which is significantly higher than the 30-40% efficiency of traditional jet engines [17]. This higher efficiency is because fuel cells do not rely on combustion, which is inherently less efficient due to heat losses. Instead, electricity is generated by fuel cells through an electrochemical reaction, resulting in less energy wasted as heat and more energy available for propulsion. Moreover, hydrogen can be used in advanced gas turbines specifically designed for its use. These turbines can be optimized to burn hydrogen more efficiently than conventional turbines. The development of these engines aims to harness hydrogen's high energy density while minimizing energy losses. As the technology evolves, the design of hydrogen-powered turbines will improve, leading to even better fuel efficiency and extended flight ranges [9].

Efficiency gains from hydrogen technology not only lower fuel costs but also contribute to the overall reduction in the environmental influence of aviation. Through the use of hydrogen more effectively, aircraft can achieve longer ranges and reduced operational costs, making hydrogen a viable alternative for future air travel.

3.2 Reducing emissions levels

Hydrogen energy technology offers significant advantages in reducing emissions, aligning closely with global sustainability goals. Unlike conventional aviation fuels that release considerable amounts of CO_2, nitrogen oxides (NOx), and other pollutants, hydrogen combustion produces only water vapor as a byproduct. This clean combustion profile positions hydrogen as a key player in efforts to lower the environmental footprint of aviation [1,4].

One of the most compelling aspects of hydrogen energy is its zero CO_2 emissions during use. When hydrogen is burned in engines or used in fuel cells, the primary byproduct is water vapor, effectively eliminating CO_2 emissions from the combustion process. This is a stark contrast to traditional jet fuels, which release substantial quantities of CO_2 and contribute significantly to global warming. The aviation industry, accountable for about 2.5% of global CO_2 emissions, stands to make a significant impact on emission reductions by adopting hydrogen technology [18].

Additionally, hydrogen combustion generates no NOx, which is another significant pollutant from traditional aviation fuels. NOx contributes to smog formation and acid rain, impacting air quality and environmental health. By using hydrogen, aviation can mitigate these pollutants, leading to cleaner air around airports and improving public health outcomes. This is especially relevant in densely populated areas where airport emissions can have a more pronounced impact on local air quality [4].

4 Aircraft design and operational integration

4.1 Aircraft design

4.1.1 Structural design

The structural design of hydrogen-powered aircraft is fundamentally different from that of traditional jet-powered aircraft due to the unique properties of hydrogen. Hydrogen is reserved either as a gas with high pressure or as a cryogenic liquid, each requiring specific structural considerations. For high-pressure hydrogen storage, the aircraft must incorporate tanks capable of withstanding pressures up to 700 bar (10,000 psi) [19]. These tanks are often made from advanced composite materials, which are both lightweight and extremely strong. The materials must resist the high pressures without deformation or failure, which is crucial for maintaining aircraft safety and performance.

When using cryogenic hydrogen, which is stored at extremely low temperatures (around -233 °C), the aircraft's structure must include highly effective thermal insulation to prevent heat transfer that could cause the hydrogen to boil off. This involves using materials that can handle such extreme cold without becoming brittle or failing [20,21]. The placement of these storage tanks within the aircraft is critical; they must be integrated in a way that maintains the aircraft's center of gravity and balance. Additionally, because hydrogen is highly flammable, safety features such as leak detection systems and reinforced tank casings are essential to prevent accidents. This intricate design process ensures that the aircraft is safe, reliable, and capable of efficient operation with hydrogen as a fuel.

4.1.2 Aerodynamic Characteristics

The aerodynamic design of hydrogen-powered aircraft must address the challenges posed by hydrogen's low density and the associated storage systems. Hydrogen's low volumetric energy density necessitates that storage tanks be either significantly larger or uniquely shaped compared to conventional jet fuel tanks. This can affect the aircraft's aerodynamic profile, potentially increasing drag and altering airflow around the aircraft [22].

To counteract these effects, engineers use advanced aerodynamic design techniques. For example, the aircraft might need to have larger wings or different fuselage shapes to accommodate the hydrogen tanks while maintaining aerodynamic efficiency. Computational Fluid Dynamics (CFD) simulations and wind tunnel testing are employed to refine these designs. CFD simulations help predict how air flows over the aircraft, allowing engineers to reduce drag and improve fuel efficiency [23]. Wind tunnel tests provide empirical data on how the design performs in real-world conditions. (See Fig. 1).

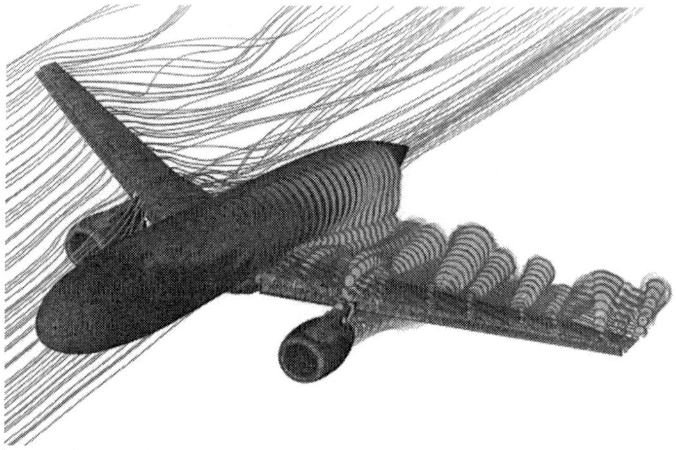

Fig. 1. Ansys Fluent simulation of an external aerodynamics study of a commercial aircraft [23]

4.2 Weight distribution

Effective weight distribution is crucial for significantly impacting the performance and safety of hydrogen-powered aircraft. Hydrogen storage systems, whether high-pressure tanks or cryogenic tanks, add significant weight to the aircraft. Proper weight distribution ensures that the acceptable limits for the aircraft's center of gravity are maintained, which is vital for stability and handling during flight. The placement of hydrogen tanks must be carefully considered to maintain balance. For instance, if the tanks are too heavy or improperly positioned, they could shift the aircraft's center of gravity, affecting its stability and control. Designers must also balance the added weight of the hydrogen storage systems with the need to maximize payload capacity, which includes passengers and cargo. Advanced materials and structural design innovations can help manage the additional weight, such as using lightweight composites to reduce the overall mass of the aircraft. Effective weight management also involves optimizing the aircraft's design to compensate for the weight of hydrogen storage. This might include adjusting wing configurations or employing innovative structural designs to ensure that the aircraft remains efficient and performs well despite the added weight. Overall, proper weight distribution is key to achieving the desired performance and safety of hydrogen-powered aircraft.

4.3 Fueling infrastructure

The development of fueling infrastructure for hydrogen-powered aircraft is a complex and essential task. Hydrogen refueling stations must be designed to handle the specific requirements of hydrogen fuel, which not only can be stored as a high-pressure gas but also as a cryogenic liquid. High-pressure refueling stations require systems to compress, store, and dispense hydrogen safely. Cryogenic refueling stations must have advanced insulation and cooling systems to keep the hydrogen in its liquid state and to prevent heat-induced vaporization. Refueling infrastructure must be integrated into existing airport facilities to ensure that hydrogen-powered aircraft can be refueled efficiently and safely [13,14]. This includes designing refueling stations that can handle the high pressures or low temperatures required for hydrogen. Efficient hydrogen distribution networks are also crucial, involving pipelines, specialized transport vehicles, and storage facilities to transport hydrogen from production sites to refueling stations.

4.4 Regulatory compliance

Stringent standards set by aviation authorities such as the International Civil Aviation Organization (ICAO) and the Federal Aviation Administration (FAA) must be met by hydrogen-powered aircraft. These standards cover safety, performance, and operational requirements to ensure that hydrogen aircraft are safe for use [3].

The certification process for hydrogen-powered aircraft involves extensive testing, including ground tests and flight trials, to verify that all systems function correctly under various conditions. This process ensures that the aircraft meets all safety and performance criteria before it is approved for operation. Additionally, hydrogen-powered aircraft must comply with environmental regulations related to emissions and fuel efficiency.

5 Environmental and economic impacts

5.1 Environmental benefits of reduced carbon emissions

The environmental benefits of hydrogen energy as an aviation fuel are significant, primarily due to its ability to dramatically reduce carbon emissions and other pollutants. The use of hydrogen fuel cells and hydrogen turbines can substantially decrease greenhouse gas emissions from aircraft during flight, thereby positively impacting the environment.

The combustion process of hydrogen fuel is extremely clean, with water vapor being the only byproduct. This means that burning hydrogen does not produce carbon dioxide (CO_2) or other harmful gases such as nitrogen oxides (NOx). This is in stark contrast to conventional aviation fuels, which release substantial amounts of CO_2 and NOx, both of which are major contributors to global warming and air pollution [6]. For instance, the International Energy Agency (IEA) reports that the aviation sector accounts for roughly 2.5% of global CO_2 emissions. [21]. Using hydrogen as a fuel can significantly reduce this percentage, helping to slow the process of global warming.

5.2 Long-term cost savings

While the initial investment for hydrogen aviation is relatively high, the potential for significant long-term operational cost savings can substantially reduce airline operating expenses.

Future fuel costs of hydrogen could become more competitive compared to traditional aviation fuels. As hydrogen production technologies advance, especially with the maturation of water electrolysis and economies of scale, the cost of hydrogen production is projected to decline significantly. The cost of producing hydrogen from renewable energy is gradually falling, thanks to technological progress and increased production scales. According to predictions from the International Hydrogen Energy Association, the production cost of hydrogen may eventually be on par with or even lower than traditional aviation fuels, thereby reducing long-term operational costs.

Maintenance costs for hydrogen fuel cell systems are generally lower than for traditional jet engines. Hydrogen fuel cell systems have a simpler structure with fewer mechanical parts, resulting in less maintenance work [23]. Moreover, the use of hydrogen reduces internal corrosion and wear in engines, further lowering long-term maintenance costs. This is particularly important in long-term operations, as maintenance costs are a significant component of airline operating expenses. Reduced maintenance needs and extended component lifespans can lead to substantial cost savings.

Although the initial investment in hydrogen aviation technology is high, significant long-term cost savings can be achieved through technological advancements and decreased production costs [24]. These savings not only help airlines reduce operational expenses but also drive the aviation industry toward a more sustainable and environmentally friendly future. Such long-term cost savings will further enhance hydrogen's economic attractiveness and practical application potential as an aviation fuel.

6 Conclusions

Hydrogen energy shows significant potential in high-power flight, especially in aerospace. As an eco-friendly and renewable energy source, hydrogen can both significantly reduce greenhouse gas emissions and aid in achieving global decarbonization goals. In terms of cost, performance and sustainability, hydrogen energy shows clear advantages over conventional aviation fuels. The application of hydrogen turbofan engines, hydrogen fuel cell power systems and hydrogen fuel cell electric fan engines can significantly improve the range and load capacity, which provides broad prospects for the future development of the aerospace field.

However, there are still some challenges to the full roll-out and application of hydrogen energy, especially in terms of hydrogen storage and distribution infrastructure. Although advances in cryogenic storage technology and lightweight tank materials have laid the foundation for hydrogen energy applications, further optimization is needed to reduce costs and improve safety. In addition, the continued advancement of hydrogen fuel cell technology will be a vital component in improving the performance of hydrogen energy applications in aviation.

Considering the advantages of hydrogen energy and its technical feasibility in the aerospace field, we believe that hydrogen energy is expected to become the main energy solution in the aerospace field in the future. Future research and development should focus on solving current technical bottlenecks and promoting the commercialization process of hydrogen energy technology to achieve its full deployment in high-power flight applications.

References

1. Yusaf, T., Fernandes, L., Talib, A. R. A., et al., Sustainable Aviation—Hydrogen is the future. Sustainability, **14**, 548 (2022)
2. Energy, B., What are the Advantages of Hydrogen Fuel Cells? Bloom Energy, 2024, retrieved on August 5, 2024. https://www.bloomenergy.com/blog/what-are-the-advantages-of-hydrogen-fuel-cells
3. IPCC, The evidence is clear: the time for action is now. We can halve emissions by 2030, April 4, 2022, retrieved on August 5, 2024, https://www.ipcc.ch/2022/04/04/ipcc-ar6-wgiii-pressrelease/

4. Adler, E. J., & Martins, J. R., Hydrogen-powered aircraft: Fundamental concepts, key technologies, and environmental impacts. Progress in Aerospace Sciences, **141**, 100922 (2023)
5. Center for Sustainable Systems, University of Michigan, Hydrogen Factsheet, 2023, retrieved on August 5, 2024, https://css.umich.edu/publications/factsheets/energy/hydrogen-factsheet
6. Rampai, M. M., Mtshali, C. B., Seroka, N. S., et al., Hydrogen production, storage, and transportation: recent advances. RSC Advances, **14**, 6699 (2024)
7. Adler, E. J., & Martins, J. R., Hydrogen-powered aircraft: Fundamental concepts, key technologies, and environmental impacts. Progress in Aerospace Sciences, **141**, 100922 (2023)
8. Boeing, Hydrogen and Sustainable Aviation, February 2023, retrieved on August 5, 2024, https://www.gnieob.com/content/dam/boeing/boeingdotcom/principles/esg/sustainability/Fact_Sheet_HydrogenProp.pdf
9. Zhang, Y., Peng, J., Qian, Y., et al., Key technologies and challenges of hydrogen powered aviation, **1**, 22 (2021)
10. Green H_2 World, Hydrogen molecular weight and molar mass, Jun 28, 2023, retrieved on August 5, 2024, https://www.greenh2world.com/post/hydrogen-molecular-weight
11. McKinsey & Company, Fuel efficiency: Why airlines need to switch to more ambitious measures, Mar 1, 2022, retrieved on August 5, 2024, https://www.mckinsey.com/industries/aerospace-and-defense/our-insights/future-air-mobility-blog/fuel-efficiency-why-airlines-need-to-switch-to-more-ambitious-measures
12. Technicas Pte. Ltd., Hydrogen Refueling Systems Technicas, 2022, retrieved on August 5, 2024, https://www.technicas.sg/hydrogen-refueling-systems/
13. She, X., Zhang, T., Meng, Y., et al., Cryogenic energy storage. Elsevier eBooks, **94**, 107 (2022)
14. Zhang, L., Jia, C., Bai, F., et al., A comprehensive review of the promising clean energy carrier: Hydrogen production, transportation, storage, and utilization (HPTSU) technologies. Fuel, **355**, 129455 (2024)
15. Zhang, W., & Xu, J., Advanced lightweight materials for automobiles: A review. Materials & Design, **221**, 110994 (2022)
16. Allendorf, M. D., Stavila, V., Snider, J. L., et al., Challenges to developing materials for the transport and storage of hydrogen. Nature Chemistry, **14**, 1214 (2022)
17. Barbir, F., Fuel cell electrochemistry. PEM fuel cells, **33**, 72 (2005)
18. Our World in Data, Ritchie, H., & Roser, M., What share of global CO_2 emissions come from aviation?, April 8, 2024, retrieved on August 5, 2024, https://ourworldindata.org/global-aviation-emissions
19. Hydrogen and Fuel Cell Technologies Office, Hydrogen Storage, n.d., retrieved on August 5, 2024, https://www.energy.gov/eere/fuelcells/hydrogen-storage
20. Ahluwalia, R., Peng, J., & Hua, T., Cryo-compressed hydrogen storage, **119**, 145 (2016)
21. IEA, Aviation, n.d., retrieved on August 5, 2024, https://www.iea.org/energy-system/transport/aviation

22. Massaro, M. C., Pramotton, S., Marocco, P., et al., Optimal design of a hydrogen-powered fuel cell system for aircraft applications. Energy Conversion and Management, **306**, 118266 (2024)
23. Ishimoto, Y., Wulf, C., Schonhoff, A., et al., Life cycle costing approaches of fuel cell and hydrogen systems: A literature review, International Journal of Hydrogen Energy, **54**, 361 (2023)
24. Energy.gov, Biden-Harris administration releases first-ever national clean hydrogen strategy and roadmap to build a clean energy future, Accelerate American Manufacturing Boom, June 5, 2023, retrieved on August 5, 2024, https://www.energy.gov/articles/biden-harris-administration-releases-first-ever-national-clean-hydrogen-strategy-and

HCCI Engines: Benefits, Challenges and Future Advancements

Yifan Fei[1]*

[1]Guanghua Cambridge International School, 200093 Shanghai, China

Abstract. Homogeneous charge compression ignition (HCCI) engines represent a significant advancement in internal combustion engine technology. This study examines the main advantages of HCCI engines, such as increased fuel flexibility, decreased pollution from nitrogen oxides (NOx) and particulate matter, and enhanced thermal efficiency. However, HCCI technology also faces challenges, such as controlling the combustion process and achieving stable ignition across diverse operating conditions. With rapid advancements in information technology and scientific methods, the automotive engine industry has made significant strides in recent years. In response to stricter environmental regulations and the growing popularity of electric vehicles, HCCI engines have gained increasing attention. Despite existing challenges, continued technological improvements are expected to enhance the feasibility and performance of HCCI engines. This paper reviews current research and technological developments, emphasizing the potential of HCCI engines to revolutionize the automotive industry while addressing critical issues that need resolution. Vital potential chances regarding this exciting field is highlighted.

1 Introduction

Along with the accommodation of oil becoming less fast, the level of the environment is souring, by the way. Many scientists have tried to find renewable resources of energy to replace conventional engines [1]. Most well-known renewable resources come from nuclear fusion energy source, geothermal energy, ocean energy, and other sources. Nevertheless, there remains a vital problem is that most renewable energies cannot be produced stably because of their availability issues. Take solar resources as an example. Although solar resource is regarded as one of the future main energy resources, they only occupy a small part of global energy requirements, which is only the whole chief energy provision's twentieth, and their low energy transfer efficiency is another insurmountable hinder to today's technology [2].

Additionally, some kinds of batteries are still the best energy storage available , while its capacity is quite restricted. Therefore, dependence on internal combustion engines which based on fossil (ICE) is necessary for many industries. With no ICE provided for transporting, it is impossible to reach such a high living standard today. Traditionally, SI and CI are two

* Corresponding author: 180904226@stu.hrbust.edu.cn

main categories of internal combustion engines. The former has drawbacks due to high knock, which are low efficiency and low compression ratio.

On the other hand, the latter was exposed to excessive amounts of harmful air pollutants, such as hydrocarbon (HC), carbon dioxide (CO_2), nitrogen monoxide (NO).These pollutants are responsible for some harmful effects like the Greenhouse Effect, etc. Incomplete combustion is one main source of primary matter pollution, as well as sea salt and dust. H_2SO_4, NH_3 and functionalized organic compounds condensed in the atmosphere formate quantities of subordinate primary matter. Human-made airborne particles are an increasing amount compared to that made in nature, PM2.5 and PM10 categorized by aerodynamic diameter size influence visibility and air quality, even making climate worse in the end [3]. One of the chief culprits to these emissions is Internal combustion engines, invisibly affecting human health.

So here comes HCCI engines, which is an unearthly solution to soar engine's efficiency, lower fuel expended and decrease the levels of emission. Compared to traditional CI and SI combustion modes, HCCI mode has a remarkably faster burning rate [4]. Here are the main advantages of HCCI engines. (i) Compared to SI engines, it is probably to reach higher fuel efficiencies, as it is managed in fuel-lean conditions and higher compression ratio. (ii) HCCI engines have remarkable fuel flexibility, and they produce less NOx and PM emissions. Nevertheless, HCCI engines have many challenges as well, including higher emissions in HC and CO emissions.

As future dominating engines instead of CI and SI engines, this paper will summarize previous related articles. This paper will mostly introduce how HCCI engines achieve these advantages, face these challenges and explain how it gains a broad prospect. This document can assist readers comprehend the characteristics of the HCCI mode, and to build the HCCI technology that aspires to produce reduced emissions.

2 Advantages of HCCI engines

2.1 High fuel efficiency

HCCI engines are capable of achieving a better fuel efficiency as compared to conventional engines (such as SI and CI engines). The main reason why HCCI has such high thermal efficiency is that HCCI engines have lean operation conditions and much higher compression ratios, which make it more likely to fit the Otto cycle approximately. According to research by Zhao et al. (2007), HCCI engines are able to reach 20 percent extra thermal efficiency compared to conventional engines, as shown in Table 1 [5]. The key factor is its auto-ignition process, which allows a more uniform, stable, complete combustion, lowering the waste of fuel and maximizing energy extraction.

Table 1. Contrast of thermal efficiency among SI, CI and HCCI

Engine Type	Thermal Efficiency (%)
SI Engine	25-30
CI Engine	35-40
HCCI Engine	45-50

https://doi.org/10.1051/e3sconf/202560601006

2.2 Low NOx and particulate emissions

The emission of NOx is significantly reduced in HCCI mode mainly achieved by its low combustion temperature, as well as PM emissions. This reduction is indispensable for ensuring emissions lower than governments' regulation standards and improving air quality. Table 2 [6,7] shows that Thring (1989) and Dec (2009) have demonstrated that NOx and PM emissions are significantly lower in HCCI engines when compared to traditional engines. NOx formation is limited because of lower combustion temperature, while the feature of homogeneous charge is useful in reducing soot formation. Research by Syed Yaser Hussaini et al.'s research also revealed that, at 2300 rpm, HCCI engines emit significantly less NOx than SI engines—roughly 10 ppm to 60 ppm, or 13.3% of the former—Fig. 1 [8].

Table 2. Emissions comparison of HCCI, CI, and SI

Emission type	SI engine emissions	CI engine emissions	HCCI engine emissions
NOx	High	Moderate	Low
Particulate matter	Low	High	Low

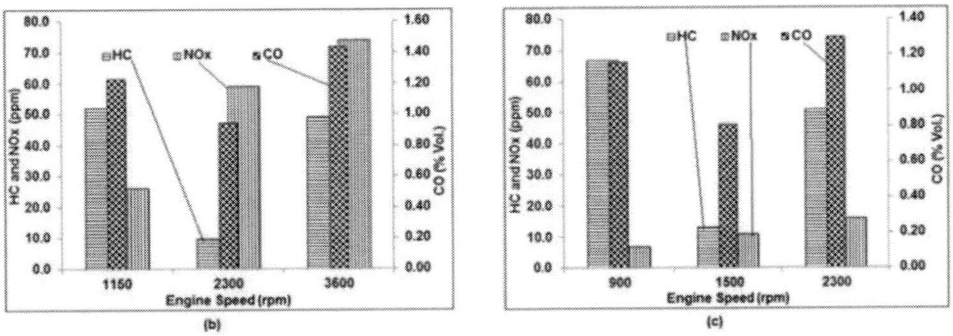

Fig. 1. Comparison of three types of emissions between HCCI and SI engines [8].

2.3 Fuel flexibility

HCCI engines have outstanding fuel flexibility, adaptable in operating various fuels, including gasoline, diesel, and alternative fuels. That is because any fuel type becomes adaptable in HCCI mode, which can be evaporated and blended with air before ignition. This adaptability makes it possible to use different types of fuels like ethanol, methanol, and biofuels in order to develop the sustainability of HCCI technology, as shown in Table 3 [9].

Table 3. Comparison of different fuel types related to their compatibility (Source: Kook et al., 2005)

Fuel Type	Compatibility
Gasoline	High
Diesel	High
Ethanol	Moderate

Methanol	Moderate
Biofuels	High

3 Challenges of HCCI engines

3.1 Combustion control

The first primary challenge of HCCI engines this paper introduced is the precise management of the combustion process. HCCI engines rely on the fuel-air mixture's chemical interactions, which makes it challenging to control the start of combustion and prevent the "knock" phenomena. This might be the biggest difference between SI and CI engines, which is well-defined in the ignition point. Onishi et al. (1979) and Dec (2009) have explored various strategies to improve combustion control in HCCI engines, including variable valve timing and advanced control algorithms, as shown in Table 4 [10,11]. They proved that accurate control over temperature, pressure, and mixture composition is critical for reliable operation.

Table 4. Different control strategies

Control Strategy	**Description**
Variable Valve Timing	Adjusting the timing of valve opening and closing to control the combustion process.
Direct Fuel Injection	delivering fuel straight into the combustion chamber in order to attain exact control.
Advanced Control Algorithms	Using sophisticated software to manage combustion parameters.

3.2 CO emissions

Studies have indicated that HCCI ones emit more CO compared to gasoline engines. This, which is clearly defined in the ignition point, may be the largest distinction between CI and SI engines. As seen in Table 4, Onishi et al. (1979) and Dec (2009) have investigated a number of approaches to enhance combustion control in HCCI engines, including variable valve timing and sophisticated control algorithms [10,11]. They demonstrated that precise regulation of mixture composition, pressure, and temperature is essential for dependable performance.

3.3 HC emissions

HC emissions generally increase when operating in HCCI mode, presenting a significant challenge for widespread adoption in vehicles. Kobayashi et al. investigated the emissions properties of a natural gas-powered HCCI engine that was converted from a gasoline engine. When compared to SI engines, they found much higher HC emissions [12]. A 25cc, four-stroke, air-cooled single-cylinder SI engine was modified by Lemberger et al. to run in HCCI mode using a variety of combustion management methods, and they recorded significant HC emissions [13]. At 3000 rpm and an equivalency ratio of 0.30, the highest HC emission detected was 950 ppm, which is almost 7 times greater than in SI mode. Milovanovic demonstrated that reducing coolant temperature could significantly decrease HC emissions [14].

3.4 Difficulty in high-load operation

HCCI engines are typically operated within a narrow range, usually under low to medium loads. When extended to high-load operations, emissions tend to increase and thermal efficiency decreases. Zhao et al. (2007) highlighted the need for further research to broaden the operational range of HCCI engines. The pressure rise rate can get quite high with high loads, which could harm the engine. Thring et al. demonstrated that the use of a specially designed fuel injector for CI engine applications was ineffective in atomizing diesel properly.

4 Future prospects and technical advancement

4.1 Composite HCCI engines

4.1.1 HCCI Gasoline engine

HCCI gasoline engine The high volatility of gasoline facilitates the formation of a homogenous fuel mixture. Researchers have focused on controlling the combustion rate and modes, allowing for mode switching under high-load conditions. Engines would then be electronically controlled to select the optimal mode for the current situation. This characteristic of modern HCCI combustion applications in gasoline engines provides superior stability during mode switching and precise fuel injection.

4.1.2 Diesel HCCI engine

Diesel features a low ignition point and a difficult time evaporating, it must be heated before it can be utilized as fuel in HCCI ICEs. High pollutant emissions and poor thermal efficiency are the outcomes of a low compression ratio. Heating aids in evaporating and mixing the diesel fuel during late injection and high swirl conditions.

4.1.3 An additional novel hybrid fuel HCCI engine

HCCI engines exhibit excellent adaption in different kinds of fuels due to their ability to accommodate a wide range of octane numbers. Besides gasoline and diesel, alternative fuels such as CH_3OH, CH_3CH_2OH, CH_4,$CH_3OCH_2CH_2OCH_3$, and H_2 have been proven exactly to be suitable for HCCI engines. In particular, using pure hydrogen fuel in HCCI engines offers significant advantages over conventional fuels.

4.2 Advanced control system

Fuel operation plays a key role in regulating the timing of ignition and combustion phases when implementing HCCI technology in engines. This involves manipulating fuel reactivity, adjusting fuel composition, and fine-tuning fuel injection strategies. The first step involved conducting experimental studies on different combinations of high- and low-reactivity fuels and non-reactive fluids to introduce impurities or distribute reactivity within the cylinder. This allowed the researchers and engineers to balance high pressure rise rates and operational range restrictions while controlling auto-ignition timing and maximizing benefit during the combustion phase. Furthermore, widely adopted and tested HCCI engines involve varying fuel compositions and sometimes adding substances to the original fuel. Another less effective method tested on HCCI engines for controlling auto-ignition timing is fuel injection strategy (Fig. 2) [15].

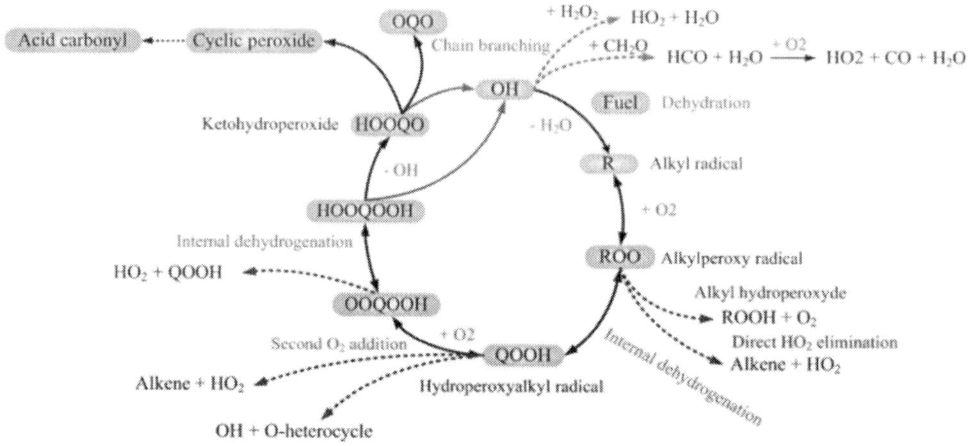

Fig. 2. Formation of different types of products in HCCI engines [15].

5 Conclusions

This article combines the HCCI engine's performance advantages and challenges to look at its prospects comprehensively. HCCI engines are a symbol of promising technology that aims to achieve cleaner and more efficient internal combustion engines. HCCI combustion mode has remarkable potential in emission features and eco-friendly fuel consumption. To cope with various environmental problems, HCCI technology might be the most suitable way for the development of internal combustion engines. HCCI engines offer innovative new ideas as well as options for pollutants and thermal efficiency issues because of their effective efficiency and low-emission characteristics, as HCCI mode has such significant advantages. Plus, HCCI engines operate smoothly and perfectly compared with traditional spark-ignition engines and direct-injection diesel engines, as it is approximately a combination of traditional SI engines and CI engines.

Moreover, HCCI engines are less reliant on scarce fuels such as oil, and their emission control systems may be simpler to manage. However, the obstacles hindering the development of HCCI engines are significant and must be addressed to unlock their full potential. To commercialize HCCI technology, several key challenges need resolution. These include precise control of injection timing, dropping of HC and CO emissions, enhancing capability to begin in cold conditions, overcoming difficulties in high-load operation, achieving homogeneous mixture preparation, and managing intake charge temperature. Given the rapid pace of technological advancements, ongoing research is likely to surmount these challenges, paving the way for practical applications of HCCI engines. Gradually, HCCI engines are expected to become a dominant force in many internal combustion engine manufacturing, offering promising and bright prospects.

References

1. R. Hilborn, C.J. Walters, D. Ludwig., Sustainable exploitation of renewable resources. Annu. Rev. Ecol. Syst. **26**, 45 (1995)

2. A. Azarpour, S. Suhaimi, G. Zahedi, et al., A review on the drawbacks of renewable energy as a promising energy source of the future. Arab J. Sci. Eng. **38**, 317 (2013)
3. F. Karagulian, C.A. Belis, C.F.C. Dora, et al., Contributions to cities' ambient particulate matter (PM): A systematic review of local source contributions at global level. Atmos. Environ. **120**, 475 (2015)
4. M. Bunce, B.G. Bunting, J. Wang, et al., Experimental and statistical comparison of engine response as a function of fuel chemistry and properties in CI and HCCI engines. SAE Int. **1**, 857 (2012)
5. H. Zhao, T. Asmus, Z. Zhang., Homogeneous charge compression ignition (HCCI) Engines: key research and development issues. SAE Int. (2007)
6. R.H. Thring., Homogeneous-charge compression-ignition (HCCI) engines. SAE Tech. Pap. **892068** (1989)
7. J.E. Dec., Advanced compression-ignition engines - understanding the in-cylinder processes. Proc. Combust. Inst. **32**, 2727 (2009)
8. S.Y. Hussaini, M.A. Noor, M.M. Rahman, et al., Procedia Technology **25**, 854 (2016)
9. S. Kook, G.J. Wachtmeister, E. Schmidt, et al., The influence of charge dilution and fuel volatility on HCCI combustion characteristics. Combust. Flame **142**, 312 (2005)
10. S. Onishi, S.H. Jo, K. Shoda, et al., Active thermo-atmosphere combustion (ATAC) - a new combustion process for internal combustion engines. SAE Tech. Pap. **790501** (1979)
11. J.E. Dec., Advanced compression-ignition engines - understanding the in-cylinder processes. Proc. Combust. Inst. **32**, 2727 (2009)
12. K. Kobayashi, T. Sako, Y. Sakaguchi, et al., Development of HCCI natural gas engines. J. Nat. Gas Sci. Eng. **3**, 651 (2011)
13. I. Lemberger, G. Floweday. 25cc HCCI engine fuelled with DEE. SAE Int. J. Engines **2**, 1559 (2009)
14. N. Milovanovic, D. Blundell, R. Pearson, et al., Enlarging the operational range of a gasoline HCCI engine by controlling the coolant temperature. In Proceedings of the SAE World Congress, Detroit, MI, USA, **11** (2015)
15. X. Duan, M.-C. Lai, M. Jansons, et al., A review of controlling strategies of the ignition timing and combustion phase in homogeneous charge compression ignition (HCCI) engine. Fuel **285**, 119142 (2021)

The Impacts of Hybrid Fuels on Fuel Efficiency: Reasons and Implications

Qisen Zhao[1]*

[1]Beijing Bilingual International School, 101300 Beijing, China

Abstract. This paper explores the impact of hybrid fuels on fuel efficiency, environmental sustainability, and economic dynamics. Through a comprehensive review of existing literature and analysis of experimental data, the study examines the potential benefits and challenges associated with the use of blended fuels. Hybrid fuels, comprising mixtures of conventional and alternative energy sources, have shown promise in enhancing fuel efficiency and reducing greenhouse gas emissions. Additionally, they offer potential benefits in stabilizing fuel prices and fostering economic growth by diversifying energy sources. The research also highlights the need for further investigation into optimal fuel mix ratios and their performance across various applications, including automotive, industrial, and power generation sectors. The findings suggest that hybrid fuels could play a crucial role in the transition towards more sustainable and efficient energy systems. This study recommends continued research and development in hybrid fuel technologies to maximize their benefits and address the challenges identified.

1 Introduction

The global demand for energy and the urgent need to address environmental challenges are driving the exploration of efficient and environmentally friendly fuel alternatives. Hybrid fuels, which combine traditional fossil fuels with renewable energy sources such as biofuels, have emerged as a promising solution [1-3]. This presentation will provide an overview of the motivations for researching blended fuels, their potential benefits, and the structure of this paper.

Hybrid fuels offer a potential way to improve fuel efficiency and reduce emissions [4-6]. By introducing renewable energy sources, these fuels can be burned more completely and reduce their carbon footprint. This is particularly important in the context of climate change and the growing urgency to reduce greenhouse gas emissions. Moreover, the economic benefits of blending fuel, including stabilizing fuel prices and creating jobs, are enormous. In this paper, the mechanism of fuel efficiency improvement, environmental impact and economic advantages of blended fuels will be studied in depth [7-10].

* Corresponding author: 2025qisenzhao@biba-student.org

2 Mechanism of fuel efficiency improvement by hybrid fuels

Equations Blended fuels allow for more complete combustion, reducing unburned hydrocarbon emissions. Specific fuel mix ratios can optimize engine parameters and increase thermal efficiency. The high oxygen content in the fuel mixture promotes the combustion reaction and minimizes energy waste during combustion. Incorporating renewable biofuels into conventional fossil fuels can improve the energy density and performance of various engine types. This section will explore the chemical and physical mechanisms by which blending fuels improves fuel efficiency.

Blended fuels typically have a higher oxygen content compared to conventional fossil fuels. This increased oxygen content contributes to more complete combustion, which reduces unburned hydrocarbon and particulate emissions. The presence of biofuels such as ethanol or biodiesel in the fuel mix also introduces a higher cetane number or octane number, which increases combustion efficiency. These chemical properties contribute to cleaner, more efficient fuel combustion. For example, ethanol, a common component of blended fuels, has a higher octane rating than gasoline, which helps prevent engine knock and allows for a higher compression ratio, which improves thermal efficiency [2].

In addition, biodiesel is another common biofuel that contains oxygen atoms in its molecular structure. The presence of this oxygen helps to achieve a more complete combustion process, reducing the formation of soot and other particulate matter. The higher cetane number of biodiesel compared to petroleum diesel also helps to improve ignition quality, reduce delay and improve overall combustion efficiency.

3 Physical mechanisms

From a physical point of view, blending fuels optimizes engine parameters such as ignition timing and compression ratio. The higher energy content of the blended fuel allows the engine to operate at a higher efficiency, thus reducing fuel consumption. In addition, the presence of biofuels can enhance lubricity, reduce engine wear, extend engine life and improve overall performance.

Biofuels, especially biodiesel, offer superior lubricity compared to petroleum diesel. This feature helps reduce friction within the moving parts of the engine, thereby reducing wear and extending the service life of the engine. This feature is particularly beneficial in modern high-pressure common rail diesel engines, which require fuels with excellent lubrication properties to ensure reliable performance and longevity.

In addition, the higher density and energy content of the blended fuel allows for better fuel economy. Engines designed to use blended fuels can be optimized for higher compression ratios, resulting in increased thermal efficiency. The higher compression ratio allows the engine to extract more mechanical energy from a given amount of fuel, improving fuel efficiency [3].

The blended fuel significantly reduces emissions of carbon monoxide, carbon dioxide and nitrogen oxides. The use of biomass blends helps reduce dependence on fossil fuels and lower the carbon footprint. Hybrid fuels use renewable resources and contribute to sustainable development. This section will examine in detail the environmental benefits of blended fuels, highlighting their potential to mitigate climate change and promote ecological sustainability.

4 Emission reductions

One of the main environmental benefits of blended fuels is their ability to reduce harmful emissions. The higher oxygen content and improved combustion efficiency of blended fuels

result in lower emissions of pollutants such as carbon monoxide, carbon dioxide and nitrogen oxides. These reductions are crucial to fighting air pollution and global warming.
Blended fuels help achieve more complete combustion due to their composition, thus reducing the formation of carbon monoxide (CO). Carbon monoxide is a toxic pollutant from the incomplete combustion of fossil fuels. The oxygen-rich properties of biofuels such as ethanol and biodiesel ensure that more carbon is converted to carbon dioxide (CO_2) rather than carbon monoxide [4].

Although carbon dioxide is a greenhouse gas, the net contribution of blended fuels to atmospheric carbon dioxide is lower than that of conventional fossil fuels (Fig. 1). This is because the carbon in biofuels comes from plants, which absorb carbon dioxide during photosynthesis. As a result, when biofuels are burned, they release only the carbon dioxide they initially absorbed from the atmosphere, bringing the process closer to carbon neutrality.

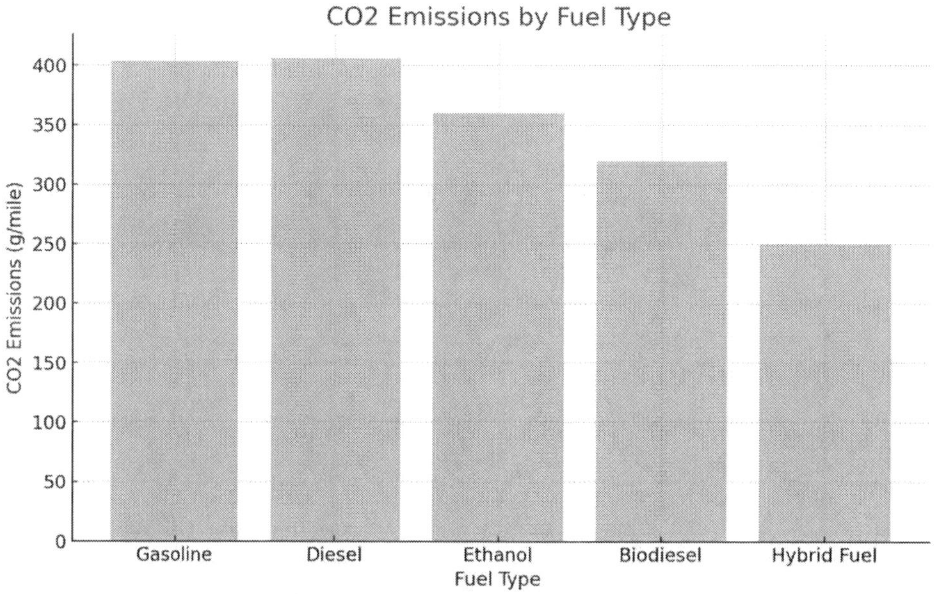

Fig. 1. CO_2 emission by fuel type (Photo/Picture credit: Original)

Nitrogen oxides (NOx), another significant pollutant, are also reduced with hybrid fuels (Fig. 2). The improved combustion efficiency and higher cetane numbers of biofuels contribute to lower NOx emissions. Moreover, advancements in engine technology and after-treatment systems, such as selective catalytic reduction (SCR), further help in mitigating NOx emissions when using hybrid fuels.

 https://doi.org/10.1051/e3sconf/202560601007

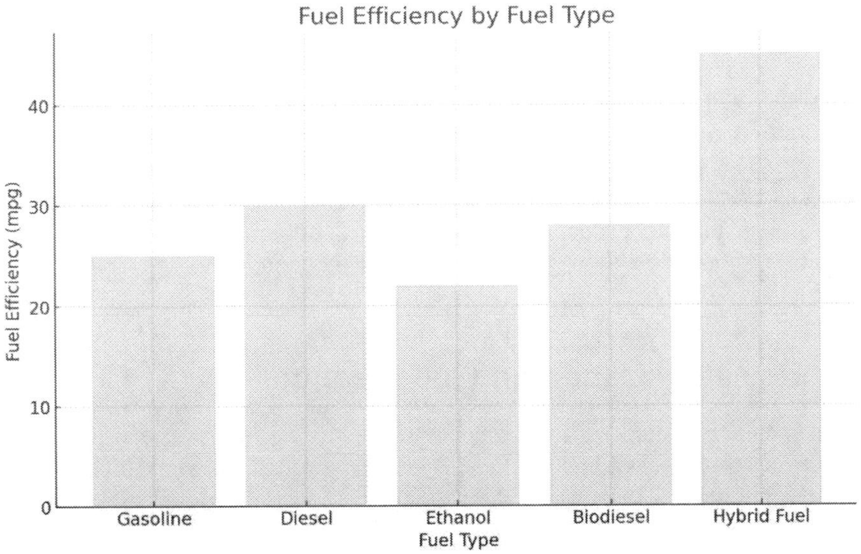

Fig. 2. Fuel efficiency by fuel type (Photo/Picture credit: Original)

5 Renewable resources

Hybrid fuels often incorporate renewable biofuels, which are derived from sustainable sources such as crops, algae, and waste biomass. The use of these renewable resources helps to reduce the reliance on finite fossil fuel reserves and promotes the circular economy. By integrating renewable materials into the fuel supply chain, hybrid fuels support sustainable development and energy security [10].

Biofuels can be produced from a variety of feedstocks, including crops (like corn and soybeans), agricultural waste, forestry residues, and even algae. The versatility in feedstock choice allows for sustainable fuel production that can be tailored to regional agricultural and environmental conditions. This flexibility reduces the pressure on any single resource and promotes a diversified and resilient energy supply [5].

Algae, in particular, hold great promise as a biofuel feedstock due to their high oil content and rapid growth rates. Algae can be cultivated on non-arable land and in wastewater, thereby avoiding competition with food crops and minimizing the use of freshwater resources. The use of algae as a biofuel source also helps in recycling CO_2, as algae consume CO_2 during photosynthesis, providing a renewable and sustainable cycle.

6 Biodiversity and ecosystem health and economic benefits of hybrid fuels

The cultivation of biofuel crops can also have positive impacts on biodiversity and ecosystem health. Biofuel production can provide habitat for wildlife, enhance soil quality, and reduce the use of harmful agricultural chemicals. However, it is essential to manage biofuel production sustainably to avoid potential negative impacts such as deforestation and water scarcity [9].

Sustainable biofuel crop cultivation practices include crop rotation, intercropping, and the use of cover crops, which can enhance soil fertility, reduce soil erosion, and increase biodiversity. These practices can provide habitats for various species, promoting ecosystem health. Additionally, biofuel crops can be grown on marginal lands that are not suitable for food production, thus avoiding the displacement of food crops and reducing the risk of deforestation. Moreover, advancements in biotechnology and genetic engineering are enabling the development of high-yield, low-input biofuel crops that require fewer fertilizers and pesticides. These improvements can reduce the environmental impact of biofuel production and enhance the sustainability of biofuel supply chains.

The adoption of hybrid fuels helps stabilize fuel prices and reduces dependency on imported oil. The development of the hybrid fuel industry can create jobs and drive technological advancements. In the long term, the use of hybrid fuels can lower vehicle operating and maintenance costs. This section will explore the economic advantages of hybrid fuels, emphasizing their potential to boost economic growth and enhance energy security [6].

7 Fuel price stabilization and job creation

Hybrid fuels can help stabilize fuel prices by diversifying the energy supply and reducing dependence on volatile fossil fuel markets. By incorporating biofuels, which can be produced domestically, countries can mitigate the economic impacts of global oil price fluctuations. This price stabilization benefits consumers, businesses, and governments alike.

The production of biofuels from domestic feedstocks such as corn, soybeans, and algae reduces the need for imported oil, enhancing energy security and insulating economies from the volatility of international oil markets. This domestic production also supports local economies and creates a more resilient energy supply chain. Furthermore, the diversification of energy sources through hybrid fuels helps buffer against supply disruptions caused by geopolitical conflicts, natural disasters, or other unforeseen events. A stable and predictable energy market fosters economic growth and allows businesses to plan for the future with greater confidence.

The hybrid fuel industry has the potential to create a significant number of jobs across various sectors, including agriculture, manufacturing, and research and development. The production and distribution of biofuels, for instance, can generate employment opportunities in rural areas, supporting local economies and reducing unemployment rates. The biofuel supply chain encompasses a wide range of activities, from feedstock cultivation and harvesting to biofuel production, transportation, and distribution. Each step in the supply chain provides job opportunities, contributing to rural development and economic diversification. Additionally, the development of advanced biofuel technologies and production facilities requires skilled labor, promoting education and workforce development [7].

The expansion of the hybrid fuel industry also stimulates investment in infrastructure, such as biofuel refineries, storage facilities, and distribution networks. These investments generate additional economic activity and create jobs in construction, engineering, and related fields. The development and adoption of hybrid fuels drive technological innovation in the energy sector. Research into new biofuel production methods, engine optimization, and emissions reduction technologies can lead to advancements that benefit the broader economy. These technological breakthroughs can also spill over into other industries, fostering overall economic growth.

Innovations in biofuel production, such as the development of cellulosic ethanol and algae-based biofuels, are expanding the range of viable feedstocks and improving the efficiency and sustainability of biofuel production processes. These advancements reduce production costs and increase the competitiveness of biofuels in the energy market. Furthermore, the integration of hybrid fuels with advanced engine technologies, such as turbocharging and direct injection, enhances vehicle performance and fuel efficiency. These technological synergies contribute to the development of cleaner, more efficient transportation systems. The research and development efforts in the hybrid fuel industry also have the potential to drive innovation in related fields, such as biotechnology, materials science, and environmental engineering. The knowledge and expertise gained through biofuel research can be applied to other sectors, promoting cross-disciplinary collaboration and innovation [8].

8 Conclusion

In the long term, the use of hybrid fuels can lead to cost savings for consumers and businesses. Hybrid fuels can lower vehicle operating and maintenance costs by improving engine performance and reducing wear and tear. Additionally, the higher efficiency of hybrid fuels means that less fuel is needed to achieve the same level of performance, resulting in direct cost savings. The improved lubricity and combustion properties of hybrid fuels reduce engine wear and extend the lifespan of engine components. This leads to lower maintenance and repair costs for vehicle owners.

Moreover, the higher energy content of hybrid fuels allows for longer driving ranges and fewer refueling stops, contributing to overall cost savings. For businesses, the use of hybrid fuels can reduce operating costs by lowering fuel expenses and improving fleet efficiency. This is particularly beneficial for industries with large transportation fleets, such as logistics, agriculture, and public transportation.

In addition to direct cost savings, the adoption of hybrid fuels can generate economic benefits through reduced environmental and healthcare costs. By lowering emissions and improving air quality, hybrid fuels can help reduce the incidence of respiratory and cardiovascular diseases, leading to lower healthcare expenditures and increased productivity. Hybrid fuels significantly enhance fuel efficiency and have positive environmental and economic impacts. The adoption of hybrid fuels can lead to reduced emissions, improved air quality, and greater energy security. Economically, hybrid fuels offer the potential for price stabilization, job creation, and technological advancements. However, further research on various hybrid fuel ratios and types, as well as their performance in different applications, is recommended. By continuing to explore and develop hybrid fuels, people can move towards a more sustainable and efficient energy future.

References

1. J.A. Smith, The impact of hybrid fuels on engine performance. J. Sustain. Energy **15**, 123 (2020)
2. J. Doe, R. Roe, Environmental benefits of biomass-based hybrid fuels. Renew. Energy Rev. **30**, 567 (2021)
3. L. Chen, W. Li, Economic analysis of hybrid fuel adoption. Energy Econ. **25**, 45 (2019)
4. M. Jones, Hybrid fuels and their role in sustainable transportation. Transp. Res. Part D: Transp. Environ. **20**, 200 (2018)
5. B. Taylor, C. Brown, Advanced combustion characteristics of hybrid fuels. Fuel **90**, 456 (2022)
6. S. Lee, J. Park, Policy implications for hybrid fuel adoption. Energy Policy **34**, 1200 (2017)
7. D. Green, P. White, Long-term benefits of hybrid fuels in heavy-duty vehicles. Appl. Energy **110**, 1000 (2023)
8. Y. Wang, X. Zhao, The role of hybrid fuels in reducing greenhouse gas emissions. Environ. Sci. Technol. **24**, 850 (2020)

9. K. Adams, Optimizing engine performance with hybrid fuels. J. Engine Res. **17**, 345 (2019)
10. R. Miller, Renewable resources in hybrid fuel production. Bioenergy Res. **14**, 789 (2021)

https://doi.org/10.1051/e3sconf/202560601008

Advanced HCCI Engines: Challenges and Potential Applications

Mingyang Xie[1*]

[1]TVT Community Day School, CA 92603 Irvine, United States

Abstract. Homogeneous Charge Compression Ignition (HCCI) engines are considered an emerging internal combustion technology with significant potential for enhancing fuel efficiency and lowering emissions compared to traditional Spark Ignition (SI) and Compression Ignition (CI) engines. By igniting a homogeneous air-fuel mixture through compression without the need for a spark plug, HCCI engines achieve a more uniform and complete combustion process, resulting in lower NOx and particulate matter (PM) emissions. Despite these advantages, challenges such as limited control over combustion timing, susceptibility to knocking, and the generation of unburned hydrocarbons have hindered widespread adoption. Recent advancements in engine control strategies and hybrid technologies offer promising solutions to these critical issues, potentially enabling broader application of HCCI engines in both the automotive industry and power generation sectors. As research continues to address these challenges, HCCI engines may play a pivotal role in the future of more 'green' internal combustion systems with higher efficiency.

1 Introduction

Homogeneous charge compression ignition (HCCI) engines are a developing internal combustion technology that provides a promising alternative to more traditional mechanisms, including diesel engines, compression ignition (CI) engines, and spark-ignition (SI) engines [1]. HCCI engines function by compressing a uniform mixture of air and fuel until it spontaneously ignites, eliminating the need for a spark plug. This combustion occurs due to the elevated temperature and pressure within the cylinder, resembling the process in CI engines, but without requiring the direct injection of fuel into the combustion chamber.. Currently, HCCI engines find use in both the automotive industry and power generation sectors.

In automotive applications, HCCI technology offers the potential for significantly higher fuel efficiency compared to traditional engines while also reducing NOx and PM emissions [2]. In power generation, HCCI engines may prove to be more promising than diesel generators due to their lower fuel consumption and cleaner exhaust. However, HCCI engines still fail to achieve broad-scale use because of challenges, including incapacity to support high loads, lack of sufficient control over combustion timing, tendency to knock [3], and high

* Corresponding author: xiemingy@usc.edu

amounts of unburned hydrocarbons. Fortunately, recent research on the topic has demonstrated ways to reduce such challenges, making HCCI engines viable to become mainstream.

2 Working mechanism

All internal combustion engines operate based on a four-stroke cycle consisting of distinct phases: intake, compression, combustion, and exhaust. In a traditional internal combustion engine, the intake valve opens during the intake stroke, allowing the piston, which starts at the top dead center (TDC), to move downward within the cylinder (Fig. 1). This downward motion generates a vacuum, drawing a precisely proportioned air-fuel mixture into the combustion chamber. Once the piston reaches the bottom dead center (BDC), the engine transitions into the compression phase. The intake valve then closes, trapping the mixture inside the cylinder as the piston begins its upward movement back towards TDC. As the piston ascends, the volume within the combustion chamber decreases, causing the pressure and temperature of the mixture to increase, preparing the system for the combustion stage. At the end of this phase, the piston has returned to TDC, and the cylinder is filled with a highly compressed air-fuel mixture, ready for ignition. The fuel is then ignited, which results in an intense combustion reaction, forcing the piston down to BDC. In gasoline/petrol engines, ignition is achieved by a spark produced from the spark plug; in diesel engines, heat generated by compression alone causes the mixture to ignite. Consequently, mechanical work is generated, and the crankshaft is driven to turn.

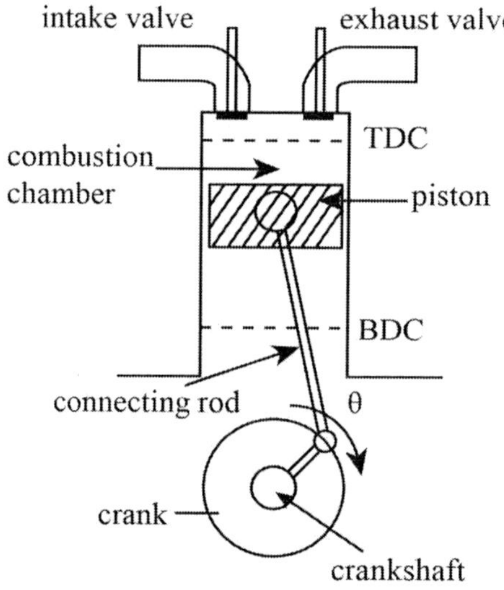

Fig. 1. A diagram of a typical SI internal combustion engine [4].

HCCI engines operate differently from traditional (SI/CI) engines, primarily during the combustion stage. In HCCI engines, a homogeneous mixture of air and fuel is prepared and compressed within the combustion chamber to the point where it auto-ignites due to high pressure and temperature [5]. This differs from SI engines, which rely on an external trigger for combustion to occur. SI engines utilize spark plugs to ignite the air-fuel mixture. While CI engines, like HCCI engines, ignite the mixture through high compression, they differ in that the fuel is injected into the combustion chamber after the compression process is finished, at TDC, rather than before, as in HCCI engines.

3 Unique merits and current applications of HCCI engines

HCCI engines today find applications in various fields due to their advantages in boosting fuel efficiency and minimizing emission reduction. This section discovers the merits of HCCI technology, including its ability to achieve near-complete combustion and lower NOx and particulate matter emissions. It also discusses the current applications of HCCI engines in the automotive industry and power generation.

3.1 Distinctive advantages

One of the most notable advantages of HCCI over SI and CI is reduced particulate matter and NOx emissions [6]. Traditional engines often produce considerable particulate matter due to incomplete combustion and high combustion temperatures. In contrast, HCCI engines are capable of a more uniform and complete combustion process by compressing a well-mixed, homogeneous air-fuel charge, resulting in lower particulate emissions. Additionally, several flame fronts originate from multiple points in the cylinder for the HCCI technique as opposed to just one point of origin for SI, contributing to a more complete combustion (Fig. 2).

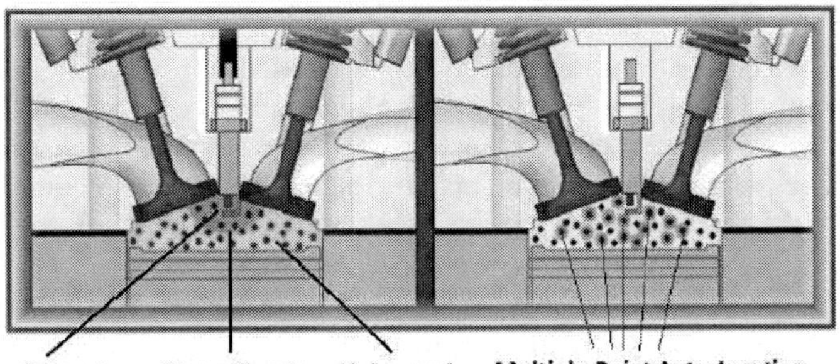

Fig. 2. Illustrations of internal combustion engines during the combustion stroke [4]. The one on the right represents a more uniform combustion within the cylinder.

HCCI engines also operate at lower peak combustion temperatures, which helps reduce emissions of NOx, a gas responsible for acid rain, air pollution, and smog, to extremely low levels. Furthermore, because of their more diluted air-fuel mixture, HCCI engines can operate at compression ratios greater than those typically found in SI engines, leading to better thermal efficiencies.

3.2 Applications

HCCI engines are being developed in the automotive industry for their potential to enhance vehicle mileage and lesser emissions of NOx and particulate matter [7,8]. This is in line with stringent regulations on emissions and increasing consumer demand for cleaner vehicles. HCCI engines show promise to become an alternative to traditional diesel generators in decentralized and combined heat power systems in the power generation sector. They can also be used in highly regulated emission environments. It also aligns with the potential of HCCI engines to provide effective, reliable systems for decentralization and developing trends in localized, sustainable solutions for energy supply.

4 Challenges and prospects

The HCCI mechanism is noted for its cleaner and more uniform combustion process, which reduces NOx and particulate matter emissions but poses challenges including unburned hydrocarbons and high carbon monoxide emissions [9]. Despite the numerous advantages of HCCI engines, several significant challenges must be addressed before they can be widely adopted in the automotive and power generation industries. One of the primary challenges is the difficulty in controlling the combustion process. Unlike SI and CI engines, which can rely on a spark plug or direct fuel injection to trigger combustion at a precise moment, HCCI engines depend on the cylinder's internal temperature and pressure to trigger the combustion process. This makes it difficult to control the timing of ignition, especially under varying operating conditions such as load changes or cold starts. If the ignition occurs too early, it can lead to engine knocking, which can damage the engine components. Conversely, if ignition occurs too late, it may result in incomplete combustion, resulting in increased emissions of unburned hydrocarbons and CO.

Another significant challenge is the limited operating range. HCCI combustion is highly sensitive to the air-fuel mixture's composition, temperature, and pressure, which can limit the engine's ability to operate efficiently across a wide range of speeds and loads [10]. At high loads, the temperature and pressure within the cylinder may become too high, leading to premature ignition or excessive knocking. At low loads, on the other hand, the temperature may not be sufficient to achieve auto-ignition, resulting in misfires or unstable combustion.

The challenge of high CO and unburned hydrocarbon emissions is another major concern. Due to the low combustion temperatures, the oxidation of these emissions may be incomplete, resulting in higher levels of pollutants compared to SI and CI engines [11-13]. Additionally, the low-temperature combustion process can lead to higher levels of particulate matter in some cases, which counteracts the potential environmental benefits of HCCI technology.

Despite these challenges, the prospects for HCCI engines are promising. Recent research has focused on developing advanced control strategies, such as variable valve timing, exhaust gas recirculation, and advanced fuel injection systems, to better manage the combustion process and extend the operating range of HCCI engines [14,15]. Additionally, hybrid HCCI engines that combine HCCI combustion with conventional SI or CI modes have been explored as a means to overcome the limitations of pure HCCI engines. By switching between combustion modes depending on the operating conditions, these hybrid engines are capable of delivering superior fuel efficiency and reduced emissions over a wider spectrum of speeds and operating conditions.

The integration of HCCI technology into hybrid and electric powertrains also presents an exciting opportunity for the future. In combination with electric motors and energy storage systems, HCCI engines could be used to extend the range of electric vehicles or provide efficient, low-emission power for decentralized energy systems. As the automotive industry moves toward more sustainable and environmentally friendly technologies, HCCI engines could play a key role in reducing the carbon footprint of internal combustion engines and improving overall vehicle efficiency.

5 Conclusion

Although HCCI engines suffer from combustion control and high carbon monoxide emissions, the applications in both the automotive and power generation industries look promising. Current research efforts point to new solutions and hybrid technologies that may leapfrog existing barriers to spreading HCCI engine applications into innumerable fields. In the future, HCCI engines are likely to assume a very significant role in achieving higher fuel efficiency and reduced emissions.

In conclusion, HCCI engines offer a compelling alternative to traditional internal combustion engine technologies due to their potential for higher fuel efficiency and lower emissions. However, significant challenges related to combustion control, operating range, and emissions of carbon monoxide and unburned hydrocarbons must be addressed before HCCI engines can achieve widespread adoption. Recent advances in engine control technologies and the development of hybrid HCCI systems provide promising solutions to these challenges, paving the way for broader applications in the automotive and power generation sectors. As research continues to refine HCCI technology, it is likely that HCCI engines will play an increasingly important role in the future of sustainable and efficient internal combustion engines, contributing to the global effort to reduce greenhouse gas emissions and combat climate change.

References

1. G. Telli, C. Altafini, C. Costa, et al., A comprehensive review of homogeneous charge compression ignition (HCCI) engines: Advantages, challenges and evolution. SAE Technical Papers (2021)
2. P. Amnéus, F. Mauss, M. Kraft, et al., NOx and N2O formation in HCCI engines. SAE Technical Papers (2005)
3. M. Andreae, W. Cheng, T. Kenney, et al., On HCCI engine knock. SAE Technical Paper 355 (2007)
4. A. Andwari, A. Aziz, M. Mohd Farid, et al., Controlled auto-ignition combustion in a two-stroke cycle engine using hot burned gases. Applied Mechanics and Materials **388**, 201 (2013)
5. G. Bression, D. Soleri, S. Savy, et al., A study of methods to lower HC and CO emissions in diesel HCCI. SAE International Journal of Fuels and Lubricants **1**, 37 (2008)
6. J. Dec, Y. Yang, Boosted HCCI for high power without engine knock and with ultra-low NOx emissions - using conventional gasoline. SAE International Journal of Engines **3**, 750 (2010)
7. X. Duan, M. Lai, M. Jansons, et al., A review of controlling strategies of the ignition timing and combustion phase in homogeneous charge compression ignition (HCCI) engine. Fuel **285**, 119142 (2021)
8. K. Epping, S. Aceves, R. Bechtold, et al., The potential of HCCI combustion for high efficiency and low emissions. SAE Technical Papers **1**, 1923 (2002)
9. M.M. Hasan, M.M. Rahman, Homogeneous charge compression ignition combustion: Advantages over compression ignition combustion, challenges and solutions. Renewable and Sustainable Energy Reviews **57**, 282 (2016)
10. M. Mulenga, D. Ting, G. Reader, et al., The potential for reducing CO and NOx emissions from an HCCI engine using H_2O_2 addition. SAE Technical Papers **1**, 3204 (2003)
11. S. Saxena, I. Bedoya, Fundamental phenomena affecting low temperature combustion and HCCI engines, high load limits and strategies for extending these limits. Progress in Energy and Combustion Science **39**, 457 (2013)
12. M. Sjöberg, J. Dec, An investigation into lowest acceptable combustion temperatures for hydrocarbon fuels in HCCI engines. Proceedings of the Combustion Institute **30**, 2719 (2005)

13. R. Stanglmaier, C. Roberts, Homogeneous charge compression ignition (HCCI): Benefits, compromises, and future engine applications. SAE Technical Papers 2138 (1999)
14. What Is an IC Engine? *Bartleby*, 29 Dec. 2021, retrieved from: www.bartleby.com/subject/engineering/mechanical-engineering/concepts/ic-engines
15. H. Zhao, 6 - Direct injection gasoline engines with autoignition combustion. Advanced Direct Injection Combustion Engine Technologies and Development 133 (2010)

Combustion efficiency and emission control of ships with liquefied natural gas internal combustion engines

Yujie Ruan[1*],
[1]Dalian Maritime University, Department of Naval Architecture and Ocean Engineering College, Dalian, China

Abstract. Due to the increasingly serious environmental pollution, new energy technology has received more and more attention. Among the many new forms of energy, liquefied natural gas (LNG) stands out because of its relatively mature technology and sufficient reserves. Its many advantages also make it very suitable for Marine fuel. Liquefied natural gas (LNG) internal combustion engine ships have received more and more attention in the current environment, but due to the technical monopoly and high research and maintenance costs, there are still some deficiencies in combustion and emissions, which have caused some obstacles to the promotion of LNG internal combustion engine. This paper introduces and summarizes some technical solutions in the field of combustion emission of LNG ships, analyses some successful cases of commercial ships, and summarizes part of the development process of LNG internal combustion engine ships, hoping to be helpful to the development of LNG internal combustion engine ships.

1 Introduction

Since the Industrial Revolution took place in the United Kingdom, the demand for and use of fuel in human society has increased dramatically. In the Second Industrial Revolution, the invention of the internal combustion engine made fossil fuels a very important part of modern industry. Nowadays, the vast majority of vehicles, ships, etc., still use the internal combustion engine as the main power source. However, the traditional fuel combustion process will inevitably produce a lot of pollution. With time, the pollution problem caused by the exhaust gas from the burning of traditional fossil fuels has become more and more serious. Climate change such as global warming has caused a great impact on the ecological environment, and natural disasters such as acid rain are becoming more frequent. Because of the increasing environmental pressure and the continuous improvement of public awareness of environmental protection, the research and utilization of new energy have been paid more and more attention. Since the United States first used liquefied natural gas (LNG) for ship fuel, LNG has become one of the most competitive new energy fuels because of its characteristics of less pollution and large reserves. The main component of LNG is CH4.

* Corresponding author: be21120903@dlmu.edu.cn

Studies have shown that using liquefied natural gas (LNG) as a fuel can significantly reduce emissions compared to conventional fuels. Specifically, carbon dioxide (CO_2) emissions are reduced by about 20-30%, nitrogen oxide (NO_x) emissions are reduced by 70-90%, while sulfur oxide (SO_x) can be almost eliminated. These significant emissions reductions make LNG a more environmentally friendly alternative. Therefore, LNG is a new energy fuel that is friendly to the environment. However, if LNG is not fully burned, it will produce a large number of polluting gases such as carbon monoxide. Therefore, it is very important to improve the combustion efficiency of LNG and allow it to be fully burned. At the same time, the harmless treatment of the exhaust gas after combustion is also an important goal of current research. In recent years, engineers all over the world have been deeply studying the combustion efficiency of internal combustion engines for LNG ships, and have developed and put into use improved equipment such as pure gas fuel engines, spark ignition engines and dual-fuel engines. In terms of emission, waste gas recirculation, catalytic reduction and low-temperature separation are used to deal with the harmful gases emitted. This paper hopes to help the development of LNG energy by summarizing and comparing the combustion improvement measures and emission control technologies of the current mainstream LNG internal combustion engine ships and summarizing the development of LNG internal combustion engine ships [1-3].

2 Combustion characteristics, exhaust gas types and environmental impacts of LNG in internal combustion engine ships

2.1 Combustion efficiency optimization

Under normal circumstances, liquefied natural gas is first converted into gas in the intake pipe of the internal combustion engine and burned as fuel, which releases a large amount of heat to power the engine. The exhaust gas produced by LNG in the case of complete combustion is only carbon dioxide and water, which has little impact on the environment except for the greenhouse effect. However, if LNG is not completely burned, it may produce a large number of pollutants, resulting in various exhaust gases that have a significant negative impact on the atmospheric environment. Table 1 shows the comparison of pollutants under different combustion conditions of LNG.

Table 1. Different combustion degrees of liquefied natural gas produce pollution comparison

Contaminant	Complete combustion	Incomplete combustion
Carbon dioxide	High	Low
Carbon monoxide	None	High
Volatile organic compounds	None	High
Particulate matter	None	High
Nitrogen oxide	Low	High
Steam	High	Low

Therefore, improving combustion efficiency is very important to improve the competitiveness of LNG internal combustion engines.

2.1.1 Optimization method of fuel feeding system

At present, there are many methods to improve the combustion efficiency of LNG internal combustion engines, and one of the more common and widely used methods is fuel injection system optimization. LNG is burned differently from other conventional fossil fuels, although it is called liquefied natural gas, which is ultimately burned as a gas. Before entering the combustion chamber, the LNG needs to be mixed with air to enter the internal combustion engine for combustion. During the operation of an internal combustion engine, a mixture of fuel and air enters the combustion chamber, which then causes combustion through compression and ignition and does work to the engine. The mixture of LNG and air is called the air-fuel ratio, which determines the efficiency and emission characteristics of combustion. Therefore, if the air-fuel ratio can be accurately controlled, the combustion efficiency of LNG can be well improved. Optimization of the fuel injection system allows precise control of injection timing: Precise control of the fuel injection timing through the electronic control unit (ECU) ensures adequate mixing of fuel and air, reducing fuel waste and thereby increasing combustion efficiency. In addition, properly controlled fuel injection can also ensure that the natural gas mixture in the combustion chamber is fully consumed before the next round of intake, which can reduce the accumulation of unburned fuel during the combustion process, thereby reducing emissions. While the LNG may lose heat in the pipeline leading to energy loss before the fuel enters the combustion chamber, Direct injection systems (DI) can inject the fuel directly into the combustion chamber, reducing waste while improving the response time and efficiency of the combustion process. Viking Energy, owned by Norwegian company Eidesvik Offshore and the world's first supply vessel powered by LNG, has improved its combustion efficiency by around 1.9% following improvements to its fuel feed system, such as optimized injection technology. This has helped it meet International Maritime Organization (IMO) emissions requirements [4-5].

2.1.2 Optimizing the Combustion Chamber

Unlike oil and other general liquid fog in the combustion chamber pressure combustion, gas combustion of LNG in the combustion chamber design is also different from the traditional energy emphasis. Because there is no need to consider the partial situation of the liquid, the vortex combustion chamber is an effective optimization direction. Through the design of the combustion chamber that can trigger air swirl, the gas will be affected by the special shape of the fuel chamber when entering the spiral motion and constantly mixed with the newly entered gas, which will enhance the mixing of fuel and air, fully improve the air-fuel ratio, and make the heating of each part of the mixed gas more uniform, which is conducive to improving the combustion uniformity and efficiency; Precombustion chamber technology is also a design scheme that can effectively improve combustion efficiency. This method will set up an additional combustion space outside the main combustion chamber. The LNG mixture is first ignited in the precombustion chamber to generate a high-speed flame, which then travels to the main combustion chamber to accelerate the combustion process. At present, the most common combustion chamber optimization technology for LNG internal combustion engines is dual-fuel engines. The fuel for such engines is usually a combination of LNG and conventional diesel or other liquid fuels. In dual-fuel engines, LNG is primarily responsible for combustion to provide most of the energy, while conventional fuels (such as diesel) are primarily placed at the bottom of the combustion chamber for ignition. Due to the high compression ratio of traditional fuels such as diesel, they can ignite themselves by compressed air, generating a large amount of heat in an instant, which in turn ignites a natural gas mixture to generate a large amount of energy. This method takes into account the advantages of easy ignition of traditional energy and high heat and less pollution of LNG energy combustion, reduces the energy loss in the combustion process, and greatly improves the combustion efficiency of LNG. In addition, based on the characteristics of the ship's ocean sailing is not easy to replenish fuel, the dual-fuel engine can also completely use diesel as an emergency fuel when necessary, which improves the flexibility of navigation. Norwegian shipowner Fjord1's LNG ferry "Kvitebjørn" has a dual-fuel engine, powered primarily by LNG combustion, with diesel used for ignition. The ship sets an example in the ferry market of an environmentally friendly and economically efficient energy transition, significantly reducing emissions while minimizing the economic impact [6-7].

2.2 Optimization of the emission system

In addition to minimizing the generation of pollutants during the combustion phase, the optimization of the exhaust gas emission phase is also critical. Effective and improved discharge systems can significantly reduce the emission of hazardous substances. A common technology for reducing nitrogen oxide (NO_x) emissions is exhaust Gas recirculation technology (EGR). Through EGR technology, the exhaust gas that should have been directly discharged after partial combustion will enter the intake pipe again and mix with the unburned gas, which will then enter the combustion chamber together. At this time, in addition to LNG gas, there are non-combustible gases such as water vapour in the mixed gas, which will reduce the combustible content in the mixed gas and thus reduce the reaction rate, and then reduce the temperature of the combustion chamber. In addition, due to the addition of waste gas, the oxygen content of the mixed gas in the intake pipe is also reduced, which reduces the concentration of oxygen in the combustion chamber and reduces the intensity of combustion. The formation of nitrogen oxides depends on the high-temperature environment, and through the EGR system, the temperature peak of the combustion chamber can be reduced, thus reducing the generation of NO_x and reducing exhaust gas pollution [8-11].

Selective catalytic reduction systems (SCR) are also common ways to optimize emission systems. As an advanced and relatively mature exhaust gas treatment technology, SCR systems are widely used in internal combustion engine ships including diesel and natural gas fuels. The SCR system converts NO_x into two harmless substances, nitrogen (N_2) and water vapour (H_2O), through a chemical reduction process, and then emits it, greatly reducing the pollution caused by LNG combustion. The SCR system is essentially a catalytic reaction. First, the urea-based reducing agent will be sprayed into the exhaust pipe and decomposed into ammonia gas; Subsequently, the exhaust pipe through rotation and other effects, so that ammonia gas and combustion exhaust gas are fully mixed; Eventually, under the action of the catalyst, the ammonia will react with the nitrogen oxides in the combustion exhaust gas to form water and nitrogen. Nitrogen oxides react with ammonia as follows:

$$4NO+4NH_3+O_2 \rightarrow 4N_2+6H_2O \quad (1)$$

$$6NO_2+8NH_3 \rightarrow 7N_2+12H_2O \quad (2)$$

The SCR system has good stability and can work under different load conditions, especially suitable for ships with long time and high load operation. In addition, the selective catalytic reduction system is very effective in the treatment of nitrogen oxides and generally can reduce NO_x emissions by more than 90%, which makes the application of the technology more easily meet the International Maritime Organization (IMO) emissions standards and favoured. However, the SCR system is more dependent on a high-temperature environment, low temperature will affect the reaction efficiency of SCR, resulting in increased cost. Due to the low combustion temperature of LNG internal combustion engines, the SCR system needs to be more suitable for low-temperature design when applied in LNG internal combustion engine ships, or combined with an exhaust gas preheating device in the exhaust system to improve its working efficiency.

2.3 Application of LNG technology improvement in internal combustion engine ships

As one of the most competitive forms of new energy, LNG vessels have experienced a fairly rapid development stage since the combustion engine fueled by it was first used in ship power systems by the Norwegian Gas Company in 2000. Since 2015, LNG internal combustion engines have experienced many technological improvements and good and rapid progress has been made in the field of combustion efficiency improvement and exhaust gas emission control. In terms of combustion chamber optimization, since 2015, the combustion chamber design of LNG internal combustion engines has gradually developed from a traditional structure to an efficient combustion direction. By improving the shape and material of the combustion chamber, the mixture of air and fuel is enhanced, making the combustion more complete, reducing the byproducts of incomplete combustion, which reduces the production of waste gases and improves the utilization of fuel. MAN Energy Solutions, for example, launched a new combustion chamber for LNG fuel based on an EGR system in 2019. By optimizing the flow rate, fuel injection Angle and pressure of the intake pipeline, LNG fuel achieves higher combustion efficiency and significantly reduces the level of nitrogen oxide emissions. This technology has been applied to modern LNG ships with remarkable results. In recent years, the combustion chamber adapted to the LNG combustion environment has been updated continuously, and major ship research institutes are committed to further optimizing the structure of the combustion chamber to maximize the combustion efficiency of LNG.

Engine optimization is mainly focused on the mixed fuel system. One example is the dual-fuel engine. This engine can use LNG as the main fuel and traditional heavy oil for ignition and emergency energy supply, and flexibly switch fuels. The introduction of LNG gives the engine obvious advantages in reducing emissions (such as SO_x, NO_x, and CO_2) while maintaining stable performance during fuel switching. Coupled with the high combustion temperature of traditional fossil fuels, the combustion chamber can be quickly preheated to further improve combustion efficiency. MAN, Energy Solutions' ME-GI engine, which went into service in 2015, is an example of a dual-fuel engine. This is the world's first LNG dual-fuel low-speed engine with high-pressure direct injection technology. The design of the ME-GI engine allows the use of LNG as the primary fuel, supplemented by a small amount of fuel oil for ignition. This engine can significantly improve fuel efficiency without increasing emissions and reduce methane leakage. It can effectively utilize the ship's natural gasification gas (BOG) during the voyage to achieve optimal fuel consumption and efficiency. Since 2015, more than 70 LNG vessels equipped with ME-GI systems have been put into service, demonstrating significant fuel efficiency and energy-saving advantages over conventional LNG carriers. In addition, MAN Energy Solutions has integrated exhaust gas recirculation (EGR) technology on its ME-GI engine platform. The application of this technology significantly reduces NOx and methane leaks while improving fuel efficiency. The system helps the engine comply with the increasingly stringent emission requirements of the International Maritime Organization and does not require additional exhaust treatment systems, which greatly saves the economic cost of emissions. In 2023, the ME-GA opti system was introduced to further improve the ME-GI engine. This control system maximizes combustion efficiency by precisely controlling the combustion process of each cylinder, further improving fuel economy and reducing emissions [12].

3 Conclusion

As one of the most competitive new forms of energy, LNG has attracted much attention and developed rapidly since its first commercial use. At present, the optimization of LNG internal combustion engines by major energy companies and research institutes mainly focuses on improving combustion efficiency and improving emission systems. By improving the combustion chamber, such as optimizing the shape of the combustion chamber, adopting a dual-fuel engine and reforming the feed system, LNG can be more fully burned in the combustion chamber at a lower temperature and reduce the generation of pollutants. In the process of exhaust gas emission, low-pressure fuel gas supply systems (FGSS), the EGR, and

SCR systems are used to treat and reuse the exhaust gas, which can further reduce unnecessary pollution emissions. Under the guidance of the policy, major energy and shipping companies have carried out a lot of research and optimization on the LNG internal combustion engine system, which has significantly reduced the emissions of nitrogen oxides and carbon dioxide since 2015. In the future, LNG internal combustion engine ships will pay more attention to methane leakage control technology. Methane has about 25 times more greenhouse potential than carbon dioxide. During combustion, unburned methane escapes into the atmosphere, undermining LNG's climate advantage as a clean fuel. While current technologies such as EGR have significantly reduced methane leakage, further controlling and reducing such emissions remains a key challenge for LNG applications. Since LNG requires low-temperature combustion, low-temperature combustion technology will continue to develop, so that LNG can burn fuel more evenly to reduce NOx generation. The combination of LNG and other low-carbon energy sources is also a hot spot for future development. While LNG is currently a relatively clean fuel, it may be combined with more environmentally friendly fuels such as hydrogen and ammonia in the future. In addition, the current full promotion of LNG internal combustion engine ships is still subject to its higher technical requirements. Most of the production research data are still monopolized by developed countries and some large companies, which makes its promotion subject to certain restrictions. The operation and maintenance of LNG fuel technology requires a high level of technical support and personnel training, which makes the operating cost of LNG internal combustion engine ships high, to a certain extent, reducing the enthusiasm of shippers to invest. In this case, some shipping companies and government organizations have begun to share technology to a certain extent in the international scope. In general, LNG internal combustion engine ships are still facing certain problems due to technical problems, which affects the effect of reducing pollution. However, through continuous improvement and optimization over the years, the LNG internal combustion engine has greatly reduced pollutant emissions and reflected huge commercial value, and LNG will continue to get more attention and development as one of the most competitive new energy methods in the future.

References

1. Paul Balcombe, Iain Staffell, Ivan Garcia Kerdan, Jamie F. Speirs, Nigel P. Brandon, Adam D. Hawkes, How can LNG-fuelled ships meet decarbonisation targets? An environmental and economic analysis, Energy, Volume **227**, (2021), https://doi.org/10.1016/j.energy.2021.120462
2. Fabio Burel, Rodolfo Taccani, Nicola Zuliani, Improving the sustainability of maritime transport through the utilization of Liquefied Natural Gas (LNG) for propulsion, Energy, Volume **57**, (2013), https://doi.org/10.1016/j.energy.2013.05.002
3. O. Schinas, M. Butler, Feasibility and commercial considerations of LNG-fueled ships, Ocean Eng., Volume **122**, (2016), https://doi.org/10.1016/j.oceaneng.2016.04.031
4. Boretti, A. Advances in Diesel-LNG Internal Combustion Engines. Appl. Sci. 2020, **10**, 1296. https://doi.org/10.3390/app10041296
5. Shou-guang Yao, Zijing Zhang, Yue Wei, Rui Liu, Integrated design and optimization research of LNG cold energy and main engine exhaust heat utilization for LNG powered ships, Case Stud. Therm. Eng., Volume **33**, (2022), https://doi.org/10.1016/j.csite.2022.101976
6. Ruiz Zardoya, A.; Oregui Bengoetxea, I.; Lopez Martinez, A.; Loroño Lucena, I.; Orosa, J.A. Methodological Design Optimization of a Marine LNG Internal Combustion Gas Engine to Burn Alternative Fuels. J. Mar. Sci. Eng. 2023, **11**, 1194. https://doi.org/10.3390/jmse11061194

7. Jiehui Li, Bingbing Wu, Gongping Mao, Research on the performance and emission characteristics of the LNG-diesel marine engine, J. Nat. Gas Sci. Eng., Volume **27**, Part 2, (2015), https://doi.org/10.1016/j.jngse.2015.09.036
8. Medhat Elkelawy, E.A. El Shenawy, Sherif A. Mohamed, Mostafa M. Elarabi, Hagar Alm-Eldin Bastawissi, Impacts of using EGR and different DI-fuels on RCCI engine emissions, performance, and combustion characteristics, ECMX, (2022), https://doi.org/10.1016/j.ecmx.2022.100236
9. Hua Xu, Dong Yang, LNG-fuelled container ship sailing on the Arctic Sea: Economic and emission assessment, Transportation Research Part D: Transport and Environment, Volume **87**, (2020), https://doi.org/10.1016/j.trd.2020.102556
10. Paul Balcombe, Dalia A. Heggo, Matthew Harrison,Total Methane and CO2 Emissions from Liquefied Natural Gas Carrier Ships: The First Primary Measurements, Environmental Science & Technology, Volume **56**, (2022), https://doi.org/10.1021/acs.est.2c01383
11. Maria Anderson, Kent Salo, Erik Fridell, Particle- and Gaseous Emissions from an LNG Powered Ship, Environmental Science & Technology, Volume **49**, (2015), https://doi.org/10.1021/acs.est.5b02678
12. E. Ekanem Attah, R. Bucknall, An analysis of the energy efficiency of LNG ships powering options using the EEDI, Ocean Eng, Volume **110**, Part B, (2015), https://doi.org/10.1016/j.oceaneng.2015.09.040

E3S Web of Conferences **606**, 01010 (2025) https://doi.org/10.1051/e3sconf/202560601010
ICNAOE 2024

Advancements and Innovations in Marine Fuel Technology

Yimeng Liu[1] *

[1]Southampton Joint Institute of Marine Engineering, Harbin Engineering University, Harbin, 150000, China

Abstract. This paper provides a comprehensive overview of the application of three clean energy engines— Liquefied Natural Gas (LNG), hydrogen, and ammonia— in the shipping industry. The shipping industry is increasingly exploring clean energy alternatives to reduce its carbon footprint. LNG is one of the most established and widely adopted options, owing to its mature technology and reduced emissions compared to conventional fuels. However, it remains a non-renewable resource and still produces some pollution, making it a less-than-ideal solution for long-term sustainability. Hydrogen, with its high combustion efficiency and zero direct emissions, holds promise as a truly clean fuel. Nonetheless, it faces significant obstacles, including high costs related to production, storage, and transportation, as well as safety concerns due to hydrogen embrittlement and flammability. Ammonia offers the potential for zero-carbon emissions and can be produced using renewable energy sources. Yet, its use is limited by poor combustion characteristics and high corrosiveness, which pose challenges for engine design and material durability. This article delves into the advantages and limitations of these three fuels, focusing on aspects such as corrosion resistance, transportation logistics, pollution levels, friction reduction, and combustion efficiency. The paper concludes by examining future development directions for each energy source within the maritime sector, highlighting the importance of continued innovation to achieve sustainable shipping practices.

1 Introduction

98% of global shipping is almost powered by fossil-derived fuels. In 2012 the shipping industry emitted about 940 million tonnes of carbon dioxide. To provide a sense of scale, if the shipping industry were considered as a country. Globally, it would be the 6th largest CO2 emitter [1]. UNFCCC has proposed a goal of reducing carbon emissions by half until 2030[2]. Therefore, as a major energy consumer, the shipping industry has to study new strategies to improve ship efficiency and reduce fuel consumption. Clean energy engines can achieve almost zero pollution when it is running, to achieve carbon peak and carbon

*Corresponding author: yl5c22@soton.ac.uk

neutrality. However, during the operation and maintenance of clean energy engines, the problems of corrosion, transportation, pollution, friction and combustion cannot be ignored.

Currently, the clean energy engines being studied and used mainly include LNG engines, hydrogen engines and ammonia engines. Compared with HFO and MGO, the actual greenhouse gas benefits of LNG are about 8 - 20%. Hydrogen and ammonia engines don't produce greenhouse gas emissions during operation [3].

In China, the research and application of marine LNG engine technology have done a lot of effective and fruitful work. Abroad, this research and development is becoming more and more mature. In China, scientists mainly focus on the research and practical application of hydrogen fuel cell systems in hydrogen engines, while developed countries have entered the experimental application stage of hydrogen-powered ships [4]. In China, the study of marine ammonia engines is still at a preliminary stage, but some foreign companies and institutions have conducted relevant simulations and experimental research.

In summary, clean energy engines have their advantages and disadvantages. LNG engines have formed a complete set of process technologies, which are relatively mature worldwide. Hydrogen engines have a high combustion efficiency, but they are easy to leak and explode and are difficult to extract [4]. Ammonia engines are clean, but they are not easy to burn and are toxic [5].

Therefore, this article aims to review the problems of corrosion, transportation, pollution, friction and combustion in marine clean energy engines which are LNG engines, hydrogen engines and ammonia engines. Finally, the parameter and their property are summarized. It is concluded that the development of ammonia engines is a key project in the future research direction. And studying its friction problems is the key.

2 Clean Energy Ship Engine

2.1 LNG internal combustion engine

Liquefied natural gas is mainly composed of CH_4 and also contains small amounts of C_2H_6, C_3H_8, C_4H_{10} and other minute amount of gas components [5]. The liquefaction of LNG reduces the volume of natural gas by about 600 times, making it easier to store and transport.

LNG has the potential to significantly reduce carbon emissions. When burning LNG, carbon dioxide emissions are approximately 20-30% lower than when burning heavy fuel oil [5]. In addition, SO_2 and NOx produced during the combustion of LNG are also significantly lower than when burning traditional fuels. It makes LNG become a cleaner fuel. LNG is essentially a fossil fuel, so its non-renewable nature is one of its greatest limitations.

2.1.1 Technical Challenges

Corrosion: The main types of marine engine corrosion include weld corrosion, stress corrosion, seawater corrosion and so on. The main reasons are the redox reaction between oxygen, water and metal and the battery reaction between oxygen and chloride ions in seawater and metal impurity elements. LNG fuel contains a small amount of sulphur and lubricating oil, so the corrosion of ship equipment will be reduced. At the same time, LNG can also reduce the generation of water and metal corrosion.

Transportation: Transportation of LNG mainly includes pipeline transportation, sea transportation and land transportation. Multimodal transportation is to combine different modes of transportation to achieve economic and environmental benefits. However, LNG has the problem of low volume density. So LNG is stored in a specially designed energy storage system to keep LNG in a liquid state.

Pollution: From Table 1, LNG ships are cleaner and more environmentally friendly than traditional petroleum fuels such as heavy oil and diesel, but they still produce pollutants during use. The generation of major pollutants produced by LNG ships is divided into two stages: the LNG combustion stage and the LNG development and transportation stage. LNG produces a small number of pollutants after combustion, such as CH4, CO2, NOx, HC, CO, PM and SOx. During the production and transportation of LNG, it is prone to leak and cause pollution. A large amount of water is used in the refrigeration process of LNG. Chemicals will pollute the water environment and soil environment. Additionally, LNG has a role of corrosion to water quality, causing water quality to deteriorate.

Therefore, reducing the emission of these pollutants is an important research direction. EGR technology can recover the heat energy and reduce the NOx content in the exhaust gas to effectively improve the combustion and emission problems of the engines. It has the greatest applicability and economy for the combustion of LNG engines.

Friction: Due to the complex working environment of LNG engines, the environment with high temperature and high pressure causes the viscosity of lubricating oil to drop sharply, resulting in direct contact and wear between parts. In addition, some organic additives will precipitate to damage the lubricating film and aggravate the friction problem [6]. To reduce friction, lubricating oil additives or surface hardening treatments are used to reduce friction loss. Using an optimized engine cooling system is a good way to avoid lubricating oil decomposition and reduce the lubrication effect.

Table 1. Using natural gas engines reduces ship exhaust emissions compared with traditional HFO engines

Air pollutant	The percentage of emissions reduction by using LNG
SOx	Over 90%
NOx	Up to 35% for Diesel cycle compression ignition engines Up to 85% for Otto cycle engines
PM	Over 85%
CO2	Up to 29%
GHG (in CO2e)	Up to 19%

Combustion: To reduce fuel consumption, improving engine performance and reducing heat loss are crucial to improve combustion efficiency. This can be achieved by changing the injector structure parameters. LNG generally needs other fuels to ignite and requires a large-aperture injector for injection [7]. This can be achieved by using fuel control strategies, the performance of natural gas substitution rate under different loads is different. So the thermal efficiency can be improved by changing the load and substitution rate [7]. This can be achieved by using injection strategies. By changing injection timing, multiple injections, post-injection and early injection, the engine performance can be effectively improved [8].

2.1.2 Application Cases

In Finland, for its European shipping business UPM company purchased seven LNG ships equipped with dual-fuel engines in 2022. The average carbon emissions per ton of cargo were reduced by 26% compared to the average carbon emissions in previous years. During the initial commissioning of these ships, the investment costs of LNG ships were higher, but they ultimately showed good long-term benefits. The use of LNG reduced port emission charges and related expenses. It also improved the public reputation and image [9].

2.1.3 Pros and Cons

Compared with traditional energy ships, the advantage of LNG ships is that the pollutants produced by combustion are significantly reduced [3]. In addition, the storage and transportation of LNG are safer than other clean energies.

However, the public still has concerns about the high cost of LNG ships in its early developed stage, energy leakage and other safety issues. In addition, with the increasing environmental pressure and the development of various new energy sources in the future, LNG ships need to continue to innovate. Therefore, researching the technologies for reducing propulsion and cooling fuel consumption, focusing on the optimization of the power system and popularizing new engines are important. In addition, the exhaust emission system needs to be improved. To improve it can use new technologies such as EGR and selective catalytic reduction systems to reduce pollution generated by ships.

2.2 Hydrogen internal combustion engine

Under standard temperature and pressure, hydrogen is a colourless, odourless and non-toxic gas. It is highly flammable. When burned, it reacts with oxygen to form water and at the same time releases a large amount of energy. Hydrogen fuel is considered a clean energy because it produces no greenhouse gases when burned and only emits water vapour. Therefore, hydrogen has significant potential to reduce carbon emissions compared to traditional fossil fuels.

2.2.1 Technical Challenges

Corrosion: The presence of hydrogen in the air will cause a chemical reaction between the surface of the metal and the air, causing the metal to produce " hydrogen embrittlement". This problem can be solved by coating [10]. In addition, in the electrolyte solution, the anode and cathode on the metal surface are prone to electrochemical reactions, forming battery corrosion. Protective silicone can be added to the seal around the turbocharger. And coating selection can be performed [11].

Transportation: Hydrogen-powered ships mostly use high-pressure gaseous hydrogen storage. A small number of ships use metal hydride hydrogen storage. However, hydrogen is a high-pressure and explosive gas. So a series of safety measures need to be taken during transportation, such as increasing the thickness of the container wall, using advanced pressure relief valves and safety valves and other technical measures to reduce transportation risks.

Pollution: The production and use of hydrogen will cause pollution. Hydrogen can be produced by a variety of methods. The most widely used method is steam methane reforming, but it will emit a large amount of carbon dioxide. Another method is coal gasification, but the production efficiency is low [10]. Therefore, it is imperative to produce "green hydrogen". The hydrogen production process uses renewable energy and adopts advanced technologies and equipment such as water electrolysis, biomass gasification, biomass-derived liquid reforming or microbial biomass conversion [12].

Friction: Hydrogen causes problems such as deterioration of the operating state of the engine friction pair, emulsification of lubricating oil and abnormal wear of the engine. Researchers at home and abroad are studying the influence of lubricating oil characteristics in a zero-carbon environment on the film formation, wear and corrosion characteristics of the interface of equivalent small sample specimens based on typical friction pair geometry and operating conditions. This research can support the improvement of lubricating oil and surface treatment of friction pairs under low-carbon and zero-carbon fuels.

Combustion: Hydrogen engines generally use hydrogen-air mixture combustion, but hydrogen has the characteristics of high flame speed, wide flammability limit, short flameout distance and instability. Therefore, it is necessary to prevent abnormal combustion and flashback. To deal with these, direct injection and special injection strategies to eliminate backflow, increase air dilution, DI combustion mode and laser ignition are generally used.

2.2.2 Application Cases

In 2015, Stena Line applied hydrogen fuel cell technology to its cargo ship " M/S Stena Germanica" in Sweden. The application of hydrogen fuel cells on ships successfully reduced emissions during navigation and demonstrated the reliability of hydrogen fuel cell systems. However, there are still problems in the application process, such as the complexity of hydrogen storage and supply, the need to build hydrogen infrastructure and compatibility with traditional power systems [13].

2.2.3 Pros and Cons

Hydrogen as a fuel can significantly reduce greenhouse gases and harmful emissions produced by ships during operation. Hydrogen has high combustion efficiency. And it is renewable. It has diverse production methods and is rich in resources. Therefore, it has significant benefits for reducing emissions and promoting sustainable shipping.

However, hydrogen is expensive to produce, store and transport, especially green hydrogen [12]. The energy efficiency of hydrogen production and combustion is lower than some traditional fuels. Hydrogen embrittlement can easily lead to corrosion damage to ship structures and components [10]. Hydrogen is highly flammable. So, it poses safety challenges.

2.3 Ammonia internal combustion engine

Ammonia is a colourless gas with a pungent odour at room temperature and pressure. It does not produce carbon dioxide during combustion. So, it has great potential for reducing carbon emissions. However, ammonia combustion may produce a small amount of nitrogen oxides [14]. In addition, by using renewable energy to produce ammonia, its renewable nature can be achieved.

2.3.1 Technical Challenges

Corrosion: Ammonia is highly acidic and alkaline. So, it can easily form corrosion layers such as nitrides and oxides on metal surfaces. Metal materials in lubrication systems are susceptible to corrosion by ammonia, resulting in a decrease in lubricant quality. In addition, the water produced by the combustion of ammonia can cause the hydrolysis of ammonium ions, making the metal susceptible to passivation, corrosion, fracture, and other problems [15]:

$$NH_4^{+}+2H_2O==NH_3{\cdot}H_2O+H3O^{+} \tag{1}$$

$$NH_4^{+}+2H_2O== NH_3{\cdot}H_2O+H^{+} \tag{2}$$

$$2H^{+}+Fe==Fe2^{+}+H_2 \tag{3}$$

To reduce ammonia corrosion, using corrosion-resistant materials is the most direct method. In addition, applying anti-corrosion coatings can provide additional protection. Ammonia fuel can also be purified to reduce impurities.

Transportation: Ammonia is processed to obtain liquid ammonia by cooling at normal pressure and liquefied under pressure, which is then stored in special pressure vessels or cryogenic tanks. The storage container must have good insulation and sealing properties. The temperature and pressure must be strictly controlled during transportation to prevent leakage and explosion risks. In addition, when storing and transporting ammonia, a leak detection system must be installed to ensure that ammonia leaks are discovered and handled timely.

Pollution: The production of Ammonia uses fossil fuels such as natural gas or coal, resulting in large amounts of CO2 and other greenhouse gas emissions, as well as other pollutants such as SOx, NOx, and nitrous oxide [16]. It can use alternative sustainable ammonia production pathways, including electrochemical and biological pathways. During use, incomplete combustion of ammonia may lead to increased NOx emissions. It can use in-cylinder reformate recirculation technology to improve the thermal efficiency of pilot-ignition diesel ammonia combustion engines and reduce unburned NH3, NOx, and N2O emissions [14].

Friction: Ammonia as a fuel does not have good lubrication properties. It is easy to produce dry combustion byproducts during combustion. And it produces highly corrosive nitrogen oxides. Even trace concentrations of ammonia can seriously affect the deposit formation characteristics and stoichiometric concentration [17].

Therefore, domestic and foreign researchers have conducted research on the composition and working conditions of lubricating oils in a zero-carbon environment [16]. At the same time, high-temperature-resistant and corrosion-resistant alloy materials are selected as the materials for key friction parts of the engine and friction and wear tests are carried out on the moisture produced by the combustion of ammonia fuel. Oil manufacturers need to consider the impact of ammonia when developing new additive packages. Moreover, when using ammonia as a fuel, appropriate chemical analysis procedures are required [17].

Combustion: The challenge of using ammonia as a marine fuel lies in the poor combustion characteristics of ammonia and the related engine combustion technology. Therefore, to overcome the inherent disadvantages of ammonia, such as high ignition energy, narrow flammability limit and slow propagation speed, HPDF and LPDF are the two main modes [18, 19]. At the same time, to solve the problem of excessive unburned ammonia emissions, EGR is usually used to control NOx emissions. The new concept of in-cylinder reformed gas recirculation (IRGR) can also be used to improve the indicated thermal efficiency and reduce unburned NH3, NOx, N2O and GHG emissions [14].

2.3.2 Application Cases

C-Job Naval Architects designed a zero-emission ship using ammonia fuel to ensure that ammonia can be stored and transported safely and efficiently, achieving the goal of zero carbon emissions. However, it also faces many difficulties. The storage and transportation of ammonia requires a low-temperature and high-pressure environment, which increases the complexity and cost of ship design. And ammonia has a greater risk of leakage in the marine environment. With the continuous advancement of technology, ammonia fuel is expected to play a key role in more fields in the future.

2.3.3 Pros and Cons

Ammonia does not produce greenhouse gas emissions during operation, which can effectively meet the pressure of the global shipping industry to reduce greenhouse gas emissions. In addition, ammonia can be produced through renewable energy, electrolyzing water to produce hydrogen, and then synthesizing it with nitrogen. The use of renewable energy to produce ammonia can achieve sustainable development. And compared with hydrogen, ammonia is relatively easier to store.

However, full decarbonization of ammonia fuels relies on low-carbon energy inputs and feedstock materials. And "green" hydrogen and ammonia from renewable sources are currently not available in the quantities required by the maritime industry [20]. Ammonia is corrosive and can corrode metal components of marine engines by used for a long time, increasing maintenance costs[15]. For ammonia-fueled shipping to become a reality, poor combustion performance and toxicity risks must be overcome, as well as infrastructure development. However, the means to overcome these challenges may be beyond the scope of the maritime sector itself [20].

3 Results and Discussion

Table 2. Performance comparison between the marine clean energy engines

	LNG	Hydrogen	Ammonia
Corrosion	LNG reduces moisture generation. It contains sulfur and lubricating properties. It has less corrosivity.	It will cause hydrogen embrittlement and easily cause battery corrosion. So, it will cause strong corrosion to the equipment.	It has a high corrosivity. Metals are prone to passivation, corrosion and fracture when contacting ammonia.
Transportation	The transportation cost is high. And the transportation pollution problem needs to be solved urgently.	High-pressure gaseous hydrogen storage is used, but a series of safety measures need to be taken. So, transportation is complicated.	It is stored in special pressure vessels or low-temperature tanks for transportation, but the risk of leakage and explosion must be prevented.
Pollution	It is clean and environmentally friendly, but during combustion and transportation, it will produce small amounts of pollutants. And the risk of methane leakage is the greatest.	The combustion products are completely environmentally friendly. However, production will cause pollution. So the production of "green hydrogen" is crucial.	It produces NOx emissions to cause pollution. So additional equipment is needed to reduce pollution.
Friction	The impact of friction is small. However, it will reduce the viscosity of the lubricating oil and cause the organic additives to precipitate, generating friction.	It causes problems such as deterioration of the operating condition of the engine friction pair, emulsification of lubricating oil and abnormal engine wear.	It does not have good lubricity. It produces dry by-products and NOx with high corrosivity.
Combustion	It has stable energy output, but the LNG combustion efficiency needs to be improved.	It uses a style of hydrogen-air mixture for combustion, which has extremely high combustion efficiency. But this style poses a safety hazard.	The poor combustion characteristics of ammonia and related engine combustion technology.

In summary, according to the table 2, LNG has high combustion efficiency, less pollution and mature technology. However, LNG engines have a high risk of methane leakage and high transportation and storage costs. Normally it is suitable for large ships. Hydrogen has high energy density and low pollutant emissions. However, it is difficult to store and transport. It has high investment costs. And there is uncertainty in fuel supply. It is suitable for short-distance and light ships. Ammonia is easy to store and distribute. And it has no carbon emissions. However, it has poor combustion performance. And there is a risk of toxicity. It is suitable for medium and long-distance transportation.

4 Summary and Conclusion

The article mainly discusses the urgency of reducing carbon emissions in the shipping industry and the application prospects and technical challenges of clean energy engines. The article points out that the use of traditional fossil fuels in ship transportation results in serious carbon emissions. Thus, this phenomenon promotes the development of clean energy engines. The article focuses on three clean energy engines: LNG engines, hydrogen engines and ammonia engines. LNG engine technology is relatively mature, which can significantly reduce carbon emissions and have low pollutant emissions. But its storage and transportation costs are high. And there is a risk of methane leakage. Hydrogen has high combustion efficiency as a fuel. And its combustion product is only water, making it extremely environmentally friendly. But production and storage costs are high. Ammonia engines have huge potential to reduce emissions and are relatively easy to store and transport. However, ammonia has poor combustion characteristics and high corrosivity. And it has certain toxicity risks. Therefore, the potential of ammonia engines has attracted widespread attention at home and abroad. It is predicted that ammonia engines will be a key project in the future ship research direction of various countries.

At present, the research about ammonia engines at home and abroad is in its infancy. And there is no mature and effective solution to solve the problems of ammonia engines. In the future, efforts should be made to develop efficient combustion technology, the safety of transportation and storage and research on mixed-use with other renewable energy sources. Among them, the friction problem is the key part of improving efficiency. The friction problem can be studied from the aspects of low-carbon environment lubricating oil analysis and testing, low-carbon environment interface tribological characteristics test research, low-carbon environment interface friction and wear detection and so on. At the same time, some parameters such as lubricating oil type, viscosity and water content also have a certain impact on the friction.

References

1. P. Balcombe, J. Brierley, C. Lewis, L. Skatvedt, J. Speirs, A. Hawkes, L. Staffell, H. Adam, S. Iain, How to decarbonise international shipping: Options for fuels, technologies and policies. Energy Convers. Manag. 182, 72-88 (2019)
2. United Nations, United Nations Framework Convention on Climate Change (United Nations, New York,1992). Available at:https://unfccc.int/resource/docs/convkp/conveng.pdf
3. Y. Wang, L.A. Wright, A comparative review of alternative fuels for the maritime sector: Economic, technology, and policy challenges for clean energy implementation. World 2(4), 456-481 (2021). https://doi.org/10.3390/world2040029

4. M.G. Sürer, H.T. Arat, Advancements and current technologies on hydrogen fuel cell applications for marine vehicles. Int. J. Hydrogen Energy (2022). https://doi.org/10.1016/j.ijhydene.2021.12.251
5. Ammonia as an energy vector: Current and future prospects for low-carbon fuel applications in internal combustion engines. J. Clean. Prod. (2021). https://doi.org/10.1016/j.jclepro.2021.126562
6. X. Di, Chuan yong fagongji qigang runhuayou ziranshuxing shiyan yanjiu, Ph.D. thesis, Dalian University of Technology (2018)
7. Y. Wang, Q. Cao, L. Liu, Y. Wu, H. Liu, Z. Gu, C. Zhu, A review of low and zero carbon fuel technologies: Achieving ship carbon reduction targets. Energy Rep. (2022). Available at: https://www.sciencedirect.com/science/article/pii/S2214166922004237
8. H. Li, R. Verschaeren, T. Beji, Investigation of evaporating sprays in a medium speed marine engine. Exp. Therm. Fluid Sci. 121, Article 110278 (2021)
9. UPM, First new sustainable LNG vessel to start shipping (2022). Available at: https://www.upm.com/articles/timber/22/first-new-sustainable-lng-vessel-to-start-shipping/
10. Z. Stepien, A comprehensive overview of hydrogen-fueled internal combustion engines: Achievements and future challenges. Energies 14, 6504 (2021)
11. S. Yuan, Y. Sun, C. Yang, Y. Zhang, C. Cong, Y. Yuan, D. Lin, L. Pei, Y. Zhu, H. Wang, A novel dual-functional epoxy-based composite coating with exceptional anti-corrosion and enhanced hydrogen gas barrier properties. Chem. Eng. J. 449, 137876 (2022)
12. B.S. Zainal, P.J. Ker, H. Mohamed, H.C. Ong, I.M.R. Fattah, S.M.A. Rahman, L.D. Nghiem, T.M.I. Mahlia, Recent advancement and assessment of green hydrogen production technologies. Renew. Sustain. Energy Rev. 189(Part A), 113941 (2024)
13. H. Bragge, C. Chantongjaroen, The future fuels of marine engines. Examensarbete på grundnivå (2021). Available at: https://odr.chalmers.se/items/6dffd334-d172-45a8-8085-5f8175022e9e
14. X. Zhou, T. Li, R. Chen, Y. Wei, X. Wang, N. Wang, S. Li, M. Kuang, W. Yang, Ammonia marine engine design for enhanced efficiency and reduced greenhouse gas emissions. Published: 07 March 2024
15. D.R. Macfarlane, P.V. Cherepanov, J. Choi, et al., A roadmap to the ammonia economy. Joule 4(6), 1186-1205 (2020)
16. M. Ozturk, I. Dincer, An integrated system for ammonia production from renewable hydrogen: A case study. Int. J. Hydrogen Energy 46, 5918-5925 (2021)
17. A. Agocs, M. Rappo, N. Obrecht, C. Schneidhofer, M. Frauscher, C. Besser, The impact of ammonia fuel on marine engine lubrication: An artificial lubricant ageing approach. Lubricants 11(4), 165 (2023). https://doi.org/10.3390/lubricants11040165
18. P. Dimitriou, R. Javaid, A review of ammonia as a compression ignition engine fuel. Int. J. Hydrogen Energy 45, 7098-7118 (2020)
19. T. Li, et al., A comparison between low- and high-pressure injection dual-fuel modes of diesel-pilot-ignition ammonia combustion engines. J. Energy Inst. 102, 362-373 (2022)
20. Y. Wang, L.A. Wright, A comparative review of alternative fuels for the maritime sector: Economic, technology, and policy challenges for clean energy implementation. World 2(4), 456-481 (2021). https://doi.org/10.3390/world2040029

Advancing Solid-State Batteries with Nanomaterials: Enhancing Safety, Performance, and Energy Efficiency

Junbo Lang [1,*]

[1]Nanjing No.13 middle school international department, Nanjing, Jiangsu, 210029, China

Abstract. Solid-state batteries (SSBs) are a promising advancement in energy storage technology, offering higher energy density and improved safety compared to traditional lithium-ion batteries. The integration of nanomaterials into SSBs has the potential to overcome key technical challenges, including low ion conductivity, interface instability, and mechanical failures. This review explores the role of nanomaterials in enhancing the performance of SSBs by improving ion transport, stabilizing interfaces, and inhibiting dendritic crystal growth. Additionally, the application of nanoscale coatings on electrodes and solid electrolytes is discussed as a strategy to further optimize battery performance. While nanomaterials significantly improve the safety, efficiency, and longevity of SSBs, challenges such as large-scale production, cost control, and long-term stability remain. Continued research and innovation are essential to fully unlock the potential of nanomaterials in SSBs, which hold significant promise for sustainable energy storage solutions, particularly in electric vehicles and renewable energy systems.

1 Introduction

Solid-state batteries represent a promising advancement in battery technology, utilizing solid electrodes and solid electrolytes, as opposed to the liquid electrolytes found in traditional lithium-ion batteries. A key distinction between solid-state and conventional lithium-ion batteries lies in the replacement of the liquid electrolyte and diaphragm with solid electrolytes. Solid-state batteries typically offer higher energy density but lower power density. Given the widely held view that lithium-ion technology has reached its theoretical limits, solid-state batteries are increasingly seen as the next evolutionary step in energy storage, especially in applications where safety, performance, and energy efficiency are paramount.*

In evaluating battery technologies, researchers focus on key factors such as safety, performance, and energy efficiency. Safety and performance are particularly critical in electric vehicles (EVs) and portable electronic devices, where they directly impact user experience. The range, charging time, and power output of EVs are influenced by battery performance—higher energy densities result in longer ranges, faster charging reduces downtime, and high-performance batteries enhance power output. These metrics are essential

*Corresponding author: lang2007@ldy.edu.rs

for the market competitiveness of EVs. Similarly, in portable electronics, batteries with extended lifespans and fast-charging capabilities greatly enhance user satisfaction.

Consequently, improving the safety and performance of batteries is essential for advancing EVs and portable electronics. The integrhold of nanomaterials into solid-state batteries holds significant promise for enhancing their performance. Nanomaterials applied to solid-state electrolytes and electrodes can significantly improve ion conductivity by optimizing the lithium-ion transport pathways, thereby increasing overall ion mobility. Additionally, nanomaterials contribute to the enhancement of interface stability between solid electrolytes and electrodes, which is crucial for maintaining long-term battery performance. By inhibiting the formation of dendritic crystals, nanomaterials mitigate the risk of short circuits, thus enhancing battery safety. Moreover, the inclusion of nanomaterials can strengthen the mechanical properties of solid electrolytes, improving their toughness and durability, which are critical for the longevity and reliability of the battery system.

This review aims to explore the utilization of nanomaterials in solid-state batteries to advance their safety, performance, and energy efficiency. It will address both the opportunities and challenges associated with nanomaterial applications, providing a comprehensive analysis of future trends and commercialization prospects. In particular, the review will consider the potential of nanomaterial-enhanced solid-state batteries to revolutionize sustainable energy storage and clean transportation, highlighting the transformative impact these technologies could have on the energy sector.

2 Technical overviews of solid-state battery

Solid-state batteries are composed of three key components: solid electrolytes, electrodes, and interfaces, each crucial to the battery's performance. The solid electrolyte serves as the core component responsible for ion transmission, replacing the liquid electrolytes used in traditional lithium-ion batteries. This transition to solid electrolytes significantly enhances the safety and stability of the battery, reducing the risks of leakage and combustion, which are common in conventional liquid electrolytes. The electrodes—comprising the positive and negative poles—are where the electrochemical reactions occur. The positive electrode typically contains lithium compounds, while the negative electrode can be made of lithium metal or other materials capable of absorbing and releasing lithium ions during charge and discharge cycles. Lastly, the interface—the contact point between the electrode and solid electrolyte—is essential for battery efficiency. Proper interface performance is critical to charge transfer and ion transport, as it directly influences the internal resistance and cycle life of the battery (Figure 1).

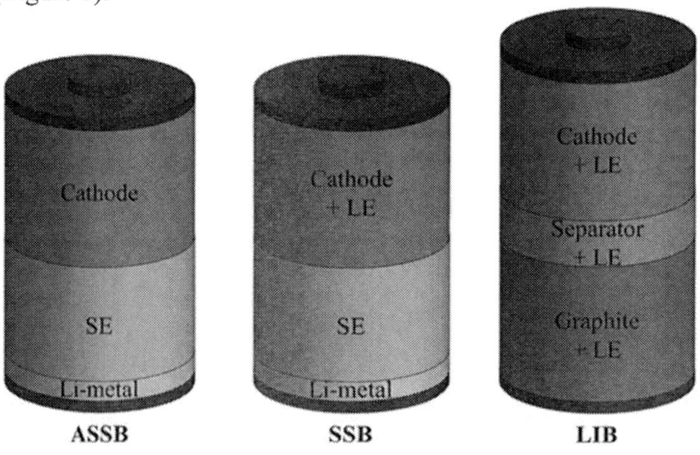

Fig. 1. Internal structure diagram of all-solid-state batteries, solid-state batteries and lithium-ion batteries [1]

A major distinction between solid-state batteries and all-solid-state batteries is the state of the electrolyte. While solid-state batteries may contain small amounts of liquid or gel electrolytes, all-solid-state batteries exclusively utilize solid electrolytes. This complete removal of liquid components improves safety by eliminating risks associated with leakage and combustion. Furthermore, all-solid-state batteries have the potential to achieve higher energy densities and longer cycle lives. However, they face significant technical challenges, including low ion conductivity in solid electrolytes and difficulties in maintaining stable interfaces between electrodes and electrolytes, which currently restrict the large-scale commercial application of all-solid-state batteries. As a result, solid-state batteries, which are somewhat less complex, are viewed as a transitional technology between conventional lithium-ion batteries and fully solid-state systems.

Compared to traditional lithium-ion batteries, solid-state batteries offer several important advantages. One of the most significant is improved safety, as the solid electrolyte reduces the risk of leakage, combustion, and other hazards associated with liquid electrolytes. This enhancement in safety makes solid-state batteries especially attractive for applications in electric vehicles (EVs) and portable electronic devices. Furthermore, higher energy density is another key benefit, as solid-state batteries can store more energy in a smaller space, extending the driving range of EVs and increasing the operational time of portable devices. Additionally, solid-state batteries are capable of operating effectively across a wider temperature range, making them more adaptable to various environmental conditions.

The introduction of nanomaterials into the development of solid-state batteries is viewed as essential to overcoming some of the key technical challenges associated with this technology. One major challenge is improving ion conductivity, which is a limiting factor in the performance of solid electrolytes. Nanomaterials can enhance the pathways for lithium-ion transport within the electrolyte, significantly increasing ion conductivity. For instance, the addition of LLZO ($Li_7La_3Zr_2O_{12}$) nanoparticles to solid electrolytes has been shown to boost ion transport and improve overall battery efficiency [2]. In addition to enhancing conductivity, nanomaterials can also play a pivotal role in stabilizing the interface between the solid electrolyte and electrodes. By incorporating polymer-based nanocomposites, the interface compatibility is improved, which reduces interface resistance and helps maintain long-term performance.

Another critical benefit of nanomaterials is their ability to inhibit the formation of dendritic crystals, which can cause short circuits in batteries, posing significant safety risks. Research shows that the introduction of ceramic nanoparticles can effectively suppress the growth of lithium dendrites, thus improving the safety and extending the cycle life of the battery [3]. Moreover, nanomaterials also improve the mechanical properties of solid electrolytes. The incorporation of materials such as nanofibers or nanoparticles into solid electrolytes enhances their mechanical strength and toughness, which increases the durability and reliability of the battery [3].

In conclusion, nanomaterials offer a promising solution to many of the technical challenges faced by solid-state batteries, including improving ion conductivity, stabilizing interfaces, inhibiting dendrite formation, and enhancing mechanical properties. These advancements not only enhance the safety and performance of solid-state batteries but also bring them closer to large-scale commercialization. As research continues, the integration of nanomaterials could play a key role in unlocking the full potential of solid-state battery technology, making it a cornerstone of future energy storage systems for applications ranging from electric vehicles to portable electronics.

3 Nanomaterials of solid electrolytes applied to solid-state batteries

3.1 Types of nanomaterials applied to solid electrolytes

Most researchers agree that the use of LLZO ($Li_7La_3Zr_2O_{12}$) solid electrolyte (SSE) in contact with the lithium metal anode can effectively reduce the formation of lithium dendrites, especially under the application of necessary pressure. This reduces the risk of short circuits and improves the battery's overall safety. However, the design of the cathode layer remains a topic of debate. Some researchers advocate for the addition of LLZO to the cathode, while others recommend using alternative electrolytes such as ionic liquids or polymers. This ongoing debate highlights the complexity of achieving the ideal configuration for lithium garnet solid-state batteries. For instance, the study by Kravchyk, K.V., Karabay, D.T., and Kovalenko, M.V. evaluates the feasibility of an all-solid-state battery using LLZO as the single electrolyte. Their analysis confirms the viability of such a configuration concerning achievable energy and power densities, as shown in Figure 2.

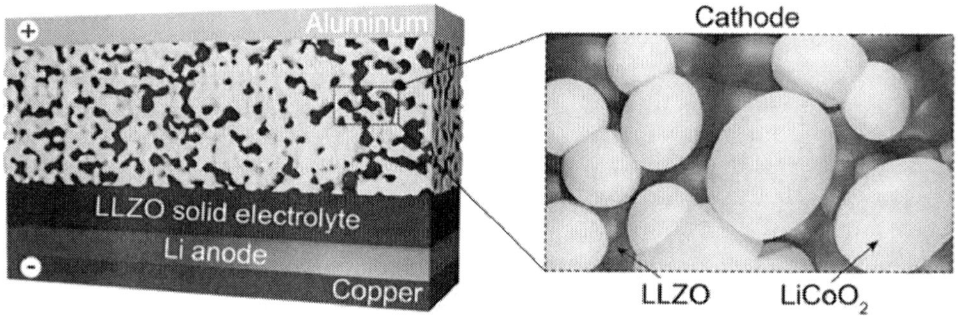

Fig. 2. Schematics of Li-garnet solid-state battery and solid-state cathode considered in this work for assessing the power and energy densities of Li-garnet SSBs [2].

In addition to LLZO, $Li_{1.5}Al_{0.5}Ge_{1.5}P_3O_{12}$ (LAGP) has also emerged as a promising material for solid-state battery applications. Recent research proposes a robust technique that involves smoothing the LAGP ceramic pellet to achieve a mirror-like surface. This method has demonstrated the capability to run a solid-state battery for an unprecedented 1,000 cycles with a capacity retention of 80% at 0.5C and 50 °C [4]. Both LLZO and LAGP are ceramic nanoparticles, and their inclusion in solid electrolytes contributes to the improved performance and longevity of solid-state batteries. Additionally, advances in polymer-based nanocomposites have shown significant potential, especially in lithium metal and "beyond lithium" (e.g., Na, K, Zn, Mg, Al) polymer-based batteries. These developments underscore the importance of ongoing research into diverse materials and configurations to enhance the performance of solid-state batteries.

3.2 The advantages of nanomaterials as solid electrolytes

Pure inorganic solid-state electrolytes (SSEs) are known for their high ionic conductivity, but they also present several challenges, such as uneven lithium deposition, low mechanical flexibility, and poor interface contact. These issues often result in high interface impedance and mechanical failure in inorganic SSE-based solid-state batteries. Nanocrystalline-based inorganic SSEs offer a promising solution, as they typically have lower interface impedance and excellent cyclic stability due to their high specific surface area and shorter lithium diffusion paths. For example, Song and his collaborators developed a dense nano-solid electrolyte, $Na_3Zr_2Si_2PO_{12}$, by calcining a nano-level ZrO_2 precursor. The solid-state symmetrical battery assembled with this electrolyte exhibited low interface impedance, indicating improved performance [5]. Additionally, Cui's research team achieved a significant breakthrough by inducing self-organized heterogeneous nanocrystalline domains in the glass ceramic-based SSE Li_2S-P_2S_5, resulting in an ultra-high lithium conductivity of 13.2 mS cm^{-1} at room temperature [6].

Applying nanoscale coatings on the surface of inorganic solid electrolytes is another common strategy to address the poor interface contact between inorganic solid electrolytes and electrode materials. For instance, insufficient contact between the inorganic solid electrolyte and the lithium metal anode can lead to rapid increases in interface resistance and uneven current density distribution. This imbalance can cause irregular lithium deposition, ultimately leading to battery failure due to the formation of lithium dendrites. Wan et al. demonstrated that coating the oxide electrolyte $Li_{6.5}La_3Zr_{1.5}Ta_{0.5}O_{12}$ (LLZT) with a pro-lithium nanofilm of Ta2O5 significantly improved the stability and reduced the interface resistance when paired with a lithium metal anode [7].

Moreover, the presence of defects, such as grain boundaries, pores, and impurity phases in inorganic SSEs, can exacerbate uneven lithium deposition and promote the growth of lithium dendrites. To mitigate these issues, Wang and his colleagues proposed a nano-confinement strategy aimed at controlling the defect-free growth of $Li_{0.33}La_{0.55}TiO_3$ (LLTO) crystals. By eliminating structural defects, they were able to significantly enhance the cycling performance of the batteries [7]. These findings underscore the critical role of nanoscale engineering in addressing the limitations of inorganic SSEs, thereby improving the reliability and efficiency of solid-state batteries.

4 Nanomaterials in electrodes

Previous studies have demonstrated that the physicochemical properties and stability of the cathode and the cathode/solid electrolyte (SE) interface have a profound influence on the performance of solid-state batteries (SSBs) [8]. Nanomaterials, with their ultra-high specific surface area, offer a significant advantage by increasing the contact area between the active material and the electrolyte. This larger contact area provides more active sites for charge transfer at the interface, which in turn enhances the electrochemical reaction kinetics. However, the increased surface area also raises the risk of side reactions, which can negatively impact the long-term performance of the battery. For instance, Wolfgang G. Zeier and his colleagues discovered that an FeS2 cathode with an average particle size of 9.7 nm delivered the highest reversible capacity (760 mAh g^{-1}) and excellent rate performance due to its reduced size. However, the capacity degradation was severe, primarily due to increased interface side reactions [9].

To mitigate this issue, without sacrificing the high capacity of nanostructured cathodes, Linda F. Nazar's group introduced a charge transfer carrier, $LiVS_2$, and designed a core-shell nanostructure $LiVS_2$@Li_2S cathode by synthesizing it on the surface of Li_2S nanotubes. This innovative design prevented direct contact between the active cathode Li_2S and the electrolyte $Li_{5.5}PS_{4.5}Cl_{1.5}$, thereby ensuring an ultra-long cycle life. The capacity retention rate was approximately 80% after 1,000 cycles, demonstrating the potential of this approach to extend battery life while maintaining performance.

Nanotechnology is also employed to protect lithium metal anodes in solid-state batteries, though the interface environment poses new challenges. The failure of solid-state lithium metal batteries is often due to the growth of lithium dendrites, which can be traced to contact failure between the lithium metal and solid electrolyte during the stripping and plating process. To address this, Jürgen Janek and his collaborators proposed a composite anode made of lithium metal and a 3D carbon nanotube (CNT) scaffold. This design maintains electrical contact between the lithium metal anode and the oxide electrolyte, $Li_{6.25}Al_{0.25}La_3Zr_2O_{12}$, thereby improving battery performance and mitigating dendrite growth [10].

One of the most promising applications of nanotechnology in SSBs is the development of functional nanoscale surface coatings for cathodes, a key step in moving from laboratory research to industrialization. The interface between the cathode and electrolyte is a critical area in SSBs, and surface coatings have emerged as an effective strategy to optimize this interface. By applying functional coatings, researchers aim to adjust the cathode/electrolyte interaction, minimize side reactions, and improve overall battery performance. In the subsequent sections, I will present experimental data to illustrate the effects of nanomaterial applications in this area, showcasing the promising advancements in interface optimization through nanotechnology.

5 Enhance safety and performance through nanomaterials

Nanomaterials significantly enhance the safety and performance of solid-state batteries by addressing critical challenges related to solid-state electrolytes (SSE), electrodes, and the interfaces between them. One of the key improvements' nanomaterials bring is the enhancement of ion conductivity. By incorporating nanoparticles such as LLZO into solid-state electrolytes, the lithium-ion transport pathways are optimized, leading to increased ion conductivity, which is essential for improving the overall efficiency of the battery. This advancement helps solid-state batteries achieve higher energy densities and better performance in various applications.

Another crucial benefit of nanomaterials is their ability to enhance interface stability. Polymer-based nanocomposites improve the compatibility between solid electrolytes and electrodes, which reduces interface resistance and ensures smoother charge transfer. This contributes to improved cycling stability and longevity of the battery, addressing one of the major challenges in solid-state battery technology. Additionally, the use of ceramic nanoparticles helps inhibit the formation of lithium dendrites, which are responsible for battery short circuits. By preventing dendrite growth, nanomaterials enhance the safety of the battery and extend its lifespan.

Nanomaterials also play a vital role in improving the mechanical properties of solid-state batteries. The inclusion of nanofibers or nanoparticles in the solid electrolyte enhances its mechanical strength and durability, making the battery more reliable, particularly in demanding applications. Furthermore, advanced designs, such as the lithium metal-3D carbon nanotube composite anode, help maintain electrical contact between the lithium metal anode and the electrolyte, preventing contact failure and improving battery performance.

Lastly, nanoscale surface coatings on the cathode provide an effective means of optimizing the interface between the cathode and the electrolyte. These coatings reduce side reactions, enhance charge transfer, and prevent electrolyte decomposition, leading to improved performance and longer cycle life for solid-state batteries. Overall, nanomaterials offer comprehensive solutions to the challenges of solid-state battery technology, significantly improving safety, efficiency, and reliability.

6 Challenges and future directions

Nanotechnology plays a pivotal role in electrochemical energy storage devices, particularly solid-state batteries (SSBs), due to the unique physicochemical properties offered by nanoscale materials. These properties, driven by the small size of nanomaterials, allow for significant improvements in the performance and interface characteristics of key materials in

SSBs. This paper reviews how structural and coated nanomaterials enhance the comprehensive performance of SSBs, focusing on improving ion conductivity, interface stability, and overall battery efficiency.

However, the large-scale production of nanomaterials for SSBs presents several challenges. The complexity of the production process, which requires precise control and specialized equipment, increases the difficulty and cost of mass production. Ensuring consistent quality is another hurdle, as even small variations in the production of nanomaterials can significantly impact the performance of batteries. Cost-effectiveness remains a crucial factor, as the high production costs associated with nanomaterials must be addressed to make them commercially viable for widespread use in SSBs.

Economic factors also influence the application of nanomaterials in commercial SSBs. A large initial investment is required to develop production facilities for nanomaterials, which can place a financial burden on companies. Additionally, the balance between cost and performance is critical nanomaterials must provide significant improvements in battery performance while keeping costs manageable to meet market demand. The success of commercial applications will also depend on the scale of production and market acceptance of SSBs, both of which are essential for driving demand.

Further research is necessary to ensure the long-term stability and reliability of nanomaterials in SSBs. This includes studying the aging mechanisms of nanomaterials to understand how they change over time during battery operation, which can help in developing strategies to extend their lifespan. Optimizing material design to improve the durability and stability of nanomaterials is also essential, as is conducting long-term testing and verification to ensure their reliability in practical applications.

In addition to technical and economic considerations, potential environmental and safety issues related to the use of nanomaterials must be addressed. It is important to conduct thorough safety assessments to ensure that nanomaterials used in battery production, use, and recycling do not pose risks to human health or the environment. Furthermore, the recycling and reuse of nanomaterials after the battery's lifecycle ends should be considered to minimize their environmental impact, making sustainability a critical aspect of future SSB development.

7 Conclusion

Nanomaterials play a critical role in enhancing the performance of SSBs. Their application improves ion conductivity, enhances interface stability, inhibits the formation of dendritic crystals, strengthens mechanical properties, protects lithium metal anodes, and optimizes interfaces. These enhancements contribute to the overall safety, efficiency, and longevity of SSBs, making them a promising technology for future energy storage solutions. Despite the significant advantages nanomaterials bring to SSBs, there are still challenges to be addressed. Issues such as large-scale production, cost control, and ensuring the long-term stability and reliability of nanomaterials must be resolved to fully realize their potential. Overcoming these challenges is essential for the successful commercialization of nanomaterial-reinforced solid-state batteries.

Continued research and innovation will be key to advancing the application of nanomaterials in SSBs. As the technology evolves, SSBs are expected to make a substantial impact on sustainable energy storage and clean transportation, improving the range and safety of electric vehicles, shortening charging times, and contributing to the broader transition away from fossil fuels. The development of SSBs will also promote the integration and utilization of renewable energy sources, aligning with global sustainable development goals. While nanotechnology has already offered promising solutions to many failure issues in SSBs, further research and investment are necessary to fully unlock its potential in creating high-energy, high-safety, and long-life batteries. The future of SSBs lies in these continued advancements, which will play a vital role in shaping the future of energy storage and sustainability.

References

1. A. M. Bates, Y. Preger, L. Torres-Castro, K. L. Harrison, S. J. Harris, & J. Hewson. Are solid-state batteries safer than lithium-ion batteries? Joule, **6**, 742-755 (2022).
2. K. V. Kravchyk, D. T. Karabay, M. V. Kovalenko. On the feasibility of all-solid-state batteries with LLZO as a single electrolyte. Scientific Reports, **12**, 1177 (2022).
3. Y. H. Zhang, et al. Nanotechnology in solid state batteries, what's next? Next Nanotechnology, **2**, 100011 (2023).
4. P. Jiang, et al. LiF involved interphase layer enabling thousand cycles of LAGP-based solid-state Li metal batteries with 80% capacity retention. Energy Storage Materials, **48**, 145-154 (2022).
5. G. Yang, A facile method for the synthesis of a sintering dense nano-grained Na3Zr2Si2PO12 Na^{+}-ion solid-state electrolyte. Chemical Communications, **57**, 4023-4026 (2021).
6. Y. Chen, T. Wang, H. Tian, D. Su, Advances in lithium–sulfur batteries: from academic research to commercial viability. Advanced Materials, **33**, 2003666 (2021).
7. W. Dong, et al. MOF lamellar membrane-derived LLTO solid state electrolyte for high lithium-ion conduction. Journal of Membrane Science, **663**, 121041 (2022).
8. K. Wang. The fractal active low-pass filter within the local fractional derivative on the Cantor set. COMPEL-The international journal for computation and mathematics in electrical and electronic engineering, **42**, 1396-1407 (2022).
9. F. Dewald, et al. Influence of Iron Sulfide Nanoparticle Sizes in Solid - State Batteries. Angewandte Chemie, **133**, 18096-18100 (2021).
10. C. Y. Kwok, et al. High-performance all-solid-state Li 2 S batteries using an interfacial redox mediator. Energy & Environmental Science, **16**, 610-618 (2021).

https://doi.org/10.1051/e3sconf/202560602002

Implementation of LFP Batteries for Energy Storage at Small hydro power station

Tianshuo Huang*

Guangzhou Huali College, Guangzhou, China

Abstract. With the application of energy storge system (ESS) in large powerplants are become increasingly mature, there's an absent of small hydropower stations. Small hydropower stations, though considered as a clean energy, are facing government regulations about restrict it outflows due to environmental protection. To help these small hydropower stations survive, ESS is needed. In this article, the battery that was used in large powerplants ESS is explored to figure out how to adopt the technology into way smaller powerplant by analysing technological, economic and construction factors. The results suggest that ESS using LFP battery can significantly improve energy consistency under outflow restriction. However, on the economic aspect, with the rapid decrease in the battery prices and different policy in different regions, the outcomes can be diverse, but the overall dropping battery prices and increasing demand of renewable energy led to a promising future. This article has illuminated the direction for the development of technology.

1 Introduction

As global environmental awareness rises, there has been a growing demand for clean and renewable energy solutions to mitigate the adverse impacts of fossil fuels. Hydropower, one of the most established renewable energy sources, has been a key contributor to clean energy production worldwide. According to the International Renewable Energy Agency (IRENA), Hydropower is an important component of power systems worldwide. It is the largest source of renewable electricity and can enables a higher penetration of variable renewables such as solar and wind by providing balancing and flexibility services. In addition to electricity, hydropower offers additional benefits like recreational activities, improved resistance to drought and flooding, and storage for drinking and irrigation water [1]. While large hydropower plants dominate the sector, small hydropower stations (SHPs), defined as stations with a capacity of less than 10 MW, are increasingly gaining attention due to their lower environmental impact and potential for providing decentralized, off-grid power solutions, particularly in remote regions.

However, small hydropower stations face unique challenges, primarily due to fluctuating water availability, seasonal variations, and strict governmental regulations that restrict outflows to protect downstream aquatic ecosystems. These regulations, often driven by

* Corresponding author: xuheng1117@tzc.edu.cn

environmental protection policies, can limit the operational flexibility of SHPs, leading to difficulties in ensuring a stable and continuous power supply. One promising solution to this problem is the integration of energy storage systems (ESS), which enable surplus energy to be stored for times when demand is low and released when required.

Although lead acid batteries continue to be the most commonly used battery technology, lithium-ion is the main battery technology for new storage applications [2]. In recent years, Lithium Iron Phosphate (LFP) batteries have emerged as a leading candidate for energy storage solutions in various renewable energy systems due to their superior thermal stability, long cycle life, and environmental safety compared to other lithium-ion technologies. Despite the extensive research on large-scale energy storage systems [3], the application of LFP batteries in small hydropower stations remains underexplored, particularly within the context of regulatory challenges that restrict water flow and environmental impacts downstream [4].

This research focuses on assessing the potential of LFP battery technology to enhance the operational efficiency of small hydropower stations under environmental constraints by analysing the following subjects including energy efficiency, energy density, duration of and energy losses during energy storage, speed of electricity generation, technology feasibility and cost-benefit analysis.

2 Technology Analysis

Compared to conventional lithium and lead acid batteries, LFP batteries are safer, last longer, need no maintenance, charge more efficiently, and discharge more quickly.

2.1 Energy Efficiency

Kang et al. [5] established that the efficiency of lithium-ion depends on current and State of Charge (SoC), and provide a method to compute energy efficiencies. In this part of we will be using the following expressions:

$$\eta_{net,ch} = \frac{W_{net}}{W_{ch}} \quad (1)$$

$$\eta_{net,dis} = \frac{W_{dis}}{W_{net}} \quad (2)$$

$$\eta_{ch,dis} = \eta_{net,ch} \cdot \eta_{net,dis} = \frac{W_{dis}}{W_{ch}} \quad (3)$$

With:

-$\eta_{net,ch}$: Net charging efficiency

-$\eta_{net,dis}$: Net discharging efficiency

-$\eta_{ch,dis}$: Combined efficiency of the charging and discharging cycle

-W_{net}: Net energy remaining after system losses

-W_{ch}: Energy input during charging process

-W_{dis}: Energy output during discharging process.

Using the equations above, the results of efficiency from discharge/charge cycles at different C-rates (C/20, 1C, 2C, 3C and 4C) were presented [6]. Charge and discharge efficiencies can be calculated by using equations (1) and (2). Afterwards, global efficiency can be calculated from half-cycle efficiencies with equation (3).

As shown in fig.1, charge efficiency and discharge efficiency are around 95% to 99%, cycle efficiency around 90% to 99%. Moreover, the efficiency remains in a high level of efficiency between 20% and 80% DoD (Depth of Discharge) and rapidly drops in two ends, which means without extreme use, the efficiency can be even higher.

Although LFP batteries may normally function in a temperature range of -20°C to 60°C, 0°C to 45°C is when they work at their best.To guarantee optimum performance, safety, and lifespan, the battery must be kept within its prescribed temperature range.

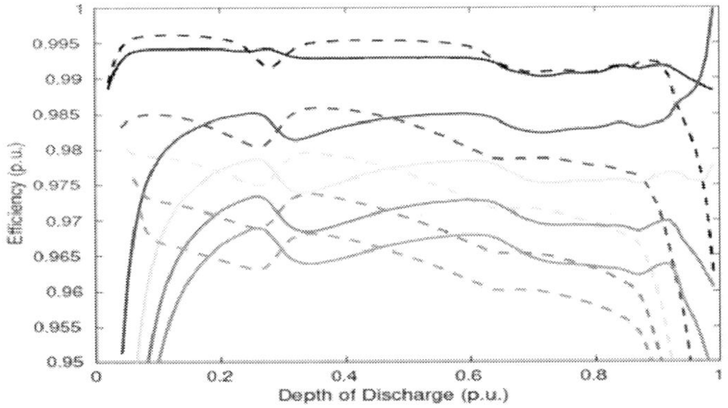

(a) Charge and discharge efficiency (respectively continuous and dashed lines).

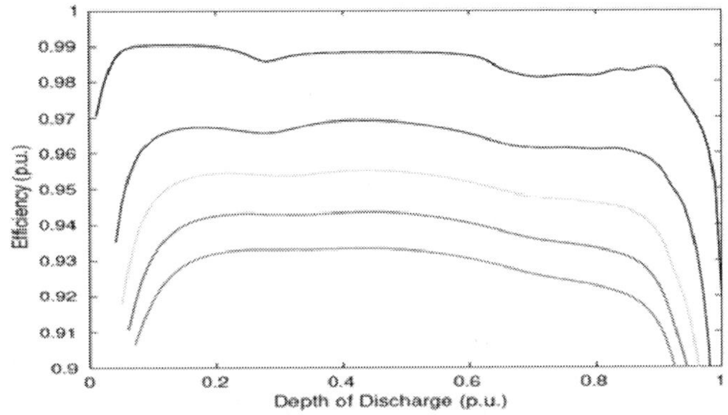

(b) Cycle efficiency.

Fig. 1. Efficiency for different C-rates [6].

2.2 Capacity Losses During Usage

During usage, discharge capacity is mainly affected by Equivalent Full Cycle (EFC), temperature, DoD, and discharge rate.

2.2.1 Equivalent Full Cycle

LFP battery remains over 80% of its initial capacity after 4000 EFC, hence we can assume that even with daily use, LFP battery can last over 10 years, as shown in fig.2.

 https://doi.org/10.1051/e3sconf/202560602002

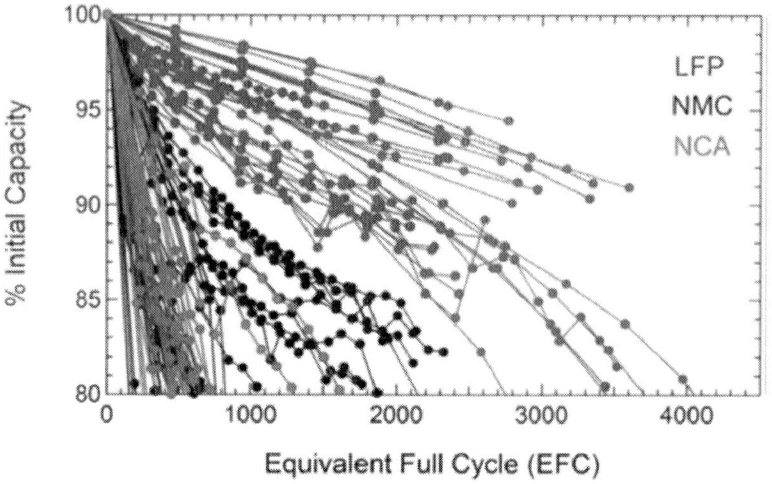

Fig. 2. Discharge capacity retention for all LFP (blue), NMC (black), and NCA (red) cells [7]

2.2.2 Temperature

The degradation of LFP battery capacity over time under different temperatures. There's roughly a 5% difference between 35 and 25 degrees, thus is vital to maintain an ideal temperature for LFP battery, as shown in fig.3.

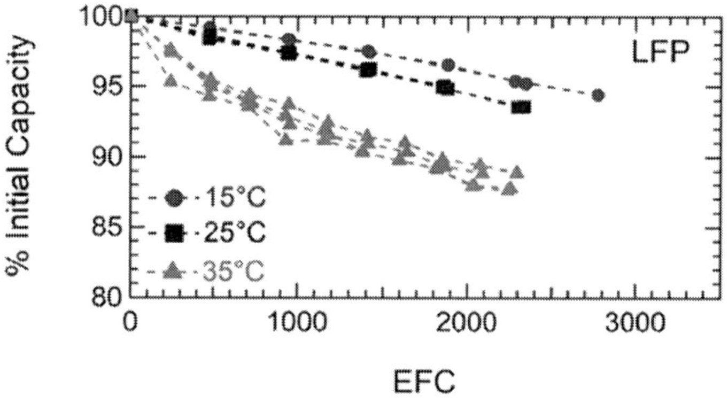

Fig. 3. Discharge capacity fade as a function of temperature [7]

2.2.3 Depth of Discharge

While maintaining excellent durability, LFP battery can be even more durable. Shallow discharge cycles (40-60%) provide the best longevity, while full discharge cycles (0-100%) lead to faster degradation, as shown in fig.4.

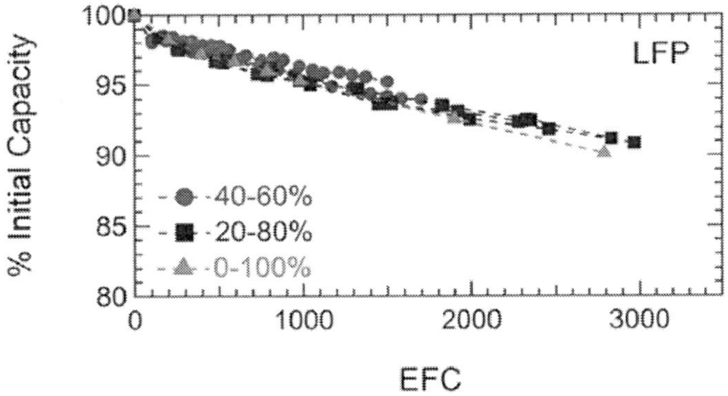

Fig. 4. Discharge capacity fade as a function of Depth of Discharge (DoD) [7]

2.2.4 Discharge Rate

Fig.5 highlights the importance of controlling the C-rate in order to maximize the lifespan of LFP batteries. Both extremely low and high C-rates negatively impact LFP battery life, while moderate C-rates like 1C and 2C are more balanced for maintaining long-term capacity retention.

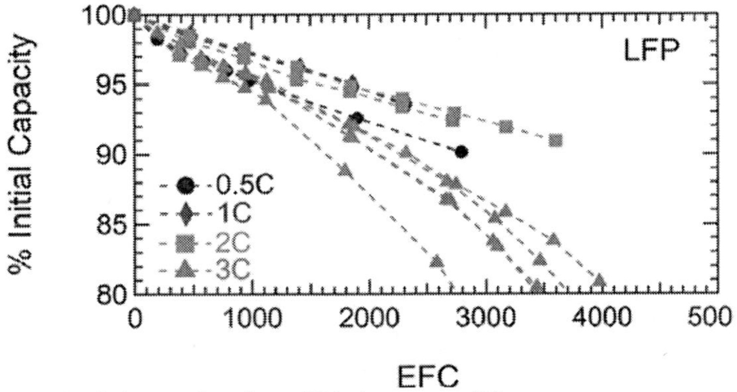

Fig. 5. Discharge capacity fade as a function of Discharge rate [7]

3 Economic Analysis

3.1 Primary Revenue Stream

The primary revenue stream in an energy storage system in China is Peak-valley arbitrage. In energy storage stations, peak-valley arbitrage is a lucrative business strategy that charges and discharges the storage system by taking advantage of the pricing differentials between peak and off-peak times in the power grid. Peak-valley arbitrage's basic concept is to maximize profits by storing energy during off-peak, low-priced times and releasing it during peak, high-priced times.

When the price differential between peak and off-peak electricity reaches 0.6 to 0.7 RMB/kWh or higher, companies using energy storage systems are considered to achieve profitability through peak-valley arbitrage.

In the case of small hydro power station, since its electricity cost is extremely low, primarily consisting of minimal operational expenses, an energy storage system will assist it to achieve peak-valley arbitrage, therefore maximizing its profits.

At the end of September 2024, provincial grid companies released the October 2024 proxy purchase electricity price list for grid enterprises. A total of 18 provinces and cities had a maximum price difference exceeding 0.7 yuan/kWh. The top three were Zhejiang, Guangdong, and Shanghai, with peak-valley price differences of 1.3522 yuan/kWh, 1.3481 yuan/kWh, and 1.2367 yuan/kWh, respectively.

3.2 Cost Analysis

Construction costs, storage battery costs, labor costs, operation and maintenance costs, disposal expenses, and other factors were taken into account when creating the energy storage power station cost model [8].

With battery being the major cost, there's two ways to obtain it. First is through one-time purchase, without any extra fees, it is supposed to be the cheapest way on average per year over the battery's lifecycle.

However, benefiting from the fierce competition in electric vehicles. The cost of LFP batteries continues to decrease rather rapidly.

Hence, with the constant dropping prices of battery, instead of purchasing battery, renting can be a cheaper way to established an ESS in a longer term. Many companies nowadays provide battery rental service. With rental battery, though costs are higher in the short term, the overall cots will be lower if battery prices continue to decrease. Moreover, with battery technology developing rapidly, renting provides access to the latest and more efficient battery.

4 Construction Analysis

4.1 Comparison of Installation Methods

4.1.1 On-site Building Installation

In this method, a dedicated structure (commonly called an energy storage building) is constructed, typically using a reinforced concrete frame. Battery stacks, power conversion systems (PCS), switchgear, transformers, and other equipment are installed inside the building. Depending on the scale, an energy storage building may contain multiple rooms for batteries, PCS, and switchgear. This approach is mainly used for large-scale energy storage stations, and most large domestic energy storage projects connected to the power grid adopt this method. The on-site building installation generally has lower upfront installation costs but entails higher long-term maintenance requirements and a longer construction period. It is also less flexible in terms of mobility.

4.1.2 Prefabricated Cabin Installation

This method uses a standard container design as a base, modified and enhanced to accommodate battery stacks, PCS, and switchgear inside the prefabricated cabin. It is commonly used for small- to medium-scale energy storage systems, such as distributed energy storage and mobile energy storage solutions. The prefabricated cabin installation offers advantages in construction speed, flexibility, and mobility. Given the project's requirements and site conditions, the prefabricated cabin installation method is selected for this project.

4.2 Analysis of Energy Storage System Efficiency

The energy storage system interfaces with the grid through PCS (Power Conversion System) and two primary circuits in the main path. During the charging and discharging processes, each stage incurs a certain amount of energy loss. Additionally, the battery system itself experiences energy loss during charge-discharge cycles.

According to current industry standards and technology, the overall efficiency of the charge-discharge process (Overall Efficiency) is determined by the PCS efficiency (~97%), battery charge-discharge efficiency (~90%), and line loss (~3%). The energy loss in the charge-discharge cycle (~10%) mostly occurs during charging. Therefore, for every 1Wh of storage capacity (with a 90% Depth of Discharge, or DOD), approximately 0.87Wh (1 * 0.9 * 0.97 * 0.997) can be discharged, and the charging process requires around 1.034Wh (0.9 / (0.9 * 0.97 * 0.997)).

5 Conclusions

The Tesla Victoria Big Battery and CATL's National Energy Storage Demonstration Project have both demonstrated the feasibility and advantages of lithium-ion Energy Storage Systems (ESS) in energy storage, as well as the gradual improvements in LFP battery technology. This paper analyzes the advantages of LFP and its application in small hydropower stations. Technically, LFP is the unequivocal first choice due to its superior performance. Economically, as the electric vehicle market develops, the cost of LFP batteries is expected to decline. Furthermore, ESS has become commercially mature, with affordable ESS rental services providing greater flexibility for private power generate applications. While battery prices decrease and the demand for clean energy is rising. Therefore, from a long-term perspective, small hydropower can achieve ecological protection and stable power output through flow control with the assistance of ESS, presenting significant future profit potential. However, the current profitability primarily depends on varying local government policies, and the costs of ESS remain relatively high.

References

1. IRENA (2023), The changing role of hydropower: Challenges and opportunities, International Renewable Energy Agency, Abu Dhabi.
2. Diaz, P. M., & El-Khozondar, H. J. Electrical energy storage technologies and the application potential in power system operation: A mini review. In 2019 IEEE 7th Palestinian International Conference on Electrical and Computer Engineering (PICECE). March, 2019.
3. Victorian Big Battery. Securing Power for victoria. Retrieved on 8th November, 2024, retrieved from https://victorianbigbattery.com.au/
4. Zhihua, C. H. E. N., Mengkai, Y. O. U., Wei, C. A. I., et al. Comprehensive Evaluation Model of Energy Storage Power Station With Full Life Cycle. Power Generation Technology, **44,** 883 (2023).
5. Kang, J., Yan, F., Zhang, P. Comparison of comprehensive properties of Ni-MH (nickel-metal hydride) and Li-ion (lithium-ion) batteries in terms of energy efficiency. Energy, **70,** 618-625 (2014).
6. Redondo-Iglesias, E., & Pelissier, S. On the Efficiency of LFP Lithium-ion Batteries. In 2022 Second International Conference on Sustainable Mobility Applications, Renewables and Technology (SMART) , 2022. IEEE.

7. James T. Frith, Matthew J. Lacey, Ulderico Ulissi. A non-academic perspective on the future of lithium-based batteries. Nature Communications, **14,** 420 (2023)
8. Preger, Y., Barkholtz, H. M., Fresquez, A., et al. Degradation of commercial lithium-ion cells as a function of chemistry and cycling conditions. Journal of The Electrochemical Society, **167**, 120532 (2020).

 https://doi.org/10.1051/e3sconf/202560602003

Development and application of pumped storage power generation system

Zhengye Liu*

School of Resources and Environment, Northeast Agricultural University, Harbin,150000, China

Abstract. As one of the most crucial energy storage facilities in modern times, pumped storage technology utilizes the principle of gravitational potential energy and mechanical energy conversion of water to pump energy facility when the eelectric load is low, and release water to generate power when it is high. The technology mainly includes pumping pump, turbine and generator and other equipment, through the two stages of pumping and power generation cycle, to realize the storage and release of electric energy. Pumped storage power generation technology has the advantages of large scale, high efficiency, clean and environmental protection, and is widely used in power systems with stability and reliability, but it is still limited by geographical factors. With the use of clean energy and the growth of electricity demand on the electricity side, pumped storage power generation technology will continue to innovate and develop, and become an important part of energy facility in the future electricity system.

1 Introduction

The earliest development and use of pumped storage technology is used for large-scale power supply to meet the electricity needs of residential and industrial production. In the 20th century, people gradually have an advanced comprehension of environmental protection, and tend to gradually replace fossil fuels with renewable and clean energy sources,which reduced the impact on climate change to a certain extent. Because renewable energy sources often exhibit variability in their energy supply, the future of energy storage technology has become particularly important. Among these technologies, pumped storage power generation has attracted much attention because of its use of water as a power generation medium and its high efficiency in power production.

The basic working rule of pumped storage technology is composed of several different modules, including the turbine, upper reservoir, lower reservoir, pump, generator, and grid [1]. The whole system of this power generation is mainly composed of two reservoirs of different heights connected to each other. When the power demand on the power side is low, the surplus power generated by pumped storage power generation will be used for water pumping between reservoirs; When the demand for electricity on the power side is high, the water is released due to the difference and through the power of turbine returnedthe reservoir [2].

* Corresponding author: lzygxx99@ldy.edu.rs

This technology first appeared in the 1890s, with the earliest invention using an individual pump impeller and group turbine generators in the Swiss, Austrian and Italian Alps. Then in the early 20th century, large hydropower stations began to appear, and during the construction process the single reversible pump turbine had become the main trend. By the middle of the 20th century, many large hydropower stations had been built, and the power sources used in the pumping process began to trend toward clean energy, such as wind, solar or a mix of energy sources. Today, pumped storage technology has been very mature, in the power supply system has played an irreplaceable role [3,4].

This paper aims to introduce the development of pumped storage technology at the present stage, give specific cases and data analysis, and make future development prediction by analyzing the advantages and disadvantages of this technology. Through such analysis, this technology can be better applied to the future large-scale power supply applications.

2 Technical principle and performance parameters

2.1 Workflow

The working process of pumped storage power generation mainly includes two stages as pumped energy storage and released water power generation.

During periods of low power demand, the entire system will be pumped for storage. This process usually uses the natural water source near the hydropower station as the raw water supply. In the pumping procedure, the pumping apparatus is actuated. With the motor serving as the power origin, water is pumped from the lower reservoir to the upper reservoir (high - level reservoir) by consuming electrical energy. In this process, electrical energy is converted into water and stored as potential energy. After pumping is completed, the water is stored for use during the power generation phase [3].

During the peak time of electricity demand or when the supply of the grid is deficient, water release power generation process starts. The stored water is released by opening the sluice gate or related control equipment at the bottom of the upper reservoir. When water flows through a turbine or turbine generator set, the kinetic energy of the water flow is converted into mechanical energy and transmitted to the generator set. At this juncture, the generator set transmutes mechanical energy into electrical energy by dint of the principle of electromagnetic induction and exports it to the grid so as to satiate the power demand [4,5].

2.2 Performance Parameters

The energy conversion efficiency is in the range of 66% to 82% [6]. Compared with other energy storage power generation methods, this power generation efficiency is already at a high level. Despite the various factors that can potentially impact the efficiency of power generation, the primary contributors to this inefficiency are likely to be energy dissipation within the unit and transformer operations, evaporation and seepage losses within the reservoir, electrical resistance encountered during the operation of the associated equipment, and fluctuations in the demand for grid power supply. These factors collectively contribute to the overall reduction in power generation efficiency.

In terms of power generation, one kilogram of water generates about 0.001 to 0.002 kWh of electricity [7]. And the gravitational potential energy generated by this water can not be all converted into electricity, there is energy loss in the process, so we multiply this data by the power generation efficiency, then 1 kg of water can be converted into electricity can be calculated. This number is so small that it's almost negligible. This is mainly because we assume a very idealized scenario in the calculation process, and the power generation efficiency is also relatively high. In reality, due to various factors, 1 kg of water can be converted into electricity will be much lower than this value.

In addition to the above data, there are also important factors such as service life, generation response time, risk response time, etc. These factors depend on the size of the hydropower station and its construction conditions [8].

3 Practical Application

Firstly, it can be used for power storage. The power of the grid in the trough is used for pumped storage, and the stored power is released in the peak time, so as to ensure the stable operation of the grid. Pumped storage technology is used to store backup power to ensure that the grid can provide stable power during peak hours or emergencies. Furthermore, it has a short - lived start - up time and a fast - paced adjustment rate. It can be utilized as an exigency power supply to promptly increase the power generation yield after the high - power loss within the system, thus guaranteeing the safe and stable functioning of the system [9].

This technology plays a significant role in regulating and optimizing power systems. Through the function of peaking and valley filling in pumped storage power station, the load fluctuation of power grid can be balanced and the energy utilization efficiency and power supply quality of power grid can be improved. In addition, the pumped storage power station can also be used as a start power supply to restore power supply in time after the occurrence of a major power outage to ensure the rapid and orderly recovery of the power system. It also play an important role in coping with the adjustment pressure brought by the continuous increase in the proportion of new energy installed capacity. Such as the Grand Guli hydropower Station in the United States, China's Fengman Hydropower Station has such a function.

The application of pumped storage power generation technology has also brought remarkable economic and social benefits. The construction and operation of pumped storage power stations will also help be conducive to local economic development, provide more jobs, protect and improve the local ecological environment. In recent years, the number of jobs that can be provided in the energy storage industry has exceeded 100,000 [3], and because of the important position of pumped storage power generation the proportion of jobs related to power stations is very large.

4 Technical Advantages and Limitations

Pumped power storage technology has been fully utilized after nearly a hundred years of development, some of its advantages are other energy storage technology can not be replaced, but there are still some shortcomings.

4.1 Technical Advantages

First, the principle of pumped storage power generation is well known. The technology has high maturity and reliability and stability, and can operate even under complex environmental conditions. Secondly, the pumped storage power station can adjust the power generation in a short time according to the demand of the grid, and has a strong adaptability. This also makes pumped storage power stations become one of the key means to solve the peak regulation of the power grid. The response speed is very fast, and the operation of regulating the power can be completed in a few seconds. Compared with other power generation methods, it has significant advantages when power grid emergencies and emergency accidents occur.

More importantly, pumped storage power generation technology has high power generation efficiency, which can save energy and improve energy utilization to a certain extent. This also means that this technology has the advantages of environmental protection and energy saving, pumped storage power stations do not produce additional carbon dioxide emissions, there is no pollution to the environment, compared to traditional thermal power stations, it is a very environmentally friendly and energy-saving way to generate electricity.

Finally, its comprehensive cost is low. Although the investment of pumped storage power stations is higher when they are initially built, the maintenance costs during operation are lower. Considering its long-term operation benefits, the comprehensive cost is low, and the investment payback period is short. And the electricity bill generated by pumped storage power generation is also less for ordinary residents [3].

4.2 Technical Limitations

These stations have high requirements for the geographical conditions of site construction, because of the particularity of its need for water generation, pumped storage power stations need suitable geographical conditions to build reservoirs and dams, which limits its wide application in some areas, and the construction period is long, which may affect the speed of its timely use. The power generation energy density of this technology is low, and its power generation is large and stable, but under the unit density or the same energy storage capacity, pumped storage power stations need to occupy more land and resources. Although the response of pumped storage power stations has developed rapidly, it is still difficult to fully keep up with the rapidly changing output of renewable energy sources such as wind power. This has limited its application in the field of renewable energy generation to a certain extent.

5 Future Outlook

As a mature and widely used energy storage technology, it will continue to maintain rapid growth in the future. Technological innovation will play a key role in the construction and operation of pumped storage power stations. With the application of digital and intelligent technology, the dispatching operation of pumped storage power stations will be more flexible and efficient, and the engineering construction technology and equipment manufacturing technology will also achieve innovation and improvement. A more complete industrial system and specialized development mode will enhance the competitiveness and sustainable development ability of the entire industry. In terms of policy, the government will continue to introduce relevant policies to support the development of pumped storage industry, improve the market mechanism, reduce electricity prices, and promote the cost reduction and high-quality development of the whole industry.

At the same time, some parts of the world are also facing challenges, some areas of pumped storage development there is not deep work, unclear demand and other problems, and some pumped storage unit production capacity is not yet able to meet the peak demand. Therefore, it is necessary to scientifically and reasonably plan the unit manufacturing and capacity transformation to avoid the problem of insufficient or excess capacity.

6 Conclusion

Pumped storage power generation is already a well-mastered energy storage technology. When the demand for electricity on the power side is low, the pumped storage power generation facility uses the extra electricity generated to draw water from the reservoir to store energy. When the demand for electricity on the power side is high, the stored water in the upper reservoir will be released for electricity generation, using the gravitational potential energy generated by the falling height of the water flow to convert into electricity release. The main advantages of pumped storage power generation are high efficiency, large energy storage scale, environmental protection, etc., but it is also limited by geographical factors and environmental impacts, and has high construction costs. Pumped storage power generation has been widely used in the world, and is an environmentally friendly, low-carbon, convenient and clean energy storage facilityy for power systems. In the future, with the rapid development of renewable energy, pumped storage will become an important means of energy storage, and government support and technological innovation will further promote the application of it. In summary, pumped storage power generation, as an efficient, reliable and economical way of energy storage, plays an important role in the power system. Although there are certain disadvantages and challenges, with the development of technology and policy support, its application prospects are still broad.

References

[1] Šćekić, L., Mujović, S., & Radulović, V. Pumped hydroelectric energy storage as a facilitator of renewable energy in liberalized electricity market. Energies, **13,** 6076 (2020)..

[2] Stocks, M., Stocks, R., Lu, B., et al. Global atlas of closed-loop pumped hydro energy storage. Joule, **5,** 270-284 (2021).

[3] Rehman, S., Al-Hadhrami, L. M., et al. Pumped hydro energy storage system: A technological review. Renewable and Sustainable Energy Reviews, **44,** 586-598 (2015).

[4] Xu B, Zhang J, Egusquiza M, et al. Exploring the Regulation Reliability of a Pumped Storage Power Plant in a Wind-Solar Hybrid Power Generation System. Water. **13,** 2548 (2021).

[5] Zhang, Z., Ding, T., Zhou, Q., et al. A review of technologies and applications on versatile energy storage systems. Renewable and Sustainable Energy Reviews, **148,** 111263 (2021).

[6] Yang, B., Li, D., Fu, X., et al. Energy and exergy analysis of a novel pumped hydro compressed air energy storage system. Energy, **294,** 130737 (2024).

[7] Cao, R., Li, W., Wang, S., et al. Comprehensive comparative study of two novel isobaric adiabatic compressed air energy storage systems coupled with pumped hydro storage. Applied Thermal Engineering, **257**, 124318 (2024).

[8] Blakers, A., Stocks, M., Lu, B., et al. A review of pumped hydro energy storage. Progress in Energy, **3,** 022003 (2021).

[9] Kong, Y., Kong, Z., Liu, Z., et al. Pumped storage power stations in China: The past, the present, and the future. Renewable and Sustainable Energy Reviews, **71**, 720-731 (2017).

Rechargeable Batteries for Renewable Energy: Current Status, Technical Challenges, and Future Directions

Qihan Hu*

Columbia International College (Canada), Hamilton, Canada

Abstract. With the increasing severity of climate change and the threat of global warming, countries are boosting the development and promotion of renewable energy sources like solar and wind to bring about the energy transition and reduce carbon emissions. However, because these energy sources are unreliable in supply, efficient energy storage technologies are required to balance energy output to satisfy daily electricity demand. Among numerous energy storage alternatives, rechargeable battery technology has received a lot of interest due to its high energy density, efficiency, and reusable nature. Currently, lithium-ion batteries, nickel-metal hydride batteries, lead-acid batteries, and other rechargeable battery types are widely utilized in consumer electronics, transportation, renewable energy storage, medical equipment, and other industries. This paper examines the basic working principle, structure, and important chemical reactions of rechargeable batteries, as well as their performance indicators such as capacity, energy density, cycle life, and so on. Furthermore, this paper addresses the technology's global application cases, as well as the status of its development and the major problems, such as safety, cost, longevity, environmental impact, and other issues. Finally, this article examines probable future technology breakthroughs, market trends, and the possibility of suitable policy support in this field.

1 Introduction

Every society depends on electrical power to meet its fundamental requirements, meanwhile, carbon dioxide emissions have drastically increased and brought about global warming. The current situation has jeopardized the environment in which people reside, leading to abnormal rainstorms, prolonged droughts, and an accelerated expansion of desert areas. Long-term variations in weather and temperature conditions, including differences in solar activity and significant volcanic eruptions, are referred to as changing the climate. Human activity has served as the primary cause of global warming since the beginning of the 19th century [1].

Carbon emissions must be reduced by nearly fifty percent by 2030 and reach net-zero levels by 2050 to avert the worst effects of climate change [2]. The creation of diverse renewable energy sources is one potential course of action. According to predictions from

* Corresponding author: 631501011112@mails.cqjtu.edu.cn

the International Renewable Energies Agency (IRENA) By 2050, 90% of the world's electricity could and ought to originate from renewable sources [3]. Utilizing renewable energy sources to the fullest can help reduce air pollution and related health problems [4]. By using the force of nature to create electricity, marine energy, wind energy, and solar energy are significant types of renewable energy that reduce carbon dioxide emissions and reduce global warming. These technologies provide energy security, lessen the negative effects on the environment and human health, and lessen reliance on fossil fuels. Renewable Energy power lead to inconsistent power output and compromise grid security. The energy storage battery system can react rapidly to variations in frequency and voltage, keeping the grid stable. The primary benefit of rechargeable battery technology is the capacity to effectively store and release electrical energy, making it appropriate for a range of applications, particularly in electric vehicle and grid regulation. Because of its adaptability, broad range of applications, and capacity to sustain power stability in an environment of significant changes in renewable energy sources, this technology is garnering attention. Rechargeable batteries are multipurpose power sources that can be repeatedly utilized and charged by an external voltage. Lead-acid batteries, nickel-metal hydride batteries, batteries with lithium ion (including lithium cobaltite and manganate), and nickel-cadmium batteries are common varieties. Rechargeable batteries have become a cost-effective and environmentally responsible choice because of their high energy density, safety, dependability, protection against the environment, and longer cycle life. They are extensively utilized in high-power, continuous-use gadgets including electric toys, power tools, and mobile phones.

Rechargeable battery power is not without its issues, though. Rechargeable batteries, for instance, may overheat when being charged or discharged, which could lead to leakage and short circuit. The battery's performance can also be lowered by erroneous operation, overcharging or draining, particularly when using a power battery, as well as by improperly timing the charging process or committing errors during operation. In this paper, principle, technical characteristics, application scenarios, problems and solutions of rechargeable battery are analyzed to elucidate how rechargeable batteries might mitigate environmental damage and significantly contribute to sustainable development.,.

2 Working principle

2.1 Definition of rechargeable battery

Rechargeable batteries are devices that can store and release electrical energy after multiple charging and draining cycles. The functioning principle is based on electrochemical processes. It stores electrical energy when charging and releases it during discharging using electrodes and electrolytes. Batteries usually consist of a positive electrode, a negative electrode, and an electrolyte. Metal oxides or phosphates are common positive electrode materials. Graphite and lithium metal are common materials for negative electrodes. Electrolytes are liquids, gels, or solids that allow ions to travel among electrodes. When a battery is charged, electrons and ions flow from the negative electrode to the positive electrode, and the reverse happens when it is discharged. This process can be repeated on multiple occasions, and it is the irreversible electrochemical reaction that enables the battery to both charge and discharge.

Rechargeable batteries use electrochemical reactions to transfer chemical energy to electrical energy. During the discharge process, an oxidation reaction happens at the anode (negative electrode), releasing electrons and delivering power over an external circuit; a reduction reaction occurs at the cathode (positive electrode), taking electrons. When charging, an external power source enables electrons to flow back, restoring these chemical processes and restoring the battery's capacity. The comparison of rechargeable battery and ordinary battery are summarized in table 1.

Table 1. The difference between rechargeable batteries and ordinary batteries [5]

Items	Rechargeable batteries	Ordinary batteries
Component	Lithium dioxide as positive platform, spongy pure lead as negative platform	Positive substance (Oxygen or hydrogen), negative substance (zinc or lead)
Application	High-energy-consuming electronic devices including digital cameras, laptop computers, and mobile phones.	Devices demanding consistent voltage and current, including torches, clocks, and others.
Rechargeable or not	Rechargeable	Nonchargeable
Strengths	Large capacity, proper for heavy energy requirements, minimizing the frequency of battery change.	Stable current and voltage, easy to transport, reliable performance, unaffected by the environment.
Weakness	Considerable cost, restricted number of charge and discharge cycles, and self-discharge.	It may not be replenished and must be replaced after usage.
Environmental impact	Reusable, minimizing the environmental impact of waste.	It must be discarded all at once, generating a vast amount of waste with protentional environmental contamination.

2.2 Types of rechargeable batteries

Nickel metal hydride batteries, lithium-ion batteries, lithium manganese oxide batteries, lithium cobalt oxide batteries, nickel cadmium batteries, and lead-acid batteries are some of the most common types of rechargeable batteries available today. Their characteristics are summarized in Table 2.

Table 2. Characteristics of various rechargeable batteries [5]

Types	Characteristics	Strengths	Weakness	Application
Lithium-ion Battery	High energy density, light weight, and high voltage	Long life, high single charge voltage, and lightweight.	Higher expense, fire risk.	Electric vehicles, such as smartphones, and laptops.
Lithium Manganate Battery	Hot temperature stability and safety, capable of sustaining higher current discharges.	Excellent safety and great temperature resistance.	Lower energy density compared to other lithium batteries.	Power tools, electric cars.
Lithium Cobalt Oxide Battery	High energy density.	High energy density and light weight.	Poor safety, high costs, and limited lifespan.	Laptops, smartphones.

Nickel Metal Hydride Batteries	Environmentally friendly, sturdy structure.	Lower cost and fewer hazardous metals.	Rapid self-discharge and limited lifespan.	Hybrid vehicles, digital cameras.
Nickel Cadmium Batteries	Low density of energy and high deep discharge capacity.	Can handle many charging and releasing cycles; suited for harsh temperatures.	Holds poisonous cadmium and has a "recall effect".	Power tools, emergency lighting.
Lead-acid Battery	Low energy density, large and bulky.	Low-cost, capable of supplying enormous amounts of current.	Heavy and short service life.	Car batteries, UPS backup power, and solar energy storage.

2.3 Main performance parameters

2.3.1 Energy density

Lithium-ion batteries typically have a power density of 150 to 200 Wh/kg, making them excellent for use in products that demand reducing weight. These products include electric vehicles (EVs), smartphones, and laptops [6]. Due to their high energy density, lithium-ion batteries keep more energy per unit weight, making them excellent for portable gadgets or applications that require a longer period of operation.

2.3.2 Charge and discharge efficiency

Lithium-ion batteries have a charge and discharge efficiency of more than 90%, which is why they are commonly employed in electronic products and electric cars. The excellent charge and discharge performance minimizes energy loss throughout the conversion process, making it perfect for systems that operate for a long amount of time or require efficient energy leadership, including solar energy storage systems and electric automobiles. Its great efficiency enables more energy to be used on each charge, extending the device's life.

2.3.3 Energy storage cost

The cost of producing lithium-ion batteries has decreased dramatically in recent years, with a current cost of approximately $137 per kilowatt-hour. With advancements in manufacturing techniques and savings of production, this cost is predicted to continue to fall in the next few years, approaching $100/kWh [7].

2.3.4 Cycle life

Lithium-ion batteries typically have 500 to 1,500 charge and drain cycles, and in some outstanding performance applications, optimized batteries can have a cycle life of over 3,000 times [8].

2.3.5 Charge and discharge speed

The charging as well as discharge speed of lithium-ion batteries can be measured using C-rate. Its rapid charge and discharge rate enables it to finish charging quickly, which is critical for everyday electronics like mobile phones and laptop computers. Furthermore, with the adoption of fast charging technology, the speed of charging of lithium-ion batteries is constantly improving, lowering users' wait times. However, high charge and discharge rates can shorten battery life, thus they are typically managed by battery management software (BMS).

2.3.6 Cost

The cost of lithium-ion batteries has been decreasing over time and is now around US$100 per kWh. As global demand for sustainable energy and electric vehicles grows, lithium-ion battery production expands, resulting in lower unit costs [8].

2.4 Applicable scenarios

2.4.1 Aerospace field

Rechargeable battery systems are essential for providing energy support in satellites, unmanned aerial vehicles （UAVs）, and other aerospace applications. In harsh settings, this sort of battery must be exceptionally reliable, long-lasting, and lightweight. Lithium-ion batteries and their new versions, among them lithium-sulfur batteries and solid- state batteries, are gaining considerable interest [9].

2.4.2 Consumer electronics

Lithium-ion batteries are commonly utilized in handheld electronics that include smartphones, laptop computers, and smart watches. Research focuses on increasing energy density, extending battery life, lowering self-discharge rates, and investigating alternate electrolyte to improve safety. Lithium-ion batteries are the chosen power source for these gadgets due to their high density of energy, lightweight design, and long service life [10].

2.4.3 Medical Equipment

Rechargeable batteries are widely used in medical devices such as pacemakers, hearing aids, and insulin units. These batteries are often compact, powerful, and long-lasting, and must operate reliably inside the bodies of people or in medical settings. The medical profession requires extremely exact standards for the safety and stability of batteries that are rechargeable. As a result, academic circles are researching the usage of solid-state batteries and bio-batteries in place of typical lithium-ion batteries to reduce dangers and increase service life [11].

2.5 Project Cases

Massive lithium-ion battery arrays are currently being used for storing wind and solar energy and offer power aid to the national grid [12]. These projects, which use the Powerwall and Powerpack storage systems for energy to provide clean energy solutions for homes and businesses, represent the commercialization of rechargeable battery technology. Tesla's

Powerwall is a popular energy storage solution for households and businesses that employs lithium-ion battery technology. Each device has 13.5 kWh of energy storage capacity, allowing users to store solar energy for use during grid disruptions or peak hours. This project demonstrates the successful integration of rechargeable batteries and renewable energy systems [13].

3 Technical challenges

3.1 Security Issues

A major safety risk for batteries made with lithium-ion is "thermal runaway," which occurs when the battery is subjected to high temperatures, charging too much, or physical damage, and the chemical reactions inside the battery become uncontrollable, causing the internal temperature to rapidly rise and cause a fire or explosion. Because of the enormous quantity of energy accumulated in high-energy-density packs of batteries, such as those used in electric vehicles, the risk of thermal runaway is very great. When the temperature within the battery exceeds a particular point, the electrolyte may dissolve and emit flammable gases, increasing the risk of a fire.

3.2 Longevity

Lithium-ion batteries have a limited life cycle, and when they are charged and discharged repeatedly, their ability steadily declines. This phenomenon is commonly known as "capacity fading" or "battery aging." After several hundred cycles, a battery may only deliver a percent of its first capacity, reducing the device's longevity.

3.3 Environmental impact issues

Lithium-ion batteries require a considerable number of rare metals, including lithium, cobalt, and nickel. The extraction of these metals has a significant impact on the environment. For example, cobalt mining is mostly centered in the Democratic Republic of Congo, which includes issues that include environmental pollution, ecological degradation, and human rights violations.

3.4 Cost challenge

Although the cost of batteries made from lithium-ion has been falling year after year in recent years, production costs stay high, particularly in large-scale applications such as electric vehicles and energy storage systems on a massive scale. Manufacturing lithium-ion batteries needs the use of high-purity elements such as cobalt, lithium, and nickel, as well as sophisticated production procedures and stringent quality control, all of which raise manufacturing costs.

4 Solutions and further trends for technical challenges

4.1 Security

4.1.1 Security Issues

Solid electrolytes provide stronger thermal stability than standard liquid electrolytes, which can considerably lower the danger of thermal runaway within the battery. Solid-state batteries not only lower the combustible gases emitted during electrolyte decomposition, but they also boost the battery's energy density.

4.1.2 New Diaphragm Technology

Higher-temperature stable separating materials can improve the reliability of the battery's inner workings and keep it from short-circuiting or failing at elevated temperatures. Ceramic separators, for example, have gradually gained popularity as a safety upgrade for lithium-ion batteries because to their exceptional thermal stability.

4.1.3 Battery Management System (BMS)

Sophisticated BMS system may track the temperature, current, voltage, and other battery characteristics in real time to prevent overcharging, over-discharging, and overheating, lowering the risk of thermal runaway. By intelligently changing the charging and discharging rates, BMS can delay the onset of thermal runaway.

4.1.4 Development trend

Solid-state batteries are seen as the future of lithium-ion battery safety. Solid-state battery technology is predicted to enter commercial production over the next ten years, with widespread use in electric vehicles, systems for storing energy, and other applications. In addition, innovative nanomaterials, coating technologies, and automated monitoring systems will improve the safety of lithium batteries.

4.2 **Lifespan Challenge**

4.2.1 Type Electrode Material

Researchers are aiming to create new silicon-based or lithium metal anodes with increased ability and cycle stability. By enhancing the structure of the materials used for electrodes, mechanical stress generated by the intercalation and deintercalation of the lithium ions can be decreased, extending the battery's service life.

4.2.2 Electrolyte improvements

High-stability electrolyte additions can prevent electrolyte decomposition under high-pressure, high-temperature circumstances while avoiding adverse reactions. The inclusion of additives (for instance, fluorinated electrolyte) can considerably increase the battery's cycle stability and delay capacity degradation.

4.2.3 Smart charging technology

Optimizing the process of charging algorithm and using advanced charging technology can help to reduce battery damage during charging and discharging. For example, several manufacturers of electric cars have begun to use rapid charging technologies and segmented recharging algorithms to alleviate battery stress during deep drain and quick charging.

4.2.4 Development trend

The life of batteries made with lithium-ion will be extended further as battery material research advances, and the battery's cycle life may eventually exceed 3,000 times [14]. At the same time, solid-state batteries are projected to play an important role in prolonging battery life due to their superior electrochemical stability. Since the need for long-life batteries in electric automobiles and energy storage increases, research into battery life improvement will become more important.

4.3 Environmental Impact Issues

4.3.1 Material Recycling Technology

To lessen dependency on rare metals, batteries recycling equipment is fast evolving. Recycling lithium, cobalt, nickel, and other metals in old batteries reduces the demand for new natural resources while also reducing the environmental impact of mining. Closed-loop recycling technology may significantly boost metal recycling efficiency while lowering the environmental impact of battery production.

4.3.2 Material Substitution

Some research aims to build cobalt-free batteries with lithium ion or replace rare elements with more abundant minerals like iron and sodium. Because of the abundance of sodium supplies, sodium-ion batteries have appeared as a potentially practical alternative to lithium batteries. In addition, the use of cobalt-free battery packs, such as iron phosphate lithium power sources, is growing, particularly in sectors where great energy density is not required.

4.3.3 Environmentally Friendly Material Development

The study and development of some novel materials is geared to minimizing environmental pollution. Water-based electrolytes, for example, are thought to be more environmentally benign than standard organic solvents. Furthermore, the research of compostable battery materials is steadily progressing to prevent environmental contamination following battery disposal.

4.3.4 Development Trend

As global environmental restrictions tighten; battery longevity will become a major focus for future improvements. Battery recycling technology is likely to advance significantly in the next years, and closed-loop logistics systems will become more widespread in the battery manufacturing industry. At the same time, the creation of green manufacturing processes and alternative resources will become a priority for the sector.

4.4 Cost Challenge

4.4.1 Scale effect and production process optimization

As the volume of lithium-ion batteries produced grows, manufacturers can save costs by improving efficiency in production and perfecting procedures. For example, automated lines of production, modern manufacturing technologies (including 3D printed electrodes), and optimized production procedures can all help to cut production costs on a big scale.

4.4.2 Material substitution and reducing dependence on rare metals

One important strategy for lowering battery prices is to limit the consumption of rare metals like cobalt and nickel. Lithium iron phosphate (LFP) batteries have appeared as the preferred low-cost batteries due to their lack of cobalt and nickel content and low material costs. Future material developments, like sodium-ion batteries, are predicted to dramatically cut battery production costs.

4.4.3 Government Subsidies and Policy Support

Some countries and regions encourage the use of electric motor vehicles and systems to store energy through government funding or tax breaks, hence boosting the growth of the lithium battery industrial chain. Furthermore, the National Strategy Reserve Program may help stabilize price variations in essential metals such as cobalt and lithium, thereby lowering the long-term expense of batteries.

4.4.4 Development Trend

The cost of producing lithium-ion batteries is likely to reduce further in the future years, driven mostly by worldwide demand for electric motor vehicles and renewable energy storage devices. With material innovation, improved production efficiency, and greater economies of scale, the cost of lithium-ion batteries is predicted to approach the critical threshold of $100/kWh, further pushing their use in a variety of situations [9].

5 Conclusion

The future growth of lithium-ion batteries confronts many hurdles, including safety, longevity, environmental impact, and cost. However, with advances in materials science, the implementation of modern technologies, and the improvement of the worldwide industrial chain, lithium battery performance and sustainability are likely to increase dramatically in the future. Innovative advancements such as solid-state batteries, renewable resources, and smart charging technology will encourage the widespread use of batteries made of lithium in electric vehicles, batteries for energy storage, and personal gadgets.

References

1. IPCC. IPCC Sixth Assessment Report. Retrieved on 5th November, 2024, retrieved from https://www.ipcc.ch/report/ar6/wg1/
2. IEA. Net Zero by 2050. Retrieved on 5th November, 2024, retrieved from https://www.iea.org/reports/net-zero-by-2050A.

3. United Nations. Renewable energy – powering a safer future. Retrieved on 5th November, 2024, retrieved from https://www.un.org/en/climatechange/raising-ambition/renewable-energy
4. World Health Organization. Billions of people still breathe unhealthy air: new WHO data. Retrieved on 5th November, 2024, retrieved from https://www.who.int/news/item/04-04-2022-billions-of-people-still-breathe-unhealthy-air-new-who-data
5. Siriyala Trilochana, C.N. Sangeetha. A Comparative study and Recent Research of Battery Technologies. The International journal of analytical and experimental modal analysis, **8**, 1183-1189 (2021)
6. Shuvajit Ghosh, Udita Bhattacharjee, Subhajit Bhowmik, et al. A Review on High-Capacity and High-Voltage Cathodes for Next-Generation Lithium-ion Batteries. Journal of Energy and Power Technology, **1**: 002 (2022)
7. China Energy Storage Network. Analysis and prospect of new energy storage technology route. Retrieved on 5th November, 2024, retrieved from https://www.escn.com.cn/20240129/18ea85be891d4a7dbd554a9c3b540624/c.html
8. NREL. Electric Vehicle Lithium-Ion Battery Life Cycle Management. Retrieved on 5th November, 2024, retrieved from https://www.nrel.gov/docs/fy23osti/84520.pdf
9. Energy Storage. Bloomberg NEF: Average battery pack prices to drop below US$100/kWh by 2024 despite near-term spikes. Retrieved on 5th November, 2024, retrieved from https://www.energy-storage.news/bloombergnef-average-battery-pack-prices-to-drop-below-us100-kwh-by-2024-despite-near-term-spikes/
10. M. Stanley Whittingham. Lithium Batteries and Cathode Materials. Chemical Review, 104:4271-4301(2004)
11. J.M. Tarascon, M. Armand. Issues and challenges facing rechargeable lithium batteries. Nature, 414: 359-367 (2001)
12. John, A.T., Nair, M. G., Morrow, M. R. Rechargeable batteries for medical implants https://pubs.acs.org/doi/10.1021/cr020731c. Journal of Power Sources, 493:229730 (2021)
13. Tesla. Tesla Power Everything. Retrieved on 5th November, 2024, retrieved from https://www.tesla.com/energy
14. Battery University. BU-808: How to Prolong Lithium-based Batteries. Retrieved on 5th November, 2024, retrieved from https://batteryuniversity.com/article/bu-808-how-to-prolong-lithium-based-batteries

Research on Current Solid Mass Energy Storage (SGES)

Yichen Su*

Beanstalk International Bilingual School, 101300 Beijing, China

Abstract. With the increasing environmental challenge posed by fossil fuel combustion, there has been a significant effort to develop alternative energy sources. In addition, hydropower, once considered as a perfect solution, has recently faced the criticism of deteriorating downstream aquatic habitat. As a result, energy sources that are susceptible to the environment, such as wind or solar, have become some of the few available options. This is accompanied by a raised demand on reliable and efficient energy storage technology. This paper examines solid mass energy storage (SGES), a newly developed energy storage technology by converting excess electricity into gravitational potential energy. The paper also lists out some of its most popular variation and relevant quantitative analysis. With a simple and straightforward mechanism, SGES has the potential to be furtherly improved and popularized for a wider range of application. However, as a newly developed concept, there are very few applicated cases of SGES. This causes a shortage in operational data and analysis on aspects such as facility lifespan, giving out a future research direction in this field.

1 Introduction

Fossil fuels have been regarded as one of the most efficient energy sources since the industrial ages in the 18th century. Pushing the turbine by using steam derived from combustion, therefore, generating electricity, fossil fuels are known and praised for their straightforward mechanism, high economic benefits, energy efficiency, and reliability. This, however, comes with serious environmental issues, namely global warming which is caused by an increase in CO_2 emission. Despite known consequences, large quantities of fossil fuels are still being used in response to the exhausting demand for energy brought by the peaked population [1]. In 2022, the global population exceeded 8 billion, resulting in a global energy consumption of 607 EJ. Among them, over 80% of the energy was contributed by fossil fuels, contributing 23.54 Gt carbon emission in the powering and industrial sectors or about 60% of the year's total CO_2 emission [2].

Driven by these issues, more and more efforts were put into the development of alternative, renewable energy sources and to replace fossil fuels with them. Hydropower, among them, must be the most mature technology. Generating over 4,200 TWh in 2023, hydropower ranked third in energy yields-just behind coal and natural gases-and outnumbered every other

* Corresponding author: Ethansu26sy@bibs.com.cn

renewable source use [3]. China, for instance, emerged as the leading force in the hydropower field. With many of its developing regions located in mountainous terrain, small-sized hydropower stations have become the most dominant energy sources for villages in southern China. However, a policy released by the Ministry of Water Resource in 2018 stated the potential disturbances of hydropower stations on the downstream aquatic habitat of the Yangtze River. The later investigations led to the dismantling of over 3,000 hydropower stations in China, with further limitations and restrictions on the usage of the plant to other 20,000 stations [4]. Such an increased awareness of protecting aquatic habitats brought a heavy blow to electricity consumption in places where hydropower stations were affected. Therefore, it is necessary to supply electricity with other energy sources, namely, solar and wind power.

Wind and solar power have boosted exponentially since the millennium, with their annual yield increasing from 31 and 1 TWh in 2000 to 2,300 and 1,600 TWh in 2023, respectively [3]. However, both technologies face the challenge of generating electricity continuously. As a natural resource, the strength and frequency of wind are completely uncontrollable. This makes wind power insufficient and extremely susceptible in places with a milder climate. Solar energy, on the other hand, is indeed abundant on a daily basis, but its peak hours (morning to noon) do not usually align with the peak of electricity demand (late afternoon to evening). If a place's power supply solely depends on such sources, it is likely for the area to undergo power shortages during peak demand hours.

The application of energy storage systems is needed to address this challenge. When supply exceeds demand, the storage system may consume the electricity generated by wind or solar plants and store it as potential energy-often gravitational potential energy. Conversely, when demand peaks but supply runs short or when under emergencies, this stored potential energy can rapidly be given out or converted back to usable electricity. This paper will provide a comprehensive analysis of solid-mass gravitational energy storage (SGES), focusing on its mechanism, quantitative analysis, state of development, and cases of implementation.

2 Qualitative and Quantitative Analysis on SGES

2.1 Mechanism and Variations of SGES

SGES systems operate in a very simple and straightforward mechanism. In short, it is to store excess energy in the form of gravitational potential energy and gives it out through a generator. This process shares similarities with hydropower plants, but in addition to it, SGES requires the system to consume energy first (therefore, storing the energy in the form of potential energy) before giving it out as usable energy. During stage 1, energy sources such as coal or solar generate excess electricity. The SGES system utilizes this excessed electricity to pull solid masses (typically composed of concrete) into the air using a crane. These masses remain above the ground until an electricity demand occurs, marking the transition to the next phase. During stage 2, the mass will be released and dropped back to the ground. While falling, the wire, with one end connected to a generator and the other to the mass, generates electricity by converting the gravitational force of the falling mass to electricity. This generated electricity is then sent back into the power grid to meet the sudden demand [5]. This means SGES is not limited by geographical or topographical factors such as the difference in elevation as required by pumped-hydro storage systems. Fig. 1 visualizes this process by showing stage 1 and 2.

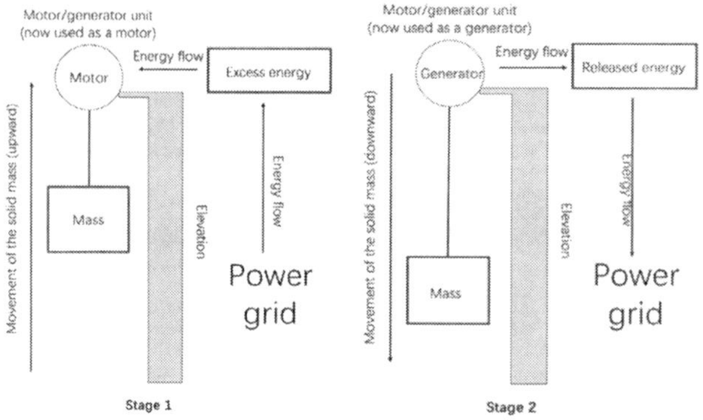

Fig. 1. SGES mechanism

There are three main SGES variations: Tower SGES (T-SGES) and Shaft SGES (S-SGES). Other variations such as Piston SGES (P-SGES) and sub-variations under it are also put to development and are relatively less mature.

T-SGES is one of the most mature systems. First developed by the Swiss-American company Energy Vault in 2017, the system comprises a modular T-shaped tower (weight-bearing tower and the cantilever), cranes on the cantilever, and a motor/generator unit on the crane. As explained in 2.1, the storage of excess energy in T-SGES is achieved by lifting masses up into the air with it. The lifted blocks are usually stacked up on different levels (stair-step configuration) to maximize space usage. The higher a block is above the ground, the more gravitational potential energy it holds, giving out more electricity when dropped. When demand occurs, the crane moves to the level of blocks that store the desired amount of electricity. The gravitational force of the falling block will allow the rope to generate electricity through the generator unit. A prominent advantage of this design is its high scalability, meaning the storage unit can be freely scaled up or down depending on the actual need. According to Energy Vault, the storage capacity of its T-SGES design may vary from 10 MWh to several GWh [5, 6]. Fig. 2 shows a schematic diagram of a T-SGES system.

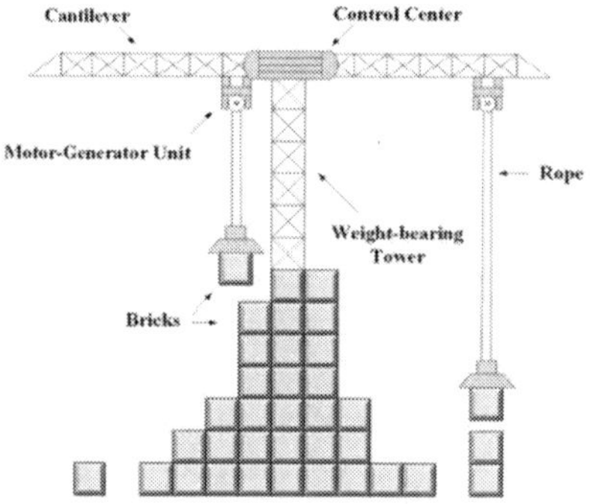

Fig. 2. Schematic diagram of T-SGES [5]

The second variation, S-SGES, has few differences compared to the previous variation. Similar to that of T-SGES, the mechanism of S-SGES also depends on the pull-drop maneuver of solid masses. The main difference between the two is the location of the system. The S-SGES system is buried underground, with a vertical hole of over 300 m dug into the ground and two motor/generator units installed on opposite sides. Two wires connected to the motor/generator units hold a single piece of solid mass in the hole through fixed pulleys and function similarly to the ropes in T-SGES [5, 7]. Fig. 3 shows a schematic diagram of a S-GES system.

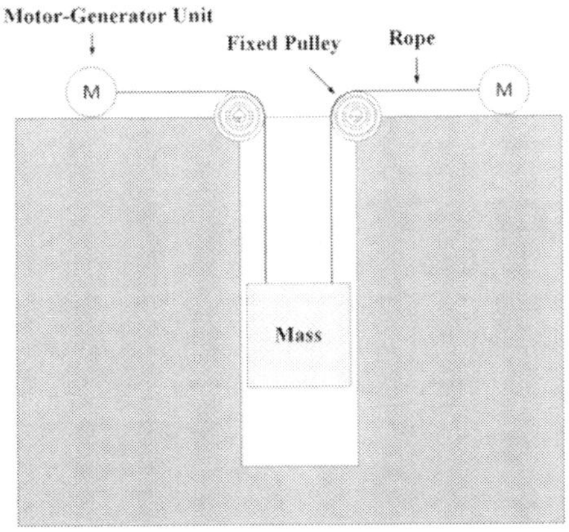

Fig. 3. Schematic diagram of S-SGES [5]

2.2 Facility Lifespan

The empirical lifespan of an SGES system is hard to determine due to lacked samples to refer to. However, theoretical estimation exists and gives an average lifespan of 30-50 years. T-SGES, according to Energy Vault, may perform for over 35 years; S-SGES with a life span of up to 50 years; and P-SGES variations for 40-60 years [5].

Factors that will affect the lifespan and reliability of an SGES system, however, can be speculated or projected. The most prominent must be the material composition of the solid masses. Considering factors such as erosion that may damage the solid blocks, the composition or material used must be taken into consideration to maximize their lifespan. Energy Vault has proposed composing these blocks out of locally sourced waste, but its feasibility and effectiveness are still yet to be tested. This factor will also affect the duration of energy storage. If the solid blocks erode or lose weight due to natural factors when they are in storage mode, they will generate or return less energy back to the grid than how much they have consumed to be lifted.

2.3 Cycle Efficiency

The efficiency cycle of energy storage technology, or the ratio between the consumable output when giving off energy and the input when converting electricity into a storable form, is another crucial factor in its evaluation. Current SGES systems are reported to have a cycle efficiency of 80%-90%, a figure that is considerably higher than the 40%-70% of compressed air energy storage and 70%-85% of hydro-pumped energy storage [5, 8, 9]. Specifically, T-SEGS operates with an efficiency of 90%; S-SGES at about 80%-90%; and P-SGES at a relatively lower 75%-80% [5]. However, most of the figures only exist on a theoretical aspect. The practical figure is still yet to be tested from application cases.

2.4 Energy Density

No available sources have directly stated the energy density of SGES in J/kg. Therefore, this report will use the T-SGES design from Energy Vault, the most typical SGES variation and the one with the most correlated data presented, as an example for calculations.

As the only design that has been put to commercial use, the company's T-SGES implemented in Rudong, China is reported to have a capacity of 100MWh. Even though no existing sources have directly stated the number of 35,000kg-concrete blocks (a standard size used by the company) used for this project, it can be speculated from an earlier prototype, which has a capacity of 35MWh and is composed of 5,000 blocks, that this project will need to include 14,000 blocks to reach its designated capacity [5]. The following calculation shows the energy density in J/kg:

$$\frac{100MWh}{14{,}000} \approx 7142.86Wh\ per\ block$$

$$\frac{7142.86Wh}{35{,}000kg} \approx 0.20Wh/kg = 720J/kg$$

Compared to other grid energy storage technologies, the energy density of T-SGES is relatively smaller. This limitation means that SGES systems depend significantly on the availability of land to reach high storage capacity. As a result, the spatial demand may become a financial challenger in areas where lands are scarce.

3 State of Development

3.1 Mechanism and Variations of SGES

Both tower SGES and shaft SGES are now in development. As pioneers of the field, Swiss-American company Energy Vault developed a modular tower SGES design in 2020. This means the storage system can be scaled either up or down depending on the actual demand of the region. Energy Vault also highlights the design's positive relationship with recycling waste. According to the company, these solid blocks can be composed of locally sourced waste materials, for instance, those from coal combustion, construction, and mining. The system is also compatible with various energy sources, namely solar and wind power, for both industrial use and community maintenance [5, 6].

However, since a tower SGES system is often over 50m high, Energy Vault stresses its possible instability under strong wind conditions. This leads to the follow-up shaft SGES design by Gravitricity from the U.K. Instead of planting the system above sea level, Gravitricity chooses to bury it underground. The company believes that apart from solving SGES instability in harsh climates, this design would also put abandoned mines in the U.K. back to use. According to the company's research, over 3,000 mines now in the U.K. are compatible with their SGES design, of which all of them could fit a storage capacity of 1.07 GWh and 350 of 6.7 GWh [10]. The company also states there are nearly 14,000 potential suitable sites worldwide [5, 7].

3.2 Cases of Application

The only commercial application was initiated in 2022 by Energy Vault in Rudong, China. With a capacity of 100 MWh and an output of 25 MW, the project was planned to be completed by the end of 2023 and the latest update in June 2024 shows that the construction is fully completed and has finished charge-discharge tests earlier in May. The facility is now waiting for the final approvals from the local authority to start operating [11].

Gravitricity, on the other hand, has completed its S-SGES prototype with a £640,000 sponsorship from the UK government. The prototype is in Leith Habour, Edinburgh, and is projected to cost £ 1 million [12]. The system, however, is much smaller in scale compared to what Gravitricity projected. It can only lift 25-50 tons, a figure hundredth of the 12,000 tons on the anticipated model. Further experimentations on a larger scale are still needed for S-SGES.

In addition to the prototype, Gravidity reached an agreement with the American Association of Bioanalysts (ABB) in December of 2023. Specialized in mine hoist for over

130 years, ABB would collaborate with gravidity by adapting their innovation on hoisting technologies into the S-SGES system. gravidity was also planning to engage with mine operators from Europe, India, and Australia, aiming to accelerate the commercialization of S-SGES in these regions [13]. A similar contract was also signed with a local company that promotes the regeneration of abandoned mines in Finland in early 2024. gravidity has received full access to the 1,400-meter-deep zinc and copper Pyhäjärvi mine, one the deepest mines in Europe, and is anticipating the project to be the first commercialized S-SGES in Europe, as shown in fig.4 [14].

Fig. 4. Pyhäjärvi mine in Finnland [14]

4 Conclusion

Solid mass energy storage (SGES) provides a promising yet challenging solution to grid energy storage under the context of global warming and increased awareness on aquatic habitat protection. With a straightforward mechanism of storing excessed electricity in the form of gravitational potential energy, SGES offers a scalable and efficient approach in grid energy storage. However, potential challenges lie in its low energy density and high land usage. Despite its high geographical and topographical adaptability, the reliance on land makes it financially costly to be implemented in expensive regions. In addition, as a newly emerged technology, SGES lacks samples of implementation. Even though now-existing prototypes or projects have proved its feasibility in storage capacity, empirical data such as lifespan and storage duration are still yet to be determined in the long run. As these technologies continue to mature and gain traction within the energy sector, they hold the potential to significantly contribute to a more resilient and sustainable energy infrastructure. By addressing current limitations and capitalizing on their strengths, SGES can play an important role in the transition towards a cleaner and more reliable energy future.

References

1. Ebhota, W. S., & Jen, T. C., Fossil fuels environmental challenges and the role of solar photovoltaic technology advances in fast tracking hybrid renewable energy system. International Journal of Precision Engineering and Manufacturing-Green Technology. **7**, 97-117 (2020).
2. IEA, CO_2 Emissions in 2023. Retrieved on 8th November, 2024, retrieved from https://www.iea.org/reports/co2-emissions-in-2023.
3. Our world in data. Ritchie, H., Roser, M., & Rosado, P. Renewable Energy. Retrieved on 8th November, 2024, retrieved from https://ourworldindata.org/renewable-energy.

4. Sixth Tone. Diao, F. China's Thousands of Small Dams Struggle to Stay Afloat. Retrieved on 8th November, 2024, retrieved from https://www.sixthtone.com/news/1007533.
5. Tong, W., Lu, Z., Chen, W., et al. Solid gravity energy storage: A Review. J. Energy Storage. **53,** 105226 (2022).
6. 6. Energy Vault, Purpose-Built Energy Storage Products. https://www.energyvault.com/products.
7. S. K. Moore, The Ups and Downs of Gravity Energy Storage: Startups are pioneering a radical new alternative to batteries for grid storage. IEEE Spectrum. **58,** 38-39 (2021).
8. Rabi, A. M., Radulovic, J., & Buick, J. M. Comprehensive review of compressed air energy storage (CAES) technologies. Thermo. **3**, 104-126 (2023).
9. GE Vernova. The most economical storage technology for long discharge duration: pumped hydro storage. Retrieved on 8th November, 2024, retrieved from https://www.gevernova.com/hydropower/hydro-pumped-storage
10. Sahoo, S., & Timmann, P. Energy storage technologies for modern power systems: A detailed analysis of functionalities, potentials, and impacts. IEEE Access. **11,** 49689-49729 (2023).
11. Nantong Government. World's first 26MW gravity energy storage project nears completion in Rudong. Retrieved on 8th November, 2024, retrieved from http://en.nantong.gov.cn/2024-06/21/c_998345.htm
12. Gravitricity. Gravitricity sets sights on leith for £1M energy storage demonstrator. Retrieved on 8th November, 2024, retrieved from https://www.engineernewsnetwork.com/blog/gravitricity-sets-sights-on-leith-for-1-million-energy-storage-demonstrator/
13. ABB. ABB and Gravitricity to collaborate on energy storage systems using end-of-life mine shafts and hoist technology. Retrieved on 8th November, 2024, retrieved from https://new.abb.com/news/detail/110473/abb-and-gravitricity-to-collaborate-on-energy-storage-systems-using-end-of-life-mine-shafts-and-hoist-technology
14. Euronews. Green, E. A Scottish company is using the Pyhäjärvi mine to build its first full-scale prototype gravity energy store. Retrieved on 8th November, 2024, retrieved from https://www.euronews.com/green/2024/02/06/this-disused-mine-in-finland-is-being-turned-into-a-gravity-battery-to-store-renewable-ene

Technical Features and Development Trends of Liquid Air Energy Storage

Yuting, Wang *

Shanghai High School International Division, Shanghai, 200231, China

Abstract. Liquid Air Energy storage (LAES), as an innovative approach to energy storage, utilizes the cryogenic properties of air to provide long duration of energy storage. Through the process of cooling air and storing it in liquid form, LAES systems can release energy when in need, expanding the air and driving turbines to generate electricity. This paper assesses LAES technology's potential, especially in grid balancing and as a back energy storage, noting its compatibility with renewable sources with a discontinuous energy supply, like wind and solar. While LAES systems provide high energy density, long operational lifetimes, and minimal environmental impact, they face challenges that include low round-trip efficiency, scalability issues, and high capital costs. Current research focuses on improving efficiency through thermal storage integration, reducing material costs, and developing hybrid systems to enhance LAES performance. Future advancements in thermal management and modular designs are expected to address existing challenges, placing LAES as a feasible solution for grid stability and renewable energy integration.

1 Introduction

Liquid air energy storage (LAES) is a type of energy storage that uses the thermodynamic properties of air for energy storage and output. In LAES systems, air is cooled down to cryogenic temperatures and then transformed into liquids. When energy is required, liquid air kept in insulated tanks at low pressures is exposed to higher ambient temperatures, which causes it to quickly expand into a gas that powers a turbine to produce electricity [1]. LAES technology provides a possibility for long-term energy storage since it can be integrated with other clean energy sources like solar and wind.

LAES technology has multiple areas of possible application, especially in large scale energy systems. One of its main areas of implantation is grid balancing in where LAES systems balance fluctuations in the supply and demands of electricity [2]. LAES can help maintain grid stability by storing excess energy during times of comparatively low consumption and releasing it during times of increased demand. This function is particularly useful in integration with other types of renewable sources, for example wind and solar energy systems, where the intermittent nature of these sources can disrupt constant energy

* Corresponding author: yuting.wang@ldy.edu.rs

https://doi.org/10.1051/e3sconf/202560602006

supply. LAES can capture and store surplus renewable energy and enable steady, long duration energy availability during periods of low energy generation.

LAES can also be used as a source of energy backup. During power outages or when other sources of energy are unavailable, LAES can be used to provide energy supply. Additionally, industrial applications of LAES, particularly in areas that require constant energy flow such as manufacturing, can benefit from LAES as it reduces energy costs and prevents operational disruptions.

LAES technology faces many obstacles despite its great potential. Among the main problems is its low energy efficiency. Air must be liquefied and then expanded to produce electricity, which takes a high amount of energy and has a low efficiency when compared to other energy storage systems [2]. There are also challenges in the scalability in LAES as these systems are typically designed for large-scale applications, making them less practical for smaller installations due to the high infrastructure costs and space requirements. Furthermore, the capital costs of building LAES systems can be large, with the need for large cryogenic storage tanks, compressors, and turbines [3]. As a new, emerging technology, LAES is still in its early stages of commercial deployment, making it essential for further development to enhance efficiency, reduce costs, and increase its competitiveness with other energy storage solutions.

The aim of this paper is to analyze and assess the feasibility and practicality of LAES technology, evaluating its advantage and disadvantage, current status, and areas of improvement.

2 Working Principles and Performance

2.1 Working principle

Based off the principle of thermodynamics, the three main phases of LAES system operation are liquefaction, the storage of energy, and energy release. During the liquefying stage, air is cooled to cryogenic temperatures-temperatures of about-196 ºC, at which it turns into a liquid-using energy, usually excess or extra energy produced from other renewable sources like wind and solar energy systems. The system first draws in air, which subsequently passes through a number of compressors. The air's temperature and pressure rise throughout this process. The compressed air is then cooled down while impurities such as carbon dioxide and water vapor are removed to prevent the equipment from freezing, which can result in damage of these equipment. After the purification process, the the air is further cooled until it reaches the desired temperatures. After that, the liquefied air is kept in low pressure, well insulated cryogenic tanks. The liquid air from the tanks is pumped into greater pressures and exposed to higher temperatures when power is needed, which causes it to quickly expand and transform back into a gas. During this process, the air can expand over 700 times its volume in the liquid state, producing a large amount of gas with high pressure. Turbines are driven by the expanding gas, which converts its mechanical energy into electrical energy [4].

2.2 Output and performance

2.2.1 Round-trip efficiency

The ratio of energy recovered to energy used and stored is known as the round-trip efficiency, or RTE. For LAES systems, the round-trip efficiency is 45% to 60%, which is relatively low compared other types of energy storage. Lithium-ion batteries, for example, can reach round-trip efficiencies over 90%. Multiple energy losses during the energy storage and release processes are the cause of LAES systems' low round-trip efficiency. First, compressing and cooling air to cryogenic temperatures requires a significant quantity of energy. Inefficiencies result from the waste of the heat power used during the compression and liquefaction process. Furthermore, inefficient use of heat sources during the gasification process can result in thermal energy losses of the liquid air [5].

2.2.2 Energy density

The quantity of stored energy per unit volume is measured by energy density. Densities of 50-200 Wh/L are found in LAES systems, which are less than lithium-ion batteries but more than compressed air energy storage (CAES), pumped hydro energy storage (PHES) systems, and pumped thermal energy storage (PTES) [1].

2.2.3 Response time

The response time measures how fast a system from switch from its inactive state to actively producing or releasing energy. LAES systems present a moderate response time of a few minutes [1]. In contrast to batteries, which can release stored electrical energy virtually instantly, the LAES system requires it to pump, heat, and expand liquid air in order to produce electricity. This process involves multiple mechanical processes and causes delay in the system.

2.2.4 Operational lifetime

The operational lifetime measure how many cycles or years that a system can continue operating efficiently before a significant degradation occurs. LAES systems are known for their long operational lifetimes, ranging from 30-40 years, with minimal performance degradation [6]. The primary contributor to this extended lifetime is the use of well-established mechanical components such as compressors and turbines which are durable and long lasting. Unlike batteries which degrade over time because of chemical reactions that occur in the cells, LAES systems do not face electric and chemical degradation. Furthermore, the components that might suffer from mechanical wear down can be easily maintained or replaced, further extending its operational lifetime.

2.3 Advantages

LAES technology serves as a potential technology for large-scale energy storage due to its many benefits. Its energy density is one of its primary benefits. The energy density offered by LAES systems, which range from 50 to 200 Wh/L, is greater than that of many other renewable energy storage options [1]. LAES systems can store a lot of energy in comparatively compact spaces attributed to their high energy density.

The minimal environmental impact that LAES technology offers is yet another significant benefit. Since there is no combustion involved in the energy storage process when using air as the medium, there are no greenhouse emissions and therefore little environmental impact.

In addition, LAES system is not limited by geographical constraints [7]. LAES systems can be deployed almost everywhere as the energy in cryogenic tanks, making it a site-free storage. This is a significant advantage over other technologies like PHES that require specific terrains and large water reservoirs. This flexibility makes it suitable for urban areas, remote locations, and places where terrain might limit the installation of traditional energy storage systems.

Moreover, LAES systems can provide storage over a long period of time. This feature is especially helpful for balancing the intermittent nature of renewable energy sources, such as solar and wind. LAES systems also possess a long operational lifetime of 30-40 years, making them a durable and cost-effective option for long periods of energy storage [6].

3 Current and future and development

LAES technology has had significant progress over the years but still remains in its early stage of commercial use. LAES systems have progressed over the conceptual and experimental phases and has pilot projects and small-scale demonstration projects. Researches are currently working on improving the efficiency, scalability, and cost reductions of LAES systems.

3.1 Current Focus Areas

3.1.1 Improving round trip efficiency

The heat energy lost during compression and liquefaction is one of the primary causes of energy loss in LAES systems. Wasted heat during the gasification process can be gathered and recycled by incorporating thermal energy storage, such as advanced thermal energy storage (TES) [8]. However, designing efficient heat exchangers and thermal storage that can retain heat for extended periods of time is a difficulty. The energy needed to reheat the liquid air can be decreased by employing latent heat storage, which uses phase-change materials to store a significant amount of heat during the liquefaction process and release it after regasification. Other techniques include sensible heat storage, which involves heating material such as rocks or molten salts and then using them to help liquid air expand.

Another significant source of energy losses is the liquefying process, which uses a large amount of energy. The efficiency of the expanders and compressors employed in the liquefaction and power generation processes is the one of the main focuses of current efforts [9]. Compression technology advancements like multi-stage compressors and cooling techniques are meant to lower the energy required for liquefaction. To improve the efficiency of the liquefaction process, research is also being done on ways to use waste heat from other sources, such as combined heat and power systems [8].

3.1.2 Reducing capital costs

The high capital costs related with LAES systems, particularly the building of the cryogenic tanks, is one of the largest barriers to commercial deployment. Researchers are finding ways to reduce the cost of material used in LAES systems, especially those of cryogenic tanks and compressors. This includes developing more affordable and also durable alloys for cryogenic storage or using new materials that can stand temperature changes without compromising system integrity. Improving the manufacturing process for the key components such as turbines, compressors, and heat exchangers can assist in reducing the capital costs.

3.1.3 Hybrid LAES systems

Hybrid LAES systems combine the use of LAES systems with other energy storage to improve the performance [10]. For example, LAES systems can be paired with other battery type storage to provide rapid and short energy discharge while supporting long duration storage. In addition, the round-trip efficiency of LAES systems can be enhanced through connection with thermal energy systems.

3.2 Future trends

In the future, with improvements in the thermal management and waste heat utilization, enhancing the efficiency and sustainability of LAES system, making them more competent in addressing situations of energy needs. LAES technology can provide long-term storage while utilizing batteries' quick response times and balancing the intermittent nature of solar and wind energy sources in hybrid LAES systems, which combine LAES systems with other storage like batteries and other renewable storage like wind and solar.

LAES systems are expected to present higher scalability in the future, enabling a larger range of application, from micro grid integration to large scale grid balancing. In addition, artificial intelligence (AI) may also be used in integration with LAES systems, where AI systems can help in meeting changing energy demands and maximize the efficiency of LAES systems. LAES systems have a large space of improvement and can continue to develop into a more resilient and mature technology.

4 Conclusion

In conclusion, LAES present a solution to long duration storage and grid balancing. Key advantages of LAES systems include its long storage duration, high energy density, environmental sustainability, and flexible applications. The goal of ongoing research is to increase round-trip efficiency, reduce capital costs, and develop hybrid LAES systems that combine the advantages of several energy storage types to enhance energy storage performance overall. Furthermore, modular LAES designs can make LAES systems even more affordable and accessible, and improvements in the materials utilized and efficient manufacturing processes should lower initial capital costs. Ultimately, LAES systems provide a solution to grid balancing and can potentially become a major energy storage technology in the future.

References

1. Vecchi, A., Li, Y., Ding, Y., et al. Liquid air energy storage (LAES): A review on technology state-of-the-art, integration pathways and future perspectives. Advances in Applied Energy, **3**, 100047 (2021)

2. Legrand, M., Rodríguez-Antón, L. M., Martinez-Arevalo, C., et al. Integration of liquid air energy storage into the spanish power grid. Energy, **187,** 115965 (20119)

3. Borri, E., Tafone, A., Romagnoli, A.,et al. A review on liquid air energy storage: History, state of the art and recent developments. Renewable and Sustainable Energy Reviews, **137,** 110572 (2021)

4. Damak, C., Leducq, D., Hoang, H. M., et al. Liquid Air Energy Storage (LAES) as a large-scale storage technology for renewable energy integration-A review of investigation studies and near perspectives of LAES. International Journal of Refrigeration, **110,** 208-218 (2020)

5. O'Callaghan, O., & Donnellan, P. Liquid air energy storage systems: A review. Renewable and Sustainable Energy Reviews, **146,** 111113 (2021)

6. Borri, E., Tafone, A., Zsembinszki, G., et al. Recent trends on liquid air energy storage: a bibliometric analysis. Applied Sciences, **10**, 2773 (2020)

7. Luo, X., Wang, J., Dooner, M., et al. Overview of current development in electrical energy storage technologies and the application potential in power system operation. Applied energy, **137,** 511-536 (2015).

8. Nabat, M. H., Zeynalian, M., Razmi, A. R., et al. Energy, exergy, and economic analyses of an innovative energy storage system; liquid air energy storage (LAES) combined with high-temperature thermal energy storage (HTES). Energy Conversion and Management, **226,** 113486 (2020).

9. Morgan, R., Nelmes, S., Gibson, E., et al. Liquid air energy storage - analysis and first results from a pilot scale demonstration plant. Applied energy, **137**, 845-853 (2015)

10. Bernagozzi, M., Panesar, A. S., & Morgan, R. Molten salt selection methodology for medium temperature liquid air energy storage application. Applied Energy, **248,** 500-511 (2019)

E3S Web of Conferences **606**, 02007 (2025) https://doi.org/10.1051/e3sconf/202560602007
ICNAOE 2024

The feasibility of conducting compressed air energy storage system

Yucheng Feng*

Yunnan University Secondary School, Kunming, 650031, China

Abstract. Energy storage has been a crucial component in human energy utilization, with the compressed air energy storage system (CAES) emerging as a significant and indispensable aspect of the conversation around energy storage solutions. Historically, the understanding of CAES was confined to theoretical models, but recent years have witnessed an increase in practical applications and constructions, bringing the technology into the public eye. These developments have yielded valuable and precise data that researchers can leverage. Utilizing these data points, this paper employs both qualitative and quantitative analysis to thoroughly evaluate the viability of implementing CAES systems. The aim of this article is to provide some insights that can significantly contribute to the advancement of CAES technology, offering a comprehensive assessment that informs future development strategies. Furthermore, the difficulty of the technology is analyzed and some solutions are provided. Finally, the further development trend of compressed air energy storage system is conducted.

1 Introduction

Energy has always been a topic closely related to the development of human beings for thousands of years, and energy has penetrated into all aspects of human life, greatly improving the convenience of human life. Energy is constantly transforming each other, and with the in-depth exploration of science, human beings have a deeper understanding of energy. From 1984 to 2004, the consumption of the energy around the world has annually increase, the primary energy has grown by 49% and the emission of the carbon dioxide has climbed by 43% [1]. The obvious climb of using energy means that investigators are supposed to focus on fixing the predicaments of generating energy and storing energy, to matching the needs of industrial electricity and household energy consumption. With the advancement of the technologies and scientific researches, people is able to invent more advanced energy storage systems, hence, researchers could find the most suitable way to balance complicated situations, such as the efficiency of operating systems, the cost of conducting systems and the pollution of maintaining systems. Over past decades, the researchers have already introduced dozens of technology choices, back in 1883, the first flywheel energy storage system has been introduced, it means people can convert the energy into various formations and easily storage them [2]. After that, the idea of using water

* Corresponding author: 278570226ljojo@ldy.edu.rs

elevation to storage energy came into view, the first pragmatic construction of pumped hydro energy storage was built in Switzerland in 1907 [2], this energy storage medium necessitates the damming of water bodies, which entails additional initial capital during project development. It is absolutely an effective way but this system has the limitation of the geographical position and constrain of the scale. In 1978, investigator conducted first compressed air energy storage system which can put into real lives in Germany through compressing air into a cavern of roughly 300,000 m^3 [3]. Until now, plenty of compressed air energy storage systems are operating to storage and regenerate the energy which utilizes in diverse facets of our lives. The basic mechanism of the CAEs is not complicated, compressed air energy storage system uses air as the carrier of the energy, large-scale compressed air energy storage used excess electricity to compress and store air in an underground structure (such as underground caves), and then mixes compressed air with natural gases when needed, burning expansion to promote gases turbine power generation. CAEs is already a relatively mature technology, but there are still many potential problems and improvements in construction and maintenance. Both the environmental impact and cost considerations still need to be comprehensively considered and reasonably solved.

More comprehensive contents are needed to generally analyze the pros and cons of the CAEs, potential application, utilization and the status of this storage technology. This study aims to collect and assemble the information that investigators got, as well as presenting the conclusion and estimating the feasibility of constructing CAEs successfully.

2 The effect of the environment

No matter what kinds of energy storage system, the consideration of the effect to the environment is a undoubtedly crucial step, not only being useful to make sure the suitability, but also guaranteeing the basic protection to the environment. Of course, the protection of the environment can be subdivided into multiple aspects, in this essay, the aim is to describe the environmental impact during the construction of the system and the environmental impact during the maintenance of the normal operation of the energy storage system. At the beginning of this technology, low efficiency has led to greater limitations of the technology, and has not received widespread social recognition and attention, one of which is because of the dependence on additional heat sources such as fossil fuels. With the invention of the Advanced Adiabatic Compressed Air Energy Storage (AA-CAES), building crews can build compression heat to indirectly reduce the use of fossil fuels that have increased the efficiency of system operation [4], this change not only effectively depict the consumption of the natural fuel, but also decrease the emission of the carbon dioxide, nitrogen dioxide, sulfur dioxide and other harmful substances, this allows the system to be built without the need for additional large-scale space to store fossil fuels and build massive conversion tools. Isothermal compressed air storage system, a new relative technology, through slow movement of the piston to keep the stability of the temperature [5]. Under this application, the system can be effectively dropping the possibility of the technological risks which caused by the fast physical friction. Thus, the system can be utilized for a long period of time.

3 feasibility of conducting this energy storage system

3.1 Cost

When calculating the cost of conducting compressed air energy storage system, there are several main facets which associated with budget. The most important one is about the capital cost, this investment includes equipment cost and installation cost. Taking a 100MW CAE system with the lifetime of 30 years as an example, the cost of equipment cost plus installation cost is 198.38 million USD without interest rate [6]. Except capital cost, another significant investment is annually operational cost, which includes normal maintenance and the consumption of the fuel. With the fixed operation cost of 5.123 million USD per year and the assumption of interest rate of 5%, so the general investment of activated system is roughly 200 million USD and maintain cost is 15.042 million USD per year. Of course, these data are based on the normal operation of the system, and do not consider the occurrence of unexpected circumstances, such as security accidents or disruption of operating funds, etc. Additionally, the estimation of the funds refers to the cost of construction and operation, and does not include the cost of land.

3.2 Policy

This section only classifies specific policy of certain countries because of the different limitation of the laws. In China, the trend of energy storage development is focusing on accelerating the construction of a clean, low-carbon, safe and efficient new energy storage system [7], providing a strong energy guarantee for economic and social development, and constantly meeting the people's growing needs for a better life. In the European Green New Deal proposed in recent years, it is mentioned that by 2050, Europe will become the world's first near-zero emission state, mainly through the provision and storage of clean, cheap and safe energy, the development of green construction and industry, and the development of sustainable transportation. Energy storage links all aspects like a chain, becoming an indispensable to all aspect of the strategy. No matter what the specific policies and restrictions are, human beings have gradually realized the important role of protecting the environment, and making changes to protect the environment has become a trend around the world. On the basis of this energy storage system is clean and environmentally friendly and relies less on fossil fuels, it has better development prospects and potential than many traditional energy storage systems.

3.3 The difficulty of construction

3.3.1 The limitation of geography

In this portion, the researcher presents a relative assumption about the volume of the cavern, the researcher hypothesizes that the shape of the cave used to store the compressed gas was an ideal oval shape, with an abc axis length ratio of 2:2:1(a being the major axis of the ellipse, b the minor axis of the ellipse, and c the thickness of the ellipse), therefore, if we need a cave with a volume of almost 300.000 cubic meters, we need to find a cave where a is about 52.4 m, b is about 52.4 m, c is about 26.2 m. Therefore, the cave has a cross-sectional area of about 8625.805 square meters and a vertical area of about 2156.451 square meters. Comparing with other conventional energy storage metrics, it does not require a very large land area, but to make sure there is enough pressure to store the gas safely, so the caves which are made by the salt, hard rock and highly permeable materials become very strong in development [8].

3.3.2 The difficulty of the technology

For the initial CAEs, the efficiency of the operator was relatively low, only 29%, with a pressure requirement of 48-66 bar, and later, with improvements to ensure higher and stable pressure, the efficiency could be increased to 36% [9]. In 1991, a large CAEs system built for commercial profit in Alabama, USA, has increased the efficiency of the advanced system to 54% by building additional heat recovery units that reduce the amount of fuel consumed during system operation [10]. In 2008, AA-CAES, a new CAES technology that reduces heat release through adiabatic systems to improve efficiency, began to be noticed and put into use, this technology successfully made the efficiency of CAES systems can be as high as 70% [11]. Until today, CAES technology has been studied and developed for nearly 27 years, and significant progress has been made in the research and improvement of compressor or caves. For this technology, the technology has been relatively mature, so there is no strong technical limitations.

4 The stage of the energy system

Based on the technology readiness level [12], the corresponded level of the compressed air energy storage system is TRL 8-9, this value means that the technology has already passes the phase of TRL 7, the system prototype demonstration under scenario, begins to construct in a relatively stable environment. Additionally, this value illustrates that the technology is under the process of real system development through handling bugs to make sure the better performance of the machine. From 1978 to the present, the traditional CAES has long lagged the trend of The Times, and it will be replaced by a more advanced and more integrated new CAES, whether it is cooperating with other energy systems, or the integration of part of the system structure, it has achieved a very significant effect in improving the performance and role of energy storage.

5 The prediction of the future development

Indeed, numerous intricate issues remain unresolved, heavily reliant on technological progress, including the expansion of system capacity and the assurance of converter safety. Until today, there are already German power companies in Germany to carry out the world's first large-scale AA-CAE project, called ADELE[10]. When many technical problems have been properly solved, I believe that large-scale and systematic work will also be carried out in the near future.

6 Conclusion

The article presents the comprehensive content of the compressed air energy storage system, collaborating the previous information with the brand technological advancement. By quantifying and defining obscure concepts, further clarification of potential issues regarding the feasibility of system construction. It is undeniable that based on the limited amount of literature read by the author, there will be some important content or new technology in the article more or less not mentioned or clearly described. In addition, the technology is still in the stage of rapid development, so this article only describes and evaluates the content before 2024. The description of the year beyond that is a forecast of the CAES technology with relatively strong uncertainty. The author expects and firmly believes that in the near future, there will be better and better technology applied to CAES, making the technology to a higher level and better serving human beings and society.

References

1. Pérez-Lombard, L., Ortiz, J., & Pout, C. A review on buildings energy consumption information. Energy and buildings, **40,** 394-398 (2008)..
2. Mitali, J., Dhinakaran, S., & Mohamad, A. A. Energy storage systems: A review. Energy Storage and Saving, **1,** 166-216 (2022).
3. Chen, H., Cong, T. N., Yang, W., et al. Progress in electrical energy storage system: A critical review. Progress in natural science, **19,** 291-312 (2009).
4. Wang, J., Lu, K., Ma, L., et al. Overview of compressed air energy storage and technology development. Energies, **10,** 991 (2017).
5. Budt, M., Wolf, D., Span, R., et al. A review on compressed air energy storage: Basic principles, past milestones and recent developments. Applied energy, **170**, 250-268 (2016).
6. 6.Lund, H., & Salgi, G. The role of compressed air energy storage (CAES) in future sustainable energy systems. Energy conversion and management, **50**, 1172-1179 (2009).
7. Zeng, M., Li, C., Zhou, L. Progress and prospective on the police system of renewable energy in China. Renewable and Sustainable Energy Reviews, **20,** 36-44 (2013)..
8. Li, L., Liang, W., Lian, H., et al. Compressed air energy storage: characteristics, basic principles, and geological considerations. Advances in Geo-Energy Research, **2**, 135-147 (2018).
9. King, M., Jain, A., Bhakar, R., et al. Overview of current compressed air energy storage projects and analysis of the potential underground storage capacity in India and the UK. Renewable and Sustainable Energy Reviews, **139,** 110705 (2021).
10. Luo, X., Wang, J., Dooner, M., et al. Overview of current development in compressed air energy storage technology. Energy Procedia, **62,** 603-611 (2014).
11. Power, R. W. E. ADELE–adiabatic compressed air energy storage for electricity supply. Retrieved on 8th November, 2024, retrieved from https://wecanfigurethisout.org/ENERGY/Web_notes/Round_Pegs/Power_Cycles_and_Energy_Storage_Supporting_Files/ADELE%20%E2%80%93%20Adiabatic%20Compressed-Air%20Energy%20Storage%20for%20Electricity%20Supply.pdf
12. NASA. Mankins, J. C. Technology readiness levels. White Paper. Retrieved on 8th November, 2024, retrieved from file:///C:/Users/Administrator/Downloads/TechnologyReadinessLevels_AWhitePaper_1995-Edited_2004.pdf

Research Progress of Solid Electrolytes in Solid-State Lithium Batteries

Nan Xia[1*]

[1] School of Materials Science and Engineering, Hubei University, Wuhan, 430062 China

Abstract. In today's world where fossil fuels are increasingly depleting, electric energy, this new type of energy, is rapidly developing, the safety of batteries is receiving more and more attention from people as well. The excellent mechanical properties of solid-state batteries significantly enhance their safety features. However, the mainstream solid-state inorganic electrolytes and gel polymer electrolytes still face issues such as low ionic conductivity, poor electrode interface performance, and inadequate environmental stability. This paper is based on years of research by predecessors on the characteristics of Solid inorganic electrolytes and gel polymer electrolytes, summarizing the recent research progress on solid-state composite electrolytes by researchers, discussing the properties of single-phase electrolytes and the improvement of composite-phase electrolytes. It discusses the improvement of solid-state battery performance by solid-state composite electrolytes from three aspects: polymer composite non-polar ceramic materials, the addition of SN catalysts, and three-dimensional skeletal structures. It analyses the problems and challenges of solid-state electrolytes, Pointing out the existing shortcomings of the current solid composite electrolytes. Finally, providing directions for the continued development of solid-state electrolytes.

1 Introduction

As technology and productivity continue to improve, the environmental pollution and energy crisis caused by fossil fuels are becoming increasingly severe, there is an essential to improve the ability of energy retention and conversion technologies. Lithium batteries are widely promoted for their efficient and green energy storage. However, in recent years, incidents of fires and explosions caused by liquid electrolyte leakage, overcharging, and high temperatures in lithium batteries have frequently occurred. In contrast, solid-state electrolyte lithium batteries have garnered widespread attention due to their superior thermal steadiness and safety. Currently, the electrolytes of solid-state batteries mainly include electrolytes made of gel-state polymers, inorganic materials, and composite materials [1]. Whereas because of the solid-solid interaction between the two electrodes and the solid electrolyte, solid-state lithium batteries' conductivity is still beneath that of conventional liquid lithium batteries [2]. Moreover, the issue of Lithium dendritic short circuits in batteries made from solid-state materials still exists, and these factors continue to impact the solid-state lithium

* Corresponding author: 202231113021001@stu.hubu.edu.cn

https://doi.org/10.1051/e3sconf/202560602008

batteries' performance in general [3]. Addressing these issues, researchers have proposed a method of combining inorganic electrolytes and gel-state polymer electrolytes to compensate for each other's shortcomings, thereby enhancing the electrolyte conductivity and interfacial performance while ensuring the safety of the battery.

This paper will respectively explore the characteristics of solid electrolytes that come in two varieties. Indicating the advantages and disadvantages of the two solid electrolytes. Summarizing the recent research progress in improving the safety, conductivity, and electrode interface performance in lithium solid-state batteries. It could supply a reference for the future research directions of composite solid-state electrolytes.

2 Solid inorganic electrolyte

Solid inorganic electrolytes (SIE) have garnered widespread attention in solid-state battery technology because of their excellent physical properties. Its excellent mechanical strength enables it to effectively deal with the battery limited circuitry issues resulting from lithium dendrites generated from the lithium battery anode, this characteristic is of great significance in enhancing battery safety [4]. Currently, there are primarily two types of solid-state inorganic electrolytes.: the electrolytes based on sulfides and oxides [5]. Sulfide-type electrolytes are gaining attention due to their high conductivity, which can reach 10^-2 S/cm [5]. However, its environmental stability and electrochemical stability are relatively poor, which directly affects its safety in high-energy-density batteries [6]. This unstable property may lead to performance degradation and potential safety hazards under prolonged use or high-load conditions. In contrast, although oxide electrolytes do not have as high conductivity as sulfide electrolytes, they possess superior thermal stability and safety [7], making oxide electrolytes more stable under high-temperature environments and extreme operating conditions. Oxide-type electrolytes have good conductivity at room temperature, it can be seen in Table 1. Moreover, the process of getting ready for oxide-type electrolytes is secure and quite mature., making them more reliable and operable in practical applications. Therefore, despite the challenge of high interfacial resistance, oxide-based electrolytes remain an important direction in current research and applications, widely favored by researchers worldwide.

Table 1. Comparison of Three Types of Inorganic Oxide Electrolytes [7-9]

Types of electrolytes	configuration	Maximum electric conductivity electrolyte chemical formula	Operating temperature	Electric conductivity
Perovskite	Tetragonal perovskite type	LLTO ($Li_{0.22}La_{0.60}TiO_3$)	25°C	4.84×10^{-4} S/cm
NASICON	Rhombus structure	$Li_{1.5}Al_{0.4}Cr_{0.1}Ge_{1.5}(PO_4)_3$ (LGP)	25°C	6.65×10^{-3} S/cm
Garnet	Three-dimensional nanostructure	$Li_{6.25}Fe_{0.25}La_3Zr_2O_{12}$	25°C	1.38×10^{-3} S/cm

3 Gel polymer electrolytes

Solid polymer electrolytes (SPE) have garnered widespread attention attributed to their excellent stability, simplicity in preparation and affordability [10]. However, the ionic conductivity of single polymer electrolytes is relatively low, usually around 5×10^-4 S/cm[11], and it is far less than that of inorganic solid-state electrolytes. Common substrate polymers for PSE contain polymethyl methacrylate (PMMA), polyethylene oxide (PEO), polyvinylidene fluoride (PVDF), polyacrylonitrile (PAN), poly (vinylidene fluoride-co-hexafluoropropylene) [PVDF-HFP]. and polyvinyl chloride (PVC) [12]. Although these polymers have good stability, their conductivity does not reach the desired level when used alone.

Gel polymer electrolytes (GPE) occupy a unique position in the classification of electrolytes. It is a special form of solid polymer electrolyte (SPE), with a state of matter that lies between liquid and solid, exhibiting characteristics of a gel-like semi-solid material. This type of electrolyte is usually formed by the combination of polymers and liquid electrolytes, resulting in a gel polymer electrolyte (GPE). GPE combines the liquid electrolytes' strong ionic conductivity with solid-state electrolytes' toughness, demonstrating excellent overall performance.

The structure of GPE endows it with excellent hardness and strength, effectively addressing the common issue of lithium dendrites in liquid electrolytes. Lithium dendrites often lead to battery short circuits and performance degradation in liquid electrolytes, whereas GPE significantly reduces this risk through its solid-state structure [3]. Furthermore, the gel-like physical properties of GPE enable it to provide excellent interfacial performance in practical battery applications. Compared to inorganic solid electrolytes, GPE can better adapt to the slight unevenness of the electrode surface when in close proximity to the cathode and anode, enhancing the battery's functionality and interfacial stability. [13].

Further research also supports the advantages of GPE in improving battery performance. For example, Liu's study on solid-state electrolytes of the garnet type points out that Solid-state electrolytes containing GPE interlayers exhibit significantly reduced interfacial resistance compared to non-gel-state solid-state electrolytes without GPE, with a reduction magnitude reaching an exponential level[14]. This improvement not only enhances the cycle life of the battery but also increases its safety and reliability.

Overall, gel-state electrolytes, because of their unique physical state and structural characteristics, offer a promising solution to address key challenges in battery technology, such as lithium dendrite issues and interface performance optimization. Therefore, further research and development of GPE will help promote the advancement of more efficient and safer battery technologies.

4 Solid composite electrolytes

The research on Solid Composite Electrolytes (SCEs) aims to explore composite solid electrolytes that possess high safety, high energy density, high conductivity, and excellent interfacial performance. Although the aforementioned solid ion electrolytes (SIEs) can achieve high conductivity, their solid-solid contact with the electrodes is not tight enough, which leads to increased interfacial resistance in the battery, causing electrons to accumulate in the circuit and directly affecting how well lithium batteries function. Nevertheless, the advantages of SIEs are also very evident, with conductivity far exceeding that of solid polymer electrolytes (SPEs) and stronger mechanical properties, it able to effectively prevent battery short-circuit issues caused by lithium dendrites, as shown in Figure. 1. Relatively speaking, SPEs have lower conductivity and insufficient mechanical strength, but their excellent interfacial properties and low interfacial resistance make their contact with the electrodes better. To solve the issues of interfacial performance and conductivity in solid-state electrolytes, researchers have proposed a strategy of combining GPE with SIE. This approach aims to overcome SIE's elevated interfacial resistance and the low conductivity of GPE, thereby promoting diversified research on solid-state composite electrolyte materials (SCEs) as well as improving solid-state batteries' overall efficiency. Tian and others' research revealed that by introducing polymer additives such as polyisobutylene (PIN) as the electrolyte matrix, conductivity can be significantly improved. The introduction of PIN provides additional weak electrostatic coordination sites to the electrolyte system, encouraging the composite SPE's ionic conductivity to increase to 5.89×10^-3 S/cm [12]. This result not only highlights the limitations of single-base polymers in terms of ionic conductivity but also emphasizes the great potential of composite electrolytes in improving conductivity and optimizing battery performance. It is evident that composite electrolytes are gradually becoming the primary path for enhancing the performance of solid-state batteries.

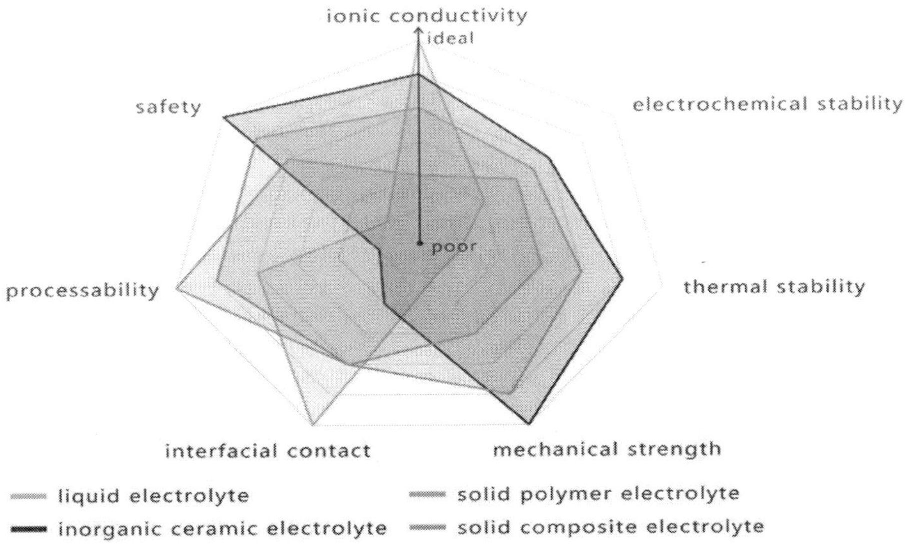

Fig. 1. Comparison of the performance of solid polymer electrolytes, inorganic ceramic electrolytes, solid composite electrolytes, and liquid electrolytes [1].

4.1 Porous inorganic particle composite polymer electrolyte

Combining small particle SIE with GPE-based polymer electrolytes is able to dramatically enhance the mechanical strength and conductivity of the electrolytes [15]. Common inorganic composite particles include LATP, LAGP, LLTO, LLZO, and LLZTaO. However, numerous studies have shown that although the initial addition of these ceramic particles does indeed enhance the conductivity of the electrolyte, the subsequent conductivity decreases. This phenomenon is believed to be related to the porous structure of the ceramic particles [15]. Isaac demonstrated that the initial increase in conductivity is mainly attributed to the dispersion of ceramic particles in the GPE, while the subsequent decrease in conductivity is due to blockage caused by poor particle contact, which is unrelated to the properties of the electrolyte itself or the concentration of the salt [15]. Tao addressed the issue of excessive thickness in SPE electrolytes by introducing nano-SiO2 particles into PVDF-based polymer electrolytes. This PVDF-based solid-state composite electrolyte (PPSE) has a thickness of only 20 μ m but a mechanical strength of up to 64 MPa. At the same time, silica nanoparticles effectively enhance the lithium-ion transference number and ionic conductivity, reaching a conductivity of 4.81 × 10^-4 S/cm, and it exhibits good system stability and sustainability [16] (Figure 2).

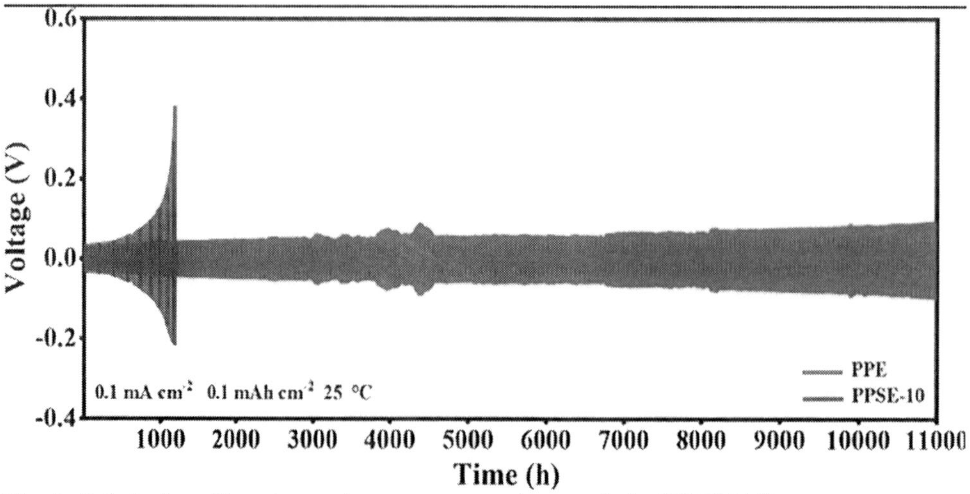

Fig. 2. Relatively stable output voltage over a very long period of PPSE [16]

4.2 Including catalysts to increase SPE's mechanical strength and conductivity

To address the issue of low conductivity in SPE, researchers have proposed doping succinonitrile (SN) as an organic catalyst in gel-state PMMA-based GPE. Experimental results show that doping with SN can greatly increase the electrolyte's conductivity, resulting in a new type of electrolyte (SN-GPE) having 2.02 × 10^-3 S/cm of conductivity, and it also possesses good mechanical properties, compensating for the deficiencies of GPE in terms of conductivity and mechanical performance [17]. However, this new method also exposes its shortcomings: the interfacial performance between SN and the lithium anode is poor due to a large number of side reactions occurring at the interface [18]. To address this, Zhao used metal-organic frameworks (MOFs) to construct the electrolyte and penetrated the MOF's multilayer channels with the LiTFSI-SN (LSN) electrolyte. This method effectively improved the battery's capacity and interfacial performance without negatively impacting ionic conductivity.

4.3 Changing the three-dimensional structure of the spatial volume to improve SCE preparation safety

Lithium batteries in solid state have not yet been put into commercial production, mainly due to issues with their interface and conductivity. Commercial preparation requires a safe and stable manufacturing process as a prerequisite for mass production. Researchers improve the structure of SCEs to enhance their stability. For example, Ruan prepared a nanofiber scaffold with a 3D rebar structure in a PEO-based composite garnet-type inorganic oxide electrolyte, which achieved a conductivity at 30° C, of 0.23 × $10^{\wedge-3}$ S/cm. This structure significantly improved compared to traditional SCEs structure electrolytes and also enhanced the stability of the battery [19]. Metal-organic frameworks (MOFs) were proposed as early as 2016, but their application regarding solid-state lithium battery technology was officially introduced by Zhou in 2023. They used bimetallic MOFs (NH3+ • SO3-@ZIFs) as fillers for PEO-based solid-state electrolytes. This composite solid-state electrolyte not only enhances conductivity but also prevents lithium dendrites from growing at the anode, which is vital for the stability and cycle efficiency of batteries [20]. On this basis, researchers further proposed a type of SPE with dual-layer asymmetry. This novel material has a mechanical strength of 16.8 MPa and exhibits good ionic conductivity (0.63 × 10^-3 S/cm), establishing a strong basis for the use of MOF in solid-state batteries [21].

5 Conclusion

Overall, in context of increasing demands for new energy storage now, The lithium battery of the future will be solid-state. This paper categorizes the two fundamental forms of solid-state electrolytes of solid-state batteries available so far, elucidating the respective advantages and disadvantages of SIE and SPE. And summarize and categorize the novel SCEs from the past three years, discussing SCEs from three aspects: the composite of inorganic ceramic particles, SN catalysts, and three-dimensional framework structures. However, these three types of SCEs are still not perfect, with the main common issue still being low conductivity, which is the biggest problem affecting the energy cycle efficiency of solid-state batteries. Exceptionally, there are also some individual issues, such as the poor interface performance of the SCEs in the composite SN catalyst and the general mechanical properties of the three-dimensional skeletal structure. This paper focuses on reviewing the two basic electrolytes of solid-state batteries, giving a reference for scholars who have not yet delved into this field but are interested in it. Using the classification models of SCEs from the past three years as a framework, it offers readers the latest developments in this topic, hoping that future readers will make progress.

At present, it seems that the conductivity and interface issues of solid-state electrolytes still require further research, and to guarantee safety, the electrolytes' mechanical strength should also be increased. The method of combining inorganic nanoparticles and MOFs can be attempted by trying to add SN catalysts to the non-polar ceramic composite polymer electrolyte to enhance conductivity, or improve the environmental stability of the sulfide electrolyte during preparation, among other things. All of them have their strengths in terms of performance enhancement.

References

1. K. Daems, et al., Advances in inorganic, polymer and composite electrolytes: Mechanisms of Lithium-ion transport and pathways to enhanced performance. RSER. **191,** 114136 (2024) .
2. H.Yang, N. Wu., Ionic conductivity and ion transport mechanisms of solid - state lithium - ion battery electrolytes: A review. ESE, **10**,51643-1671 (2022).
3. Y. Chen, et al., Understanding the lithium dendrites growth in garnet-based solid-state lithium metal batteries. JPS, **521,** 230921 (2022).
4. Q. Lv, et al., Suppressing lithium dendrites within inorganic solid-state electrolytes. CRPS. **3**,1 (2022).
5. Q. Liu, et al. Recent advances in stability issues of inorganic solid electrolytes and composite solid electrolytes for all - solid - state batteries. TCR. **22**,10 (2022)
6. L. Lilu, et al. Advances in electrochemical stability of sulfide solid electrolytes. JCC **47**,10 (2019): 1367-1385.
7. K.J. Kim, et al.. Solid - state Li - metal batteries: challenges and horizons of oxide and sulfide solid electrolytes and their interfaces. AEM. **11**.1 (2021): 2002689.
8. M. Illbeigi, A. Fazlali, M. Kazazi, A.H. Mohammadi. SSI. , **289**, 180.2016
9. R. Wagner, GJ Redhammer, D Rettenwander, et al. Fast Li-ion-conducting garnet-related Li7-3xFexLa3Zr2O12 with uncommon I4-3d structure. CM. **28**, 16, 5943-5951 (2016).
10. G. Yuan, MK Wang, L Long. Research advances on gel electrolytes for lithium-ion batteries. ANRE. **8**, 4 331-338 (2020).

11. H. Yang, et al. PDOL-based solid electrolyte toward Practical application: Opportunities and challenges. NML. **16**,1, 12,7 (2024).
12. X. Tian, Y Yi, P Yang, et al. High-charge density polymerized ionic networks boosting high ionic conductivity as quasi-solid electrolytes for high-voltage batteries. AMI. **11**, 4, 4001-4010 (2019).
13. L.X u, et al. Garnet solid electrolyte for advanced all - solid - state Li batteries. AEM. **11**, 2 (2021)
14. B. Liu, et al. Garnet solid electrolyte protected Li-metal batteries. AMI .**9**,22, 18809-18815 (2017)
15. J.A. Isaac, D Devaux, R Bouchet. Dense inorganic electrolyte particles as a lever to promote composite electrolyte conductivity. NM. **21**, 12, 1412-1418(2022) .
16. Y. Ma, C Wang, K Yang. Ultrathin and Robust Composite Electrolyte for Stable Solid-State Lithium Metal Batteries. AMI. (2023).
17. Z. Zhou, et al. PMMA-Based Composite Gel Polymer Electrolyte with Plastic Crystal Adopted for High-Performance Solid ECDs. POL. **15** (2023).
18. T. Zhao, et al. Laminar composite solid electrolyte with succinonitrile-penetrating metal-organic framework (MOF) for stable anode interface in solid-state lithium metal battery. JPS. (2023).
19. Y. Ruan, et al. A novel reinforced concrete-like composite solid-state electrolyte with enhanced performance for all-solid-state lithium batteries. JSEE. **28**, 8(2024)
20. X. Zhou, et al. Difunctional MOF for dendrite-free all-solid-state lithium metal batteries by the synergistic effect of hydrogen bond and electrostatic interaction. NE. **108** (2023)
21. Q. Cheng, et al. 3D interconnected MOF-derived asymmetric bilayer solid-state electrolyte for enabling homogeneous Li deposition of all-solid-state lithium metal batteries. JSEE. **28**, 8, 2631-2642 (2024)

A Review on Element Doping in Li-Rich Cathode Materials for Lithium-Ion Batteries

Zhenyu Luo[1,*]

College of Chemistry and Environmental Engineering, Shenzhen University, Shenzhen, 518055, China

Abstract. With the growing demand for clean energy, lithium-ion batteries (LIBs) have become essential in applications like electric vehicles due to their high energy density and efficiency. However, conventional cathode materials such as $LiCoO_2$ and $LiFePO_4$ are limited by relatively low specific capacities. To address this, Li-rich Mn-based cathode materials (LRMs) have been proposed as promising alternatives, offering greater specific capacity. Despite their potential, LRMs suffer from challenges like voltage decay, poor rate performance, and safety problems during cycling. This review discusses the use of element doping, such as the introduction of Al, Cr, and Ti, to improve the structural stability and electrochemical properties of LRMs. While significant progress has been made in enhancing capacity and cycle life, issues like voltage fade and long-term stability persist. Future research should focus on optimizing doping strategies, enhancing structural integrity, and reducing costs to facilitate the widespread use of LRMs in energy storage and electric vehicle systems.

1 Introduction

As the demand for clean energy rises, particularly for electric vehicles and sustainable technologies, lithium-ion batteries (LIBs) have become increasingly crucial. LIBs are widely used in portable devices like smartphones and laptops [1], as well as in large-scale applications due to their high energy density and lightweight nature. Nonetheless, the rapid advancements in the energy sector place higher demands on LIBs, including increased capacity, faster discharge rates, and improved safety. Compared to cathodes, anodes are less expensive, with graphite being a particularly cost-effective and commercially viable material. Consequently, the high cost of cathode materials hinders the progress in enhancing specific capacity and reducing overall LIB costs. Traditional cathode materials, such as $LiCoO_2$, $LiMn_2O_4$, and $LiFePO_4$, although commercially successful, have limitations due to their relatively low specific capacity [2-4]

To overcome these limitations, LRMs have emerged as promising alternatives, offering higher specific capacities [5] and relatively lower costs. In these materials, surplus lithium ions within the Li_2MnO_3 phase occupy one-third of the transition metal layer sites, resulting in the development of a superstructure. Despite their potential, LRMs face challenges such

* Corresponding author: 2022142010@email.szu.edu.cn

as poor rate performance, voltage decay, high initial irreversible capacity, low Coulombic efficiency, and structural transformations from layered to spinel structures [6].

Among various enhancement strategies—including surface coating, ion doping, as well as structural optimization—doping has demonstrated significant promise [7]. This technique involves substituting elements like Al [8], Cr [9], or Ti [10] into the cathode material's crystal lattice, thereby improving structural stability, electronic conductivity, and electrochemical performance.

An overview of recent advancements in doping LIBs will be provided in this review, focusing on improvements in specific capacity, discharge capacity, and cycling life achieved through the introduction of various metals into LRMs.

2 Lithium-rich materials

2.1 Composition and structure

Lithium-rich materials (LRMs) consist of two components: Li_2MnO_3 and $LiMO_2$. Li_2MnO_3, which is part of the monoclinic system with the C/2m space group, shares similarities with $LiMO_2$, characterized by the α-$NaFeO_2$ structure in the hexagonal system (R-3m space group). Both exhibit a similar lithium layer, albeit with some distortion. The key difference lies in the transition metal layer: in Li_2MnO_3, one-third of the transition metal is substituted by lithium ions, whereas $LiMO_2$ contains only transition metals, as illustrated in Fig. 1. The structure of LRMs remains debated; Dr. Haijun Yu and etc propose a two-phase composite of monoclinic and hexagonal crystals [11]. By using atomic-resolution scanning transmission electron microscopy (STEM), their experimental results directly unveil the two-phase coexistence of LiTMO2 and Li2MnO3-like structures. While Karalee A. Jarvis and etc advocate for a single-phase solid solution structure [12]. With the help of XRD and STEM, they conclude that Li [Li0.2Ni0.2Mn0.6]2 composition is integral, but other compositions especially with excess lithium may lead to two-phase coexistence.

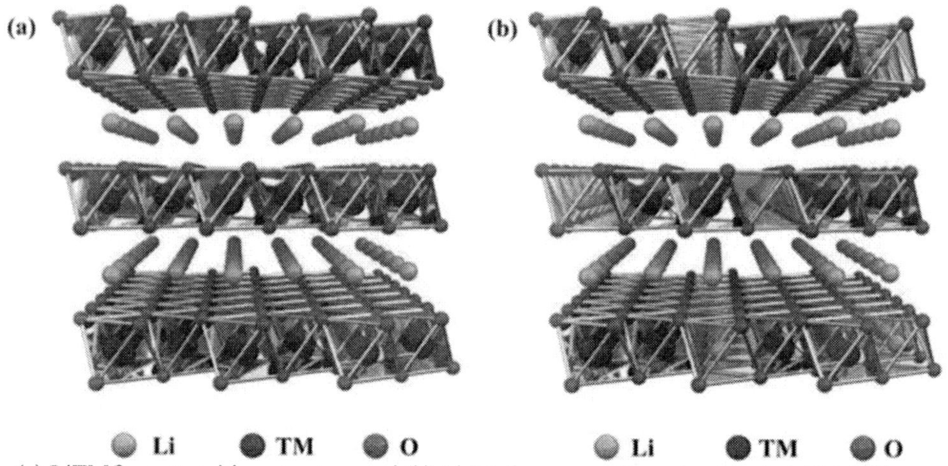

Fig. 1. (a) $LiTMO_2$ composition structure and (b) Li_2MnO_3 composition structure in LRM [13].

2.2 Unique situation

Fig. 2 depicts the lithium-rich oxide material PLLO with a P2-type layered structure, featuring two distinct platforms in its charge-discharge curve. The initial platform, occurring before 4.5 V, resultss from the oxidation of transition metals as lithium ions are deintercalated, with Co^{3+} and Ni^{2+} oxidizing to Co^{4+} and Ni^{4+}, respectively. The platform at 4.5 V appears only during the first charge, leading to debates regarding its origin. The reversible oxygen redox hypothesis suggests that at this voltage, oxygen ions may partially oxidize (from O^{2-} to O^- or O_2^-), contributing additional capacity. Furthermore, the migration of transition metals may cause voltage decay. The low initial efficiency of lithium-rich manganese is attributed to an irreversible phase change that produces Li_2O.

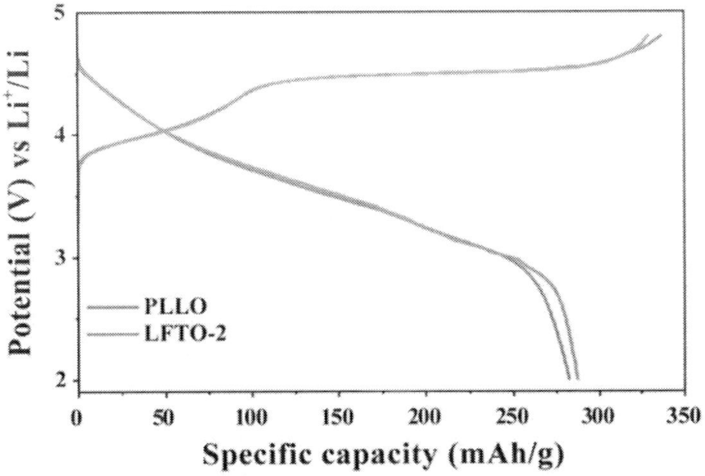

Fig. 2. The initial charge-discharge profiles for PLLO and LFTO-2 at 0.2C (where 1C equals 250 mA/g) [14].

3 Challenge

3.1 Capacity and voltage decay

A primary challenge with lithium-rich manganese (LRM) materials is the degradation of capacity and voltage over time. Voltage decay directly results in capacity decay, making battery life much lower than our requirements. As shown in Figure 3, after 100 cycles, the capacity of LLNMC decreased by 41 mAh/g, retaining only 81.7% of its initial capacity, while the average voltage also dropped significantly. Current research attributes this voltage decay to O_2 release, migration of transition metal ions, and the development of spinel-like phases. So far, this problem still can't be solved perfectly. Despite attempting various new materials, after long-time cycles, either the voltage and capacity decay remains apparent or the cost is high.

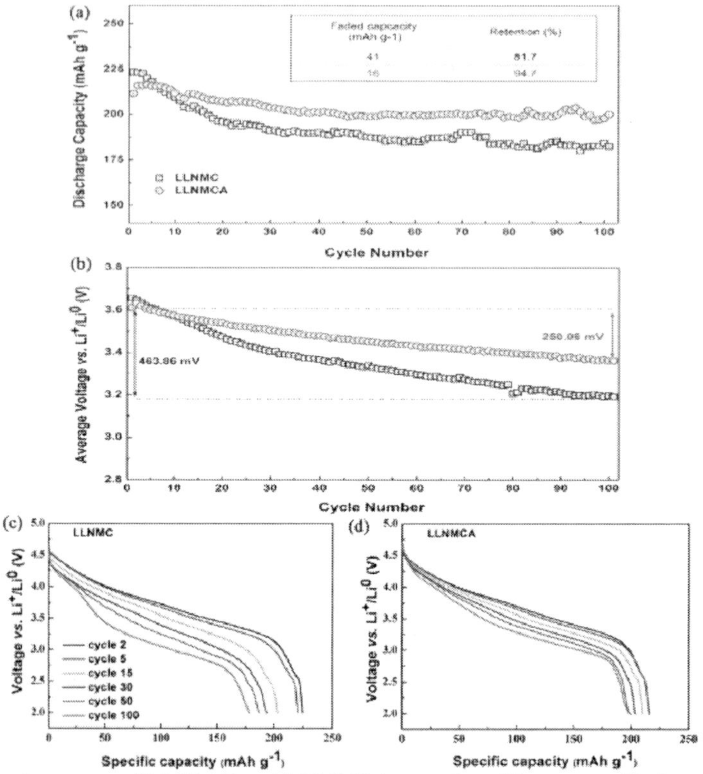

Fig.3. (a) Cycle performance of LLNMC and LLNMCA samples; (b) average discharge voltage as a function of cycle number; (c, d) specific discharge curves for LLNMC and LLNMCA from the second to the one-hundredth cycle[15].

3.2 Low magnification

The rate capability of a battery, defined as the ratio of discharge current to its rated capacity (usually represented as "C"), is critical for performance. Figure 4 illustrates the rate performance of LRMs. Notably, as the discharge rate increases, the specific capacity of the material decreases. The poor rate performance is conditional on various factors. One of them is the intrinsic structure of Li2MnO3 which limits the conductivity of Li+. The irreversible phase change that results in the production of LiMn2O4 further impedes the diffusion of L^{i+} over extended cycle operations. Thus, the secret to enhancing the subpar rate performance is maintaining the integrity of the inner structure.

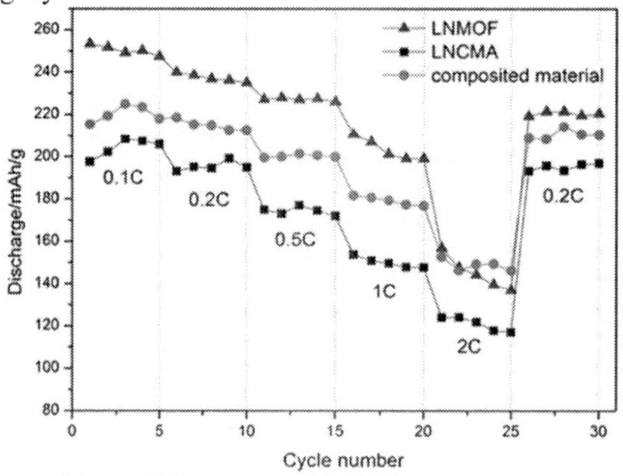

Fig.4. Rate performances of three different materials [16].

3.3 Safety

Lithium-ion batteries pose safety risks, including fire and explosion due to exothermic reactions. The main safety concerns are caused by the use of flammable carbonate and carboxylate solvents in electrolytes, which can break down under high temperature and pressure, leading to thermal runaway. Additionally, lithium dendrites can form on the anode during long-term cycling. Some dendrites may break off as "dead lithium," while others can pierce the separator, causing short-circuits [17].

4 Methods of Improvement

4.1 Surface coating

Active materials in lithium-rich systems are prone to corrosion by the electrolyte during electrochemical reactions, leading to structural degradation. To mitigate this, surface coatings create a protective layer that minimizes electrolyte contact, effectively reducing corrosion under high-voltage conditions. The performance of lithium manganese-based materials can vary based on the coating material, including carbon coating, oxide surface coating (common choices include Al_2O_3[18] and ZnO[19]) and phosphate surface coating. According to studies, these coatings greatly improve the lithium-rich materials' electrochemical characteristics.

4.2 Doping

In LRMs, doping is a commonly used technique to improve conductivity, decrease voltage decay, and improve structural stability. The integrity of the layered structure is directly related to the stability of the voltage and capacity. Structural stability, catalytic activity, and interactions with the electrolyte are all lowered by adding electrochemically inert ions to the lattice, especially at high charge states. This method enhances long-term cycling performance and successfully prevents voltage decay. Depending on the charge characteristics, doping can be classified as cation doping, anion doping, or anion-cation co-doping. Table 1 shows how several common dopants affect particular electrochemical properties.

Table 1. Improvement through the use of doping strategies [15,20,21].

Materials	Elements	Initial discharge capacity (mAh g^{-1}/C)	Initial coulombic efficiency	Capacity retention percentage (%/cycles/C)
$Li_{1.14}(Ni_{0.136}Co_{0.136}Mn_{0.544})O_2$	Al	225/0.1	72%	82%/100/0.1
		212/0.1(Al)	73%	95%/100/0.1
	Mg	286/0.1	84%	61%/100/0.1
$Li_{1.2}Ni_{0.16}Mn_{0.56}Co_{0.08}O_2$		273/0.1(Mg)	78%	82%/100/0.1
		160/0.1(Mg)	64%	150%/100/0.1

$Li_{1.20}Mn_{0.54}Ni_{0.13}Co_{0.13}O_2$		261/0.1	75%	52%/300/1
	Na	276/0.1	77%	81%/300/1
	F	282/0.1	78%	76%/300/1
	Na+F	291/0.1	82%	85%/300/1

Even if the modification technique helps lithium-rich batteries operate better electrochemically, there are currently many difficulties in the widespread application of lithium-rich batteries. Therefore, further innovation and breakthroughs are still needed. For example, when using the doping modification method, researchers found that only doping with a certain element can only improve the electrochemical performance of one aspect, and another element can improve the electrochemical performance of another aspect, so a co-doping method was adopted. From a microscopic point of view, for positive electrode materials, how to increase the diffusion rate of lithium ions and maintain the stability of the layered structure is crucial. Therefore, a deeper understanding of the reasons for the weakening of electrochemical performance and using the existing technologies are needed. After understanding the mechanism of the modification method, there will be more direction and research efficiency for further.

5 Conclusion

LRMs are regarded highly promising for next-generation lithium-ion batteries due to their high energy density and unique electrochemical properties. However, they encounter major challenges, including low initial Coulombic efficiency, poor rate performance, and decay of capacity and voltage. Additionally, large-scale industrial use is limited by issues like oxygen release and structural transformations during charge and discharge. This review systematically examines the structure, working mechanisms, and electrochemical behavior of Li-rich Mn-based cathodes, offering modification strategies like doping and surface treatments to address these challenges.

Looking ahead, future research should focus on optimizing these strategies and developing new techniques, such as co-doping and multi-layer coatings, to improve performance. This work will guide future studies and support industrial applications, particularly in electric vehicles and energy storage systems. Collaboration among chemists, physicists, and materials scientists will be crucial for addressing interface and surface issues, ensuring that these materials meet the growing demands for high-efficiency energy storage.

References

1. P. Peng, Y. Chen, Q. Zhou, L. Shen, Y. Wen, F. Du, Y. Chen, J. Zheng, J. Tructure Engineering with Sodium Doping for Cobalt-Free Li-Rich Layered Oxide toward Improving Electrochemical Stability. Colloid Interface Sci. **676**, 847-858 (2024)
2. Y. Zhang, H. Xie,H. Jin, Q. Zhang, Y. Li, X.Li, K. Li, C.Research Status of Spinel Limn2o4 Cathode Materials for Lithium Ion Batteries. Bao, IOP Conf. Ser.: Earth Environ. Sci. **603**, 12-51 (2020)

3. Y. Ding, D. Mu, B. Wu, R. Wang, Z. Zhao, F. Wu, Recent Progresses on Nickel-Rich Layered Oxide Positive Electrode Materials Used in Lithium-Ion Batteries for Electric Vehicles. Appl. Energy **195**, 586-599(2017)
4. Y. Wang, P. He, H. Zhou, Olivine Lifepo 4: Development and Future. Energy Environ. Sci. **4** (3), 805-817 (2011)
5. M. M. Thackeray, S. H. Kang, C. S. Johnson, J. T. Vaughey, R. Benedek, S. A. Hackney, Li2mno3-Stabilized Limo2 (M = Mn, Ni, Co) Electrodes for Lithium-Ion Batteries. J. Mater. Chem. **17** (30), 3112 (2007)
6. Y. Li, Z. Li, C. Chen, K. Yang, B. Cao, S. Xu, N. Yang, W. Zhao, H. Chen, M. Zhang, Recent Progress in Li and Mn Rich Layered Oxide Cathodes for Li-Ion Batteries. Energy Chem. **61**, 368-385 (2021)
7. P. Oh, J. Yun, S. Park, G. Nam, M. Liu, J. Cho, Recent Advances and Prospects of Atomic Substitution on Layered Positive Materials for Lithium-Ion Battery Adv. Energy Mater. **11** (15), 2003197 (2021)
8. W. Yan, C.; Y. Xie, J. C. Jiang, D. Y. Sun, X. D. Ma, Z. G. Lan, Y. C.Enhanced Rate Performance of Al-Doped Li-Rich Layered Cathode Material Via Nucleation and Post-Solvothermal Method. Jin, ACS Sustain. Chem. Eng. **6** (4), 4625 (2018)
9. B. Song, C. Zhou, H. Wang, H. Liu, Z. Liu, M. O. Lai, L. Lu, Advances in Sustain Stable Voltage of Cr-Doped Li-Rich Layered Cathodes for Lithium Ion Batteries. Electrochem. Soc. 2014, **161** (10), A1723-A1730 (2014)
10. Y. Wang, S. Liu, X. Guo, B. Wang, Q. Zhang, Y. Li, Y. Wang, G. Wang, L. Gu, H. Yu, Elements Gradient Doping in Mn-Based Li-Rich Layered Oxides for Long-Life Lithium-Ion Batteries. Mater. Sci. Technol. **207**, 266-273 (2025)
11. H. Yu, R. Ishikawa, Y.-G. So, N. Shibata, T. Kudo, H. Zhou, Y. Ikuhara, Cathode Material for Lithium-Ion Batteries. Angew. Chem. Int. Ed. **52** (23), 5969-5973 (2013)
12. K. A. Jarvis, Z. Deng, L. F. Allard, A. Manthiram, P. J. Ferreira, Atomic Structure of a Lithium-Rich Layered Oxide Material for Lithium-Ion Batteries: Evidence of a Solid Solution. Chem. Mater. **23** (16), 3614-3621 (2011)
13. H. Zheng, X. Han, W. Guo, L. Lin, Q. Xie, P. Liu, W. He, L. Wang, D. -L. Peng, Mater. Recent Developments and Challenges of Li-Rich Mn-Based Cathode Materials for High-Energy Lithium-Ion Batteries. Today Energy. **18**, 100518 (2020)
14. S. Wang, J. Suo, Y. Liu, W. Guo, G. Gao, X. Han, M. Fan, R. Wu, D.-L. Peng, Q. Xie, Enhancing the Electrochemical Properties of Li-Rich Layered Oxide Cathodes by a Facile Fe/Ti Integrated Modification Strategy. Chem. Eng. J. **497**, 154387 (2024)
15. H. Guo, Y. Xia, H. Zhao, C. Yin, K. Jia, F. Zhao, Z. Liu, Stabilization Effects of Al Doping for Enhanced Cycling Performances of Li-Rich Layered Oxides. Ceram. Int. **43** (16), 13845-13852 (2017)
16. H. Zhu, Y, Xiao, J. Dong, D. Zhang, C. Chang, New Composite Li1.4mn0.61ni0.18co0.18al0.03o2.4 and Lini0.5mn1.5o3.9f0.1 Cathode Material with Higher Specific Capacity and Better Capacity Retention. Ionics **26** (8), 3749-3760 (2020)
17. H. K. Hu, W. D. Xue, Y. Li, P. Jiang, Safety Protection Measures for Lithium Ion Batteries: An Overview and Outlook. Acta Polym. Sin. **53** (5), 457-473 (2022)
18. Y. Zhou, P. Bai, H. Tang, J. Zhu, Z. Tang,Chemical Deposition Synthesis of Desirable High-Rate Capability Al2o3-Coated Li1.2mn0.54ni0.13co0.13o2 as a Lithium Ion Battery Cathode Material. Electroanal. Chem. **782**, 256-263 (2016)

19. G. Singh, R. Thomas, A. Kumar, R. S. Katiyar, A. Manivannan, Electrochemical and Structural Investigations on Zno Treated 0.5 Li 2mno 3-0.5limn 0.5ni 0.5o 2 Layered Composite Cathode Material for Lithium Ion Battery. Electrochem. Soc. **159** (4), A470-A478 (2012)
20. P. K. Nayak, J. Grinblat, E. Levi, M. Levi, B. Markovsky, D. Aurbach, Understanding the Influence of Mg Doping for the Stabilization of Capacity and Higher Discharge Voltage of Li-and Mn-Rich Cathodes for Li-Ion Batteries. Phys. Chem. Chem. Phys. **19** (8), 6142-6152 (2017)
21. P. Zhang, X. Zhai, H. Huang, J. Zhou, X. Li, Y. He, Z. Guo, Synergistic Na+ and F− Co-Doping Modification Strategy to Improve the Electrochemical Performance of Li-Rich Li1·20mn0·54ni0·13co0·13o2 Cathode. Ceram. Int. **46** (15), 24723-24736 (2020)

E3S Web of Conferences **606**, 02010 (2025) https://doi.org/10.1051/e3sconf/202560602010
ICNAOE 2024

An Overview of Lithium-Ion Battery Safety: Existing Problems and Potential Solutions

Wenduo Fu[1], *Shiqi* Sun[2*], and *Bo* Wang[3]

[1]School of Materials Science and Engineering, Hebei University of Technology, 300401 Tianjin, China
[2] School of Resources and Environmental Engineering, East China University of Science and Technology, 200237 Shanghai, China
[3] School of Chemical Engineering and Technology, Dalian University of Technology, 300401 Dalian, China

Abstract. Lithium-ion batteries (LiBs) are extensively utilised in the energy sector owing to their superior performance. In this paper, thermal runaway (TR) is introduced as one of the safety dangers, analysed the mechanism and characteristics of it. This paper focuses on the optimization of Lib, which can reduce the TR caused by mechanical, electrical and TR. The study suggests improvements in three crucial areas: electrode materials, electrolytes, and separators, aiming to reduce the dangers of TR in LiBs. Through doping, coating, nanoparticle design, and metal atom replacement, electrode materials can be improved to increase cathode safety, LiF coatings on NCM811 cathodes are one prominent example. By adding chemicals like flame retardants and creating solid-state electrolytes, electrolyte optimization greatly increases battery safety. Furthermore, improving separator stability by applying ceramic coatings, high-temperature-resistant polymers, and surface modifications fortifies mechanical and thermal endurance, lowering the risk of TR and short circuits. When combined, these methods make the LiB system safer and more dependable.

1 Introduction

The worldwide transition to sustainable energy has expedited the uptake of electric vehicles (EVs), with lithium-ion batteries (LiBs) becoming the favored energy storage choice owing to their high energy density and comparatively minimal environmental impact.
[1]. However, as battery technology advances—particularly with the shift from NMC111 to NCM811 ternary electrode materials—new safety concerns have arisen [2].

Increasing nickel content while reducing cobalt levels in these batteries can lead to significant safety risks, including susceptibility to short circuits, thermal runaway, and potential fires or explosions [2]. Resolving these difficulties is essential for guaranteeing the safe functioning of LiBs in EVs.

This report offers an in-depth review of safety concerns associated with LiBs. It highlights the primary danger that might threaten the function and security of batteries. In order to

* Corresponding author: 21014003@mail.ecust.edu.cn

address these safety concerns, the study also looks into a number of potential solutions. The research seeks to increase the general safety of lithium-ion devices by suggesting enhancements. By lowering the chance of breaks down, these improvements aim to increase battery reliability for general utilization.

2 Lithium-ion battery working principle

A lithium-ion battery (LiB) is a rechargeable energy storage device that functions by moving lithium ions between the cathode and anode during charging and discharging processes. [1]. It consists of four main components: the cathode, anode, electrolyte, and separator. The cathode is typically made of oxides like $LiFePO4$ and NMC, used in electric vehicles, while the anode is primarily composed of carbon or silicon materials such as graphite and SiOX [3]. The electrolyte includes organic solvents or inorganic lithium salts, and the separator is a microporous insulating material designed to prevent short circuits between the electrodes [1].

3 Thermal runaway mechanism

Thermal runaway (TR) in LiBs occurs when an unexpected rise in internal temperature triggers a series of chain reactions that can pose serious safety risks. TR is often caused by mechanical, electrical, or thermal abuse [4]. This process can lead to significant energy release, sometimes resulting in smoke, fire, or explosions. Ouyang employed Accelerating Rate Calorimetry (ARC) to investigate temperature variations during TR, identifying three critical temperatures: T1, T2, and T3 [5]. T1 signifies the initiation of self-heating, generally linked to the degradation of the solid electrolyte interface (SEI) layer; T2 indicates the commencement of TR; and T3 represents the maximum temperature attained during TR, reflecting the peak energy release. [5]. Consequently, improving battery safety necessitates the elevation of T1 and T2 while diminishing T3 [6].

3.1 Mechanical abuse

Mechanical abuse, including damage from impacts, is a significant contributor to TR in LiBs, particularly in electric vehicles. This abuse can distort the battery, impairing the separator and resulting in a short circuit between the electrodes.[3]. This leads to a rapid temperature rise to T1, breaking down the SEI film and triggering TR [6].

3.2 Electrical abuse

Electrical abuse is a common trigger for TR in LiBs. This abuse includes overcharging, over-discharging, and external short circuits [4]. External short circuits may occur due to events such as vehicle collisions, water submersion, or contamination with conductive substances. This type of short circuit leads to a rapid discharge, causing a significant temperature increase, gas production, and eventual battery expansion.

Overcharging occurs when the battery voltage exceeds its maximum limit, often due to a malfunction in the battery management system (BMS). This condition generates substantial heat, promotes lithium dendrite growth at the anode, and can damage the cathode. Conversely, over-discharging occurs when the battery voltage decreases excessively, resulting in significant heat production, probable copper dissolution in the current collector, interior short circuits, and the initiation of thermal runaway.

3.3 Thermal abuse

Thermal abuse denotes a condition in which a LiB is exposed to excessive temperatures in a high-heat environment. Such conditions can provoke chemical reactions within the battery, resulting in an immediate rise in internal temperature and pressure. This immediately induces TR, and if a single cell undergoes TR, it can initiate a chain of events that threatens the security of the whole battery group.

4 Thermal runaway characteristics

TR in LiBs generally progresses through four stages: Usually, the process starts with gas emission and venting, then TR, which is frequently accompanied by flames. An explosion could result from this intensifying into a TR. The TR event comes to an end when the flames are extinguished, completing the process. [6]. This section focuses on lithium iron phosphate (LFP) and ternary lithium batteries, the predominant types used in electric vehicles (EVs) today.

Table 1 compares the TR parameters for both battery types. LFP batteries are known for their superior heat stability due to the robust covalent P=O bond within the molecule and the octahedral structure of (PO4)3- [6]. As a result, LFP batteries have a longer ignition time compared to ternary lithium batteries and demonstrate enhanced safety and thermal stability.

Table 1. The comparation of the TR parameters [7-9].

	Lithium Iron Phosphate Battery	Ternary Lithium Battery		
	Electrical abuse	Thermal abuse	Electrical abuse	Thermal abuse
Ignition time/s	3040	916	810	431
Thermal runaway duration/s	186	181	670	400
Maximum temperature/°C	423	372.1	736.8	1180
Temperature rate/ (°C/s)	0.12	0.32	2.33	29.50

5 Optimization of thermal runaway

This study aims to enhance safety in lithium batteries by optimising electrode materials, electrolytes, and separators to decrease the danger of thermal runaway induced by mechanical, thermal, and electrical abuse.

5.1 Optimization of electrode materials

Contemporary suggestions for enhancing the safety of cathode materials include doping, coating, micro-particle design, and metal atom substitution [10]. Doping and coating with chemically inert materials can reduce reactions between anode and electrolyte, lowering TR risks. For example, applying thermal-responsive polymers to electrodes and separators can effectively prevent short circuits and control temperature increases [11]. Coating cathode materials, such as LiNi0.8Co0.1Mn0.1O2 (NCM811), with LiF, enhances thermal and structural stability, minimizes interface side reactions, and reduces oxygen release during charging, thereby improving battery safety [12]. Moreover, silicon/carbon nanotube composites provide enhanced lithium storage capacity and increased electrode stability. [13].

Xu et al. have proposed a technique employing a conductive polymer (Poly(3,4-ethylenedioxythiophene)) at the microscopic level, which improves electrochemical and thermal stability while mitigating internal reactions, hence decreasing temperature elevation. [14].

These techniques, including LiF coatings, thermal-responsive polymers, and silicon/carbon nanotube composites, effectively optimize LiB electrode materials, significantly mitigating the risk of TR.

5.2 Optimization of electrolytes

The performance of the electrolyte is crucial for the safety of lithium-ion batteries, mainly improved by electrolyte additives and the advancement of novel solid-state electrolytes.

5.2.1 Electrolyte additives

Electrolyte additives, including film-forming agents, overcharge protection substances, and flame retardants, improve the thermal stability of LiBs. Film-forming additives facilitate the development of an SEI on the anode's surface during initial charge-discharge cycles. This passivation layer facilitates lithium ion intercalation while preventing adverse reactions between the electrolyte and electrode materials, thereby reducing electrolyte decomposition, minimizing irreversible capacity loss, and decreasing the risk of TR. Notable examples include fluoroethylene carbonate (FEC) and vinylene carbonate (VC) [15,16].

Overcharge protection additives, which encompass redox and polymeric monomer additives, remain inert under normal conditions but engage in redox reactions or electro polymerization during overcharging to prevent irreversible damage [17]. Commonly used compounds include 1,3,5-trihydroxybenzene, 2,5-difluoro-1,4-dimethoxybenzene, cyclohexylbenzene, alkyl 3,3,3-trifluoropropionate, and 4-bromoanisole [18,19].

Flame retardant additives function through mechanisms such as chemical interference (capturing free radicals), physical barriers (isolating oxygen and heat), and heat absorption, effectively reducing the risk of TR. These additives are typically phosphate esters or fluorinated organic solvents [17]. However, their interaction with other electrolyte components can negatively impact battery performance. K. Liu et al. proposed encapsulating triphenyl phosphate flame retardants in micro-sized shells via electrospinning, allowing the retardant to be released only at elevated temperatures, thus mitigating potential adverse effects [20].

5.2.2 Solid-state electrolytes

Advancements in solid-state electrolytes for LiBs are critical, as they offer a fundamental solution to fire risks. These electrolytes are categorized into three types: polymer, oxide, and sulfide. Polymer solid-state electrolytes provide moderate lithium-ion conductivity, stability, and ease of manufacturing. Oxide solid-state electrolytes excel in conductivity but are costlier to produce. Sulfide electrolytes offer high ionic conductivity but are sensitive to oxygen, necessitating strict environmental controls. To address the limitations of individual electrolytes, inorganic fillers can be integrated into polymer matrices, combining the benefits of both inorganic and organic materials to enhance ionic conductivity and electrochemical stability [21].

In summary, optimizing LiB electrolytes involves the use of film-forming additives, overcharge protection additives, flame retardants, and the development of solid-state electrolytes. These strategies significantly improve battery safety, with solid-state electrolytes representing a promising direction for mitigating TR.

5.3 Optimization of separator materials

The separator in LiBs functions as an ion transport channel while preventing short circuits between the anode and cathode. Traditional polyolefin separators can shrink under high temperatures, raising the risk of TR. Effective separators must also possess mechanical strength to withstand impacts and punctures, as well as chemical stability to resist electrolyte corrosion. Notably, lithium dendrites can grow through separators, potentially causing short circuits. Therefore, developing separators that inhibit dendrite growth is crucial.

Ceramic-coated separators address these challenges by maintaining structural stability at elevated temperatures and reducing pore size, thus minimizing dendrite penetration [22-27]. High-temperature-resistant polymer separators, such as polyimide (PI) and polybenzimidazole (PBI), offer excellent thermal and chemical stability, preserving integrity under heat and preventing TR. Additionally, surface modifications of polyolefin separators—through plasma treatment, UV irradiation, or chemical grafting—can enhance wettability, ionic transport, and overall mechanical strength [28]. Collectively, these strategies improve separator stability and mitigate TR risks in LiBs.

6 Conclusion

This article examines the causes of TR in LiBs and outlines strategies for prevention. Key triggers include mechanical, thermal, and electrical abuse, with different battery types exhibiting varying behaviours during TRs.

To mitigate these risks, this study propose optimizations across three main areas: electrode materials, electrolytes, and separator materials. Enhancements in cathode safety can be achieved through doping, coating, micro-particle design, and metal atom substitution. Electrolyte optimization involves using additives and developing solid-state alternatives. Additionally, employing high-temperature-resistant polymer separators, ceramic-coated separators, and surface modifications can significantly bolster separator stability, reducing TR risks.

Notably, solid-state electrolytes offer a promising solution to the safety challenges of LiBs and represent a critical area for future research.

Authors Contribution

All the authors contributed equally and their names were listed in alphabetical order.

References

1. Y. Tang. A review of the application status and development of lithium-ion battery in electric vehicles. Environ. Technol. **41**, 94–100 (2023).
2. W. Zhang, W. Xiao, Y. Yi, et al. Research progress on safety modification strategies of lithium-ion battery. Energy Storage Sci. Technol., 1–18 (2024).
3. X. Pan, C. Shao, X. Wang. Analysis of thermal runaway characteristics and control methods of lithium batteries for vehicles. Times Automob. **17**, 145-147 (2024).
4. X. Feng, M. Ouyang, X. Liu, L. Lu, Y. Xia, X. He. Thermal runaway mechanism of lithium-ion battery for electric vehicles: A review. Energy Storage Mater. **10**, 246–267 (2018).
5. X. Feng, S. Zheng, D. Ren, X. He, L. Wang, X. Liu, et al. Key characteristics for thermal runaway of Li-ion batteries. Energy Procedia **158**, 4684–4689 (2019).

6. Y. Zhi, R. Lu, Q. Yu, & F. Yan. Analysis of the characteristics and prevention technologies of thermal runaway in power batteries. Automot. Eng. **46**, 139–150 (2024).
7. Z. Jia, S. Wang, P. Qin, C. Li, L. Song, Z. Cheng, et al. Comparative investigation of the thermal runaway and gas venting behaviors of large-format LiFePO4 batteries caused by overcharging and overheating. J. Energy Storage **61**, 106791 (2023).
8. X. Zhu, Z. Wang, Y. Wang, H. Wang, C. Wang, L. Tong, M. Yi. Overcharge investigation of large-format lithium-ion pouch cells with Li(Ni0.6Co0.2Mn0.2)O2 cathode for electric vehicles: Thermal runaway features and safety management method. Energy. **169**, 868–880 (2019).
9. C. Jin, Y. Sun, H. Wang, Y. Zheng, S. Wang, X. Rui, et al. Heating power and heating energy effect on the thermal runaway propagation characteristics of lithium-ion battery module: Experiments and modeling. Appl. Energy **312**, 118760 (2022).
10. X. Zhu, Z. Wang, H. Wang, W. Hsin. Research review on thermal runaway and safety management of lithium-ion power batteries. J. Mech. Eng. **56**, 91–118 (2020).
11. Y. Zheng, N. Lin, Y. Zhao, W. Wang, Y. Qian, Y. Zhu, Y. Qian. A flexible micro/nanostructured Si microsphere cross-linked by highly-elastic carbon nanotubes toward enhanced lithium-ion battery anodes. Energy Storage Mater. **17**, 93–100 (2019).
12. J. Lang, Y. Long, Q. Jiale, X. Luo, H. Wei, K. Huang, H. Zhang, L. Qi, Q. Zhang, Z. Li, W. Hui. One-pot solution coating of high-quality LiF layer to stabilize Li metal anode. Energy Storage Mater**.** **16**, 85–90 (2019).
13. J. Park, J. Kim, C. Lee, S. Lee. Mixed ion/electron-conductive protective soft nanomatter-based conformal surface modification of lithium-ion battery cathode materials. J. Power Sources **263**, 209–216 (2014).
14. G. Xu, Q. Liu, K. K. Lau, Y. Liu, X. Liu, H. Gao, X. Zhou, M. Zhuang, Y. Ren, J. Li, M. Shao, M. Ouyang, F. Pan, Z. Chen, K. Amine, G. Chen. Building ultra conformal protective layers on both secondary and primary particles of layered lithium transition metal oxide cathodes. Nat. Energy **4**, 484–494 (2019).
15. H. Zhao, Y. Xueqing, J. Li, B. Li, H. Shao, L. Li, & Y. Deng. Film-forming electrolyte additives for rechargeable lithium-ion batteries: Progress and outlook. J. Mater. Chem. A **7**, 8700–8722 (2019).
16. H. Zhao, Y. Xueqing, J. Li, B. Li, H. Shao, L. Li, & Y. Deng. Film-forming electrolyte additives for rechargeable lithium-ion batteries: Progress and outlook. J. Mater. Chem. A **7**, 8700–8722 (2019).
17. D. He, J. Wang, Y. Peng, B. Li, C. Feng, L. Shen, S. Ma. Research advances on thermal runaway mechanism of lithium-ion batteries and safety improvement. Sustain. Mater. Technol. **41**, e01017 (2024).
18. C. Zhu, J. Liu, Y. Xiaohua, Y. Zhang, X. Jiang, P. Dong, Y. Zhang. 4-Bromoanisole (4BA) as additive for overcharge protection of lithium-ion batteries. Int. Electrochem. Sci. **14**, 4571-4579 (2019).
19. S. Wang, J. Liu, Y. Xiaohua, J. Liu, C. Zhu, J. Rong, Q. Wang, Z. Yuan, Y. Zhang. Electrochemical behavior of 3,4-ethylenedioxythiophene as an anti-overcharge additive for lithium-ion batteries. Int. J. Electrochem. Sci. **14**, 9527–9536 (2019).
20. S. P. Herle. Ceramic coating on battery separator (2022).

21. Z. Xu, L. Zheng, B. Chen, T. Zhang, X. Chang, S. Wei, & Z. Dai. Overview of research on composite electrolytes for solid-state batteries. Energy Storage Sci. Technol. **10**, 2117–2126 (2021).
22. P. Bai, J. Li, F. R. Brushett, & M. Z. Bazant. Transition of lithium growth mechanisms in liquid electrolytes. Energy Environ. Sci. **9**, 3221–3229 (2016).
23. S. P. Herle. Ceramic coating on battery separator (2022).
24. C. Kim, J. Yoo, K. Jeong, K. Kim, & C. Yi. Investigation on internal short circuits of lithium polymer batteries with a ceramic-coated separator during nail penetration. J. Power Sources. **289**, 41–49 (2015).
25. C. Shi, P. Zhang, L. Chen, P. Yang, & J. Zhao. Effect of a thin ceramic-coating layer on thermal and electrochemical properties of polyethylene separator for lithium-ion batteries. J. Power Sources **270**, 547–553 (2014).
26. W. Shin & D. Kim. High-performance ceramic-coated separators prepared with lithium ion-containing SiO2 particles for lithium-ion batteries. J. Power Sources. **226**, 54–60 (2013).
27. J. Choi, S. Kim, & D. Kim. Enhancement of thermal stability and cycling performance in lithium-ion cells through the use of ceramic-coated separators. J. Power Sources. **195**, 6192–6196 (2010).
28. Z. Wang, D. Peng, & K. Sun. Research progress of separator materials for lithium-ion batteries. CIESC J. **69,** 282–294 (2018).

Based on the form of new energy, the development prospect of supercapacitors in electric vehicles is discussed

Kangcheng Dong [1,*]

[1]School of Chemical and Environmental Engineering, Wuhan Polytechnic University, 430023 Wuhan, China

Abstract. With the increasing global emphasis on environmental protection and sustainable development, new forms of energy have gradually become a key development direction in the energy sector. Electric vehicles, as an important application area of new energy technology, have their performance improvement and technological innovation as the focal points of industry attention. Supercapacitors, as an emerging energy storage device, have shown great potential in the field of electric vehicles due to their fast charging and discharging, high power density, and long cycle life. This article analyzes the development trend of electric vehicle technology under new energy forms and deeply explores the basic principles of supercapacitors and their specific applications in electric vehicles. Furthermore, this article looks forward to the development prospects of supercapacitors in electric vehicles, pointing out their enormous potential in improving electric vehicle performance and promoting efficient energy utilization. However, the development of supercapacitors also faces challenges such as high costs, limited energy density, and insufficient technological maturity. Addressing these issues, this paper puts forward suggestions such as strengthening technological research and development and optimizing joint usage strategies. The aim is to achieve positive interaction and collaborative development between supercapacitors and electric vehicles.

1 Introduction

With the heightened global awareness of environmental protection and the prevalence of sustainable development, new energy has emerged as a significant development direction in the global energy sector. In this context, electric vehicles (EVs), as the core carriers of new energy applications, exhibit increasingly obvious and irreversible development trends. The promotion of EVs not only helps significantly reduce tailpipe emissions from traditional fuel vehicles, thereby improving air quality and alleviating environmental pressure, but also significantly enhances energy utilization efficiency, driving the optimization and upgrading of the energy structure toward a cleaner and more efficient direction[1]. In the development

* Corresponding author： ailuzhang@ldy.edu.rs

of electric vehicles (EVs), the application of supercapacitor technology has garnered widespread attention. Currently, lead-acid batteries are widely used in multiple industries due to their low cost. However, their inherent drawbacks such as low power density, inability to charge and discharge rapidly, and short cycle life can easily lead to inefficiencies in application processes, thereby increasing the application costs. However, Supercapacitors, with their high power density, rapid charging and discharging capabilities, long lifespan, and environmental friendliness, are playing an increasingly important role in EV energy systems. Especially in urban traffic environments, where frequent starts and stops demand high power requirements for EVs, supercapacitors can quickly absorb and release energy, meeting instantaneous response needs while also demonstrating significant advantages in energy recovery [2].

Furthermore, the combined use of supercapacitors and batteries further enhances the overall performance of electric vehicles. By reasonably pairing supercapacitors and batteries, the complementary advantages of both in terms of energy density and power density can be fully utilized. This hybrid energy storage system not only meets the demand for long driving ranges in electric vehicles but also provides sufficient power support in high-power output scenarios such as acceleration and hill climbing [3,4].

The emergence of intelligent supercapacitors has injected new vitality into the development of electric vehicles. Intelligent supercapacitors integrate smart functions such as electrochromism, self-healing, and shape memory, not only enhancing the performance of energy storage devices but also broadening their application fields. In electric vehicles, intelligent supercapacitors are expected to further elevate the level of vehicle intelligence and meet consumers' demands for personalized and convenient transportation [5].

This article will delve into the basic concepts, working principles, and specific applications of supercapacitors in electric vehicles. Supercapacitors, with their unique electrochemical properties, exhibit significant advantages in rapid charging and discharging, long cycle life, and high-power density, offering new possibilities for enhancing the performance of electric vehicles. Their applications are not limited to the power systems of electric vehicles but can also be extended to energy recovery, auxiliary power supplies, and other aspects. Based on an understanding of the fundamentals of supercapacitors, this article will further analyze their application prospects and potential challenges in the field of electric vehicles.

2 Overview of the supercapacitors

The working principle of supercapacitors is primarily based on the interfacial double-layer effect between the electrode material and the electrolyte, enabling them to complete the charging and discharging process within an extremely short period of time, and this process is highly reversible. This characteristic grants supercapacitor significant advantages in high-power output situations, particularly suitable for applications requiring frequent and rapid charging and discharging.

Structurally, supercapacitors are typically composed of electrodes, electrolytes, separators, and casing, among other components. Among them, the choice of electrode material is crucial to the performance of supercapacitors. Currently, commonly used electrode materials include activated carbon, metal oxides, and conductive polymers. These materials possess high specific surface area, good electrical conductivity, and chemical stability, which can effectively enhance the energy storage capacity and cycling stability of supercapacitors.

In general, supercapacitors, as a new type of energy storage device, exhibit broad application prospects in the field of electric vehicles. With continuous technological advancements and cost reductions, supercapacitors are expected to play a greater role in electric vehicles, driving innovation and development in electric vehicle technology. Meanwhile, as people's focus on environmental protection and sustainable development deepens, supercapacitors, as a clean and efficient energy storage technology, are predicted to occupy an important position in the future energy sector.

2.1 Application of supercapacitors in electric vehicles

With their unique performance advantages, supercapacitors are gradually exhibiting broad application prospects in the field of electric vehicles. As efficient energy storage components, supercapacitors bring new possibilities for performance enhancement and energy management in electric vehicles.

In the power system of electric vehicles, supercapacitors can serve as an auxiliary power source, working in conjunction with the battery pack. Vehicles require a large amount of electricity instantaneously during startup, and supercapacitors offer better short-term electrical energy impulse capabilities compared to conventional power sources, along with a relatively longer lifespan [6].

A supercapacitor is a circuit structure composed of multiple parallel converters, which offers higher efficiency, stable output, a broader range, and smaller stable fluctuations compared to traditional single converters, effectively enhancing the overall efficiency of the circuit [6].

2.2 Development trend of supercapacitor technology

The development trend of supercapacitor technology is closely related to the demand for new energy products such as electric vehicles. In the pursuit of higher energy density, researchers are exploring new electrode materials and electrolytes, hoping to store more energy within a limited volume. This not only helps improve the driving range of electric vehicles but also achieves superior power performance while maintaining vehicle lightweightness.

In terms of cost reduction, with improvements in production processes and the realization of large-scale production, the manufacturing cost of supercapacitors is expected to further decrease. This will enhance the market competitiveness of electric vehicles and promote their wider adoption and application. At the same time, the cost reduction will also provide strong support for the expansion of supercapacitor applications in other energy storage fields.

3 Application of supercapacitors in electric vehicles

3.1 Description of the supercapacitor function

The most remarkable features of supercapacitors are their high power density and rapid charge and discharge capability. This means that during electric vehicles' acceleration and braking processes, supercapacitors can quickly provide or absorb a large amount of electrical energy to meet the demand for instantaneous high-power output. In contrast, traditional batteries such as lithium-ion batteries, although having high energy density, are often limited by their charging and discharging speeds when it comes to high-power output. Supercapacitors can charge to over 90% of their rated capacity within seconds to minutes, a feature that is particularly important in urban traffic environments where electric vehicles require rapid energy replenishment or frequent starts and stops.

 https://doi.org/10.1051/e3sconf/202560602011

Supercapacitors also exhibit an extremely long cycle life, capable of undergoing tens of thousands to even millions of charging and discharging cycles with minimal performance degradation. This characteristic greatly reduces the maintenance costs and replacement frequency of electric vehicles over their service life. Furthermore, supercapacitors require no complicated maintenance procedures in daily use, further saving time and resources.
Electric vehicles (EVs) have a wide range of application scenarios, from cold northern regions to hot southern regions, posing high demands on the adaptability of energy storage components. Supercapacitors maintain good performance within a wide temperature range (such as -40°C to 85°C), far surpassing many traditional batteries. This makes supercapacitors more advantageous in EV applications under extreme climatic conditions, ensuring stable vehicle operation in different environments.

Supercapacitors do not pollute the environment throughout their entire lifecycle, from production, and use to dismantling, and are considered a green and environmentally friendly energy storage solution. They do not contain heavy metals or harmful chemicals, thus preventing safety hazards such as leaks or explosions during use. As society's emphasis on environmental protection and safety continues to increase, the application prospects for supercapacitors in the EV field will become even broader.

3.2 Current research results

3.2.1 Auxiliary power system and energy recovery system

In pure electric vehicles (EVs), a typical application of supercapacitors is in auxiliary power systems and energy recovery systems. For example, a certain brand of pure electric bus uses supercapacitors as an auxiliary energy storage device, allowing for rapid charging at designated charging stations every 2-3 miles, with the entire process taking only a few minutes. This efficient charging method not only shortens the vehicle's charging time but also significantly improves the operational efficiency of the bus. Additionally, supercapacitors can recover energy from the braking system, converting kinetic energy that would otherwise be wasted into electrical energy for storage, further reducing the vehicle's energy consumption. According to statistics, these buses consume 40% less electricity than trolleybuses and only one-third of the energy of fuel-powered vehicles, truly realizing green transportation.

3.2.2 Energy management of hybrid electric vehicles

In this model's application, supercapacitors serve as an auxiliary component in the engine. When the car is started, the engine converts kinetic energy into electrical energy through the generator, which is then converted into direct current by an inverter and output to the electric motor. If the vehicle speed does not reach the set value after startup, the electric motor assists in propulsion. When the vehicle encounters poor road conditions during operation, the supercapacitor can be activated to provide electrical energy to the electric motor. During driving, when the speed reaches the set value, the electric motor stops operating, and the vehicle is powered by the engine. In this process, excess power is used to charge the supercapacitor. After the vehicle stops, the supercapacitor can automatically discharge into the battery . When the vehicle's power demand is low, this energy storage system is charged, enabling the recycling of energy. The rapid charging and discharging characteristics of supercapacitors allow hybrid electric vehicles to manage energy more effectively during acceleration and braking, improving energy utilization efficiency. According to tests, hybrid electric vehicles can save 30%-50% of fuel and reduce pollution by 70%-90%, making significant contributions to energy conservation and emission reduction.

3.2.3 To improve the vehicle performance

Supercapacitors also play a significant role in enhancing the performance of electric vehicles. During startup in cold weather, electric vehicles may face difficulties due to the decreased performance of batteries. However, supercapacitors can be used in parallel with batteries in low-temperature environments of -40°C, ensuring successful ignition in one attempt and significantly improving the vehicle's low-temperature starting performance. During the instant of starting, acceleration phase, and climbing processes of electric vehicles, high power output is provided to reduce the impact of instantaneous high current on a single battery, and to compensate for the low energy density of a single supercapacitor. Additionally, supercapacitors can provide instant high current for electric vehicles, helping the vehicles reach their maximum speed in a short period of time and enhancing the driving experience.

3.2.4 Optimization of hybrid energy storage system

In the overall design of electric vehicles, supercapacitors are often used in conjunction with traditional batteries such as lithium-ion batteries to form a hybrid energy storage system. This combination fully leverages the advantages of supercapacitors, including rapid charging and discharging and high power density, as well as the high energy density of lithium-ion batteries. By optimizing energy management strategies, the hybrid energy storage system can enhance the electric vehicle's acceleration performance, braking energy recovery efficiency, and overall service life while ensuring driving range. Additionally, the incorporation of supercapacitors helps reduce the number of charging and discharging cycles of lithium-ion batteries, thereby extending the lifespan of the battery pack and lowering the overall vehicle cost.

4 Challenges and development direction

4.1 The current research challenges

The energy density of supercapacitors is relatively low compared to other battery forms, one of the reasons for which is their low rated voltage (<2.7V). Currently, due to the gap between the energy density of supercapacitors and batteries versus actual needs, a major challenge in the field of supercapacitors remains improving their energy density [7]. Secondly, the production cost of supercapacitors is relatively high, which increases their application cost in new energy-electric vehicles. Reducing costs and improving cost-effectiveness are key to promoting their large-scale application.

Moreover, supercapacitor technology is not yet fully mature, and certain key technologies still need to be broken through. At the same time, the industry lacks unified standards and specifications, which has somewhat impacted their market promotion and application.

To promote and develop supercapacitors in the market, their performance needs to be enhanced to possess characteristics superior to other capacitors. The performance of supercapacitors largely depends on the choice of electrode materials and electrolytes. Therefore, developing new high-performance materials and improving their stability and cycle life are important directions for current research.

4.2 Future development direction

By optimizing electrode structures, developing new electrode materials and electrolytes, and improving production processes, it is expected that the energy density of supercapacitors will be significantly increased in the coming years, making it closer to the level of lithium-ion batteries. Increasingly more research is being conducted to develop new materials with high surface areas and the ability to withstand larger voltage windows. If these expansions are achieved, the energy density of supercapacitors will be comparable to that of batteries [8].

With the expansion of production scale and continuous technological advancements, the production cost of supercapacitors is expected to gradually decrease. Simultaneously, optimizing production processes and improving production efficiency can also reduce costs to a certain extent.

Strengthening research and development, as well as innovation in supercapacitor technology, is crucial for achieving breakthroughs in key technologies. Additionally, establishing unified industry standards and specifications to promote the standardization and modularization of technology will help improve product quality and reduce production costs.

The application of supercapacitors in new energy electric vehicles is not limited to auxiliary energy storage and energy recovery systems but can also be expanded to areas such as vehicle starting, acceleration assistance, and braking energy recovery. In the future, with continuous technological advancements and expanding application fields, the application prospects of supercapacitors in new energy electric vehicles will become even broader[9,10].

Supercapacitors can be combined with other energy storage technologies such as lithium-ion batteries to form hybrid energy storage systems. This type of system can fully leverage the advantages of various energy storage technologies, improving overall energy storage efficiency and performance. In the future, hybrid energy storage systems are expected to become an important development direction for energy storage systems in new energy electric vehicles.

5 Conclusion

In the context of new energy sources, this paper delves into the development prospects of supercapacitors in the field of electric vehicles (EVs). Through comprehensive analysis, the following main conclusions are drawn: The rise of new energy sources has provided significant impetus for the technological advancement of electric vehicles, as well as promoting environmental protection and sustainable development. Technological innovations in EVs have not only reduced environmental pollution but also optimized the energy structure. As a new type of energy storage device, supercapacitors demonstrate remarkable potential in enhancing EV performance due to their rapid charging and discharging capabilities, long cycle life, and high power density. Despite gradually increasing market acceptance, the application of supercapacitors in EVs still faces challenges in terms of cost, performance, and technology maturity. Therefore, it is necessary to strengthen technology research and development, enhance energy density, reduce costs, and optimize their combined use strategies with battery packs. In the future, the coordinated development of supercapacitors and EVs will become an important trend in the development of the new energy industry, contributing to technological innovation and sustainable development.

With technological advancements and the booming new energy market, the application prospects of supercapacitors in the EV field are broad. Future technology research and development should focus on improving the energy density of supercapacitors by exploring new electrode materials, electrolytes, and energy storage mechanisms to meet the diverse needs of EVs. Meanwhile, combining Internet and artificial intelligence technologies to develop innovative application scenarios will be a research hotspot. From the perspective of industrial development, improving the supercapacitor industry chain, strengthening raw material research and development, enhancing production processes, and expanding markets are crucial for achieving widespread application. Additionally, promoting international cooperation and technological exchanges will accelerate the application process of supercapacitors in the EV field. Despite these challenges, supercapacitors remain worthy of belief, as they will play a significant role in the field of electric vehicles and inject new vitality into the sustainable development of the new energy industry.

References

1. L. Tian, Supercapacitor and its application in electric vehicle, Micro-motor **8**,3 (2008)
2. Q. Xiong, Research progress of supercapacitors in hybrid electric vehicles, J. Sun Yat-sen University (Natural Science Edition) **S1**, 4 (2004)
3. D. Zhang, An energy storage device combining a supercapacitor and an exchangeable battery pack and an electric vehicle driven thereof, 1-2 (2019).
4. D. Zhang, An energy storage device combining the supercapacitor and the battery pack and related electric vehicles,(2020)
5. C. Xiong, Recent progress in the development of smart supercapacitors, Smart Mat **4**,2 (2022)
6. X. Chen, K. Xiao, L. Liang, G. Yuan, Research on application of supercapacitor energy storage in electric vehicle, Digital communication world **06**,281 (2019)
7. S. Cao, The challenge countermeasure and development of supercapacitor, Water conservancy technical supervision, **06**, 210-216 (2023)
8. H. Shifei, Z. Xianglin, S. Samrat, Z. Yufeng,Challenges and opportunities for supercapacitors, APL Materials, **10**, 7 (2019)
9. Q. Si, Study on multi-objective optimization of power allocation in EV hybrid energy storage system, J. Auto. Indus (2024)
10. F. Han, B. Bin, H. Xu, D. Yang, The automobile industry: Electric power industry. Electro. **19,** 6-8 (2023)

E3S Web of Conferences **606**, 02012 (2025) https://doi.org/10.1051/e3sconf/202560602012
ICNAOE 2024

A Review of Research on Potential Solutions for Dendrite Growth in Solid State Cells

Yang Liu*

Department of Materials, Chemistry and Chemical Engineering, Soochow University, 215123 Soochow, China

Abstract. The formation of lithium dendrites can lead to irreversible capacity loss and pose safety risks in lithium batteries. One proposed model suggests that when the current density is too high, a depletion layer of lithium ions forms near the anode, promoting dendrite growth. Lithium dendrites may puncture the membrane of the battery and reach the positive electrode, causing the battery to short circuit, causing the battery to overheat and possibly explode. This growth can result in short circuits in conventional lithium batteries. Several strategies can be employed to inhibit the growth of lithium dendrites. These include coating the lithium metal surface with durable layers or alloy compounds, micro-modulating the solid-state electrolyte to reduce dendrite nucleation, and managing nano-cracks in the electrolyte. These approaches have shown promising results in enhancing the stability and cycling performance of lithium metal anodes. However, technological constraints still limit the widespread implementation of these strategies, and more advanced characterization techniques are needed to better understand and address the issue of dendrite growth in solid-state batteries.

1 Introduction

In light of the global trend toward decarbonization, efficient and environmentally friendly energy storage technologies have become a central focus of research [1]. Among these, lithium batteries hold a prominent position. Despite their potential to achieve high energy densities, existing lithium-ion batteries do not meet the increasing demands for performance in the future energy market. For instance, electric vehicles are expected to travel approximately 500 kilometers on a single charge, yet most current models equipped with lithium-ion batteries can only manage about 200 to 300 kilometers per charge. This limitation is largely due to the capacity constraints of commercial lithium-ion batteries, which typically offer around 250 Wh/kg and 770 Wh/L [2].

To address future energy needs, the development of solid-state batteries—known for their higher energy density, superior cycling performance, and enhanced safety—has become essential. Solid-state batteries are considered an ideal solution because of their exceptional performance [3]. However, in practical applications, solid-state electrolytes still face challenges, particularly with the issue of short circuits caused by the growth of lithium

* Corresponding author : 2109404061@stu.suda.edu.cn

dendrites. For example, the grain boundaries in ceramic solid-state electrolytes often provide a pathway for these dendrites to grow, which can ultimately lead to battery failure [4].

The purpose of this paper is to review recent research advancements and the challenges faced by various solid-state electrolytes in addressing the issue of lithium dendrite growth, which can lead to short circuits. Additionally, it aims to refine strategies for mitigating this problem. This work is essential for enhancing the safety and stability of solid-state electrolytes, thereby supporting the commercialization of solid-state batteries and facilitating updates and improvements in the energy market.

The title is set in bold 16-point Arial, justified. The first letter of the title should be capitalised with the rest in lower case. You should leave 22 mm of space above the title and 6 mm after the title.

2 Lithium dendrite formation mechanism

The formation of lithium dendrites is primarily explained by the physical model proposed by Chazalviel. As shown in Figure 1, the deposition of lithium dendrites on the surface of the lithium anode occurs in two main phases. When the current density is below a critical value, the ion concentration near the lithium anode is similar to that of the electrolyte. In this scenario, the growth of lithium dendrites is hindered. However, when the current density exceeds this critical value, lithium ions near the anode are rapidly consumed, creating a depletion layer of lithium ions. This results in a localized space charge that encourages the reduction of lithium atoms and promotes the formation of dendrites [5].

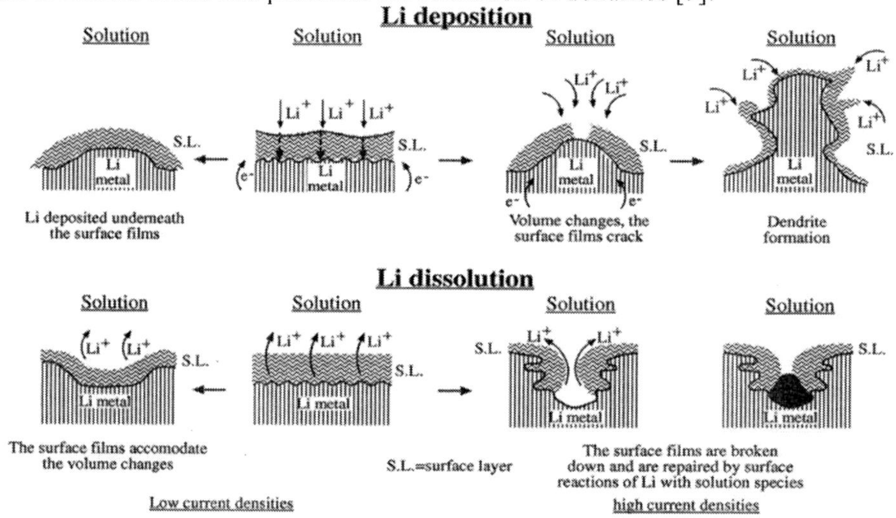

Fig. 1. the deposition mechanism of lithium dendrites on the lithium anode [6].

Chazalviel explains that the lithium-ion deposition process is crucially situated beneath the solid electrolyte interface of the battery [7]. Here, the reduced lithium atoms are subjected to internal stresses and interfacial surface tension and move along the grain boundaries, leading to increasing internal stresses at the lithium anode. This condition significantly increases the internal stress at the lithium anode. When this stress surpasses a critical threshold, lithium atoms exit the pores and form lithium dendrites. As these dendrites expand and reach a certain size, they become detached from the root, resulting in dead lithium. This phenomenon leads to irreversible capacity loss, emphasizing the need for improved battery management strategies [8].

The growth of lithium dendrites is an irreversible process. In conventional lithium batteries, if lithium dendrites penetrate the separator, they can cause the battery to short-circuit. In contrast, solid-state batteries are less susceptible to puncture by lithium dendrites due to the dense nature of their electrolyte. However, their operation depends on lithium penetration and ion transport, which can lead to the formation of lithium dendrites within the electrolyte, potentially resulting in short circuits.

The location and growth patterns of lithium dendrites are significantly influenced by the quality of the electrolyte material. These factors are closely related to the morphology of the crystal boundaries and defects in the electrolyte[6]. Ceramic electrolytes are generally more likely to promote the growth of lithium dendrites compared to polymer electrolytes, primarily due to the presence of nano-cracks.

3 Modification strategies

Recent studies indicate that lithium dendrite growth can be inhibited through various methods, including the coating of anode lithium metal surfaces, micro-modulation of solid-state electrolytes, and control of nano-cracks.

3.1 The coating of anode lithium metal surfaces

The short-circuit principle highlights a critical issue: the diffusion of lithium atoms along grain boundaries beneath the solid electrolyte interface (SEI) is constrained, resulting in stress and dendrite formation at the anode. By employing a robust SEI layer with exceptional insulating properties, we can effectively curb the growth of lithium dendrites while preventing direct contact between the anode and the electrolyte. Research efforts have increasingly concentrated on both organic and inorganic coatings, alongside promising polymer and alloy compound coatings that have produced impressive results.

Chang et al. have made significant advancements by creating an artificial SEI using planar polycyclic aromatic hydrocarbons on lithium metal electrodes through an innovative in-situ formation technique. Their product, the dihydroxy violet anthracene (DHV) layer, acts as a powerful protective shield, enhancing the stability of the SEI. Remarkably, the DHV layer has been shown to enhance the Coulombic efficiency of the lithium metal anode significantly, achieving outstanding cycling stability of over 1,000 hours at a rate of 4 mAh cm^{-2}. Furthermore, lithium-metal anodes shielded with the DHV layer demonstrated an impressive Coulombic efficiency of 99.8% and sustained excellent long-term cycling stability after 300 cycles in $LiFePO_4$/Li batteries. This innovation marks a pivotal step forward in battery technology [9].

Inorganic coatings play a significant role in this field. Liu et al. employed an innovative technique where they alternately sprayed negatively charged graphene oxide and positively charged polylysine solutions onto a copper substrate using spray coating. They then utilized microwave plasma chemical vapor deposition (CVD) to grow a nanodiamond film on this copper substrate, creating a high-strength protective layer. This layer boasts a modulus of elasticity exceeding 200 GPa, effectively preventing the propagation of lithium dendrites. Moreover, it demonstrates exceptional stability during cycling tests.

A key aspect of Liu's innovation is the two-layer design. By repeating the processes of seed-layer spraying and CVD, they were able to grow a second layer of nanodiamond thin film. This approach significantly enhances the defect tolerance of the interface [10].

In 2017, Liu proposed the use of dynamically crosslinked polymers, such as Play-Doh, as an interfacial layer for lithium metal anodes. This polymer exhibits viscoelastic characteristics, providing it with a unique "solid-liquid" hybrid property [11].

As shown in Figure 2, (a) illustrates the structural formula of the material, while (b) and (c) display comparison graphs for the scenarios without and with the coating, respectively. The results indicate that the crosslinked polymer coating effectively prevents the growth of lithium dendrites. This efficiency arises from the material's ability to reversibly transition between "liquid" and "solid" properties based on the lithium growth rate. Consequently, it maintains good contact even as the volume and morphology of the lithium metal change. This innovation provides a new perspective on stable operation and uniform deposition/stripping of lithium metal anodes.

In addition, alloying the lithium metal surface by magnetron sputtering at a specific temperature provides higher conductivity efficiency by lower resistance of the alloy compounds compared to other coating materials.

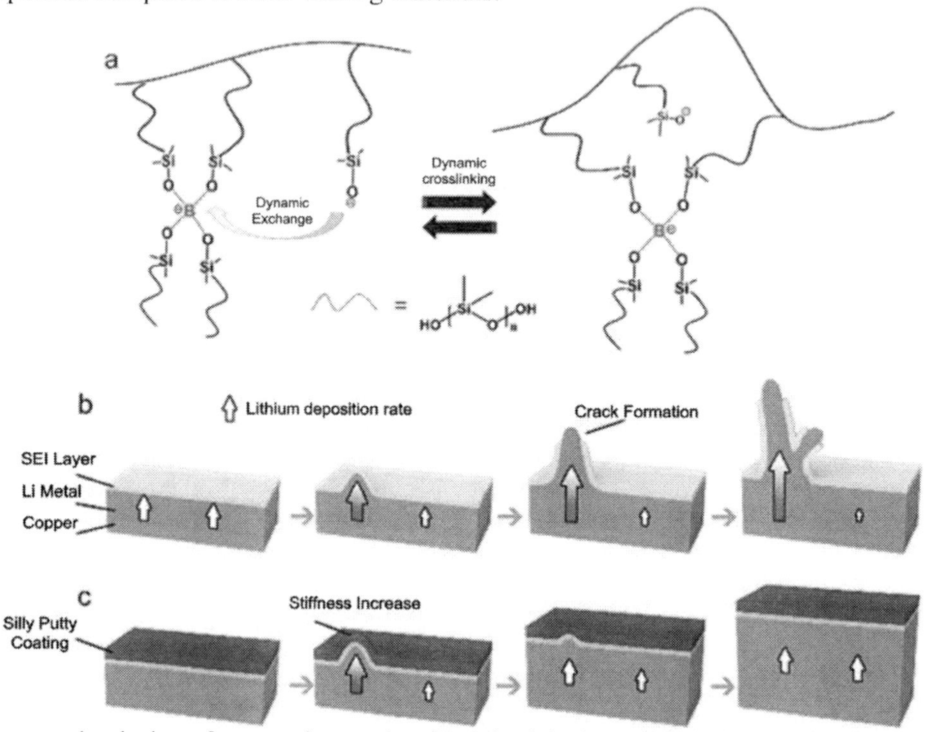

Fig. 2. the design of composite coating SP (plasticine) modified Li anode [11].

3.2 Micro-modulation of solid-state electrolytes

At high current densities, the ion concentration exceeds that of the electrolyte, leading to the formation of a space charge in the dilute electrolyte. This space charge accelerates the nucleation and growth of lithium dendrites. To effectively mitigate this dendrite nucleation process, it is beneficial to use electrolytes with either a zero anion transfer number or a high concentration of supporting electrolyte. Additionally, a single ionic conductor can enhance lithium ion mobility, meaning that lithium ions can move through the electrolyte much more efficiently than other ions, thus reducing ion loss [12].

The performance of the electrolyte can be enhanced by incorporating additives without altering its fundamental properties. For instance, Chae et al. studied a novel electrolyte additive called lithium cyano tris(2,2,2-trifluoroethyl)borate (LCTFEB). Their characterization of lithium metal morphology and the solid-state electrolyte interface used techniques such as field emission scanning electron microscopy (FE-SEM), cryo-transmission electron microscopy (cryo-TEM), and X-ray photoelectron spectroscopy (XPS). The findings demonstrated that the additive could form a solid-electrolyte interphase (SEI) layer on the lithium metal surface, significantly improving the electrochemical performance of both the NCM523/Li battery and the symmetric Li/Li battery [13].

3.3 Control of nano-cracks

Compared to traditional electrolytes, solid electrolytes possess a high shear modulus, which theoretically makes them better at resisting lithium metal penetration. However, a study found that the oxide solid electrolyte LLZO, with a shear modulus of approximately 60 GPa, still experiences intrusion even at a surface-averaged current density significantly lower than 1 mA cm^{-2} [14].

Based on this, Geoff McConohy and his colleagues conducted lithium metal plating experiments on the surface of LLZO using in situ microprobe scanning electron microscopy. They controlled the contact force between the microprobe and the LLZO to observe how lithium penetration initiates and grows. The results showed that increasing the contact force significantly enhanced the likelihood of lithium penetration. This occurs because higher contact forces create new nano-cracks on the electrolyte surface, thereby confirming the positive relationship between nano-crack formation and lithium penetration[15].

To mitigate the risk of lithium metal permeation, it is essential to carefully manage the contact force, as this can help minimize crack formation caused by excessive pressure. Additionally, the study indicated that small compressive strains can effectively guide lithium penetration along the direction of strain, which means that applying external stresses can also be a viable method for influencing crack growth and lithium penetration. Consequently, McConohy suggested using external compressive strains to promote crack closure, thereby increasing the effective fracture toughness of the cracks. This crack closure effect can ultimately reduce the likelihood of lithium metal penetration.

4 Conclusion

Traditional lithium-ion batteries have struggled to meet the increasing demand for new energy solutions, making the practical application of solid-state batteries urgently necessary. A key challenge for the commercialization of solid-state batteries is addressing the growth of lithium dendrites within solid-state electrolytes. Unlike traditional lithium batteries, where lithium dendrite countermeasures focus on different approaches, this paper outlines three methods to mitigate dendrite growth in solid-state electrolytes: the coating of anode lithium metal surfaces, micro-modulation of the electrolyte, and control of nano-cracking.

However, these advancements are still limited by technological constraints. In the future, employing more advanced characterization techniques such as scanning electron microscopy (SEM), transmission electron microscopy (TEM), and X-ray diffraction (XRD) could provide a clearer understanding of lithium dendrite growth on anodes, thereby enabling more targeted solutions to this issue.

References

1. L. Wang, J. Li, G. Lu, W. Li, Q. Tao, C. Shi, H. Jin, G. Chen, S. Wang, Wang. Fundamentals of Electrolytes for Solid-State Batteries: Challenges and Perspectives. Front. Mater. **7**, (2020)
2. J. Xu, X. Cai, S. Cai, Y. Shao, C. Hu, S. Lu, S. Ding, High-Energy Lithium-Ion Batteries: Recent Progress and a Promising Future in Applications. Energy Environ. Mater. **6**, (2022)
3. X. Wang, Preparation and characterization of PEO-based solid composite electrolyte, Xi'an Technological University, (2019)

4. A. M. Bates, Y. Preger, L. T. Castro, K. L. Harrison, S. J. Harris, J. Hewson, Loraine Torres-Castro, Katharine L. Harrison, Stephen J. Harris, John Hewson. Are solid-state batteries safer than lithium-ion batteries. Joule. **6**, 742-755 (2022)
5. Y. S. Cohen, Y. Cohen, D. Aurbach, Micromorphological studies of lithium electrodes in alkyl carbonate solutions using in situ atomic force microscopy J. Phys. Chem. B. **104**, 12282–12291 (2000)
6. X. Cheng, R. Zhang, C. Zhao, F. Wei, J. Zhang, Q. Zhang, A review of solid electrolyte interphases on lithium metal anode, Adv. Sci. **3**, 1500213 (2016)
7. T. Foroozan, S. S. Asl, R. S. Yassar, Recent advances in dendrite-free lithium metal anodes for high-performance batteries, Jps. **461** (2020)
8. X. Zhang, C. Sun, A review of solid electrolyte interphases on lithium metal anode, Phys. Chem. Chem. Phys. **24**, 19996-20011 (2022)
9. S. Chang, X. Jin, Q. He, T. Liu, J. Fang, Z. Shen, Z. Li, S. Zhang, M. Dahbi, J. Alami, K. Amine, A. Li, H. Zhang, J. Lu. In Situ Formation of Polycyclic Aromatic Hydrocarbons as an Artificial Hybrid Layer for Lithium Metal Anodes, Nano Lett. **22**, 263-270 (2022)
10. Y. Liu, Y. Tzeng, D. Lin, A. Pei, H. Lu, N. A. Melosh, Z. Shen, S. Chu, Y. Cui, Lithium Metal Anodes with an Adaptive "Solid-Liquid" Interfacial Protective Layer, Joule. **2**, 1595-1609 (2018)
11. K. Liu, A. Pei, H. R. Lee, B. Kong, N. Liu, D. Lin, Y. Liu, C. Liu, P. Hsu, Z. Bao, Y. Cui, Lithium Metal Anodes with an Adaptive "Solid-Liquid" Interfacial Protective Layer, J. Am. Chem. Soc. **139**, 4815−4820 (2017)
12. M. D. Tikekar, S. Choudhury, Z. Tu, L. A. Archer. Archer. Design principles for electrolytes and interfaces for stable lithium-metal batteries, Nat. Energy. **1**, 1-7 (2016)
13. O. B. Chae, V. A. K. Adiraju, B. L. Lucht, Borate as a Multifunctional Electrolyte Additive for High-Performance Lithium Metal Batteries, ACS Energy Lett. **6**, 3851−3857 (2021)
14. L. Frenck, G. K. Sethi, J. A. Maslyn, N. P. Balsara,Factors That Control the Formation of Dendrites and Other Morphologies on Lithium Metal Anodes, Front. Energy Res, **7**, (2019)
15. G. McConohy, X. Xu, T. Cui, E. Barks, S. Wang, E. Kaeli, C. Melamed, X. W. Gu, W. C. Chueh, Mechanical regulation of lithium intrusion probability in garnet solid electrolytes, Nat. Energy. **8**, 241-250 (2023)

Improvement of the Conductive Ability of Polymer Electrolytes

Hailiang Xu*

College of Materials Science and Engineering, Beijing University of Chemical Thechnology,102200 Beijing, China

Abstract. In current applied research, traditional electrolytes face significant challenges. The emergence and application of polymer electrolytes provide effective solutions to these problems. This paper will introduce the types and characteristics of polymer electrolytes based on three aspects. The star-shaped compound based on PEGMA, i.e., poly(ethylene glycol) methacrylate, can significantly enhance the mechanical and electrical properties of polymer electrolytes; the gel electrolyte prepared by compounding highly polar β -polyvinylidene fluoride (β -PVDF) with nylon 6 shows significant application potential in improving ion conduction efficiency and enhancing battery safety. Brush-shaped polymers have good performance in avoiding the crystallization of linear polymers and inhibiting the growth of lithium dendrites and improving mechanical strength. These three types of polymer electrolytes each have their characteristics and good application prospects. Still, they all face problems such as high cost and difficulties in industrialization at the present stage and need further development to better benefit human society. This paper aims to provide those who want to explore and do some decent research on this topic with some basic information as well as fundamental advice.

1 Introduction

Electrolytes are crucial in electrochemical energy storage. However, the safety issues of liquid electrolytes and the performance degradation caused by dendrite growth have seriously limited the practical application of lithium-ion batteries. Therefore, polymer electrolytes (PE) are expected to replace liquid electrolytes (LE). Although solid polymer electrolytes (SPE) have good safety and mechanical properties, they are limited by temperature, have low ionic conductivity, and have poor contact with electrodes and insufficient cycling performance. For example, the high crystallinity of linear polyethylene oxide (PEO) leads to a low ionic conductivity of SPE, and the strong complexation between PEO and lithium ions (Li^+) limits the conduction of Li^+, resulting in a generally low lithium-ion transference number of the electrolyte. The scientific community is improving polymer electrolytes in various ways to achieve better performance.

This paper classifies and lists the main types and characteristics of polymer electrolytes, describes their preparation methods, and provides a preliminary overview of the research and

* Corresponding author :2022020333@buct.edu.cn

application progress of polymer electrolytes. For the problems existing in the currently widely used electrolytes, polymer electrolytes are used in a targeted manner and their properties are precisely utilized to solve the problems. It is our hope to provide some inspiration in the direction of the further development of polymer electrolytes by subsequent researchers.

2 Star-shaped compounds

2.1 Structural optimization of star-shaped compounds

The structure of star-shaped compounds can be optimized by adjusting cross-linking agents in varieties of means and the relative molecular mass of PEGMA, i.e., poly(ethylene glycol) methacrylate. Incorporating star-shaped compounds into cross-linked solid polymer electrolytes significantly improves not only the mechanical but also electrochemical properties of the solid electrolyte. Studies have shown that adding poly(ethylene glycol) methacrylate (6 mmol), trimethylolpropane tris(3-mercaptopropionate) (2 mmol), and benzophenone (mass fraction 5%) into a solvent-free system and vigorously stirring for 4 hours to promote uniform mixing. Subsequently, the slurry was dropped into a polytetrafluoroethylene mold and irradiated with 365 nm UV light for 30 seconds to obtain a star-shaped compound (T475). Compared with the initial solid polymer electrolyte, after incorporating the star-shaped compound, the tensile toughness of the sample is significantly improved, indicating that incorporating star-shaped compounds into cross-linked polymer electrolytes in a solid state can significantly improve the mechanical properties of the solid electrolyte. [1]

2.2 Ionic conductivity and conductance performance

Ionic conductivity is one of the most effective indicators for measuring the applicability of SPEs. The mobility of polymer segments increases with increasing temperature, thereby accelerating the complexation and decomplexation processes between lithium ions and polymer segments and promoting the conduction of lithium ions.

The longer chain length of the PEGMA arm in the introduced T950 star-shaped plasticizer can further disrupt the regularity of polymer segments, increase the movement rate, and thus promote the conduction of lithium ions. Thanks to its better flexibility, it promotes the stable interface contact between the solid electrolyte film and the electrode, ensuring the smooth cross-interface transmission of lithium ions inside the battery [1].

The introduction of star-shaped compounds promotes the sufficient relaxation of polymer segments in the SPE - T950 structure, ensuring that it has both better ionic conductivity and better mechanical flexibility. In addition, the good flexibility of SPE - T950 promotes good interface stability between the SPE electrodes, providing a smooth channel for the cross-interface migration of Li^+, which ensures that the Coulomb efficiency of the battery is maintained above 96% during the charge-discharge cycle[2].

3 Gel electrolytes

3.1 Improvement of battery safety performance

Highly polar β -polyvinylidene fluoride (β -PVDF) has all-trans conformations, with its Fluorine and Hydrogenium atoms located on opposite sides of the polymer backbone, and has been proven to be a promising artificial solid electrolyte interface coating on copper and lithium metal anodes, enabling dendrite-free lithium deposition/stripping and enhanced cycling performance [3]. Molecules contain a large number of carbon - fluorine functional groups (-C—F), which can promote the movement of lithium ions in the electrolyte.

However, the polyvinylidene fluoride matrix has low mechanical strength and poor thermal stability and cannot be used alone as a polymer electrolyte, especially in a working environment that requires high-rate battery discharge [4]. The method of making a composite diaphragm from other polyvinylidene fluorides with good mechanical properties and thermal stability can effectively improve the performance of the polymer electrolyte matrix. Since nylon 6 (PA6) has a high melting point, is resistant to organic solvent corrosion, and is inexpensive, it has been widely used in various fields in the past few decades. Therefore, compounding nylon-6 - 6 with polyvinylidene fluoride to prepare a gel polymer electrolyte and applying it to lithium batteries has good prospects.

Dissolve polyvinylidene fluoride and nylon 6 in a formic acid - acetic acid mixed solution respectively, adjust the mass fraction, and then perform electrospinning for three layers. The first layer is a nylon layer, the second layer is polyvinylidene fluoride, and the third layer is again a nylon layer. Use the composite diaphragm to absorb a small amount of lithium hexafluorophosphate electrolyte (1 mol/L lithium hexafluorophosphate solution) to obtain a GPE as the battery electrolyte and lithium metal as the battery negative electrode. After that, prepare a positive electrode sheet by mixing commercial lithium iron phosphate, carbon black, and a binder in a mass ratio of 8:1:1. Finally, assemble a CR2025 type lithium-ion battery in an argon atmosphere-protected glove box, and then use a blue electricity tester to test the various properties of the battery in the end.

Tests show that if the battery experiences a thermal runaway due to an accidental local short circuit when the temperature rises above 160 °C, the polyvinylidene fluoride layer begins to melt, while the nylon layers on both sides remain stable at this time. The melted polyvinylidene fluoride will enter the pores of the nylon layer and block the pores, completely blocking the positive and negative electrodes of the battery. At this time, the circuit is interrupted at the power source, and the electrolyte is equivalent to an open circuit, with no current passing through. Therefore, the PA6/PVDF/PA6 composite diaphragm has good thermal stability, and when used as a gel polymer electrolyte, it can effectively improve battery safety [5].

4 Brush-shaped polymers

During the charging and discharging process of LMBs, the growth of lithium dendrites at the electrode/electrolyte interface can lead to a battery short circuit and even cause serious safety accidents [6]. The main strategies for inhibiting the growth of lithium dendrites include: constructing an artificial solid electrolyte interface (SEI) layer, constructing a nano-composite electrolyte [7], preparing a high-mechanical-strength 3D network polymer, and introducing new additives [8], etc. Polyacrylic acid (PAA) can react with lithium metal on the negative electrode surface in situ to form lithium polyacrylate (LiPAA), and the formed SEI layer can effectively inhibit the growth of lithium dendrites [9]. In addition, the brush-shaped polymer segments have strong mobility and can avoid the crystallization problem of linear polymers, making them an ideal solid polymer electrolyte matrix material. However, the grafting density of brush-shaped polymers currently used for SPE is low, and the synthesis steps are cumbersome. In addition, the irregular growth of dendrites on the lithium negative electrode surface is a major challenge for lithium metal batteries, and lithium dendrites may pierce the electrolyte membrane and cause a battery short circuit.

It has been reported that using poly2-((2-bromopropoxy) methyl) acrylate (PBMA) as a raw material, the PEG side chain is connected to the PBMA main chain by the "grafting-to"

 https://doi.org/10.1051/e3sconf/202560602013

strategy. This method obtains a PBMA - g - PEG brush-shaped polymer by dialysis and freeze-drying. Based on this, various brush-shaped polymers with different topological structures are constructed to inhibit the crystallization behavior of PEO, enhance the mobility of polymer segments, and improve the ion-conducting performance. At the same time, by designing a single-ion structure and introducing polyester or polycarbonate segments, the content of free Li^+ in the electrolyte is increased, and the lithium ion transference number is improved.

4.1 Inhibition of lithium dendrite growth

The interaction between the polymer matrix or lithium ions and the polymer can effectively inhibit the crystallization behavior of PEG, reducing the crystallinity of the electrolyte. The rigidity of the PS base combined with the flexibility of PEG reduces the brittleness of the material while avoiding the problem that PEG is difficult to form. The low crystallinity creates good conditions for ion movement. SPEs with high mechanical strength can migrate the growth of lithium dendrites on the lithium negative electrode surface as a way improve battery safety. The brush-shaped structure can promote the movement of polymer segments, enhance the ion-conducting ability of the electrolyte, and simultaneously improve the thermal and electrochemical stabilities of the electrolyte. When the main chain length of the brush-shaped polymer is the same, the lower the Tg (glass transition temperature) of the brush-shaped polymer electrolyte as the longer the side chain is, the more efficient the complexation - decomplexation process with lithium ions, and the higher the ionic conductivity. In addition, the single-ion structure can increase the content of free lithium ions moving through the electrolyte and improve the lithium-ion transference number [10].

4.2 Improvement of mechanical properties

Using a polycaprolactone (PCL)-based brush-shaped polymer electrolyte. A macromolecular initiator containing hydroxyl and carboxyl groups can initiate and catalyze the ring-opening process of the polymerization of ε -caprolactone monomers to obtain a brush-shaped polymer. Based on a series of PHMA macromolecular initiators with various polymerization degrees. Polyester-based brush-shaped polymers with the same PCL side chain length and different main chain lengths can be obtained by changing the consistency of reaction conditions, and how different main chain lengths affect the performance of SPEs can be studied. Mixing PH - PCL and PEO in a certain proportion and preparing SPEs with different brush-shaped polymer contents or different main chain lengths by solution casting method. That is, pouring the solution of the prepared PH - PCL and PEO mixture onto a clean and dry polytetrafluoroethylene mold, and then placing it in a refrigerator, repeatedly freezing and thawing to evaporate the water, thus obtaining a solid product with a coexistence of crystalline and amorphous states. Linear PCL is similar to PEO and is a semi-crystalline polymer, which is not conducive to the conduction of Li^+. However, studies have shown that the brush-shaped structure helps to inhibit the crystallization of the polymer. This is because the PCL has a strong influence between itself and PEO can inhibit each other's behavior of crystallization. XRD results show that the brush-shaped structure can initially inhibit the crystallization behavior of linear PCL, and there are only a few crystalline regions in the polyester-based brush-shaped SPEs, which is obviously beneficial to improving its electrochemical performance. The strong complexation between the polyether matrix and Li^i limits the migration movement of Li significantly, so the tLi+ of traditional PEO - based SPEs is low (0.20, 60 oC). In comparison, the tLi+ of 0.6PCL is increased to 0.52 (60 oC), because the complexation between the carbonyl of the polyester matrix and Li^i is weak, allowing Li^i to migrate more efficiently in the electrolyte.

4.3 Improvement of interface stability

The interface stability between SPEs and the lithium metal anode significantly affects the cycling stability of LMBs. Studies have shown that the presence of an appropriate amount of carboxylic groups in the electrolyte helps to inhibit the growth of lithium dendrites on the lithium anode surface [11]. The carboxylic acid can react in situ with lithium metal to form a stable SEI layer as long as the amount is proper, which can have tremendous influence on the growth of lithium dendrites. Since the conduction of lithium ions in SPEs depends on the waving polymer chains, and general polymer chains lack sufficient movement ability at room temperature, LMBs are difficult to work at room temperature [7]. PCL has ideal chain flexibility and thanks to the brush-shaped structure, PCL can further enhance the movement ability of polymer chains by destroying the regularity of polymer chains.

4.4 Limitations and future development

The biggest problem with polymer electrolytes is that their manufacturing cost is relatively high, and the manufacturing process is complex, which is not conducive to large-scale industrial applications. Moreover, the volume of the electrolyte is too large, which goes against the current pursuit of smaller battery volumes. It is possible to further explore changes in polymerization methods such as emulsion polymerization to obtain higher molecular weight main chain polymers and increase the alcoholysis degree to obtain more grafts to enhance their electrochemical and mechanical properties.

5 Conclusion

This paper explores the incorporation of star-shaped compounds into polymer electrolytes to improve their ion passage performance and mechanical properties; it also explores and introduces that in gel electrolytes, the fluorine atoms in the main chain play an extremely important role in the transmission of lithium ions, and in addition, the gel electrolyte can protect the circuit well in case of a short circuit due to its characteristic of melting at high temperatures; furthermore, it introduces brush-shaped compounds for polymer electrolytes. Due to their unique characteristics of graft copolymers, they can create polymer electrolytes with multiple functional groups coexisting and thereby improve the ion transmission efficiency between the electrode and the electrolyte.

Whether it is star-shaped compounds, gel electrolytes, or brush-shaped polymers, the problems they all have lie in the low synthesis yield and high cost of high-molecular solid electrolytes at the present stage, and the unclear industrial application scenarios. However, they all demonstrate that high-molecular materials have a solid theoretical foundation and bright application prospects in reducing lithium salt crystallization and improving the migration ability of lithium ions in electrolytes in solid electrolytes. For example, the mechanical properties and the proportion of high-molecular amorphous states of solid electrolytes can be adjusted by regulating parameters such as polymerization degree and alcoholysis degree to achieve quantitative control of physical properties and improve SPEs.

In the future, polymer electrolytes will play a key role in human exploration of batteries.

References

1. W.Wang, C. Wang, J. Hu, Construction of Star-shaped Compounds to Improve the Performance of Cross-linked Network Solid Polymer Electrolytes. J. South China Normal Univ. (Nat. Sci. Ed.) **56**, 25-31 (2024).

2. S. Li, Z. Xue, Design and Electrochemical Performance of Brush-shaped Solid Polymer Electrolytes for Lithium Metal Batteries. (2024).
3. J. Luo, C.C. Fang, N.L. Wu, High Polarity Poly(vinylidene difluoride) Thin Coating for Dendrite-free and High-performance Lithium Metal Anodes. Adv. Energy Mater. **8**, 1701482 (2018).
4. L.G. Shen, S.S. Feng, J.X. Li, et al., Surface Modification of Polyvinylidene Fluoride (PVDF) Membrane via Radiation Grafting: Novel Mechanisms Underlying the Interesting Enhanced Membrane Performance. Sci. Rep. **7**, 2721 (2017).
5. X. Wu, X. Hou, Gel Polymer Electrolyte Based on Nylon 6 and Polyvinylidene Fluoride Composite Diaphragm. J. South China Normal Univ. (Nat. Sci. Ed.) **54**, 36-41 (2022).
6. X. Li, Y. Tian, L. Shen, Z. Qu, T. Ma, F. Sun, et al., Electrolyte Interphase Built from Anionic Covalent Organic Frameworks for Lithium Dendrite Suppression. Adv. Funct. Mater. **31**, 2009718 (2021).
7. H. Huo, J. Gao, N. Zhao, D. Zhang, N.G. Holmes, X. Li, et al., A Flexible Electron-blocking Interfacial Shield for Dendrite-free Solid Lithium Metal Batteries. Nat. Commun. **12**, 176 (2021).
8. S.H. Lee, J.Y. Hwang, J. Ming, Z. Cao, H.A. Nguyen, H.G. Jung, et al., Toward the Sustainable Lithium Metal Batteries with a New Electrolyte Solvation Chemistry. Adv. Energy Mater. **10**, 2000567 (2020).
9. P. Parikh, M. Sina, A. Banerjee, X. Wang, M.S. Souza, J.-M. Doux, et al., Role of Polyacrylic Acid (PAA) Binder on the Solid Electrolyte Interphase in Silicon Anodes. Chem. Mater. **31**, 2535-2544 (2019).
10. B. Gupta, Geeta, A.R. Ray, Preparation of Poly(ε -caprolactone)/Poly(ε -caprolactone-co-lactide) (PCL/PLCL) Blend Filament by Melt Spinning. J. Appl. Polym. Sci. **123**, 1944-1950 (2012).
11. D. Rosenbach, N. Mödl, M. Hahn, J. Petry, M.A. Danzer, M. Thelakkat, Synthesis and Comparative Studies of Solvent-free Brush Polymer Electrolytes for Lithium Batteries. ACS Appl. Energy Mater. **2,** 3373-3388 (2019).

https://doi.org/10.1051/e3sconf/202560602014

A review of metal catalysts for lithium-sulfur battery applications

Li Shuai *

School of Energy and Chemical Engineering (SECE), Xiamen University Malaysia, 43900 Sepang, Malaysia

Abstract. When compared to ordinary lithium batteries now on the market, lithium-sulfur batteries stand out as the future research direction for energy storage systems owing to their higher volumetric energy density, theoretical specific energy, and gravimetric energy density. However, Li-S batteries' drawbacks have significantly impeded their use and commercialization. A series of problems such as shuttle effects and volume expansion can remarkably reduce the service lifespan of Li-S batteries. By adding metal catalysts to the positive electrode of Li-S batteries, it is possible to improve the reaction kinetics of these batteries and efficiently address issues like the shuttle effect. During the paper, metal catalysts are separated into three groups based on the many ways they express themselves: metal single-atom catalysts, metal nanocatalysts, and metal compound catalysts. Metal single atom catalysts have catalytic active sites that are shaped like a single metal atom. Metal nanocatalysts are catalysts composed of metal particles with a size of 1-100 nm. Metal compound catalysts refer to catalysts composed of compounds formed by metals and other elements (such as oxygen, nitrogen, sulfur, etc.), including metal oxides, metal nitrides, metal sulfides, etc. It also lists and summarises the research progress of previous generations for the reference of related personnel.

1 Introduction

In this day and age, the continued consumption of fossil energy sources and the environmental problems they bring have led to a growing focus on new and clean energy sources. High-performance energy storage systems are becoming more and more in demand, and the battery field is attracting much attention and developing rapidly. Among them, lithium-ion batteries are widely used commercially by virtue of their unique advantages. However, the existing commercial lithium-ion batteries' energy density (200–300 Wh kg^{-1}) is still inadequate, and the high production cost (> $300 kW-1h-1) has failed to meet current demand within the market, significantly restricting their promotion. Therefore, the development of the elevated energy density batteries as a new generation of energy storage systems has become an urgent task. Lithium-sulfur batteries are gradually gaining focus because of their high gravimetric energy density (2600 Wh kg^{-1}), elevated theoretical specific capacity (1675 mA h g^{-1}) and high volumetric energy density (2800 Wh L^{-1} ,assuming a

* Corresponding author : EGE2109330@xmu.edu.my

2.15V average voltage). Nevertheless, the commercialisation of lithium-sulfur batteries has not yet occurred, mainly due to problems with electron/ion insulation, shuttle effects as well as volume growth of the cathode material of monomeric sulfur and its product sulfide. Li2S, the discharge product, and sulphur are both electrical insulators, so the reaction must be facilitated by the introduction of a conductive substrate to promote a smooth reaction. Secondly, the conversion reaction of sulphur is multistep and results in the development of a variety of soluble polysulphide intermediates (Li2Sx,4 ≤ x ≤ 8), which have the opportunity to traverse the separator to the exterior of the lithium metal anode as a soluble substance and react with lithium on the lithium metal anode's surface to be reduced to a solid deposit, leading to a self-discharge. In addition, the lithium precipitates will return to the cathode surface to undergo oxidation again, causing the active sulphur and lithium metal to be lost.. Since Li_2S_2 and Li_2S occur in solid form, the interconversion of the two is very slow, resulting in slow overall kinetics. It has been found that the efficient conversion of polysulfides can be effectively promoted by introducing various catalytic substances at different parts of the Li-S battery. Among them, boosting the reaction kinetics of Li-S batteries by introducing metal catalysts at the anode is a major research hotspot. The introduction of metal catalysts can impede the lithium polysulphide shuttle, improve the sulfur utilisation as much as possible by speeding up the sulfur redox process in addition to enhance the lithium-sulfur batteries' overall performance.

This paper classifies and describes the metal catalysts used in lithium-sulfur batteries. Additionally, prior research advancements in this area are presented for the benefit of researchers in related domains.

2 Metal monoatomic catalysts

While an increase in the bulk metal-based materials' weight ratio can improve the utilisation of sulphur in lithium-sulfur batteries, the introduction of inactive components reduces the sulphur loading rate, thus decreasing polysulphide conversion and battery energy density. Single-atom catalysts (SACs), which comprise uniformly scattered metal sites with excellent catalytic performance and close to 100% atom utilization, can effectively alleviate this problem. The introduction of single atom catalysts does not sacrifice the cathode's sulphur ratio, thus increasing the energy density of lithium-sulfur batteries. The unsaturated coordination environment and distinct electronic structure of single-atom catalysts endow them with obvious catalytic advantages, which make them more competitive than traditional catalysts. Since single atoms have high free energy, in order to preserve structural stability, they are typically coordinated with other atoms (such as N and O). By anchoring atomically dispersed metal atoms on various carbon substrates, the redox reaction of sulfur can be catalyzed efficiently. SACs were first reported in 2011, and scientists have developed and designed a number of different high-quality SACs through many different synthesis methods. Currently, a number of SACs have been synthesized that can exhibit electrocatalytic activities better than those of metal nanoparticles. Their use in Li-S batteries is currently in the investigation stage, though[1]. Currently, the mainstream metal monatomic catalysts for lithium-sulfur batteries include Fe, Co, and other monatomic metals such as Ni, Zn, Mo, V, W, and Ru.

Because of their exceptional cost-effectiveness and environmental friendliness, atomic iron electrocatalysts with Fe-Nx activity centres have garnered a lot of interest. Wang prepared FeN4 /CN-SAC consisting of nitrogen-doped Fe-N4 active sites. Despite the comparatively low iron loading (1.14 wt %), FeN4 /CN-SAC exhibited low overpotentials at different scanning speeds were able to exhibit lower overpotentials, suggesting an enhancement of the kinetics of the polysulfide conversion reaction [2].

https://doi.org/10.1051/e3sconf/202560602014

Density Functional Theory (DFT) calculations have shown that single-atom sites can lower the energy barrier for polysulfide conversion and enhance the cell multiplicity performance [3]. The coordinating environment has a major impact on the catalytic activity of individual atoms. The tetra-coordinated structure of M-N4 (M stands for metal) has great stability and is often used as the active site for sulfur reduction reactions. Even though the structure is stable, the tetra-coordinated configuration may not exhibit the best catalytic performance, and thus it is an effective strategy to enhance solitary atoms' catalytic activity by modulating the atmosphere of coordination. For example, the adsorption capacity of Fe-N5-C is significantly superior to that of Fe-N4-C, and the migration energy barrier of lithium ions on the surface of Fe-N5-C is significantly reduced thanks to the increase of N coordination number.

Wang uniformly embedded single iron atoms (SAFe) on a porous nitrogen-rich carbon matrix (NC) and subsequently placed it at the anode of a Li-S battery, and a series of electrochemical analyses and theoretical simulations and spectroscopic analyses showed that the uniformly distributed single iron atoms (SAFe) could create a lithium-sulfur battery with low activation voltage without sacrificing the current rate due to their sufficiently high catalytic activity. At the same time, they also created ultra-high-speed lithium-sulfur batteries by using single-atom catalytic cathodes, which were tested to be excellent in terms of long cycle life at 5 C and excellent rate performance at 12 C. The results showed that the long cycle life lithium-sulfur batteries at 5 C and excellent rate performance at 12 C were very good. The aforementioned evidence clearly demonstrated that single-atom catalysts are capable of efficiently accelerating the kinetics of high-performance batteries' ultrafast conversion reactions [2].

Chen et al. integrated Co complexes into conjugated microporous polymers (CMP-CoN_2) as monatomic metal catalysts for Li-S batteries. Due to the enhancement of hybridisation between the 3d orbitals of the cobalt atoms and the 3p orbitals of the sulfur atoms in the polysulphides, the monatomic catalysts possessed superior catalytic activity in facilitating polysulphide conversion in Li-S batteries when in contrast to the saturated analogues, and the experimental The obtained experimental data show that the single-atom catalyst has a great multiplicative capacity of 673.2 mA h g-1 at 6 C and a high specific capacity of 1393 mA h g-1 at 0.1 C. At 2 C after 1000 cycles, it likewise exhibits a very low capacity decay rate of 0.045. [4].

3 Metal nanocatalysts

Pt was the first metal to be used in metal nanoresearch, and as the field has grown, Co, Fe, and nano-alloys have been widely developed. Co is the most widely studied metal, and its synergistic interaction with carbon-doped materials can quicken the reaction of sulphur reduction, raise the quantity of active sites and enhance interactions with polysulfides, and boost the rate of the reaction. Fe nanometals have likewise attracted attention, and Ye have reported a highly graphitised Fe-enriched and N-site-rich carbon tubes for catalysing the electrochemical reaction of lithium-sulphur batteries [5], while Ru nanometal has been shown to quicken Li2S2's transformation into Li2S. In order to integrate the advantages of single metals, the combination of nano-alloys is highly valued, specifically such as CoFe, FeNi, ZnPd alloys and high-entropy alloys, which are all supported by carbon substrates with a view to enhancing the catalytic effect.

Chen synthesised and used Co-Fe mixed phosphide nanotubes as lithium-sulfur cathode electrodes, which have a network of interconnected, very porous pores, together with excellent metal conductivity, and which can significantly enhance the Li-S batteries' redox reaction kinetics. Chen found that Co-Fe-P nanotubes have a strong influence on polysulfide by using UV-visible absorption spectroscopy.

LiPS with strong chemisorption ability, and also Co-Fe-P nanotubes with strong chemotrapping ability for LiPS as shown by DFT studies, and the conversion reaction's kinetics were increased by these interactions between LiPS and cathode materials. The battery's capacity was 678 mAh g-1 following 500 cycles of continuous cycling, with a 0.043% cycle-average capacity decrease [6].

Yang studied basic problems such as the in-situ conversion process of monometallic catalysts in Li-S batteries and the chemical state of the resulting catalytically active centers in addition to examining the reaction kinetics of lithium polysulphide in Li-S batteries. In order to convey sulfur, they uniformly loaded nickel nanoparticles into a flexible material made of a carbon-based bacterial cellulose framework (CBC/Ni). They then used in situ XRD and in situ Raman experiments to investigate the precise chemical processes taking place in this material. They confirmed that during battery operation, the nanoscale nickel catalyst in the S@CBC/Ni electrodes changes into NiS2 through interactions with lithium ions, serving as the new active center to aid in the conversion of lithium ions. After 500 cycles, the average capacity decay rate is only 0.064%, and the lithium-sulfur battery with the nickel nanocatalyst exhibits a high multiplicative capacity of 798.7 mAh g^{-1} at 2C., which can be considered an excellent performance. The experimental data are satisfactory[7].

4 Metal compound catalysts

After a comparative study of the abundance and price of each element in the periodic table, compared to the previous two, there are numerous benefits to employing metal oxides in Li-S batteries as cathode catalysts. Thus, the application of metal compounds—more especially, metal carbides, metal nitrides, metal phosphides, metal sulfides, metal selenides, and metal oxides—as sulfur redox catalysts is a significant area of scientific interest. The catalysts of different metal compounds have different electronic structures and surface properties, so their catalytic activities are also very different. There is a wide variety of metal oxides, which have a better catalytic effect than non-polar carbon and can react with polysulphides to form thiosulphates and polysulphates to enhance sulfur utilisation. The polar bonding between the metal cation and oxygen anion in metal oxides provides sufficient active sites to anchor the polysulfides. Nevertheless, the lack of electrical conductivity of metal oxides remains a drawback. As a result, future designs ought to concentrate on enhancing Li-S batteries' performance by boosting their conductivity through the addition of carbon or conductive polymers. Metal sulfides are well anchored to polysulphides due to their better conductivity and strong sulphophilicity. Selenides are similar to metal sulfides but exhibit better catalytic activity due to the high conductivity of selenium. Because of their high conductivity, metal nitrides have garnered a lot of interest, but are expensive to prepare and optimising their cost has become an important research direction. Metal phosphides are more mild and economical to prepare, and possess high electrical conductivity and excellent catalytic effect, but are easily oxidized at room temperature, and there is an urgent need to explore the one-step synthesis of metal phosphides with excellent performance.As shown in fig 1, by introducing modification strategies such as internal and surface defects, micromorphology, and elemental doping, more active sites can be generated to improve the catalyst activity and achieve rapid conversion of polysulfides.

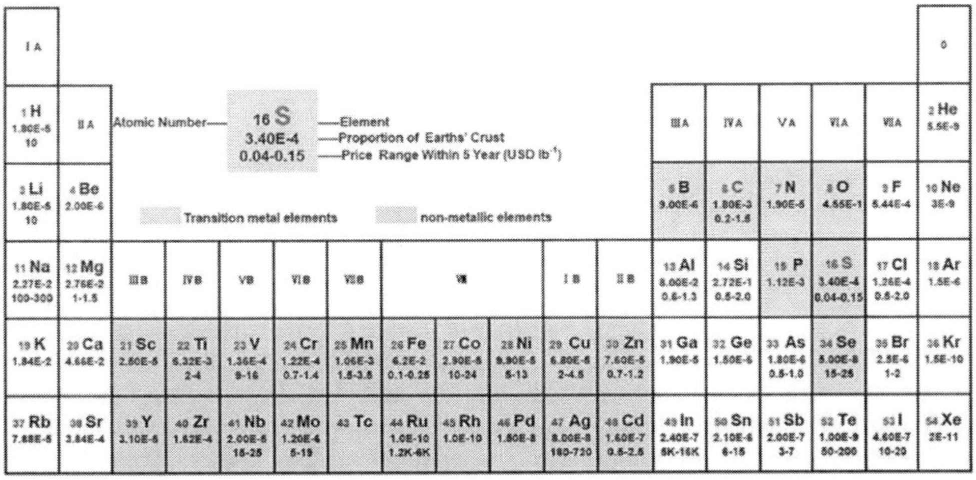

Fig. 1. Prices and abundance of elements related to lithium battery catalysts [1].

Iron oxide was studied as a catalyst for use in the cathode region of Li-S batteries beginning as soon as 2014, marking the beginning of research on metal oxides. Zhao prepared porous iron oxide by wet chemistry, and then the prepared porous iron oxide was embedded in the cathode section of Li-S batteries, where sulphur from the cathode portion was combined with it. The shuttle effect was lessened by using the porous iron oxide as a reservoir for polysulfides. The experimental results show that the lithium-sulfur battery containing 5% iron oxide has improved cycling performance and greater battery stability, having a 574 mAh g-1 charging capacity that remains usable after 50 cycles at 0.5°C ambient temperature. [8]

Taking metal sulphides as an example, metal sulphide catalysts have been improved through continuous research and exploration to obtain catalysts with better catalytic effects.Lin introduced MoS2 as an efficient electrocatalyst in lithium-sulphur batteries. The CV result plots of the symmetric cell containing MoS2-x showed that the narrow inter-peak distance and sharp peaks of the MoS2−x electrode indicate excellent electrode electrochemical reversibility, which suggests that the LiPs conversion reaction is easy to carry out. This improves the use of sulphur by significantly and successfully reducing the buildup of discharge products on the cathode surface. A little amount of MoS2 combined with sulfur may cycle 1600 times at 5 C rate with a capacity degradation of 0.083% per cycle thanks to this kinetic effect, which is already an exceptional cycle life performance. [9]

The immobilisation of polysulfides in hollow conducting carbon is a major research direction to limit the shuttle reaction, and a variety of carbon materials including graphene, carbon nanotubes, graphene oxide, and microcarbon/medium carbon have been extensively explored and investigated by the previous researchers with a view to turn them into cathodic sulphur hosts to capture polysulfides to inhibit the shuttle reaction. Iron carbide (Fe_3C) stands out due to its excellent magnetic and catalytic activity. Its excellent ability to capture sulfides and catalytic activity for lithium-sulphur battery cathodes is worth exploring in research. Gao et al. proposed to prepare ACT@Fe/Fe_3 C/graphene sulphur hosts by embedding Fe/Fe_3C into graphene shells and placing them on flexible activated cotton-twisted (ACT) fibres, which were put into lithium-sulfur battery cathodes to be used as catalysts. They discovered that this catalyst's superior magnetic field-enhanced trapping mechanism allows it to efficiently prevent the diffusion and dissolution of polysulfides, opening up a new avenue to lessen the polysulphide shuttle effect. In contrast to the 100-cycle cycle life that pure ACT/S can accomplish, the cycle life with ACT@Fe/Fe3 C/S can be greatly enhanced to 600 cycles, which is about six times more than that without iron carbide, and it has a 764 mAh g^{-1} initial discharge capacity [10].

He and colleagues looked on the use of nickel metal boride (NiB) in Li-S batteries as a catalyst. According to the theoretical study, it can be concluded that NiB can effectively enhance the charge transfer and accelerate the sulfur redox kinetics, thus enhancing battery performance considerably. Lithium-sulfur battery's discharge capacity with NiB catalyst under cycling conditions at a sulfur loading of 5 mg cm^{-2}, 0.5 C, is 1239 mAh g^{-1}, which is retained by 83.2% after 150 cycles, indicating that the theory is highly accurate based on the analysis of the experimental results. Additionally, it maintains a 590 mAh g^{-1} capacity with a 14.89 mA cm-2 current density with a degradation rate of just 0.07%. This study has significant research value and significance since it demonstrates how metal boride catalysts can enhance lithium-sulfur batteries' performance [11].

The excellent electron orbital assignment ability of phosphorus can enhance the activity of phosphide catalysts, and Huang proposed the use of iron phosphide nanocrystals as catalysts uniformly embedded in 3D porous graphene-carbon nanotubes scaffolds (FeP/rGO/CNTs) for Li-S batteries. According to the DFT theory, FeP has the ability to efficiently adsorb LiPSs and encourage their conversion, preventing their dissolution and diffusion, thus alleviating the shuttle effect, enhancing the redox kinetics, and boosting the battery performance. On the basis of this study, the research on FeP catalysts has never stopped; Ma constructed hollow sea urchin FeP by phosphorylating self-assembled sea urchin FeOOH precursors in a solvothermal procedure in order to provide enough active surfaces for LiPSs; Xia et al. used gradient porous conductive carbon skeletons as FeP catalysts in order to maximize the active Li-poly(II) sulfide binding sites and catalytic sites to ensure high sulfur utilisation and catalytic efficiency; and Xia et al. used gradient porous conductive carbon skeletons as FeP catalysts in order to maximize the active Li-poly(II) sulfide binding sites and catalytic sites. In order to maximise the active lithium polysulphide binding sites and catalytic sites to ensure high sulfur utilisation and catalytic efficiency, Xia et al. used a gradient-conducting carbon skeleton as a carrier for FeP nanoparticles to fabricate a lithium-sulfur battery cathode body, PCM/FeP [12].

5 Conclusion

Because of their great theoretical capacity, Li-S batteries are very competitive and have garnered a lot of interest in current energy storage device development. However, there are still a number of issues with Li-S batteries that need to be resolved, including the shuttle effect, volume expansion, and the low conductivity of polysulfides. Solving these problems is a must to promote the commercialisation of Li-S batteries. The aforementioned issues can be successfully improved and resolved by adding metal catalysts to the Li-S battery's cathode.

Currently, the research on metal catalysts for lithium-sulfur batteries is primarily separated into three groups, namely metal single atom, metal nano and metal compounds, each with its own advantages and disadvantages. All things considered, metal catalysts can significantly enhance lithium-sulfur batteries' performance by lowering the polysulphide shuttling phenomenon and increasing the electrode reaction kinetics. However, there is a certain impact on the sulphur loading when metal catalysts are introduced. The research on metal catalysts for Li-S batteries is still going on steadily. Maintaining the sulfur loading in order to attain a high sulphur loading while enhancing the Li-S batteries' performance through the introduction of metal catalysts may become a major research direction to promote the realisation of metal catalysts in practical applications.

The criteria for evaluating the performance of metal catalysts should be multidimensional, primarily in lithium-sulfur batteries' electrochemical performance, including their specific capacity, multiplicity, and cycling stability, among other aspects. However, the potential mechanism of action of the catalysts still needs to be studied in depth. A deeper understanding of catalysts through the development of advanced characterisation methods can greatly facilitate the advancements in research and practical applications of lithium-sulphur batteries.

References

1. S. Qian. Research Progress of Metal Catalysts for Sulfur Reduction of Lithium Sulfur Batteries. Journal of Organic Chemistry Research, **11(04),** 314-333 (2023).
2. J. Wang, L. Jia, J. Zhong, Q. Xiao, C. Wang, K. Zang, H. Liu, H. Zheng, J. Luo, J. Yang, H. Fan, W. Duan, Y. Wu, H. Lin, & Y. Zhang. Single-atom catalyst boosts electrochemical conversion reactions in batteries. Energy Storage Materials, **18**, 246–252 (2018).
3. Y. Li, J. Wu, B. Zhang, W. Wang, G. Zhang, Z. Seh, N. Zhang, J. Sun, L. Huang, J. Jiang, J. Zhou, & Y. Sun. Fast Conversion and Controlled Deposition of Lithium (Poly)sulfides in Lithium-Sulfur Batteries Using High-Loading Cobalt Single Atoms. Energy Storage Materials, **30**, 250-259 (2020).
4. L. Chen, J. Xia, Z. Lai, D. Wu, J. Zhou, S. Chen, X. Meng, Z. Wang, H. Wang, L. Zheng, L. Xu, X. Lv, C. W. Bielawski, & J. Geng. Coordinatively Unsaturated Co Single-Atom Catalysts Enhance the Performance of Lithium-Sulfur Batteries by Triggering Strong d -p Orbital Hybridisation. ACS Nano (2024).
5. H. Ye, J. Sun, S. Zhang, H. Lin, T. Zhang, Q. Yao, & J. Lee. Stepwise Electrocatalysis as a Strategy Against Polysulfide Shuttling in Li-S Batteries. ACS Nano, **13**, 14208-14216 (2019).
6. Y. Chen, W. Zhang, D. Zhou, H. Tian, D. Su, C. Wang, D. Stockdale, F. Kang, B. Li, & G. Wang. Co–Fe Mixed Metal Phosphide Nanocubes with Highly Interconnected-Pore Architecture as an Efficient Polysulfide Mediator for Lithium–Sulfur Batteries. ACS Nano, **13(4),** 4731–4741 (2019).
7. C. Yang, F. Wang, D. You, W. Yang, Y. Wang, W. Han, Y. Zhang, Z. Zhu, & X. Li. In-situ chemical state transition of Ni nano-metal catalytic site promotes the reaction kinetics of lithium-sulfur battery. Chemical Engineering Journal, **496,** 153812 (2024).
8. C. Zhao, C. Shen, F. Xin, Z. Sun, & W. Han. Prussian blue-derived Fe2O3/sulfur composite cathode for lithium–sulfur batteries. Materials Letters, **137,** 52–55 (2014)
9. H. Lin, L. Yang, X. Jiang, G. Li, T. Zhang, Q. Yao, G. W. Zheng, & J. Y. Lee. Electrocatalysis of polysulfide conversion by sulfur-deficient MoS2 nanoflakes for lithium–sulfur batteries. Energy & Environmental Science, **10(6),** 1476–1486 (2017).
10. Z. Gao, Y. Schwab, Y. Zhang, N. Song, & X. Li. Ferromagnetic Nanoparticle–Assisted Polysulfide Trapping for Enhanced Lithium–Sulfur Batteries. Advanced Functional Materials, **28(20)** (2018).
11. X. He, Z. Xie, W. Zhang, Z. Gao, Y. Cheng, X. Zhang, Y. He, F. Kang, & L. Peng. Operationally Robust Li–S Batteries at High Current Density Enabled by Metallic, Dual Sulfurphilic Nickel Boride. ACS Materials Letters, 5307–5315 (2024).
12. X. Liu. Research Status of Iron-Based Compounds as Catalysis for Lithium-Sulfur Batteries. Material Sciences, **14(02),** 130–141 (2024).

Research Progress on Oxygen Reduction Catalyst Materials for Proton Exchange Membrane Fuel Cells

Beiheng Cao[1],*

[1]College of Chemical Engineering, Sichuan University of Arts and Science, 63500 Dazhou, China

Abstract. The sustainable development of society gives rise to the increasing energy demand, making environmental pollution a problem that cannot be ignored. Among them, fuel cells are attracting attention as efficient and environmentally friendly. Proton exchange membrane fuel cells (PEMFCs) represent an advanced and eco-friendly energy conversion technology with substantial potential in transportation, distributed power generation, and other applications, while their performance can be remarkably impeded by the cathodic oxygen reduction reaction (ORR) which exhibits slow kinetics. Pt-based catalysts are the preferred materials for catalysis but are detrimental to the practical application of fuel cells due to their high cost and instability. Therefore, it is essential to develop stable and low-cost cathode catalysts. This paper reviews the advancements in cathode catalysts for PEMFCs, encompassing traditional platinum-based catalysts and their enhancement strategies, as well as the current status of emerging non-platinum-based catalysts. The paper also outlines future research directions for these cathode catalysts.

1 Introduction

Against the backdrop that energy demand grows rapidly and environmental problems deteriorate a lot, developing clean and renewable energy is of primary importance. Proton exchange membrane fuel cell (PEMFC) exhibits high energy conversion efficiency, high energy density, and environmental protection properties, capable of ensuring chemical energy is directly converted into electrical energy. It is widely used in vehicles (such as cars and airplanes), power generation stations, and military fields. However, the relevant cathodic oxygen reduction reaction (ORR) significantly affects fuel cells' overall performance because of the involvement of multiple proton-electron transfers and inherent slow kinetics. Although Pt-based catalysts exhibit excellent catalytic performance, their high cost, scarcity, and strict requirements for oxygen purity limit their applications. In terms of cost reduction, Deng et al. prepared an iron nanocatalyst encapsulated in pod-shaped carbon nanotubes, which can reduce the contact between the catalyst and air pollutants and also reduce the consumption of precious metals [1]. In addition, most PEMFCs use air as the raw material gas, and impurities such as SO_2 and NO_2 present in the air can easily poison the catalyst, significantly

* Corresponding author: jiqianchang@ldy.edu.rs

reducing the efficiency and service life of the fuel cell. Misz et al.summarized the impact of some pollutants in the air on PEMFC and found that a small amount of toxic substances can significantly reduce the performance of fuel cells [2]. Therefore, improving the anti-toxicity of catalysts, and ensuring their stability and activity, has become a key research direction at present. With the continuous advancement of PEMFC towards practical applications, researchers are required to pay more attention to anti-toxicity weakening and cost reduction while ensuring stability.

This article introduces several catalysts developed in recent years and explores the current problems, research directions, and corresponding solutions of traditional PEMFC oxygen reduction electrocatalysts to promote the development of PEMFC technology.

2 Cathodic reaction principles for PEMFC

PEMFCs have the function of generating electrical energy by the hydrogen-oxygen electrochemical reaction, enabling the transfer of electrons from the anode to the cathode relying on the ORR and the acquisition of electrical energy from fuel cells by external electrical appliances. Oxygen enters the fuel cell's cathode inlet and reaches the surface of the cathode catalyst through the gas diffusion layer, providing sufficient oxygen for the reaction. The hydrogen gas at the anode decomposes into protons and electrons, with the former reaching the cathode through the proton exchange membrane (PEM) and the latter flowing toward the cathode along the external circuit. When oxygen, protons, and electrons meet on the surface of the cathode catalyst, an oxygen reduction reaction occurs, producing water.

3 Pt-based catalysts

3.1 Traditional Pt-based catalysts

Platinum (Pt) has long been the preferred material for cathode catalysts in PEMFC relying on the outstanding catalytic activity towards ORR. Nevertheless, the scarcity of Pt resources results in high catalyst costs and insufficient stability under actual working conditions. In traditional Pt-based catalysts, excessive Pt dosage has greatly hindered their large-scale production applications. At the same time, Pt nanoparticles are loaded on carbon supports, which provides a larger specific surface area (SSA). However, corrosion is prone to occur at high potentials of carbon supports, and exposing more active sites also makes Pt-based catalysts susceptible to poisoning. Because the gas used for the cathode is usually air, rather than high-purity oxygen, pollutants such as SO_2 and NO_2 contained in the air will adsorb onto the surface of Pt-based catalysts, thereby reducing their catalytic activity. The problem of cathode catalyst poisoning can be solved from four perspectives: adding air filters, ozone treatment, electrochemical recovery, and developing new catalyst materials [3].

3.2 Pt-based alloy catalysts

Although Pt exhibits excellent catalytic performance in ORR, its scarce resources, high cost, and other issues greatly prevent PEMFCs from being applied. According to research, alloying Pt with other transition metals of Co and Nican improves the catalyst performance [4]. For example, Pt_3M (where M is another transition metal) alloy catalysts present improved catalytic mechanisms, thereby being well-concerned in the field of fuel cells. Alloying can adjust Pt atoms in terms of the electronic structure and optimize their ability to adsorb and dissociate oxygen molecules. For example, Pt-Ni alloy catalysts shift the d-band center of the catalyst through the modulation of the ratio of Pt and Ni, thereby enhancing the catalytic activity. Zhou et al. added a small amount of Ag to PtCu and found an obvious improvement in the catalytic activity after adding Ag [5]. The current common method for preparing alloy catalysts is a solvothermal method, but this method has difficulties in removing organic residues when preparing end-capping agents, and the process flow is complex and costly [6]. Therefore, researchers are improving solvothermal methods (such as polyol methods) and exploring other preparation methods (such as electrodeposition and atomic layer deposition).

3.3 Pt-based core-shell catalysts

The Pt-based catalyst with a core-shell structure usually uses nonprecious metals (such as Co, Ni) as the core and Pt as the shell. The important reason why Pt-based catalysts with core-shell structure have attracted attention is not only because they can reduce the amount of Pt used, but also because there are two important adjustable interactions at the core-shell interface, namely strain effect and ligand effect, which can effectively improve the activity of electrocatalysts [7]. Upon the alloying of Pt with a transition metal, the electrons of the transition metal will interact with the electrons of Pt. If the electronegativity of the two is different, it will cause electron transfer, which will change the electron cloud density distribution of Pt atoms and thus affect their adsorption capacity for the reaction substrate. The strain effect originates from the change in the distance between atoms on the catalyst surface. Lattice distortion can occur because of the different sizes of atoms, resulting in strain. Under compressive strain, the d-band center of Pt presents a downward shift, reducing the adsorption energy for reaction intermediates and promoting the desorption of reaction products, thereby increasing the chemical reaction rate. Despite the role of the core-shell structure in remarkably decreasing the amount of Pt in PEMFCs, there are still some issues that need to be addressed, among which durability remains an important indicator to be considered. Regarding this, the Lv Yang team Pd3Au@Pt/C study was conducted on the C core-shell structure electrocatalyst, and it was concluded that adding a small amount of high redox potential precious metals to the original catalyst core can effectively enhance durability [8]. Although Pt-based core-shell catalysts have great advantages, there are still many shortcomings that need to be improved.

3.4 Pt-based single-atom catalysts

Pt-based single-atom catalysts refer to those in which a single Pt atom is uniformly distributed on a specific carrier. The smaller the radius of Pt particles, the higher their utilization rate. Single-atom catalysts can maximize the participation of active sites, therefore single-atom catalysts are currently developing into highly active electrochemical reaction catalysts. Liu et al. synthesized a Pt1/NBP catalyst using a low-temperature hydrothermal ethanol reduction method to develop an efficient ORR catalyst [9]. The catalyst was strongly selective for the 2-electron pathway of ORR. Further high-temperature heat treatment was performed to reconstruct the coordination environment for Pt atoms, forming a single-atom catalyst fixed with pyridine-N Pt1@Pt /NBP, thus obtaining an efficient 4-electron pathway [10]. Although single-atom Pt has a unique electronic structure and exhibits special catalytic activity and selectivity, it has great research potential. However, commonly used characterization methods make it difficult to verify the intrinsic relationship between single-atom size and its catalytic performance. Therefore, to study its reaction mechanism, it is necessary to develop new characterization methods.

4 Non-Pt-based catalysts

4.1 Transition metal nitrides

Transition metal nitrides exhibit excellent catalytic properties in oxygen reduction reactions. Specifically, Fe-N-C material has an initial potential approaching to Pt/C catalyst, and this catalytic property mainly results from the synergistic effect between nitrogen atoms and transition metals [11]. In the ORR reaction, nitrogen atoms can adsorb oxygen molecules, while transition metals promote the dissociation of oxygen molecules through electron transfer, forming adsorbed oxygen atoms that react with protons and electrons to generate water. By adjusting the electronic structure of transition metals, their adsorption and activation abilities can be optimized, thereby improving catalytic performance. Compared to ordinary C materials, adding N to the catalyst can activate more oxygen molecules, and the atomic radii of N and C are close, which can effectively solve the problem of lattice mismatch [12]. Therefore, the performance of nitrogen-doped carbon catalysts will be better than other catalysts. Jia et al. synthesized nitrogen-doped nanocages with porous self-supporting structures and found that they have dual catalytic effects, the ORR performance of which approaches that of commercial Pt/C [13]. According to research results, among nitrogen-doped carbon materials with different N configurations, pyridine-N has the best catalytic activity.

4.2 Transition metal chalcogenides

Transition metal chalcogenides are compounds composed of transition metal elements (Fe, Mo, Ni, etc.) and chalcogenides, divided into binary and ternary compounds, with Ru-based compounds being the main ones [14]. There are two structural types in ternary compounds: the Chevrel phase and the amorphous phase [15]. These compounds typically have a layered structure, with transition metal atoms and chalcogen atoms connected by chemical bonds within the layers, and weaker van der Waals forces interacting between the layers. This structure gives it a high SSA and rich active sites. For example, MoS_2 has high ORR activity in alkaline media, and its active sites are mainly distributed at the edge of MoS_2. By modifying and regulating the edge position, its catalytic performance can be improved. During the ORR process, metal atoms at the edge of transition metal chalcogenide can adsorb oxygen molecules, while sulfur atoms can alter metal atoms' electronic structure, promoting oxygen molecules to be activated.

4.3 Non-metal carbon-based catalysts

Despite the significant progress in Pt-based catalysts and non-noble metal catalysts, there are still shortcomings. Carbon-based materials exhibit certain catalytic activity in ORR reactions, such as graphene and carbon nanotubes. Carbon-based materials, with massive sources and low prices, can well resist methanol and remain stable in acidic and alkaline media. However, the catalytic activity of a single carbon material as a catalyst is not ideal. To solve this problem, its catalytic activity can be further improved by changing its porous structure or doping with heteroatoms [16]. Nitrogen-doped carbon materials are active, and doping nitrogen atoms can alter the carbon materials' electronic structure, introducing defect sites on their surfaces to be active sites for ORR. During ORR, nitrogen-doped carbon materials such as pyridine nitrogen and pyrrole nitrogen can adsorb oxygen molecules and activate them to generate water through electron transfer. Meanwhile, the pore structure of carbon materials also has a significant impact on ORR performance, and an appropriate pore structure can promote the diffusion and adsorption of reactants.

5 Conclusion

With the acceleration of social development, the contradiction between social development and the natural environment has become increasingly prominent. Therefore, developing new energy sources has become an urgent task, and PEMFCs, as a zero-pollution and zero-emission new energy source that uses hydrogen as fuel and water as a product, have enormous potential. However, there are still many issues to be addressed before its widespread application. Because most oxygen reduction catalysts currently rely on Pt, its scarce resources and high cost have primarily restricted the research progress of PEMFCs. Therefore, it is particularly important to enhance the performance of Pt-based catalysts and develop new non-Pt-based catalysts. This article summarizes the cathodic reaction principle of PEMFCs, explores the problems of traditional catalysts, and introduces the research directions of Pt-based catalysts (traditional Pt-based catalysts, Pt-based alloy catalysts, Pt-based core-shell catalysts, and Pt-based single-atom catalysts) and non-Pt-based catalysts (transition metal nitrides, transition metal chalcogenide, and Non-metal catalysts). The research summary is as follows: Further explore new cathode catalyst materials, especially high activity, high stability, and low-cost non-Pt-based catalysts; Conduct in-depth research on catalyst active sites and catalytic mechanisms to theoretically benefit catalyst design; Develop methods to more effectively prepare reusable catalysts, to meet the large-scale application needs of PEMFCs.

In summary, although Pt-based catalysts and non-Pt-based catalysts have made some progress in research, they still face many challenges. With the deepening of research and the development of technology, it is expected to develop new catalysts with high performance and cost-effectiveness for benefiting the achievement of the national carbon peak and carbon neutrality strategic goals.

References

1. D. Deng, L. Yu, X. Chen, et al., Iron encapsulated within pod-like carbon nanotubes for oxygen reduction reaction. Angew. Chem. **125(1)** (2013).
2. U. Misz, A. Talke, A. Heinzel, et al., Sensitivity analysis of the impact of air pollutants on automotive fuel cells. Fuel, Cell. **16(4)**, 444-462 (2016).
3. Q. Zhang, H. Zhao, J. Rong, et al., Research progress on anti-toxic electrocatalysts for oxygen reduction reaction in proton exchange membrane fuel cells. Chem. Prog. **42(09)**, 4677-4691 (2023).
4. Q. Kang, G. Zhang, Z. Zhang, et al., Research progress on non-noble metal catalyst for proton exchange membrane fuel cells. Prog. New Energy. **6(01)**, 55-61 (2018).
5. Y. Zhou, D. Zhang, Nano PtCu binary and PtCuAg ternary alloy catalysts for oxygen reduction reaction in proton exchange membrane fuel cells. J. Power Sources **278**, 396-403 (2015).
6. B. Wang, J. Han, X. Yu, et al., Solvothermal synthesis of nickel cobalt oxide and its electrochemical properties. Nonf. Met. Sci. Eng. 1-13 (2014).
7. Z. Li, T. Shen, Y. Hu, et al., Progress in the application of ordered intermetallic compound electrocatalysts in fuel cells. J. Phys. Chem. **37(09)**, 187-208 (2021).
8. Y. Lyu, Y. Song, H. Liu, et al., Pt-based core-shell electrocatalysts with Pd in the core. J. Phys. Chem. **33(02)**, 283-294 (2017).
9. J. Liu, J. Bak, J. Roh, et al., Reconstructing the coordination environment of platinum single-atom active sites for boosting oxygen reduction reaction. ACS Cat. **11(1)**, 466-475 (2020).

10. M. Zhang, J. Wei, J. Ouyang, et al., Construction and application of Pt single-atom catalyst. Anal. Lab. **41(12)**, 1400-1410 (2022).
11. X. Li, Q. Feng, C. Li, et al., Research progress of Fe/N/C materials as cathode catalysts in proton exchange membrane fuel cells. Appl. Chem. **53(04)**, 991-995 (2024).
12. J. Li, Preparation and electrocatalytic properties of Fe-N-C based materials. Three Gorges University (2023).
13. N. Jia, Q. Weng, Y. Shi, et al., N-doped carbon nanocages: Bifunctional electrocatalysts for the oxygen reduction and evolution reactions. Nano Res. **11**, 1905-1916 (2018).
14. Y. Wang, L.Wang, H. Fu, Research progress of Fe-NC catalysts for the electrocatalytic oxygen reduction reaction. Sci. China Mater. **65(7)**, 1701-1722 (2022).
15. D. Zhao, G. Yin, J. Wei, Polymer membrane fuel cell cathode non-Pt catalyst. Chem. Prog. **21(12)**, 2753-2759 (2019).
16. H. Sun, J. Wang, Z. Zhu, et al., Porous organic polymer derived carbon-based electrocatalysts for oxygen reduction reactions in fuel cells. Chem. Prog. **35(11)**, 1638-1654 (2023).

Development trend of cathode materials commonly used in lithium batteries

Gongsebaimu*

Zhejiang University Chemistry Department, Zhejiang University, 310058 Hangzhou, Zhejiang, China

Abstract. With the profound scientific and economic advances, people's demand for electronic products is increasing daily. Lithium-ion batteries are broadly applied due to their significant advantages in weight, volume, and lifespan. The anode of a lithium-ion battery is the main component in the electronic reaction of lithium ions. The materials currently used for the anode of lithium-ion batteries are mainly $LiFePO_4$, ternary lithium batteries, and $LiMnO_2$. Sometimes lithium titanate and $LiCoO_2$ are also used. Different anode materials share their benefits and drawbacks regarding various aspects such as energy density, cost, cycle life, safety, and environmental effects. Material modification or surface coating of the positive electrode of lithium-ion batteries is currently a popular solution, which also has a significant influence on changing the market share of different positive electrode lithium-ion materials. At last, this paper discusses the modification schemes for different types of lithium-ion batteries in the future by reviewing the literature.

1 Introduction

As the scientific and economic advances are increasingly prosperous, people's demand for electronic products is increasing day by day. Lithium-ion batteries are broadly applied in a wide range of industries like consumer electronics, electric vehicles, renewable energy storage, medical equipment, drones, and robotics due to their significant advantages in terms of weight, size and service life. However, we currently still face various problems when using lithium-ion battery anode materials regarding energy density, cost, cycle life, safety and environmental influence. It is inevitable that lithium-ion battery technology will be improved in the future production development.

The basic structure of a lithium-ion battery comprises four main sections: the positive electrode material, the negative electrode material, the electrolyte and the separator. During the charging process, lithium ions in the positive electrode material are precipitated, entering the electrolyte, and passing through the separator to embed in the negative electrode. During the discharging process, lithium in the negative electrode material loses electrons, is converted into lithium ions, returns to the electrolyte and passes through the separator to return to the positive electrode [1-3]. This cyclic process ensures that lithium-ion batteries can be repeatedly used. Currently, the production technology of lithium-ion batteries has

* Corresponding author :3220106321@zju.edu.cn

https://doi.org/10.1051/e3sconf/202560602016

basically reached 300 Wh/kg. Under the development of various fields in the future, the energy amount of lithium-ion batteries may reach 500Wh/kg

In the 1990s, Japanese companies were the first to successfully produce lithium-ion batteries, and their rapid development followed. Over the past three decades, the application of lithium-ion batteries has gradually expanded, and they have become the main type of battery in consumer electronics, electric vehicles, renewable energy storage, medical equipment and other fields. In the future, with the global trend of the Internet of Everything, lithium-ion batteries will play an essential part in various fields.

At present, the main cathode materials for lithium-ion batteries on the market are mainly $LiFePO_4$, ternary lithium batteries, and $LiMnO_2$. Sometimes lithium titanate and $LiCoO_2$ are also used. The research on lithium-ion batteries significantly contributes to promoting the popularization of renewable energy storage, power interruption (solar energy storage), and the development of portable electronic devices. Moreover, its application in smart grids, energy storage systems, and microgrids has improved energy utilization efficiency and system stability.

2 Lithium iron phosphate material

Lithium iron phosphates ($LiFePO_4$) is the earliest material adopted in lithium-ion batteries and is usually called 'ferroelectric'. At present, the main methods for preparing $LiFePO_4$ anode materials include the high-temperature solid-phase method, the carbon thermal reduction method, and the hydrothermal synthesis method. LiFePO4 batteries are widely applied in solar and wind energy storage systems. They can provide stable power output and extend the cycle life, which is ideal for large-scale power storage needs. Their durability and safety make them ideal for home and industrial energy storage [7]. Moreover, $LiFePO_4$ is one of the battery materials chosen by many electric vehicle manufacturers. In 2022, the installed capacity of power batteries in China will be 302.3 GWh, of which the installed capacity of $LiFePO_4$ power batteries will be 184.5 GWh, accounting for 61%. A large number of popular models such as the entire range of BYD vehicles, the Model Y/3, the Xiaopeng P7/P5, the GAC Aion S/Y, etc. all use $LiFePO_4$ power batteries [6]. LiFePO4 batteries are also available for a wide range of portable electronic devices, whose safety and longevity make them widely used in specific portable devices. However, $LiFePO_4$ faces problems as a cathode material, like low energy density, poor low-temperature performance, poor electrical conductivity and bulky size. There are currently several solutions to these challenges.

First, the nano treatment of this material can extend specific surface areas, shorten the lithium-ion diffusion path, and effectively control both the shape and size of the $LiFePO_4$ particles, thereby improving conductivity and energy density. Second, doping modification, which involves introducing appropriate metal or non-metal elements into $LiFePO_4$, can improve the lithium-ion diffusion performance at low temperatures, enhance its low-temperature discharge capacity, and optimise the crystal structure to improve the lithium-ion diffusion rate and electronic conductivity. In addition, the addition of highly conductive additives can improve the overall conductivity of the electrode, or it can be compounded with high energy density materials to balance safety and energy density. In addition to optimizing the cathode material, the recycling of cathode materials also needs to be considered in order to achieve the goal of sustainable development. At present, direct regeneration has more advantages in terms of economy and environmental protection, but this technology is still in the preliminary research stage, while indirect regeneration is suitable for situations where the raw materials are highly complex or where high-value resource reserves are required[5].

Although the above solutions are highly feasible, nanoprocessing is difficult and costly to implement on a large scale. Currently, the market demand for LiFePO4 batteries is continuing to expand, providing market space and economic incentives for the industrial development of LiFePO4 battery recycling. However, the raw materials of LiFePO4 batteries are not highly valuable, and the economic benefits of recycling them are limited. The costs and benefits of the entire life cycle of LiFePO4 batteries need to be comprehensively considered in order to clarify the economics of industrialisation and enhance the driving force for sustainable development[5].

3 Ternary lithium battery material

Ternary lithium batteries refer to a series of oxides composed of nickel, cobalt and manganese, and the ratio of each element directly affects the performance of the battery. Studies have shown that increasing the nickel content significantly improves the energy density, indicating that nickel plays a key capacity maintenance part in ternary lithium-ion batteries [4]. As a ternary cathode material for lithium-ion batteries, NCM combines the high specific capacity of lithium nickelate ($LiNiO_2$), the profound thermal stability and the lower price of lithium manganese oxide ($LiMnO_2$), and the advantages of lithium cobalt oxide ($LiCoO_2$) with stable electrochemical properties, it has demonstrated excellent electrochemical performance. It has long been the main cathode material used in power batteries due to its high energy density [8,9]. However, its disadvantages should not be ignored, which mainly include high cost (scarce and expensive cobalt resources), low cycle life, and the risk of oxygen molecules being released during charging and discharging, which may lead to combustion or deflagration under high-temperature conditions [8]. Therefore, corresponding control and early warning measures must be taken when using ternary lithium-ion batteries to prevent extreme situations [2]. To solve these problems, high-nickel ternary cathode materials have been optimised in recent years by means of ion doping, surface coating, co-modification, concentration gradients, electrolyte modification and structural regulation.

Ion-doped cathode materials can expand the interlayer spacing of the material, increase the diffusion rate of Li, reduce the Li/Ni mixed row, and stabilize the lattice structure. Furthermore, ion doping forms higher energy TM-O bonds, which inhibit structural distortion and the generation of microcracks, thereby stabilising the material structure and improving the battery's lifespan. Coating the NCA cathode material with different substances hinders direct contact between the electrolyte and the material, inhibiting the erosion of the electrolyte and thereby improving the battery's electrochemical performance. Copolymerization combines the advantages of different modified components to more effectively enhance the electrochemical properties of the material. Modification of the microstructure of the concentration gradient can enhance the stability of the outer layer of the material, and keep the original electrochemical characteristics of the cathode material at the same time. Electrolyte modification can improve the chemical stability of the electrolyte to inhibit decomposition and form a reactive protective film on the cathode material surface, extending the cycle life of the battery. Structural regulation improves the electrochemical performance of high-nickel ternary cathode materials by adjusting the lattice structure and particle morphology, thereby enhancing mechanical stability and Li^+ diffusion performance [8]. The above optimisation solutions are currently mainly being studied in laboratories. In terms of industrialization, some of these technologies have matured and are highly feasible, promoting the future development of ternary lithium batteries.

4 Lithium manganese oxide material

In the domestic market and research field, the share of $LiMnO_2$ batteries compared to $LiFePO_4$ batteries and ternary lithium batteries is extremely limited, and their production and research and development are mainly concentrated in Japan. $LiMnO_2$ batteries have good safety performance, good low-temperature performance, resistance to overcharging or over-discharging, and are very inexpensive. They are seen as one of the most prosperous cathode materials, and their main application area is plug-in hybrid vehicles. It is also important to note that there are 36 pure electric buses with $LiMnO_2$ as the main application, accounting for as high as 17% [10]. However, there are still significant problems with its high-temperature performance and cycle stability.

Several improvement schemes have been proposed to address these deficiencies. First, by coating the surface of the material, a layer of inert elements or compounds can be coated on the surface of the lithium nickel manganese oxide material, which can effectively decrease the contacted area between the electrolyte and the lithium nickel manganese oxide, thereby reducing the occurrence of side reactions and protecting the active material. Alternatively, a layer of fast ion conductors can be coated on the surface of the lithium nickel manganese oxide material, which can protect the positive electrode material while enhancing the lithium-ion transport rate, thereby improving the electrochemical performance [11]. Secondly, replacing the nickel-manganese ions with cations with stronger chemical bonds replacing the oxygen ions with anions, or filling the oxygen defects can make the material structure more stable, so that it can withstand more stress without collapsing [11]. Finally, structural optimisation, such as nanotreatment or porous structure design, can improve battery performance. At present, these modification methods are already being used in production and laboratories, but the technical aspects of surface coating are still widely under-explored and need further studies.

5 Lithium cobalt oxide material

In addition to the three common lithium-ion batteries mentioned above, types such as LiCoO2 batteries are still reasonable in specific application fields, although they are rare in production and research. Due to their ease of processing, unseen volumetric energy density and high working potential, lithium cobalt oxide (LCO)-based battery materials are dominant in lithium-ion batteries (LIBs) based on 3C (computers, communications and consumer electronics) [12]. However, despite their superior energy density, their other performance indicators are still inferior to those of other types of batteries. Therefore, the main solutions currently focus on surface coating technology, and the following coating methods are commonly used. (1) Carbon itself is an electronic conductor material. Some foreign scholars used the low-temperature liquid phase method to coat carbon on the surface of Li_2CO_3 material and found that carbon can enhance the cycle rate as well as high-temperature storage performance of the material. (2) Ionic conductor coating: LATP is an excellent ionic conductor material. Coating it on Li2CO3 can improve the rate and cycle performance of the material at 4.5V. (3) Electronic-ionic double conductor coating: The electrode itself is a good electronic-ionic conductor. Existing research has confirmed that AlWxFy is an excellent electronic-ionic conductor. Coating it on Li_2CO_3 can make it exhibit good electrochemical properties at 4.5V. (4) Double electronic and ionic insulation coating: oxides such as Ti, Mg, Al and Zr are often used. The research found that mechanised coating of Li_2CO_3 material with TiO helps to enhance its stability at high voltages and significantly improves its chemical and electrical properties [13]. Surface coating using these methods can solve most problems, but due to the scarcity of cobalt in the raw material of $LiCoO_2$, $LiCoO_2$ materials still cannot become a relatively large proportion of lithium-ion materials on the market. Finding suitable metal ions and a suitable ratio to replace them is also the main development direction of $LiCoO_2$ materials at present.

6 Conclusion

The main battery types on the market today are $LiFePO_4$ batteries and ternary lithium batteries, while $LiMnO_2$ batteries and $LiCoO_2$ batteries are mainly used in specific fields. Different battery types have different advantages and are suitable for different applications. $LiFePO_4$ batteries can be applied in various fields such as electric vehicles (EVs) and grid energy storage that require high safety and long life and lower energy density, due to their profound performance in these aspects. Ternary lithium batteries with high energy density are ideal components for these high-performance electric vehicles and portable electronic devices, but are relatively expensive and have relatively poor thermal stability. $LiMnO_2$ is low-cost and safe, and is suitable for small and medium-sized devices with short-term use, such as electric bicycles and power tools. $LiCoO_2$ is widely used in consumer electronics like cellphones and laptops, owing to its high energy density, but it is expensive and relatively unsafe. There are some commonalities in the ways in which different materials enhance the cathode materials performance. The main technologies that have been put into production are surface coating, structural optimization and doping modification, but the materials used in the modification process are quite different. Some materials have not yet been mass-produced due to high technical barriers or costs.

In the future, if the relevant technical problems can be solved, the performance of various types of batteries will be significantly improved. At the same time, the doping modification solutions under research also have the potential to bring about significant changes in market share.

References

1. L. Wang, Analysis of the Development Trend of Positive Electrode Materials for Lithium-ion Batteries. J. Changji Univ. **04**, 118-122 (2020).
2. Y. Dong, Y. Liu, Y. Hu, et al., Research on the Reaction Dynamics and Reversibility Enhancement Mechanism of Lithium Storage Based on VS_2/MoS_2 Schottky Heterojunction. Sci. Bull. **65**, 1470-1478 (2020).
3. P. Zhu, Experimental Analysis of Charging and Discharging Characteristics of Lithium-ion Power Batteries. Sci. Technol. Innov. **19**, 24-25 (2020).
4. S. Liu, H. He, C. Chang, Discussing the Past, Present, and Future of Lithium Batteries from the 2019 Nobel Prize in Chemistry. J. Appl. Technol. **20**, 135-139 (2020).
5. Y. Wang, X. Zheng, D. Ruan, et al., Recycling of Lithium Iron Phosphate Batteries: From Basic Research to Industrialization. Chin. Eng. Sci. 1-14 (2022).
6. D. Liu, F. Xiao, The Current Status of Patent Technology for Lithium Iron Phosphate Power Batteries. China Sci. Technol. Inf. **17**, 15-17 (2024).
7. L.-X. Yuan, Z.-H. Wang, W.-X. Zhang, X.-L. Hu, J.-T. Chen, Y.-H. Huang, J.B. Goodenough, Energy Environ. Sci. **4**, 269-284 (2011).
8. L. Zhao, J. Wang, Y. Zhang, et al., Research Progress on Modification of High Nickel Ternary Cathode Materials for Lithium-ion Batteries. Funct. Mater. **55**, 9064-9070 (2024).
9. P. Li, W. Li, C. Xu, et al., Research Progress on Doping Modification of Ternary Positive Electrode Materials for Lithium-ion Batteries. J. Zhengzhou Univ. (Sci. Ed.) **56**, 80-87 (2024).
10. S. Zeng, Y. Chen, J. Tang, et al., Comparison of Electrochemical Performance Based on Ternary, Lithium Iron Phosphate, and Lithium Manganese Oxide Cathodes. J. Guangdong Norm. Univ. Technol. **44**, 53-58 (2023).

11. J. Yan, Preparation and Modification of Lithium Nickel Manganese Oxide Cathode Material. Central South Univ. (2023).
12. J. Gu, J. Ji, W. Zhou, et al., Research Progress on Lithium Cobalt Oxide as Cathode Material for Lithium-ion Batteries. Appl. Chem. 1-4 (2022).
13. J. Xie, On the Development of High Voltage Lithium Cobalt Oxide Cathode Materials. Technol. News **21**, 44-47 (2023).

Physical and Chemical Methods for Mitigating Carbon Dioxide

Yuhang Wu[1*]

[1]Department of Chemistry, University of Nottingham Ningbo China, Ningbo, 315100, China

Abstract. The greenhouse effect caused by greenhouse gases, especially carbon dioxide emissions, has become a central issue in global environmental governance. Traditional emission reduction strategies, although effective, are difficult to meet global climate goals. In response, the scientific community has begun to explore new technologies for capturing, separating, storing and utilizing carbon dioxide through physical and chemical means. These technologies offer potential avenues for greenhouse gas (GHG) management, but still face technical, economic, and environmental challenges before large-scale application. This study systematically evaluates the progress of the application of these physical and chemical means in carbon dioxide (CO_2) treatment, and discusses in depth the technical bottlenecks and future improvement directions by analyzing the advantages and disadvantages of each technology. It is hoped that this study provides a valuable theoretical basis and practical reference for the development and application of future large-scale CO_2 abatement technologies.

1 Introduction

With the increasing global climate change [1], greenhouse gas emissions, especially CO_2 emissions, have become the core issue of global environmental governance. CO_2 is the main factor leading to the greenhouse effect. The continuous rise of CO_2 concentration has a profound impact on the earth's climate system, resulting in a series of serious consequences [2]. Therefore, how to effectively slow down CO_2 emissions, or even remove them from the atmosphere, has become the focus of academia, industry and government policymakers.

Traditional CO_2 emission reduction strategies mainly focus on energy structure adjustment and improving energy efficiency. However, these methods have effectively reduced CO_2 emissions to a certain extent, more is needed to meet the global climate goals. To this end, in recent years, the scientific community has begun to explore a more direct and technical solution, namely the use of chemical and physical means of CO_2 capture, separation, storage and utilisation. These technologies offer a potential pathway to greenhouse gas governance by capturing CO_2 from point sources or the air and converting it into valuable chemicals or long-term storage in geological formations.*

* Corresponding author: shyyw20@nottingham.edu.cn

Although chemical and physical methods show great promise in the laboratory and small-scale applications, there are technical, economic, and environmental challenges before they can be adopted on a large scale. At present, most of the research focuses on the development of a single technology, and there needs to be more systematic evaluation of multi-technology integration and its practical application potential. In addition, the energy efficiency, stability and economy of CO_2 treatment technology need to be further optimised, and the solution to these problems is crucial to achieving large-scale CO_2 emission reduction in the future.

This study systematically evaluates the current progress in the application of physical and chemical methods in CO_2 treatment, and analyse its technical bottlenecks and potential directions for improvement. By combining physical and chemical methods, this paper attempts to provide theoretical support and a practical path for large-scale CO_2 emission reduction in the future. Through this research, we can promote the practical application of physical and chemical methods in coping with climate change and contribute new ideas and technical support to achieving global Sustainable development goals.

2 Physical method

Physical methods are a significantly important part of people's usual deal with CO_2. Common physical methods of CO_2 include physical adsorption, membrane separation, and cryogenic condensation. These methods exploit the physical properties of CO_2 to achieve its capture and separation through different mechanisms. The following is a detailed description of each method.

2.1 Physical adsorption

Among CO_2 capture technologies, adsorption has garnered significant attention due to its advantageous characteristics revealed in recent years. Physical adsorption is a process where gas molecules are adsorbed into the surface of solid materials through physical forces such as van der Waals forces and electrostatic attraction. This process involves weak intermolecular interactions without forming or breaking chemical bonds. Porous carbon-based materials have become one of the most versatile CO_2 adsorbents [3]. Porous carbon-based materials are a class of adsorbents characterised by their high surface area, tunable pore structures, and excellent adsorption properties. These materials include activated carbon, carbon nanotubes, and graphene. They can effectively capture and store CO_2 because of their high porosity and surface reactivity. The process of physisorption (CO_2 adheres to the adsorbent without reacting) is shown in Figure 1.

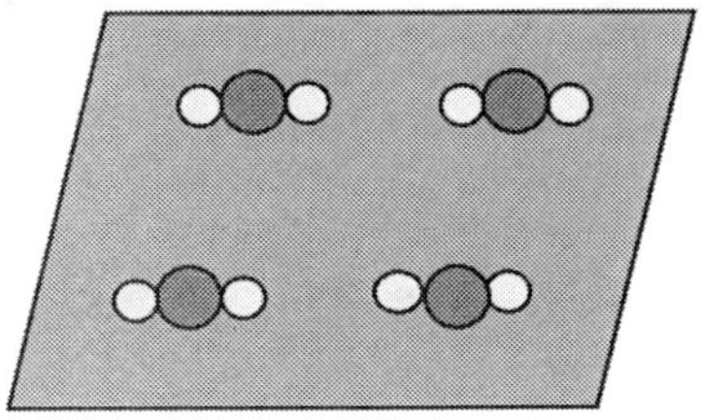

Fig. 1. Schematic diagram of CO_2 physical adsorption process [4].

2.2 Membrane separation

Membrane separation is a process that utilises selectively permeable membranes to separate components of a gas mixture based on their different permeation rates. This method is becoming more and more popular for capturing CO_2 because it is environmentally friendly, highly efficient and energy-saving. CO_2 molecules have the capability to penetrate more easily in specific membrane materials due to their lower molecular size and higher solubility. Membrane separation is divided into 5 categories: Polymeric membrane separation, inorganic membrane separation, mixed matrix membrane separation, composite membrane separation and liquid membrane separation. As a membrane in liquid Membrane Separation, the Water membrane can quickly and effectively absorb CO_2 [5]. What is more, water can not only be easily gotten but also be more stable than other membranes and eco-friendly. CO_2 membrane separation technology is important promoting clean energy production.

2.3 Cryogenic condensation

The cryogenic condensation method is a technology that condenses CO_2 from gas to liquid by reducing the temperature so as to achieve gas separation and capture. When the temperature of CO_2 gas is lowered below its condensation point, the CO_2 gas molecules condense into a liquid state, which can be separated from the gas mixture. Because CO_2 has a relatively high condensation point, condensing and separating at lower temperatures is easy. Using this principle, CO_2 can be extracted from a variety of mixed gases. On the one hand, it can be used in commercial, industrial, medical, and other fields, such as carbonated beverages, dry ice, and other uses. Moreover, it can reduce CO_2 in the atmosphere and reduce the greenhouse effect.

3 Chemical method

Chemical methods play a vital role in reducing CO_2. These methods are not only efficient and applicable, but also convert CO_2 into valuable products, driving resource recycling and sustainable development. On a global scale, using chemical methods to capture, fix, and convert CO_2 has become an important strategy to combat climate change and the greenhouse effect. The following will discuss several major chemical methods, including biochemical methods, solvent absorption, carbonation reaction, and catalytic conversion, and analyze their principles, applications, advantages, and challenges.

3.1 Physical adsorption

Biochemical Methods use microorganisms or enzymes to catalyze the fixation and conversion of CO_2 to convert CO_2 into valuable chemicals or biomass through biochemical reactions. These methods rely primarily on the ability of organisms to fix and transform CO_2 in natural processes. Biochemical methods are mainly divided into two methods: photosynthetic organisms and microbial electrochemical systems.

3.1.1 Photosynthetic organisms

Photosynthetic organisms fix CO_2 through photosynthesis. The main process of photosynthesis is (Eq.1):

$$6CO_2 + 6H_2O + \text{lignt energy} \rightarrow C_6H_{12}O_6 + 6O_2 \tag{1}$$

Photosynthesis is divided into two stages: the light reaction and the dark reaction. In the light reaction, light is absorbed by chlorophyll and converted into chemical energy. In the dark reaction, CO_2 is fixed and converted into glucose through a series of reactions. According to this principle, CO_2 can be effectively converted into glucose and oxygen, which not only reduces CO_2 in the atmosphere, but also makes it a more useful substance. After learning the principle of photosynthetic organisms, here are two practical parts of that principle.

First, biological sequestration of CO_2 is used. Algae's high growth rate and ability to adapt to a variety of environments make them ideal for bioenergy production and atmospheric CO_2 emission reduction. Microalgae are highly efficient in solar energy conversion and biomass production. Certain species exhibit high CO_2 fixation rates, making them very effective at reducing atmospheric CO_2 concentrations. These microorganisms can trap carbon from the atmosphere and flue gas. The fixed CO_2 is converted into biomass and can be used for a variety of purposes, including biofuel production. The efficiency of microalgae in photosynthesis ensures that a large amount of CO_2 is removed from the atmosphere. What is more, seaweed or Marine macroalgae also play an important role in CO_2 sequestration. They participate in the carbon cycle by converting CO_2 into organic matter through photosynthesis. Schemes like the Ocean Macroalgae Afforestation (OMA) aim to use seaweed farming to offset man-made CO_2 emissions. Seaweed farming in coastal areas enhances CO_2 capture. This process involves the growth of seaweed, which absorbs CO_2 and thus reduces the CO_2 concentration in the atmosphere. The harvested seaweed biomass can be further processed into biofuels, and the resulting biomass can be used for bioenergy, providing a sustainable way to manage carbon and thus achieve negative carbon emissions [6].

Second, Microalgae bioreactors can be used. An algal bioreactor is a device for cultivating microalgae on a large scale, using the photosynthesis of microalgae to fix and convert CO_2. These reactors maximize the biomass yield and CO_2 capture efficiency of microalgae by optimizing growth conditions such as light, temperature, and nutrient supply [7].

3.1.2 Microbial electrochemical systems

Microbial Electrochemical Systems (MES) use electroactive microorganisms to catalyze the reduction of CO_2 on electrodes [8]. Electrons can be transferred from the electrode to CO_2 through the metabolic activity of these specific microbes, facilitating their conversion into valuable chemicals such as formic acid, carbon monoxide, or acetic acid. The key reactions typically include:

$$CO_2 + 2H^+ + 2e^- \rightarrow CO + H_2O \tag{2}$$

Or

$$CO_2 + 4H^+ + 4e^- \rightarrow HCOOH \tag{3}$$

The electrons and protons produced by the anode microorganisms oxidize the organic substrate, and the electrons are transmitted to the cathode through an external circuit, where they are combined with CO_2 under the action of electroactive microorganisms to achieve efficient CO_2 reduction. MES can not only effectively reduce greenhouse gas emissions, but also capture and reduce CO_2 while treating organic wastewater, which has significant environmental and economic benefits. The system, combined with renewable energy, provides a low energy consumption and efficient CO_2 reduction solution. However, further research is needed regarding microbial activity optimization, electrode material durability, and large-scale applications.

3.2 Solvent adsorption

The solvent absorption method has been widely studied and applied due to its mature and efficient CO_2 trapping technology. The solvent absorption method is a chemical reaction between chemical solvents and CO_2: the CO_2 in the gas is absorbed and fixed, and then the pure CO_2 is released through the regeneration process. The method uses the chemical properties of the solvent to form reversible compounds with CO_2 under specific conditions, thereby achieving the capture and release of CO_2. The following is the basic principle and process of solvent absorption.

Solvent absorption primarily relies on the chemical reaction between the solvent and CO_2. The most common solvents are amine-based solvents, such as monoethanolamine (MEA), diethanolamine (DEA), and methyldiethanolamine (MDEA) [9]. These amine-based solvents react with CO_2 through the following reactions:

MEA and CO_2 reaction:

$$RNH_2 + CO_2 + H_2O \rightarrow RNH_3^+ + HCO_3^- \quad (4)$$

Where R stands for hydroxyethyl

DEA and CO_2 reaction:

$$R_2NH + CO_2 + H_2O \rightarrow R_2NH_2^+ + HCO_3^- \quad (5)$$

These reactions produce bicarbonate (HCO_3^-) and amine salts (RNH_3^+). During the regeneration process, these reactions are reversible. By heating or reducing the pressure, the absorbed CO_2 can be released, and the solvent is regenerated.

A schematic diagram of an absorber/stripper system for separating carbon dioxide from exhaust gases is shown in Figure 2. The absorption process usually takes place in an absorption tower. Gas containing CO_2 enters from the bottom of the absorption tower and comes into countercurrent contact with the solvent sprayed down from the top. In this process, CO_2 is absorbed by the solvent, and a chemical reaction occurs. The process can be divided into the following steps: First, the gas enters the absorption tower. The exhaust gas containing a high concentration of CO_2 is introduced from the bottom of the absorption tower and is in contact with the solvent through the packing layer or plate tower. The second step is a solvent spray. Absorb the spray solvent at the top of the tower, make it flow down the tower wall, and countercurrent contact with the rising gas. The solvent can be an amine solvent (MEA, DEA, or MDEA) or other solvents with absorptive capacity. Then, there is the absorption reaction. CO_2 molecules react chemically with solvent molecules at the gas-liquid interface to form stable compounds such as bicarbonate and amine salt. Finally, the purification gas is discharged. The purified gas is discharged from the top of the absorption tower, and the CO_2 concentration is significantly reduced to achieve CO_2 capture and reduction.

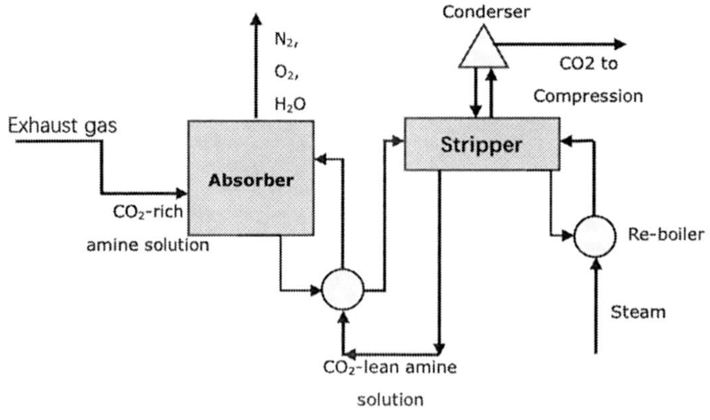

Fig. 2. Simplified PFD of an absorber/stripper system for CO_2 separation from exhaust gas [10].

The solvent absorption method has several advantages over other CO_2 capture technologies. First, the absorption efficiency of the solvent absorption method is high, and a large amount of CO_2 can be captured, especially for high-concentration CO_2 emission sources. Secondly, the method has good selectivity and can effectively distinguish and preferentially absorb CO_2 without interference from other gases. In addition, the solvent absorption technology is mature, proven by years of industrial practice, with reliable operating performance and controllable operating parameters. Therefore, the solvent absorption method is widely used in large industrial facilities such as power plants and chemical plants and has become one of the key technologies for CO_2 emission reduction.

3.3 Carbonation reaction

Carbonation reaction, as an emerging CO_2 capture and fixation technology, is a process that converts CO_2 into stable carbonates by using the chemical reaction between CO_2 and alkaline minerals or industrial wastes. This process not only enables the permanent fixation of CO_2, but also converts industrial by-products into valuable materials. The following is the basic principle of carbonation reaction.

The carbonation reaction consists of three basic steps:

The first step is the dissolution of CO_2: in the presence of water, CO_2 is first dissolved in water to form carbonic acid (H_2CO_3). This process can be expressed as:

$$CO_2 + H_2O \rightarrow H_2CO_3 \tag{6}$$

The second step is the dissociation of carbonic acid: Carbonic acid dissociates in an aqueous solution into hydrogen ions (H^+) and bicarbonate ions (HCO_3^-), and further dissociates into hydrogen ions and carbonate ions (CO_3^{2-}). These dissociation reactions can be expressed as:

$$H_2CO_3 \rightarrow H^+ + HCO_3^- \tag{7}$$
$$HCO_3^- \rightarrow H^+ + CO_3^{2-} \tag{8}$$

The third is the precipitation of carbonate ions: the reaction of carbonate ions with metal cations (e.g. Ca^2, Mg^{2+}) produces an insoluble carbonate precipitate. Taking the formation of calcium carbonate as an example, the reaction can be expressed as:

$$Ca^{2+} + CO_3^{2-} \rightarrow CaCO_3 \downarrow \tag{9}$$

CO_2 reacts directly with metal oxides to form carbonates in an anhydrous environment. This direct reaction usually occurs at higher temperatures and sometimes requires a catalyst to increase the reaction rate. Taking magnesium oxide (MgO) as an example, the reaction is as follows:

$$MgO + CO_2 \rightarrow MgCO_3 \tag{10}$$

The energy change in this reaction is negative (i.e., exothermic) because the carbonate formation is thermodynamically stable. This means that the resulting carbonate is difficult to reverse back into CO_2 and metal oxides under natural conditions so that the CO_2 can be fixed for a long time [11].

The common carbonation is mineral carbonation. Mineral carbonation is a process in which CO_2 reacts with minerals containing metal ions such as magnesium and calcium (such as serpentine and olivine) to form stable carbonates. This process is an exothermic reaction and can occur naturally or through accelerated reactions in industrial Settings. Mineral carbonation is a long-term form of CO_2 sequestration because the resulting carbonate is very stable under environmental conditions and does not decompose easily, which can effectively fix CO_2 and prevent it from re-entering the atmosphere. This method can not only utilize the abundant mineral resources in nature but also achieve the permanent storage of CO_2 by treating industrial by-products. Mineral carbonation technology is considered one of the safest and most feasible methods of carbondioxide capture and storage and is particularly suitable for implementation in areas rich in these minerals, such as the northern regions of Pakistan [12].

3.4 Catalytic conversion

Catalytic Conversion refers to the process of using catalysts to accelerate chemical reactions. A catalyst is a substance that increases the rate of a chemical reaction without being consumed by itself. In the catalytic conversion process, the catalyst enables the reactant to be converted to the target product more quickly and efficiently by reducing the activation energy of the reaction [13]. The principle of catalytic conversion in the treatment of CO_2 mainly involves accelerating a chemical reaction through a catalyst, thereby efficiently converting CO_2 into other chemicals or fuels. Here are the main principles:

3.4.1 Catalytic hydrogenation reaction

In catalytic hydrogenation reactions, the catalyst is usually a metal (such as nickel, copper, iron [14], etc.) whose role is to reduce the activation energy of the reaction, causing CO_2 to react with hydrogen (H_2) to form chemicals such as methane (CH_4) and methanol (CH_3OH). The reaction formula is as follows:

$$CO_2 + 4H_2 \rightarrow CH_4 + 2H_2O \tag{11}$$
$$CO_2 + 3H_2 \rightarrow CH_3OH + H_2O \tag{12}$$

These reactions are usually carried out under high temperature and pressure conditions, and the catalyst can increase the reaction rate and the product's selectivity.

3.4.2 Photocatalytic reduction reaction

Photocatalytic reduction uses a semiconductor material (such as titanium dioxide TiO_2 [15]) as a catalyst to excite electrons to jump from the valence band to the conduction band under light, forming an electron-hole pair. Electrons can reduce CO_2 to form carbon monoxide (CO) or methanol (CH_3), while holes are decomposed by water to produce O_2. The reaction process can be simplified as:

$$CO_2 + 2H_2O + light\ energy \rightarrow CH_3OH + O_2 \tag{13}$$

3.4.3 Dry Reforming of Methane （DRM）

In the DRM process, CO_2 and CH_4 react in the presence of a catalyst to produce syngas, a mixture of H_2 and CO [16]. The chemical reaction can be represented as:

$$CH_4 + CO_2 \rightarrow 2H_2 + 2CO \tag{14}$$

Syngas can be further used to produce liquid fuels, chemicals, and other energy products. This process not only converts greenhouse gases, but also provides raw materials for industrial production.

4 Conclusions

This study reveals the advantages and shortcomings of physical and chemical means in CO_2 treatment by systematically evaluating the application of these technologies. Physical methods such as physical adsorption, membrane separation and low-temperature condensation show good application prospects and are suitable for CO_2 capture and separation, while chemical methods such as biochemical methods, solvent absorption, carbonation reactions and catalytic conversion can convert CO_2 into valuable products, thus promoting resource recycling. Although these technologies have shown great potential in laboratory and small-scale applications, they still need to be further optimized in terms of energy efficiency, stability and economics. Future research should focus on the systematic assessment and practical application potential of multi-technology integration in order to realize large-scale CO_2 emission reduction and provide theoretical support and practical pathways for addressing climate change.

References

1. R. Lindsey, L. Dahlman. Climate Change: Global Temperature. NOAA Climate.gov. **116**, 058303 (2024)
2. K.C. Nadeau, I. Agache, M. Jutel, et al. Climate change: A call to action for the United Nations. Allergy. **77**(4), 1087-1090 (2022)
3. O.H.P. Gunawardene, C.A. Gunathilake, K. Vikrant, et al. Carbon Dioxide Capture through Physical and Chemical Adsorption Using Porous Carbon Materials: A Review. Atmosphere. **13**(3), 397 (2022)
4. Patel, H.A., Byun, J., Yavez, C.T. Carbon dioxide capture adsorbents: Chemistry and Methods. *ChemSusChem*. **10**, 1303–1317 (2017).
5. W. Jung, J.S. Lee, H. Yoon, et al. Water membrane for carbon dioxide separation. Separation and Purification Technology. **210**, 540-548 (2018)
6. D. Moreira, J.C.M. Pires. Atmospheric CO_2 capture by algae: Negative carbon dioxide emission path. Bioresource Technology. **215**, 371-379 (2016).
7. B. Wang, Y. Li, N. Wu, et al. CO_2 bio-mitigation using microalgae. Applied Microbiology and Biotechnology. **79**(5), 707-718 (2008)
8. B.E. Logan, K. Rabaey. Conversion of Wastes into Bioelectricity and Chemicals by Using Microbial Electrochemical Technologies. Science. **337**, 686-690 (2012)
9. A. Cachaza, D. Gómez-Díaz, A. Montáns, et al. Carbon dioxide chemical absorption by solvents based on diamine and amines blend. AIChE Journal. **64**(7), 2702-2710 (2018)
10. H.M. Kvamsdal, J.P. Jakobsen, K.A. Hoff. Dynamic modeling and simulation of a CO_2 absorber column for post-combustion CO_2 capture. Chemical Engineering and Processing: Process Intensification. **48**(1), 135-144 (2009)
11. S.J. Gerdemann, W.K. O'Connor, D.C. Dahlin. Carbon dioxide sequestration by aqueous mineral carbonation of magnesium silicate minerals. Environmental Science & Technology. **41**(7), 2587-2593 (2007)
12. M.I. Rashid, Z. Yaqoob, M.A. Mujtaba, et al. Carbon capture, utilization and storage opportunities to mitigate greenhouse gases. Heliyon. **10**(3), e25419 (2024)

13. B. Taleb, R. Jahjah, D. Cornu, et al. Exploring Hydrogen Sources in Catalytic Transfer Hydrogenation: A Review of Unsaturated Compound Reduction. *Molecules*. **28**(22), 7541 (2023)
14. H. Tang, T. Qiu, X. Wang, et al. A Brief Review of Recent Theoretical Advances in Fe-Based Catalysts for CO_2 Hydrogenation. Molecules. **29**(6), 1194 (2024)
15. A. Bratovčić, V. Tomašić. Design and Development of Photocatalytic Systems for Reduction of CO_2 into Valuable Chemicals and Fuels. Processes. **11**(5), 1433 (2023)
16. A.G.S. Hussien, K. Polychronopoulou. A Review on the Different Aspects and Challenges of the Dry Reforming of Methane (DRM) Reaction. Nanomaterials. **12**(19), 3400 (2022)

https://doi.org/10.1051/e3sconf/202560603002

Research progress on the source, adsorption treatment and conversion technology of CO2

Yingfei Wang[1]*
[1]School of Chemistry and Chemical Engineering, China University of Petroleum (East China), Qingdao, 266580, China

Abstract. As an important greenhouse gas, the increase in emissions of carbon dioxide (CO_2) has become one of the main drivers of global climate change. This paper systematically discusses the source, adsorption treatment method and conversion technology of CO_2. First of all, the sources of CO_2 are divided into two categories: natural processes and human activities. Among them, CO_2 generated by human activities is the main cause of environmental problems such as global warming. Subsequently, the article introduces the adsorption method as an important means of CO_2 treatment, covering the adsorption principle, common adsorbent types, adsorption process and adsorption regeneration methods. Finally, this paper focuses on the conversion technology of CO_2, including thermal catalytic conversion, photocatalytic conversion and electrocatalytic conversion. Each method has unique advantages and challenges, and provides different ideas and solutions for the effective conversion and utilization of CO_2.

1 Introduction

With the acceleration of global industrialization and the continuous growth of energy demand, the problem of CO_2 emissions is becoming more and more serious. As a major greenhouse gas, excessive CO_2 emissions not only lead to global warming, but also cause a series of environmental problems and ecological disasters. Therefore, finding efficient and sustainable CO_2 conversion technology and converting CO_2 into valuable chemicals or fuels has become a hot topic for global researchers. This paper aims to explore the way of catalytic conversion of CO_2, and to provide new ideas and methods for the efficient conversion and utilization of CO_2.

2 Sources of CO_2

CO_2 comes from a variety of sources, covering two aspects: natural processes and human activities.

2.1 Natural processes

* Corresponding author: jicong@ldy.edu.rs

There are a wide range of sources of CO_2 in nature. Animal and plant activity is the main cause of CO_2. The respiration process of animals and plants produces CO_2. The respiration of underground microorganisms will release CO_2, which may dissolve in groundwater and penetrate into the surface or atmosphere. Organic matter such as animal and plant carcasses, feces and humic acid will release CO_2 under the decomposition of microorganisms. In addition, some geological activities also produce CO_2. Natural phenomena such as volcanic eruptions will release a large amount of CO_2 into the atmosphere [1].

2.2 Human activities

CO_2 generated by human activities is the root cause of some environmental problems such as global warming [2]. In the fields of electric power production, industrial manufacturing, transportation and residential life, the combustion of fossil fuels such as coal, oil and natural gas produces a large amount of CO_2. In addition to the direct combustion of fossil fuels, some industrial production processes (such as cement manufacturing, steel smelting, etc.) also release a large amount of CO_2. In addition, deforestation, agricultural activities (such as the oxidation of methane from rice cultivation to CO_2) and urbanization will change the way of land use, thus affecting the emission and absorption of CO_2.

From the perspective of human behavior, the main sources of CO_2 emissions can be further refined into the following sectors. The first is the power and thermal energy production sectors, whose total CO_2 emissions account for a large proportion of global greenhouse gas emissions, mainly due to the combustion of fossil fuels. Secondly, in various industrial manufacturing processes, the fuel combustion of automobiles, aircraft, ships and other vehicles is an important source of CO_2 emissions in the transportation sector. The CO_2 emissions in the construction industry mainly come from the production of building materials, energy consumption in the process of building operation and the treatment of construction waste [3].

3 CO_2 adsorption treatment

3.1 Adsorption principle

The adsorption method selectively absorbs CO_2 molecules to the surface of another material through weak van der Waals force (physical adsorption) or strong covalent bonding force (chemical adsorption), thus achieving the purpose of enriching CO_2. Physical adsorption is mainly based on the physical interaction between the adsorbent and CO_2 molecules, such as van der Waals force, while chemical adsorption involves the chemical reaction between the adsorbent and the CO_2 molecule to form a chemical bond.

3.2 Type of adsorbent

Adsorbents are the core of adsorption technology. Common adsorbents include activated carbon, molecular sieve, hydrotalc, cage hydrate, metal organic skeletons (MOFs), etc. These adsorbents have different physical and chemical properties and are suitable for different adsorption scenarios. For example, activated carbon is widely used for the adsorption of CO_2 due to its large specific surface area and microporous structure; molecular sieve is favored for its specific pore size and shape selectivity [4].

3.3 Adsorption process

The adsorption process can be carried out in many ways, the two most common of which are the filling bed and the fluidized bed. In the filling bed, the adsorbent is loaded into the tower, and the gas mixture flows through the gap between the fixed adsorbent particles [5]. In the fluidized bed, the gas flows at a higher speed, causing the adsorbent particles to be suspended in the airflow. As the adsorption process progresses, the adsorbent gradually saturates. At this time, regenerative measures need to be taken to release the adsorbed CO_2 and restore the adsorption capacity of the adsorbent [6].

3.4 Adsorbent regeneration method

According to different regeneration methods, the adsorption method is often divided into variable pressure adsorption (PSA), variable vacuum adsorption (VSA) and variable temperature adsorption (TSA). PSA refers to the regeneration of adsorbents by changing the pressure of the system. In the process of adsorption, the system maintains a high pressure to promote the adsorption of CO_2. While in the process of regeneration, the pressure of the system is reduced, so that the adsorbed CO_2 is detached and released. VSA is similar to PSA, but it is usually vacuumed to further reduce the system pressure, thus improving the regeneration efficiency. TSA is usually regenerated by changing the temperature of the adsorbent. During the adsorption process, the adsorbent is kept at a low temperature to adsorb CO_2. In the process of regeneration, the adsorbent is heated to a higher temperature, so that the adsorbed CO_2 is detached and released [7].

4 CO_2 conversion

CO_2 conversion is an important research field. It involves a variety of methods and approaches aimed at transforming CO_2, a greenhouse gas, into a valuable chemical or energy source. The following is a detailed introduction of the main methods of CO_2 conversion.

4.1Thermal catalytic conversion

The thermal catalytic conversion of CO_2 is an important chemical process. This method uses thermal energy to drive the catalyst to promote the chemical reaction between CO_2 and reactants (such as hydrogen) to generate valuable chemicals or fuels.

The thermal catalytic conversion method activates CO_2 with thermal energy, and converts it into CO, CH_4, methanol and other high-value-added products under the conditions of the presence of hydrogen sources or other reducing agents, which is easier to achieve. Typical technologies include CO_2-CH_4 reforming reaction and thermal hydrogenation and reduction of CO_2. CO_2-CH_4 reforming syngas is a typical CO_2 utilization technology, which makes CH_4 and CO_2 CO and H_2 under the action of high-temperature systems, water vapor, oxygen and catalysts. Syngas can be used to produce various products, for example, the production of advanced alkanes and oxygen compounds through the Feto synthesis method. The method is simple and low-cost, but it requires an operating temperature of 800 ~ 1,000 °C to achieve the high-equilibrium conversion of methane and CO_2 to H_2 and CO, and carbon deposition will cause catalyst inactivation [8]. Thermal hydrogenation and reduction of CO_2 also requires the activation of CO_2 in a high-temperature environment. At present, the research mainly focuses on the selection and utilization of high-efficiency catalysts. The catalysts of Fe, Co, Cu, Zn and other metals are loaded on carriers with large specific surface area and high thermal stability. They are used for CO_2 hydrogenation to produce low-carbon olefins [9], and the

 https://doi.org/10.1051/e3sconf/202560603002

preparation process is continuously improved. Cu-based catalysts and Pd, Pt, Au and other precious metal catalysts are widely used in CO_2 addition. Hydrogen to methanol [10], methanol is more than 80% selective. At present, CO_2 hydrogenation to methanol technology has entered the stage of industrialization.

4.1.1 Photocatalytic transformation

CO_2 photocatalytic conversion is a technology that uses photocatalysts to promote the conversion of CO_2 into other valuable chemicals under light conditions. This technology has the advantages of green environmental protection and sustainability, so it has received widespread attention and research.

The photocatalytic conversion method uses photosensitive semiconductors to generate electrons and holes under the action of light, and electrons reduce CO_2 to CO, formic acid and other products, which is also known as artificial photosynthesis. The study found that finding high-efficiency photocatalysts is the key to strengthening artificial photosynthesis.

The basic principle of semiconductor photocatalytic reduction CO_2 is shown in Figure 1 [11]. As can be seen from Figure 1, (1) under the irradiation of a certain wavelength of light ($h\nu \geq Eg$), the electrons on the semiconductor catalytic material price band (VB) are excited to jump to the conduction band (CB), producing photoelectron-hole pair. (2) Photoelectron-hole separation and migration to the surface of the photocatalytic material. At the same time, photoelectron-holes may be compounded in the body center or on the surface. (3) Photoelectrons and holes migrate to the surface to participate in the redox reaction respectively, where CO_2 reacts with photoelectrons and H^+ in water to form organic compounds ($CO_2+H^+ +e^- \rightarrow$ organic compounds Matter), the photohole on the surface of the photocatalytic material reacts with water to generate $\cdot OH$ and H^+, that is: $H^+ + H_2O \rightarrow \cdot OH+H^+$, thus completing the photocatalytic reduction CO_2 reaction. In the reduction process, the reaction conditions and catalytic materials are different, and the reduction products obtained are also different, such as HCOOH, CO, HCHO, CH_3OH and other hydrocarbons. The reaction formula and reaction electrode potential are as follows:

$$2H_2O+4H^+ \rightarrow O_2+4H^+ \quad +1.23\ eV \tag{1}$$
$$2H^++2e^- \rightarrow H_2 \quad 0\ eV \tag{2}$$
$$CO_2+e- \rightarrow CO_2^- \quad -1.49\ eV \tag{3}$$
$$CO_2+2H^++2e^- \rightarrow HCOOH \quad -0.19\ eV \tag{4}$$
$$CO_2+2H^+ +2e^- \rightarrow CO+H_2O \quad -0.19\ eV \tag{5}$$
$$CO_2+4H^+ +4e^- \rightarrow HCHO+H_2O \quad -0.06\ eV \tag{6}$$
$$CO_2+6H^+ +6e^- \rightarrow CH_3OH+H_2O \quad +0.03\ eV \tag{7}$$
$$CO_2+8H^+ +8e^- \rightarrow CH_4+H_2O \quad +0.18\ eV \tag{8}$$

From the above principle, it can be found that photocatalytic reduction of CO_2 needs to meet two basic conditions: first, the photon energy must be greater than or equal to the band gap. Second, the conductive band potential is more negative than the surface electron receptor potential, and the valence band potential is more correct than the surface electron body potential. In this way, the reaction process of photocatalytic reduction of CO_2 can be realized [11]. Compared with the thermal catalytic conversion method, photocatalytic reduction CO_2 can be carried out at room temperature and pressure, which obviously has a wider application prospect. However, the cheap and efficient photocatalyst has not been developed. At present, there are still problems of low light utilization efficiency and low yield [12].

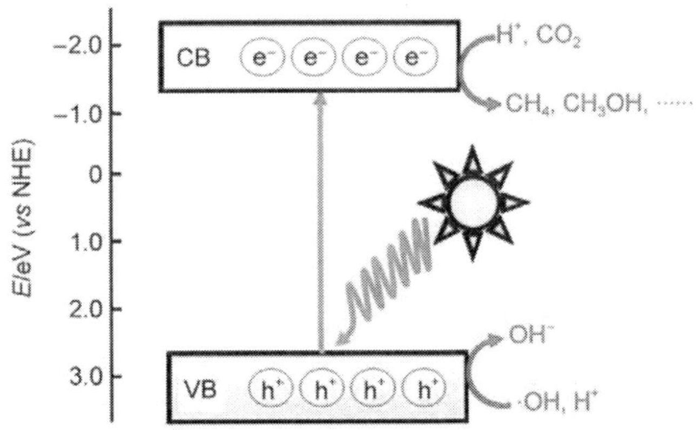

Fig. 1. The basic principle of semiconductor photocatalytic reduction CO_2.

4.1.2 Electrocatalysis

The electrocatalytic conversion method converts CO_2 into a high-value-added product through the potential difference between the two electrodes. This method has low requirements for the reaction medium (gas and liquid phase conditions can occur). It can be carried out at room temperature and pressure. The products are rich, and the selective synthesis of products can be achieved by controlling the application of electrocatalytic conditions such as potential and electrode materials, which makes the electrocatalytic reduction of CO_2 become the forefront of today's CO_2 transformation research.

Electrocatalytic CO_2 reduction is the process of reducing electrons obtained from the surface of the electrode by CO_2 molecules or CO_2 solventized ions in solution. From the perspective of thermodynamics, CO_2 conversion will undergo different multi-electron transfer reactions at different potentials. Water or solvents (electrocatalytic processes in non-water environments) also undergo redox reactions on the electrode surface, providing protons for the CO_2 reduction process. At present, the main products of reported CO_2 electrochemical reduction include carbon monoxide (CO), formic acid (HCOOH), methane (CH4), ethylene (CH_2CH_2), ethanol (CH_3CH_2OH) [13], methanol (CH_3OH) [14], formaldehyde (HCHO) [15], acetate (CH_3COO^-) [16] and other products. The electrocatalytic CO_2 reduction in aqueous solution is generally accompanied by the hydrogenation reaction of the side reaction. At present, the catalyst for CO_2 electrochemical reduction reaction has expanded from traditional metals and metal oxides to a wide range of nanomaterials. These new nanomaterial electrocatalysts show a unique ability to adsorb and activate CO_2, and promote CO_2 electrochemical reduction at a high rate under the condition of low energy supply (low over potential).

Electrode materials, catalysts, electrolytes, etc. of electrocatalytic conversion will affect the selectivity of reaction products. If it is to be widely used in CO_2 conversion, it is necessary to develop efficient electrical catalysts to reduce the overelectric potential of the reaction and improve the efficiency of electric energy utilization. Otherwise, excessive power consumption will not be worth the loss.

5 Conclusion

The treatment and utilization of CO_2 has become a key issue in dealing with global climate change. Through the systematic analysis of the source of CO_2, adsorption treatment technology and its conversion methods, this paper emphasizes the importance of multi-channel response to CO_2 emissions. As an effective means of CO_2 capture, the adsorption method shows a variety of application scenarios. Activated carbon, molecular sieve and other adsorbents show excellent adsorption performance under specific conditions. However, the regeneration process of adsorbents still has problems such as high energy consumption and low efficiency. In the future, research should be committed to developing more efficient and renewable adsorption materials and optimizing the regeneration process to improve the overall processing efficiency. The conversion technology of CO_2 provides new possibilities for its resource utilization. The thermal catalytic conversion method can react CO_2 with reducing agents such as hydrogen to produce high-value-added products such as syngas or methanol, but it is limited by high-temperature operation and catalyst inactivation. Photocatalysis and electrocatalytic conversion technologies are carried out at room temperature and pressure, with broader application prospects. However, at present, the light utilization efficiency of the photocatalyst is low, the yield is not high, and the power utilization efficiency of the electrocatalytic process also needs to be improved.

Generally speaking, the effective treatment and transformation of CO_2 requires the coordinated development of multiple technologies. Future research should focus on the following aspects: developing efficient and low-cost adsorbents and catalysts; improving the energy utilization efficiency of the conversion process; and combining a variety of technologies to realize the integration of CO_2 capture and conversion. Through these efforts, we can achieve the efficient use of CO_2 resources while reducing CO_2 emissions and help achieve the global sustainable development goals.

Reference

1. L. Q. Xu, L. J. Zhu, R. D. Yang, et al. Mineralization of CO_2 from different sources and the way it occurs. Guizhou Geology. **27(01),** 22-26, (2010)
2. B. Q. Lin, L. J. Jiang. Environmental Kuznets curve prediction of carbon dioxide in China and analysis of influencing factors. Management World, **04,** 27-36, (2009)
3. A. Ning. Characterization of Atmospheric CO_2 Column Concentration and Anthropogenic Emission Distribution in China and Pakistan. Nanjing University of Information Engineering, China. (2023)
4. W. Z. Tan, Q. F. Song, X. Q. Chen, et al. Advances in carbon dioxide solid adsorbent materials. Colloids and Polymers, **38(02),** 90-94 (2020)
5. M. P. Sheng. Study on carbon dioxide removal by rotary filled bed enhanced organic amine absorber and its mass transfer process. Beijing University of Chemical Technology, Beijing, China. (2019)
6. Y. F. Han. Numerical simulation study of CO_2 adsorption on fluidized bed based on shrinkage kernel mode, Xiangtan University. (2021)
7. Z. G. Xie. Experimental study on the performance of adsorbent for CO_2 removal device for space station and development of principle prototype. Lanzhou University of Science and Technology. (2009)
8. D. Pakhare， A. Spivey. A review of dry (CO_2) reforming of methane over noble metal catalysts. Chemical Society Reviews, **22,** 7813- 7837 (2014)
9. J. G. Shi, Z. J. Liu, C. S. Liu. Progress of carbon dioxide hydrogenation to low-carbon olefins. Chinese and Foreign Energy. **2,** 59-73 (2019)
10. Y. Zhang. Study on the Performance of Copper-Based Catalysts for CO_2 Hydrotreating and Methanol Synthesis Using Acetylacetonate as Precursor. Kunming University of Science and Technology. 11-12 (2018)

11. B. Y. Lin, H. F. Shi. A review of photocatalytic CO_2 conversion to hydrocarbon fuel systems. Journal of Physical Chemistry. **30(12),** 2177-2196 (2014)
12. P. F. Lu, S. J. Ruan. Progress of carbon dioxide catalytic conversion technology under dual-carbon target. Comprehensive Utilization of Resources in China. **42(01),** 100-102 (2024)
13. Y. B. Zhen, Q. Zhang, J. Shi, et al. Progress of catalysts for electrocatalytic reduction of CO_2 to multiple products. Progress in Chemical Engineering. **41(03),** 1209-1223 (2022,)
14. K. Kuhl, T. Hatsukade, E. R. Cave, et al. J Am Chem Soc. **136,** 14107-14113 (4014)
15. Jianping Q, Juntao T, Jie S, et al. ElectroChim Acta. **203,** 99-108 (2016,)
16. Liu Y, Chen S, Quan X, et al. J Am Chem Soc. **137,** 11631 – 11636 (2015)

Research on carbon dioxide treatment and utilization technology

Ketong Chen[1]*

[1]Shenzhen Foreign Language School, Shenzhen, China

Abstract. This paper summarizes the various types of carbon dioxide treatment technology, the current situation, and the development trend, and describes the progress and challenges in the application of carbon dioxide treatment technology at the present stage. It is found that the focus of research and development should be on improving carbon capture efficiency and reducing carbon capture costs. CO_2 utilization technology is currently in the industrial demonstration stage, and breaking through the bottleneck of high temperature and high pressure environment, and searching for suitable catalysts to improve the carbon utilization efficiency are the key research directions in the next stage of CO_2 utilization technology. CO_2 treatment technology needs to overcome the challenges of economic profitability, technological innovation, cost reduction and efficiency enhancement, and policy subsidies and incentives. CO_2 bioprocessing technology will become a new mode of CO_2 treatment promotion in the future. This study is of reference significance for accurately grasping the research direction of CO_2 treatment technology, promoting the progress and innovation of CO_2 treatment technology, and accelerating the leapfrog development of CO_2 treatment technology.

1 Introduction

Carbon dioxide treatment (CDT) is a technology that captures carbon dioxide from the atmosphere by direct or indirect means and stores it permanently. In response to increasing global climate change, the Paris Agreement proposes to limit the increase in global average temperature to 2°C above pre-industrial levels and to work towards limiting the increase to 1.5°C [1]. In its 2017 Energy Technology Perspectives Report, the International Energy Agency (IEA) states that in order to achieve the goal of limiting global warming to 2°C by the end of the century, the energy sector will need to achieve net-zero emissions by 2060, or earlier if the goal is a lower warming (e.g., 1.75°C) [2]. The Intergovernmental Panel on Climate Change (IPCC), in its Special Report on Global Warming of 1.5°C, emphasized the critical importance of achieving net-zero carbon dioxide emissions by mid-century [3]. Decarbonization is not only necessary to achieve net-zero emissions, it is also a key means of compensating for the net-negative emissions required for over-temperature (temperatures exceeding 1.5°C) or critical temperature rise control (temperatures close to, but not exceeding,

* Corresponding author: chenketong2008@icloud.com

1.5°C), all of which relies on the large-scale use of a variety of carbon dioxide treatment methods.

Different CO_2 treatment methods differ significantly in terms of sequestration potential, implementation difficulty and socio-economic benefits. With the continuous development of CO_2 treatment technologies and the deepening of their understanding, it is no longer possible to effectively achieve the current GHG emission reduction targets by solely relying on traditional means of emission reduction. Therefore, based on the background of actively promoting carbon neutrality, systematically sorting out and analyzing the path of post-combustion carbon dioxide treatment technology, as well as synthesizing and summarizing the status quo, progress and typical cases of domestic and international carbon dioxide treatment technology under the "dual-carbon" goal, will help accelerate the pace of development of China's carbon dioxide treatment technology, and achieve carbon peak in 2030 and carbon peak in 2060 at the earliest possible date [4]. Therefore, the study and promotion of carbon dioxide treatment technology is of great significance to the realization of global climate goals.

2 Physical methods

Physical methods to realize carbon dioxide emission reduction mainly include carbon dioxide capture and storage (CCS), physical adsorption and supercritical CO_2 technology [5]. Among them, CCS captures CO_2 from industrial emission sources or air through adsorption, absorption or membrane separation, and compresses and stores the captured CO_2 in geological structures to prevent it from re-entering the atmosphere. Physical adsorption utilizes porous materials to efficiently adsorb carbon dioxide, while membrane separation technology separates carbon dioxide from gas mixtures by selectively transmitting the membrane. Direct air capture technology, on the other hand, captures CO_2 directly from the air, which is captured by the action of adsorbents or absorbers and then compressed and stored, and is suitable for widely distributed CO_2 sources with low concentrations. In addition, supercritical CO_2 technology utilizes the special properties of carbon dioxide in a supercritical state for capture or storage. These physical methods excel in reducing CO_2 emissions from industrial processes and provide important technical support for achieving carbon neutrality goals.

3 Biocapture

Among the many carbon utilization technologies, artificial bioconversion of CO_2 technology integrating synthetic biology means plays a particularly important role: on the one hand, it can effectively reduce the emission of greenhouse gases and accelerate the pace of advancement towards carbon neutral conversion; on the other hand, it also provides a key idea for solving the major topics such as food security and space exploration [6].

Some scholars have reported the development of a new "genomic scalpel" that rapidly prunes and knocks out the genome of microalgae to form a highly efficient cellular factory, which can then use CO_2 and sunlight to customize the production of biomolecules such as biofuels or bioplastics [7]. Designing and modifying E. coli to use hydrogen (H_2) to convert carbon dioxide (CO_2) into formic acid, thereby capturing atmospheric CO_2 and helping to curb the greenhouse effect. Replacing molybdenum atoms with tungsten atoms in a formic acid hydrogen cleaving enzyme in E. coli to create a tungsten-rich environment for E. coli to grow [8]. A special pressurized bioreactor was then designed that was filled with H_2 and CO_2 that could be used by the E. coli to capture the carbon dioxide and produce formic acid from it, which could then be used to make fuels, plastics, or chemicals. A scholar has reported for the first time the route to synthesize artificial starch from carbon dioxide, and the first time to realize the ab initio synthesis of carbon dioxide to starch in the laboratory, which is a major disruptive and original breakthrough in the field of artificial starch synthesis by human beings.

4 Chemical method

4.1 chemical absorption

According to the different principles of CO_2 separation, chemical methods can be categorized into solvent absorption, adsorption, membrane absorption, electrochemical method and hydrate method.

4.2 chemical transformation method

In addition to adsorption, catalytic conversion of CO_2 is another promising technology for emission reduction. The catalytic conversion method is driven by a catalyst to convert CO_2 into other valuable chemicals or energy carriers for resource utilization [9]. Based on the type of catalyst and reaction conditions, catalytic conversion can be categorized into thermo-catalytic conversion, electrocatalytic conversion and photocatalytic conversion.

4.2.1 Thermocatalytic Conversion

Thermocatalytic conversion involves the conversion of carbon dioxide into methane, carbon monoxide, methanol and other chemicals by means of a catalyst under high temperature conditions. Common thermocatalytic reactions include methanation (also known as the Sabatier reaction) and CO_2 hydrogenation. In the methanation reaction, CO_2 reacts with hydrogen in the presence of a nickel-based catalyst to produce methane and water. The reaction is favorable due to its high conversion and relatively mild reaction conditions [10]. However, it is energy intensive and requires large amounts of hydrogen.

4.2.2 Electrocatalytic Conversion

Electrocatalytic conversion is a method of reducing CO_2 using electrical energy. In this process, CO_2 is reduced to carbon monoxide, formic acid, ethylene and other organic compounds with the help of an electrocatalyst [11]. Advantages of electrocatalytic conversion include mild reaction conditions, a wide variety of product types, and flexible reaction control. However, the efficiency of electrocatalytic conversion is usually limited by the selectivity and stability of the electrocatalyst. In addition, large-scale electrocatalytic conversion requires a large amount of electrical energy, placing a higher demand on the development of renewable energy sources.

4.2.3 Photocatalytic Conversion

Photocatalytic conversion is a method that utilizes light energy to reduce carbon dioxide. Common photocatalysts include titanium dioxide, graphene and metal-organic frameworks. In photocatalytic conversion, the photocatalyst absorbs light energy to produce electron-hole pairs, and the electrons react with the CO_2 molecules to produce reduction products [12]. The advantage of photocatalytic conversion is its ability to utilize solar energy directly for the reaction, making it environmentally friendly and available in a wide range of products.

However, the overall efficiency of photocatalytic conversion is generally low, and the selectivity and stability of photocatalysts need to be further improved.

5 Conclusion

Excessive CO_2 emissions are a primary driver of global climate change. To address this challenge, technologies such as adsorption and catalytic conversion have emerged as promising solutions. The adsorption method captures CO_2 using various types of adsorbents, offering advantages in efficiency and renewability, though it faces challenges related to selectivity and cost. Catalytic conversion methods, including thermal catalytic, electrocatalytic, and photocatalytic processes, transform CO_2 into valuable chemicals or energy carriers, presenting significant potential. However, issues such as efficiency, stability, and economic feasibility still require further research and improvement.

As scientific and technological advancements continue, these methods are expected to play an increasingly vital role in reducing CO_2 emissions and mitigating global climate change. The successful implementation of these technologies, combined with global efforts to reduce fossil fuel dependency and enhance renewable energy adoption, will be crucial in the ongoing battle against climate change.

References

1. P. FriedlingsteinA. Global carbon budgets: determining limits on fossil fuel emissions. Weather, **75,** 7 (2020)
2. H. Takashi, N. Hoseok , C. Andrew. Multilateral energy technology cooperation: Improving collaboration effectiveness through evidence from International Energy Agency Technology Collaboration Programmes. Energy Strategy Reviews, **43,** 100920 (2022)
3. L. Rolf, S. Göran. Lost in transformation: The Paris Agreement, the IPCC and the quest for national transformative change. Frontiers in Climate. **4,** (2022)
4. K. Amith, E. Ahmed , M. A. Aliyu. An Assessment of CO2 Capture Technologies towards Global Carbon Net Neutrality. Energies, **17(6),** (2024)
5. Q. Liu, J. Xiao, H. Yu. Research progress of variable pressure adsorption capture CO_2 technology and its application case study in petrochemical industry. Southern Energy Construction, **1-13,** (2024)
6. Z. Shuping, L. Zhenrong. Advances in the biological fixation of carbon dioxide by microalgae. Journal of Chemical Technology and Biotechnology, **96,** 1475-1495 (2021)
7. K. P. Vikas, D. Anima, K. Ritu, et al. Recent progress and challenges in CRISPR-Cas9 engineered algae and cyanobacteria. Algal Research, **71,** 103068 (2023)
8. R. Magali, B. Fraser, G. William, et al. Efficient Hydrogen-Dependent Carbon Dioxide Reduction by Escherichia coli. Current Biology, **28,** 140-145, (2018)
9. P. Sangwon, G. L. Min, P. Jinwon. CO_2 (carbon dioxide) fixation by applying new chemical absorption-precipitation methods. Energy. **59,** 737-742 (2013)
10. S. F. Zhou; Z. Z. Guan, G. Chen, et al. Advances and strategies in the electrocatalytic conversion of carbon dioxide to valuable chemicals and fuels. Fuel, **377,** 132539 (2024)
11. Z. X. Zhang, Y. W. Zhang, X. J. Zhang, et al. Advances in technologies related to the catalytic conversion of carbon dioxide for the production of chemicals. Refining Technology and Engineering, **54(07),** 9-12 (2024)

12. R. Galeboe, J. Kalala, K. Phathutshedzo. Carbon dioxide conversion to fuel over alumina-supported ruthenium catalysts. Heliyon, **10,** 313492024 (2024)

https://doi.org/10.1051/e3sconf/202560603004

Progress in Carbon Dioxide Capture and Storage (CCS) and Conversion Utilization Research

Junteng Ye[1*]

[1]Shenzhen Foreign Language School, Shenzhen, China

Abstract. The issue of climate change caused by excessive carbon dioxide (CO_2) emissions has garnered widespread attention. This paper reviews the current major technologies for CO_2 treatment, including carbon capture and storage (CCS) and CO_2 utilization. CCS technology effectively reduces atmospheric CO_2 concentrations by injecting captured CO_2 underground for long-term storage, thereby mitigating climate change. Additionally, improvements in technology can enhance oil recovery rates. CO_2 utilization transforms CO_2 into high-value chemicals such as methane and carbon monoxide (CO) through thermal catalysis, photocatalysis, and electrocatalysis. Despite challenges such as low energy efficiency and high costs, these technologies show significant potential in reducing carbon emissions and promoting carbon recycling.

1 Introduction

Climate change is the most serious global challenges, with the fundamental cause being the large-scale release of greenhouse gases, especially CO_2. Since the Industrial Revolution, with the widespread use of fossil fuels and the acceleration of industrialization, the amount of CO_2 emissions has increased rapidly. These excessive CO_2 emissions have led to a significant intensification of the greenhouse effect, causing profound changes in the Earth's natural climate system. The greenhouse effect not only causes a continuous rise in the global average temperature, but also triggers a series of chain reactions, such as the accelerated melting of polar glaciers, a significant rise in sea levels, and the frequent occurrence of extreme weather events. These changes seriously threaten the balance of the ecosystem, the stability of the global food supply chain, and the safety of human living environments. Therefore, reducing CO_2 emissions has become the primary environmental issue that the global community must urgently address.*

To meet this challenge, countries and research institutions around the world are actively exploring various technological means to mitigate or even reverse the impact of excessive CO_2 in the atmosphere. This article will explore in detail several major technological approaches to addressing the issue of excessive CO_2 in the atmosphere, including CCS and

* Corresponding author: 1812010715@stu.hrbust.edu.cn

CO_2 utilization. CO_2 capture and storage technology can effectively reduce the concentration of CO_2 in the atmosphere and mitigate the process of climate change by capturing CO_2 from industrial emission sources and safely storing it underground. CO_2 utilization technology, on the other hand, can convert CO_2 into useful chemicals or fuels through advanced catalysis and chemical reactions, achieving efficient resource utilization.

2 Physical Methods for CO_2 Treatment

2.1 Carbon Dioxide Storage

CCS is a crucial technology for combating climate change, aimed at reducing greenhouse gas emissions, especially CO_2 concentrations, to mitigate global warming.

Deep geological storage is currently the most mature technology, achieving long-term storage by injecting CO_2 into abandoned oil and gas fields, salt cavern formations, or coal seams. Depleted oil fields are one of the preferred locations for geological CCS. These oil fields have stored large amounts of oil and gas throughout geological history, proving their good sealing and storage capabilities. Injecting CO_2 into depleted oil fields not only achieves long-term storage but can also enhance oil recovery (Enhanced Oil Recovery, EOR) to further extract residual oil and gas. EOR technologies mainly include three categories: water flooding, gas flooding, and chemical flooding. Water flooding increases oil recovery by injecting water into the reservoir, potentially improving oil recovery by 10%-20% [1]. Gas flooding improves reservoir displacement efficiency by injecting gases such as natural gas or CO_2, with CO_2 injection in EOR potentially increasing recovery by 10%-15% [1]. Chemical flooding alters the physical and chemical properties of oil by injecting surfactants, polymers, and other chemicals, which can enhance recovery by 8%-15% [1]. Each of these EOR technologies has its characteristics and can be selected based on specific reservoir conditions to improve recovery.

2.2 Advantages of Carbon Dioxide Storage

CCS technology has significant advantages as an important measure for greenhouse gas reduction. Its main advantage lies in effectively mitigating climate change. Permanently storing captured CO_2 rather than releasing it into the atmosphere can significantly reduce greenhouse gas emissions and is significant in combating climate change [2]. Additionally, injecting CO_2 into reservoirs can increase oil recovery and extend the lifespan of oil fields, as confirmed by research by Ampomah et al. [3]. Furthermore, the processes of capturing, transporting, and storing CO_2 can generate carbon offset credits, providing new business opportunities for companies participating in carbon trading markets and becoming an important means for companies to address climate change [4].

2.3 Disadvantages of Carbon Dioxide Storage

CCS technology also faces potential drawbacks and challenges. Its construction and operational costs are relatively high, which may impose certain pressures and obstacles on the project's economic viability [5]. Moreover, long-term storage of CO_2 may lead to accidental leaks, posing potential safety risks to the environment and surrounding residents, necessitating strict monitoring and management measures [6]. The processes of capturing, compressing, and transporting CO_2 also require significant energy, which can reduce the overall energy efficiency of power plants or factories, necessitating further advancements in technology to improve efficiency [7]. Overall, governments and enterprises need to continuously optimize technologies and improve policies and regulations to further promote the development and application of CCS technologies, achieving the goal of greenhouse gas reduction.

Overall, governments and enterprises need to continuously optimize technology and improve policies and regulations to further promote the development and application of CCS technology, achieving greenhouse gas reduction goals.

3 Chemical Conversion of CO_2

3.1 Carbon Dioxide to Methane

3.1.1 Thermocatalytically Conversion of Carbon Dioxide to Methane

The Sabatier reaction is a thermocatalytic method for converting CO_2 to methane. This reaction can be represented as: $CO_2 + 4H_2 \rightarrow CH_4 + 2H_2O$. In the study by Li et al. [8], Ni/Al_2O_3 was used as a catalyst for the hydrogenation of CO_2 to produce methane. According to their research [8], at higher reaction temperatures, CO_2 molecules first adsorb on the surface of the Ni/Al_2O_3 catalyst and are thermally activated. The activated CO_2 reacts with hydrogen in the reaction atmosphere to generate methane and water. The produced methane desorbs from the catalyst surface, and water also desorbs, exposing the active sites on the catalyst surface again. Although this reaction efficiently converts CO_2 into the clean fuel methane, achieving both capture and utilization of greenhouse gases, the required hydrogen can come from clean hydrogen production technologies, providing a renewable source of raw materials for the process. However, the Sabatier reaction requires high temperature and pressure conditions, which increases equipment and energy consumption, raising the cost of the process. Additionally, the preparation process of composite catalysts like Ni/Al_2O_3 is relatively complex, making large-scale production and commercialization difficult.

3.1.2 Photocatalytic Conversion of Carbon Dioxide to Methane

Photocatalysis is another method for converting CO_2 into methane. Under visible light or sunlight, various semiconductor photocatalysts, such as TiO_2 or cadmium sulfide (CdS), are excited to generate electron-hole pairs. These photo-generated charge carriers can activate CO_2 molecules adsorbed on the catalyst surface and undergo a series of reduction reactions with water molecules or hydrogen, ultimately producing fuels such as methane, methanol, and ethanol [9]. This method has the advantages of utilizing renewable solar energy resources, not consuming fossil fuels, and having a relatively simple process. However, the activity and selectivity of existing photocatalysts are low, and production efficiency needs improvement.

3.1.3 Electrocatalytic Conversion of Carbon Dioxide to Methane

Electrocatalytic conversion of CO_2 to methane is another method that has attracted attention. Hori et al. [10] indicated that in the electrocatalytic process, an electric current drives CO_2 to be reduced to CH_4 on the surface of the electrocatalyst. Birdja et al. [11] noted that electrochemical CO_2 reduction has advantages such as high energy density and high flexibility for storing renewable electricity, allowing for low overpotential and efficient reversible conversion of CO_2 into useful products through suitable electrocatalysts. However, this technology still faces challenges such as low energy efficiency, the need for improved product selectivity, low overall conversion rates, and insufficient technological maturity.

3.2 Carbon Dioxide to Carbon Monoxide

Converting CO_2 to CO is an important carbon capture and utilization technology. The target product of this process is the more reactive CO, which can be used as a chemical raw material and fuel. One of the main methods for converting CO_2 to CO is the reverse water-gas shift (RWGS) reaction. The RWGS reaction is the opposite of the water-gas shift reaction, which is typically used in industrial processes to produce hydrogen (H_2) from CO and water (H_2O) [12]. The RWGS reaction can be expressed by the following chemical equation: $CO_2 + H_2 \rightleftharpoons CO + H_2O$. In this reaction, CO_2 and H_2 are converted into CO and H_2O. This is an endothermic reaction that requires input of thermal energy to proceed. The typical working temperature range for the RWGS reaction is 600 °C to 900 °C, depending on the specific catalyst and reactor design. The conversion of CO_2 to methane mainly uses metals as catalysts, primarily consisting of Cu, Pt, and Rh fixed on various supports [12]. The choice of catalyst materials and supports very impact on the activity, selectivity, and stability of the RWGS reaction. The RWGS reaction can be conducted at lower temperatures (~165 °C) [12] with minimal production of by-products such as methane [13]. A promising method to enhance the performance of the RWGS reaction is the use of nanostructured catalysts. For example, MD Porosoff et al. developed a RWGS catalyst based on platinum nanoparticles that achieved 95% CO selectivity and approximately 50% CO_2 conversion rate at 550 °C. They found that compared to traditional bulk platinum catalysts, nanoscale platinum particles could better activate CO_2 molecules and selectively generate CO [14]. Overall, converting CO_2 to CO through the RWGS reaction is a promising method for utilizing CO_2 as a valuable chemical feedstock. However, research and development of catalyst design, reactor engineering, and process optimization remain key to improving the efficiency and cost-effectiveness of this technology.

3.3 Conversion of Carbon Dioxide to Other Chemicals

Converting CO_2 to other products, such as alcohols or alkenes, is an effective way to reduce carbon emissions and achieve carbon neutrality. Ethylene is a product that can be obtained by catalytically converting CO_2. The electrochemical method is a way to convert CO_2 to ethylene through electrolysis. This general reaction can be represented as: $2CO_2 + 4H_2 \rightarrow C_2H_4 + 2H_2O$. The electrochemical reduction of CO_2 to ethylene is a complex process involving multiple steps and intermediates. First, CO_2 adsorbs on the copper surface, then it is reduced to CO as an intermediate, and finally further reduced to products like CH_4 and C_2H_4 [15]. Converting CO_2 to ethanol is also a feasible method. The electrochemical reduction of CO_2 to ethanol is a multi-electron, multi-proton reaction, which can be represented as: $2\ CO_2 + 6\ H^+ + 6\ e^- \rightarrow CH_3CH_2OH + 2H_2O$. A class of promising catalysts is based on Cu materials. Research shows that a single copper catalyst can produce ethanol and other C^{2+} products. Optimizing the surface structure and active sites of copper is key to improving ethanol selectivity. For example, controlling the orientation of copper crystal planes can significantly impact product distribution [16]. Introducing a second metal element into copper-based bimetallic electrocatalysts can further enhance the yield and selectivity of ethanol. For instance, bimetallic catalysts such as Cu-Pd and Cu-Sn benefit from the introduction of a second metal to modulate electronic structure and catalytic active sites [16]. Furthermore, in the study by Jiang et al. [17], they developed a Cu-based electrocatalyst that achieved high Faradaic efficiency (>60%) for the selective reduction of CO_2 to C^{2+} products, including ethanol. They attributed this performance enhancement to the metal-ion battery cycling method that exposes more favorable (100) and (211) crystal planes for C-C coupling.

4 Conclusion

The technologies for capturing, storing, and converting CO_2 play a crucial role in addressing global climate change. CCS technologies effectively reduce the concentration of CO_2 by capturing CO_2 produced during industrial and energy generation processes and safely storing it underground, thereby mitigating the negative impacts of the greenhouse effect and climate change. In addition, CCS technology can be combined with enhanced oil recovery to achieve a win-win situation for energy production and environmental protection. Meanwhile, the conversion and utilization of CO_2 technologies provide new solutions for reducing carbon emissions. Through thermal catalysis, photocatalysis, and electrocatalysis, CO_2 can be converted into fuels such as methane and CO or other high-value-added chemicals, which not only helps to reduce CO_2 emissions but also provides raw materials for the chemical industry, achieving carbon recycling. However, these technologies still face challenges in energy efficiency, cost, and catalyst stability, requiring further research and technological breakthroughs. In the future, to achieve the global carbon neutrality goal, it is necessary to promote the development of CO_2 treatment technologies at multiple levels. It is hoped that this article can contribute to the development of CO_2 capture, storage, and conversion technologies, reduce global carbon emissions, and address climate change.

References

1. V. Alvarado, E. Manrique. Enhanced Oil Recovery: An Update Review. Energies, **3,** 1529-1575 (2010)
2. Masson-Delmotte, Valérie, et al. Global warming of 1.5 C. An IPCC Special Report on the impacts of global warming, **1,** 93-174 (2019)
3. W. Ampomah, et al. Performance assessment of CO_2-enhanced oil recovery and storage in the Morrow reservoir. Geomechanics and Geophysics for Geo-Energy and Geo-Resources, **3,** 245-263 (2017)
4. DOE, US. Carbon capture, utilization and storage: Climate change, economic competitiveness, and energy security. 1-12 (2016)
5. Leeson, Duncan, et al. A Techno-economic analysis and systematic review of carbon capture and storage (CCS) applied to the iron and steel, cement, oil refining and pulp and paper industries, as well as other high purity sources. International Journal of Greenhouse Gas Control, **61,** 71-84 (2017)
6. Koornneef, Joris, et al. The environmental impact and risk assessment of CO_2 capture, transport and storage - An evaluation of the knowledge base. Progress in Energy and Combustion Science, **38(1),** 62-86 (2012)
7. Turan, Guloren, et al. Global status of CCS 2021. CCS Accelerating to net zero (2021).
8. K. N. Li, K. Yasutaka, Y. Hiromi. Hollow carbon-based materials for electrocatalytic and thermocatalytic CO_2 conversion. Chemical Science, (2024)
9. Izumi, Yasuo. Recent advances in the photocatalytic conversion of carbon dioxide to fuels with water and/or hydrogen using solar energy and beyond. Coordination Chemistry Reviews, **257(1),** 171-186 (2013)
10. Hori, Y. I. Electrochemical CO_2 reduction on metal electrodes. Modern aspects of electrochemistry, 89-189 (2008)
11. Birdja, Y. Yuvraj et al. Advances and challenges in understanding the electrocatalytic conversion of carbon dioxide to fuels. Nature Energy, **4(9),** 732-745 (2019)
12. CO_2 conversion by reverse water gas shift catalysis: comparison of catalysts, mechanisms and their consequences for CO_2 conversion to liquid fuels

13. Rodriguez, A. Joséet al. CO_2 hydrogenation on Au/TiC, Cu/TiC, and Ni/TiC catalysts: Production of CO, methanol, and methane. Journal of catalysis, **307,** 162-169 (2013):
14. Porosoff, D. Marc et al. Identifying trends and descriptors for selective CO_2 conversion to CO over transition metal carbides. Chemical Communications, **51(32),** 6988-6991. (2015)
15. Kuhl, P. Kendra et al. New insights into the electrochemical reduction of carbon dioxide on metallic copper surfaces. Energy & Environmental Science, **5(5),** 7050-7059 (2012)
16. J. L. Qiao, et al. A review of catalysts for the electroreduction of carbon dioxide to produce low-carbon fuels. Chemical Society Reviews, **43(2),** 631-675 (2014)
17. K. Jiang, R.B. Sandberg, A.J. Akey, et al. Metal ion cycling of Cu foil for selective C-C coupling in electrochemical CO_2 reduction. Nat Catal, **1,** 111-119 (2018)

 https://doi.org/10.1051/e3sconf/202560603005

Analysis of aerosol climate effects under global warming

Guo Li [1,*]

[1]Jiangsu Normal University, School of Geography, Geomatics and Planning, 221116 Xuzhou, China

Abstract. Global warming is a global environmental problem that has a profound impact on natural ecosystems, human socio-economic activities and the future fate of the planet. Aerosols, as tiny particles that are widespread in the atmosphere Its climate effects are complex and uncertain, and it is a key factor in global climate change research. This essay first introduces the definition and classification of aerosols, including natural source aerosols and anthropogenic source aerosols. Then it analyzed the interrelationship between global warming and aerosols. Then it explored the differences in aerosol impacts in various regions, using North China, Northeast China and the Qaidam Basin as examples. It was found that the differences in the spatial and temporal distribution of aerosols can be summarized from the seasonal and interannual variations of Aerosol Optical Depth (AOD) as well as the characteristics of the spatial distribution. The higher AOD values in spring and summer are mainly affected by sand and dust aerosols, localized pollution, and the EI Niño-Southern Oscillation (ENSO) phenomenon. The spatial distribution of AOD showed heterogeneity. The research provides an important scientific basis for the development of effective climate policy and response measures.

1 Introduction

In the context of global climate change, global warming has become a serious challenge that cannot be ignored. Global warming has far-reaching influences on natural ecosystems, human socio-economic activities and the future fate of the earth. Global warming is mainly driven by a combination of factors, including increased greenhouse gas emissions, land-use change, deforestation and accelerated industrialization. Among the many factors exploring climate warming, aerosols, as tiny particulate matter widely existing in the atmosphere, have been the most uncertain factor in anthropogenic climate forcing for a long time, and their complex climate effects have gradually received extensive attention from the scientific community [1].

Aerosols are referred to a gaseous dispersion system consisting of solid or liquid particles suspended in a gaseous medium. Aerosols not only directly affect the Earth's energy balance, but also indirectly influence cloud formation, development, and dispersion through a series of complex physical and chemical processes, thus having a significant impact on the

* Corresponding author: 3020220308@jsnu.edu.cn

climate system. In the study of aerosol climate effects, cloud-aerosol interactions have become a hot topic of research for a wide range of scientists due to their large uncertainty factors and their non-negligible moderating effect on the global climate system [2]. The climate effects of aerosols are particularly important in the context of global warming. On the one hand, aerosols can directly cool the atmosphere and the earth's surface by scattering and absorbing solar radiation, which has a certain mitigating effect on global warming, the so-called "aerosol cooling effect". On the other hand, aerosols can also indirectly alter the energy distribution and water cycle of the climate system by affecting the radiative properties of clouds and precipitation efficiency. Such indirect effects are often more complex and difficult to predict than direct effects [3]. Therefore, an in-depth analysis of the climate effects of aerosols in the context of global warming not only helps to understand more comprehensively the driving mechanisms and evolutionary patterns of climate change but also provides an important scientific basis for the formulation of effective climate policies and response measures.

This paper is to study the climate effects of aerosols in the context of global warming, specifically including the classification of aerosols and their characteristics, the interrelationship between global warming and aerosols, and the difference in aerosol impacts in different regions. Analyse the effects of global warming on aerosol generation, propagation and removal, and the direct and indirect radiative feedback of aerosols on global warming by elaborating on the classification of natural and anthropogenic sources of aerosols. Meanwhile, taking North China, Northeast China and Qaidam Basin as examples, this paper will explore the seasonal distribution characteristics, spatial distribution characteristics and influencing factors of aerosol optical thickness in different regions, to provide a scientific basis for comprehensively grasping the driving mechanism and evolutionary law of climate change.

2 Classification of aerosols

2.1 Natural source aerosols

Natural source aerosols, as an important component of atmospheric aerosols, have a wide range of sources and are complex. Natural source aerosols mainly include sea salt aerosols, sand and dust aerosols, biomass burning aerosols, and biogenic aerosols. These aerosols are produced primarily by physical, chemical and biological processes in nature without direct human involvement and are therefore widespread globally [2].

2.2 Anthropogenic source aerosols

Aerosols of anthropogenic origin are aerosol particles produced directly or indirectly by human activities. These particles are widespread in the atmosphere and have significant impacts on air quality, climate change and human health. With accelerated industrialization and urbanization, aerosol emissions from anthropogenic sources are increasing and have become an important part of the global environmental problem. Aerosols of anthropogenic origin come from a variety of sources, including fossil fuel combustion, industrial processes, agricultural activities, urban buildings and road construction. For example, the combustion of fossil fuels in large quantities releases large amounts of sulphur dioxide, which increases the concentration of sulphate aerosols in the atmosphere, thus contributing to global climate change. In the case of China, the direct radiative effect of sulphate aerosols causes a generalized cooling of inland China [4].

3 Interrelationships between global warming and aerosols

3.1 Impact of global warming on aerosols

Global warming affects aerosol generation, propagation and removal processes in a variety of ways. Elevated temperatures act directly on the aerosol generation process. For example, the rates of some chemical reactions may change as temperatures rise. In terrestrial ecosystems, higher temperatures can affect the physiological activity of vegetation. Volatile organic compounds (VOCs) emitted by plants are precursors to secondary aerosol formation, and temperature changes can alter the rate of VOCs emitted by plants. For some biogenic aerosols, elevated temperatures may induce enhanced microbial metabolic activity, which in turn increases the release of microbial aerosols.

In terms of transmission, warmer temperatures may alter the thermal structure and stability of the atmosphere. A warmer air layer may increase the height of the atmospheric boundary layer, which could affect the aerosol propagation range in the vertical direction. Diffusion and mixing of aerosols in the vertical direction can be altered by changes in the height of the atmospheric boundary layer, which may lead to a more complex distribution of aerosols in the atmosphere. In terms of removal processes, temperature affects chemical and physical processes in the atmosphere. For example, increased temperatures may accelerate the evaporation process of aerosols, especially for some semi-volatile aerosol components. In addition, increased temperatures may change the characteristics of precipitation, indirectly affecting the removal efficiency of aerosols by wet deposition.

Changes in source areas due to climate change have a non-negligible indirect impact on aerosol sources. In the case of the oceans, as global warming increases ocean temperatures, sea ice melts, which can alter marine ecosystems. Phytoplankton are affected by the environment in which they grow, and their abundance and distribution change. Phytoplankton are one of the important sources of marine-sourced aerosols, and the dimethyl sulphur (DMS) they release oxidizes in the atmosphere to form sulphate aerosols. Therefore, changes in marine ecosystems indirectly affect the generation of marine-sourced aerosols.

In terrestrial ecosystems, climate change may lead to the expansion of arid and semi-arid areas and a reduction in vegetation cover. This will increase the release of dust from the surface and expand the source area for dust aerosols. At the same time, the frequency and intensity of forest fires may increase as a result of climate change, and forest fires release large amounts of soot and aerosol precursors, thus becoming an important temporary source area for aerosols. These climate change-induced changes in source regions have complex implications for the overall source strength of aerosols, further affecting their role in the global climate system.

3.2 Aerosol feedback to global warming

3.2.1 Direct effect of radiation

Aerosols directly affect solar radiation through scattering and absorbing, altering the radiative balance of the geothermal air system, which affects the regional climate. It has been found that changes in surface albedo affect aerosol radiative effects and the Earth's radiation balance and that aerosols are more susceptible to radiative warming effects when surface albedo is high. Annan Chen et al. combined satellite and ground observation data to study aerosol and surface albedo under different aerosol characteristics and pointed out that the interaction of aerosol direct radiative effect and surface albedo is important for balancing the Earth's energy.

As surface albedo decreases, the aerosol cooling effect at the top boundary of the atmosphere increases and the warming effect decreases [5].

The scattering of solar radiation by aerosol particles is an important radiative process in the atmosphere. Sunlight encountering suspended aerosol particles is scattered and part of the radiation does not reach the surface. Scattering aerosols, such as sulphates, depend on particle size and other factors, such as sulphate and sea salt aerosols can reduce the amount of solar radiation reaching the surface to produce a cooling effect. Some aerosol particles such as black carbon absorb solar radiation directly, and black carbon aerosol is a strong absorbing component of the atmosphere and has a warming effect on the surface. A study by Huijuan Lin et al. in Regional Climate Effects of Black Carbon in Asia showed that the concentration of black carbon had an extremely significant effect on the shortwave heating rate, especially in the region of large values of black carbon concentration (Fig. 1) [6].

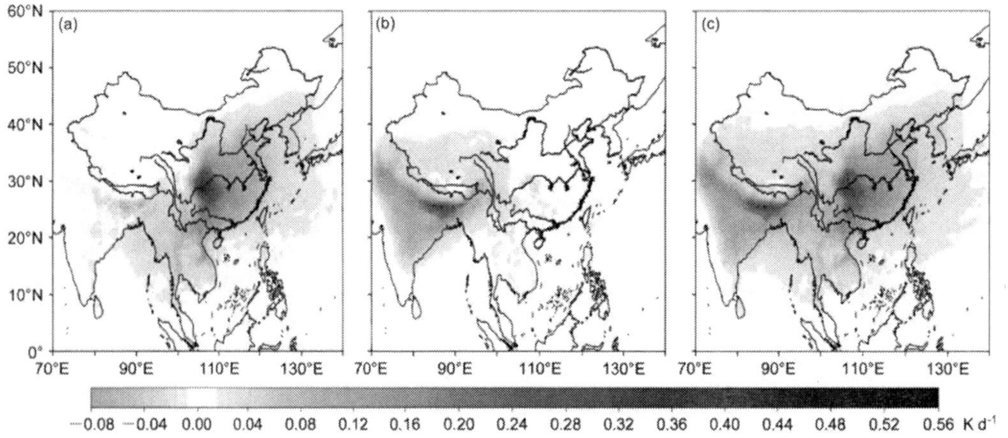

Fig. 1. Changes in shortwave radiation heating rate (SWHR; units: K d-1) near 850 hPa for (a) Chinese b lack carb on, (b) Indian b lack carb on, and (c) Chinese and Indian b lack carb on overwinters (1996 - 2006). The black dots indicate the student's t-test 90% confidence level. [6]

3.2.2 Indirect radiation effect

Cloud formation occurs mainly through the process of heterogeneous nucleation that occurs when aerosols are used as cloud condensation nuclei and water vapor accumulates on them. The interaction between aerosols and clouds is an important source of uncertainty in future climate predictions. Graham Simpkins studied how increased aerosol concentrations affect mid-latitude storm clouds by using high-resolution imagery and a global convective licensing model. Studies have shown that aerosols increase the liquid water content of polluted systems, and cloudiness, and thus enhance the albedo of storms, and the outgoing flux of shortwave radiation. These findings help to limit estimates of anthropogenic radiative forcing and improve climate projections [7].

Aerosols have three important types of indirect effects on clouds and the climate system. The first type of indirect effect is the Twomey effect, where an increase in aerosol number concentration leads to an increase in cloud particle number concentration, a decrease in radius, and an increase in cloud albedo [8]. In the oceans, anthropogenic emissions of aerosols increase the concentration of cloud particle numbers, resulting in whiter clouds and increased albedo. The second type of indirect effect, the Albrecht effect or cloud lifecycle effect, is that an increase in anthropogenic aerosols reduces the radius of cloud particles, inhibiting precipitation and altering the cloud lifecycle. In and around large cities, human activities generate aerosols that reduce stratocumulus cloud particle radii, inhibit precipitation, and lengthen cloud duration. Semi-direct effect refers to the absorption of solar energy by light-absorbing aerosols to heat the atmosphere and evaporate clouds, such as soot, black carbon, and other aerosols can heat the atmosphere and clouds to evaporate the cloud droplets, reducing the number of clouds, shortening the cloud lifetime, and reducing the average albedo of the cloud body [9]. In turn, clouds can be cleared of aerosols by precipitation, affecting their effect on clouds. Convective cloud formation begins with the rise of the surface of a heated planet, and the magnitude of the rise depends on the change in temperature with altitude or the rate of descent of the troposphere.

4 Differences in aerosol impacts in different regions

4.1 Impact of aerosols in North China

Aerosol Optical Thickness (AOD) is a physical quantity that describes the light-cutting effect of aerosols. It is a key indicator of the degree of atmospheric turbidity and the most important optical property parameter in determining the climatic effects of aerosols. Spatial averaging of the parameter Column_Optical_Depth_Aerosols_532 in the CALIPSO satellite Level 2 5km aerosol layer product from June 2006-October 2015 by Xingxing Gao et al. Based on multi-source satellite remote sensing data and ground observation data, it is found that the seasonal distribution characteristics of aerosol optical thickness in North China are obvious [10]. As can be seen from Fig. 2, the 532 nm AOD value was the largest in North China in summer, followed by autumn and winter in that order, and the smallest in spring. The corresponding 532 nm AOD values were 0.74, 0.58, 0.55 and 0.49 in that order. The high AOD areas are concentrated in the densely populated and highly urbanized plains of North China, while the low AOD areas are concentrated along the hilly mountains at higher elevations. It can be seen that the spatially varying properties of atmospheric aerosols are strongly influenced by human activities.

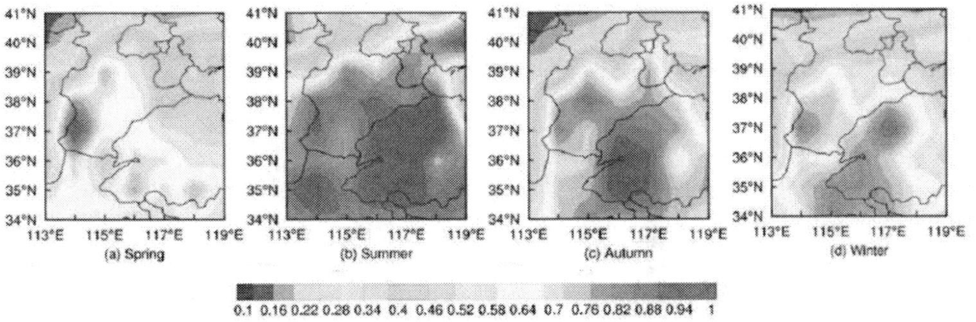

Fig. 2. Spatial distribution of mean 532 nm AOD values in North China, 2006-2015 (a. spring; b. summer; c. fall; d. winter) [10]

4.2 Impact of aerosols in Northeast China

Li Wan et al. analysed the spatial distribution characteristics and inter-annual trends in Northeast China using MODIS atmospheric aerosol optical thickness data and the China Multiscale Emission Inventory Model (CMEIM) from 2003 to 2022 [11]. It was found that all high values of AOD in Northeast China occurred in spring and summer, with a decreasing trend of AOD in fall and an increase in AOD in winter (Fig. 3) The reason is that the burning of straw in spring and the strong radiation and high humidity in summer favor the acceleration of the air-particle transformation process, which has a greater impact on the atmospheric extinction in the Northeast. Figure 4 shows the regional average values of AOD in the northeast region. As can be seen from the figure, the annual average AOD value is highest in Liaoning. And the AOD values in Liaoning were higher than those in Jilin and Heilongjiang in spring, summer, autumn and winter. The increase of AOD in summer in Northeast China is mainly related to the environmental humidity, and the boundary layer meteorological conditions have some influence on AOD in winter.

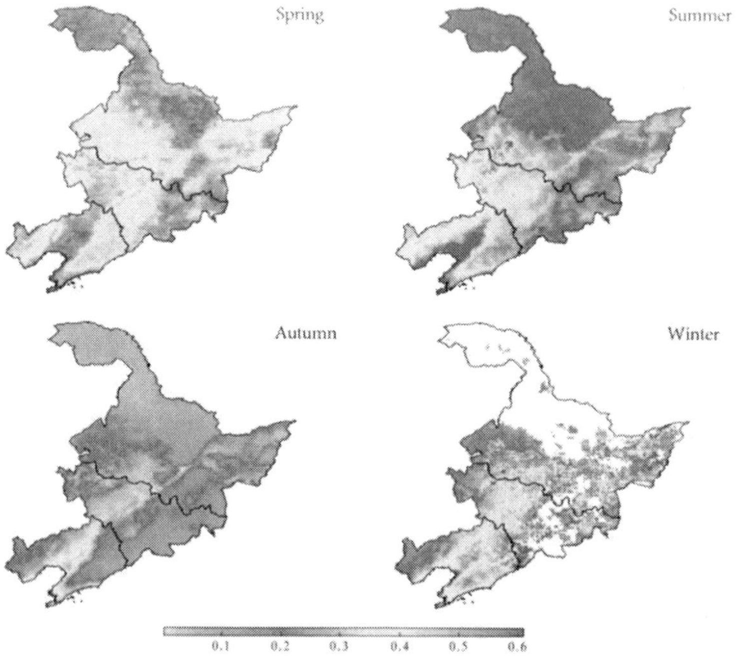

Fig. 3. Seasonal distribution of AOD in Northeast China from Mar 2003 to Feb 2023 [11]

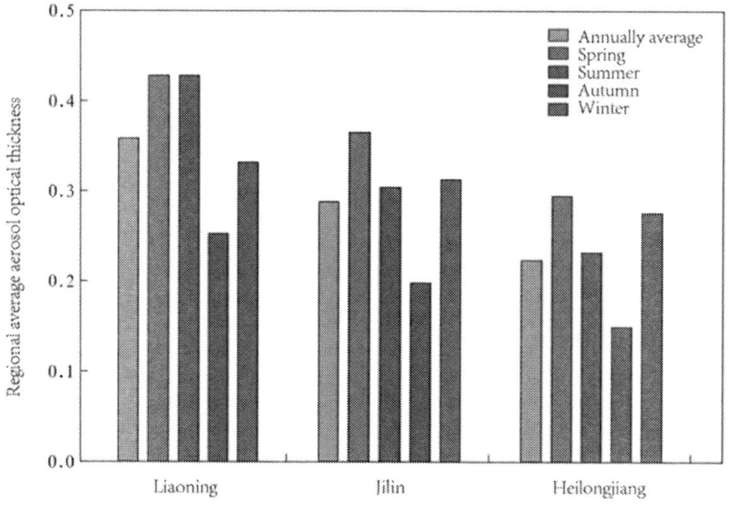

Fig. 4. Regional average of AOD in Northeast China from Mar 2003 to Feb 2023 [11]

https://doi.org/10.1051/e3sconf/202560603005

4.3 Impact of aerosols in Qaidam basin

Xiao Hongdan et al. explored the spatial and temporal distribution characteristics and influencing factors of AOD in the Qaidam Basin from 2001 to 2021 based on the MODIS MCD19A2 aerosol dataset using linear trend, Spearman's correlation analysis, and Ångström's exponential interpolation method [12]. As can be seen in Figure 5, it indicates that AOD in the Qaidam Basin has obvious seasonal variations, with the area of high AOD values being the largest in the spring, gradually decreasing in the summer, and dissipating in the autumn and winter. Since the basin is mostly desert, dust storms are frequent in the spring and are accompanied by drastic weather changes and the accumulation of large amounts of dust aerosols, resulting in the formation of high-value areas. Figure 6 shows the characteristics of seasonal changes of AOD in the Qaidam region, and it can be seen that the value of AOD is spring > summer > autumn > winter. Due to the topographic constraints of the Qaidam Basin, air convection is poor. The Qaidam Desert is the main source area of sand and dust in the basin, and its AOD reaches its highest level in the spring, which makes it difficult for sand and dust aerosols to diffuse, and thus leads to the emergence of the "Basin Effect".

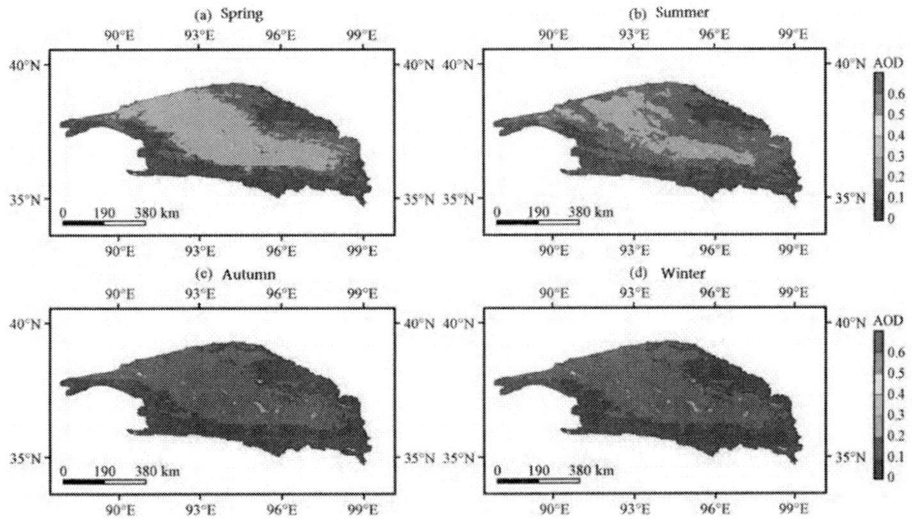

Fig. 5. Seasonal mean spatial distribution of AOD in the Tsaidam Basin during 2001 -2021 [12]

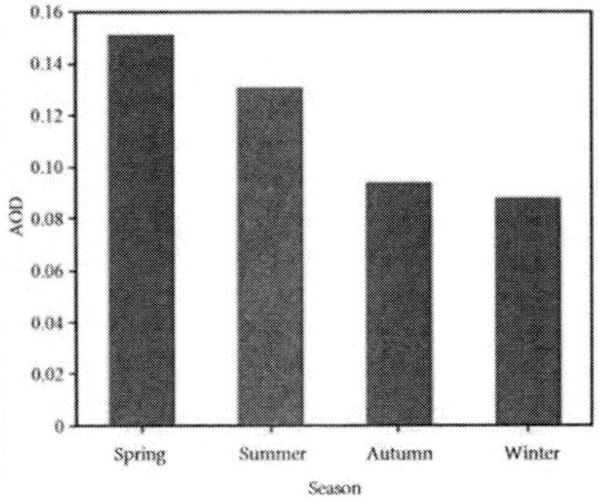

Fig. 6. Characteristics of seasonal changes in AOD in the Qaidam Basin during 2001 -2021 [12]

In terms of seasonal and interannual variations in AOD and spatial distribution characteristics, a case study can summarize the differences in aerosol impacts in different regions found that the differences in the spatial and temporal distribution of aerosols. One reason for the higher AOD values during the spring and summer seasons is the effect of dust aerosols. Dust aerosols are blown up and spread over longer distances in the spring due to dry air and favorable wind conditions, resulting in a greater optical thickness of dust aerosols in the spring. The second is the effect of localized pollution. Emissions due to human activities increase aerosols, which can lead to an increase in the optical thickness of aerosols. Especially during the spring and summer months, when human activities are more frequent due to higher temperatures and increased sunlight hours, the aerosol optical thickness increases. The third is the impact of the El Niño-Southern Oscillation (ENSO) phenomenon. In the spring of the year following El Niño, dry air and low precipitation favor biomass-burning activities in northern continental Southeast Asia, leading to an increase in carbonaceous aerosol emissions, and thus an increase in aerosol optical thickness. On the contrary, the La Niña phenomenon leads to a decrease in the aerosol optical thickness. The spatial distribution of AOD is characterized by inhomogeneity.

5 Conclusion

This paper provides the climatic effects of aerosols in the context of global warming and the differences in impacts in different regions. Global warming and aerosols interact in a complex way, with global warming influencing aerosol production, propagation and removal, and aerosols in turn feeding back on global warming through direct and indirect radiative effects. Aerosol optical thickness in different regions shows distinct seasonal and spatial distribution characteristics. The high AOD areas in North China are concentrated in the densely populated and highly urbanized North China Plain. The high values of AOD in Northeast China were in spring and summer, and the AOD values in Liaoning were higher than those in Jilin and Heilongjiang. The AOD in the Qaidam Basin formed a high-value area in the spring due to the sandstorm weather, and the "basin effect" appeared due to the limitation of the topography. Overall, the differences in the spatial and temporal distribution of aerosols are influenced by several factors, including dust aerosols, localized pollution, and the ENSO phenomenon. The understanding of the climate effects of aerosols and regional differences is of great significance to future climate change research and response efforts and can provide a scientific basis for the formulation of effective climate policies and response measures.

References

1. IPCC. Climate Change 2013-The Physical Science Basis. Intergovernmental Panel on Climate Change,1535 (2013)
2. Yu Yong, Wu Ziyue, Brief analysis on climate effects of aerosols. Technology Wind. 34, 174-175 (2020)
3. Li Zhanqing, Overview of the impact of aerosols on China's weather, climate and environment. Trans Atmos Sci. 43,76-92 (2020)
4. Yu Yong, Brief analysis on the impact and feedback of sulfate aersosls on climate change. Technology Wind. 34,128-130 (2021)
5. Annan Chen, Chuanfeng Zhao, Haotian Zhang1, Surface albedo regulates aerosol direct climate effect. Nat Commun. 15, 7816 (2024)

6. LIN Huijuan, LI Muyang, ZHUANG Bingliang, Climate Effects of Black Carb on Aerosol from the Mainland of China and India in East Asia in Winter. Chinese J. Atmospheric Sci. 47,1113-1130 (2023)
7. Simpkins, G, Aerosol - cloud interactions. Nature Climate Change. 8, 457 (2018)
8. Twomey S, Pollution and the planetary albedo. Atmos Environ. 8,1251-1256 (1967)
9. Mingjie Liang, Zhiwei Han, Jiawei Li, Yue Li, Aerosol effects during heat waves in summer 2022 and responses to emission change over China. Clim Atmos Sci. 7,190 (2024).
10. Gao Xingxing, Aerosol optical characteristics and climate effects in typical regions of China. Lanzhou University (2018)
11. Li Wan, Zhao Hujia, Wang Changshuang, Spatial Distribution Characteristics and Interannual Variation Trend of Aerosol Optical Depth in Northeast China from 2003 to 2022. J. Appl. Meteorol. 35, 211-224 (2024)
12. Xiao Hongdan, He Qing, Yuan Shujie, Analysis of the Spatial and Temporal Distribution Characteristics of Aerosol Optical Thickness and the Impact of Meteorological Factors in the Qaidam Basin. Plateau Meteorology. 43,762-774 (2024)

Aerosol Effects on the Atmosphere: Types, Mechanisms and Future Perspectives

Yuanjie Wang[1*]

[1]Nanjing University of Information Science and Technology, Reading academy, 210000 Nanjing, China

Abstract. Aerosols, composed of solid or liquid particles suspended in the atmosphere, with complex sources and mechanisms of action, are widely present in both natural environments and human activities. Aerosols directly affect the Earth's energy balance by scattering and absorbing solar radiation and indirectly affect climate and precipitation patterns by acting as cloud condensation nuclei, altering cloud formation and microphysical properties. This paper discussed the different types of aerosols, including natural, anthropogenic, and hybrid aerosols, and analyzed their physical and chemical properties. This paper also analyzed the direct and indirect climate effects of aerosols in the atmosphere and explored future research directions, including technological innovations, interdisciplinary collaborations and international policy drivers. Future research on aerosols as an important factor in climate change will contribute to a better understanding of its complexity and to strategies to address global climate change.

1 Introduction

Aerosols are gaseous dispersed systems of solid or liquid particles suspended in a gaseous medium, with particle diameters ranging from a few nanometres to tens of microns, which are widely distributed in the natural environment and human activities. As the problem of climate change becomes more serious, aerosols are appearing more frequently in various articles, giving rise to a wide range of concerns about them.

Atmospheric aerosols have the capacity to directly influence the Earth's energy balance through their scattering and absorption of solar radiation. Additionally, they can indirectly impact the Earth's climate by serving as cloud condensation nuclei and modifying cloud formation as well as microphysical properties.[1]. Therefore, the international community attaches considerable importance to the study of the sources and formation mechanisms of aerosols in the atmosphere. With the development of the national economy, atmospheric particulate pollution has become urban and regional, and organic aerosols have also become a significant challenge for air pollution control [2]. Through long-term atmospheric monitoring, some research teams in the United States have identified in detail the major sources of anthropogenic aerosol (e.g., sulfate and nitrate aerosols) emissions and their concentration distribution patterns in the atmosphere in different regions. In the study of the

* Corresponding author: nw808469@student.reading.ac.uk

mechanism of action of aerosols on the atmosphere, European scientific research groups have used advanced satellite remote sensing technology and a network of ground-based observatories to investigate in detail the scattering and absorption properties of aerosols in response to solar radiation. The research team discovered that black carbon aerosols significantly contributed to the absorption of solar radiation, with this absorption having a notable impact on regional climate patterns. In the field of aerosol action mechanisms, China has conducted comprehensive research on the physical, chemical and climatic effects of aerosols, utilizing self-developed observational equipment and numerical models. For instance, research into cloud physics has demonstrated that aerosols, acting as water vapor condensation nuclei, exert a considerable influence on precipitation distribution in China, particularly in arid and semi-arid regions, where alterations in aerosol concentration may result in shifts in precipitation patterns. Nevertheless, despite the notable advancements in aerosol research at home and abroad, there remain some challenges and issues that necessitate further experiments and research.

This paper is designed to explore the sources and characteristics of different types of aerosols in the atmosphere and their mechanisms of influence on climate. By classifying and analysing natural aerosols, anthropogenic aerosols and hybrid aerosols, the role of aerosols in radiation, cloud and surface interactions is revealed in conjunction with the findings of international research.

2 Types and characteristics of aerosols

2.1 Natural aerosols

Natural aerosols are defined as aerosols emitted to the atmosphere through a variety of physical, chemical and biological processes occurring in the natural environment [3]. The sources of natural aerosols are numerous and diverse, encompassing volcanic eruptions, forest fires, sea salt particles, wind and sand. The concentration and distribution of these aerosols in the atmosphere are subject to seasonal, geographical and climatic influences. The release of sulfate aerosols into the atmosphere as a result of volcanic eruptions has the potential to exert a significant influence on the global radiation balance. In contrast, sea salt aerosols present in the oceans play a pivotal role in cloud formation. Natural aerosols can accelerate certain chemical processes within the atmosphere, and cause changes in atmospheric colour and a range of other effects.

2.2 Anthropogenic aerosols

Anthropogenic aerosols are aerosols generated through human activities, mainly from industrial emissions, fossil fuel combustion, and agricultural activities. Sulfate, nitrate and organic carbon aerosols are the most common types of anthropogenic aerosols. These aerosols not only influence the climate system by scattering and absorbing solar radiation but also have significant implications for air quality. These include reduced visibility, respiratory diseases, acid rain, and other adverse effects [4].

2.3 Mixed aerosols

Mixed aerosols are a combination of natural and anthropogenic aerosols, and typically possess a complex chemical composition and physical properties [5]. For instance, dust aerosols can adsorb pollutants emitted by anthropogenic activities as they traverse industrial zones, thereby forming mixed aerosols. Such mixing properties contribute to the intricacy and diversity of aerosol climate impacts.

2.4 Physical and chemical properties of aerosols

The physical characteristics of aerosols encompass particle diameter, shape and surface properties, whereas the chemical properties pertain to their composition and chemical reactivity. Particle size is a pivotal factor in regulating the behaviour of aerosols in the atmosphere, influencing their atmospheric residence time, optical properties and settling rate. The physical properties of aerosols encompass particle diameter, shape and surface properties, whereas the chemical properties pertain to their composition and chemical reactivity. Particle size is a pivotal factor in regulating the behaviour of aerosols in the atmosphere, influencing their atmospheric residence time, optical properties and settling rate [6].

3 Mechanisms of aerosol impact on the atmosphere

3.1 Physical effects

3.1.1 Scattering and absorption of solar radiation

The scattering of sunlight by aerosol particles is one of their most significant physical effects [7]. As sunlight traverses the atmosphere, it is subject to scattering by aerosol particles, resulting in the light being dispersed in all directions. This scattering effect not only reduces the intensity of solar radiation reaching the ground but also affects the optical properties and visibility of the atmosphere. The scattering effect of aerosol particles is related to their size, shape, refractive index and other factors. Aerosol particles of small size undergo mainly Rayleigh scattering, resulting in a blue colouration of the sky, while particles of larger size may undergo Mie scattering, causing optical turbidity in the atmosphere. Some aerosol particles are capable of absorbing energy directly from solar radiation, such as black carbon. This absorption increases the heat content of the atmosphere and affects the Earth's energy balance. Concurrently, compounds attached to the surface of aerosols may also absorb solar radiation, thus enhancing the overall absorption effect of aerosols. In East Asia, for example, summer temperatures and pressures are influenced by regional sulphate aerosols in China, with widespread cooling and pressure increases north of about 25°N inland (Fig. 1) [8]. Black carbon aerosols at 500 hPa cause a decrease in temperature north of 45°N and an increase in temperature south of 45°N in China. At 850 hPa, black carbon aerosols cause a significant increase in temperature in the eastern part of China and its coastal areas, and a decrease in temperature in the western and northern parts of the country (Fig. 2) [9].

 https://doi.org/10.1051/e3sconf/202560603006

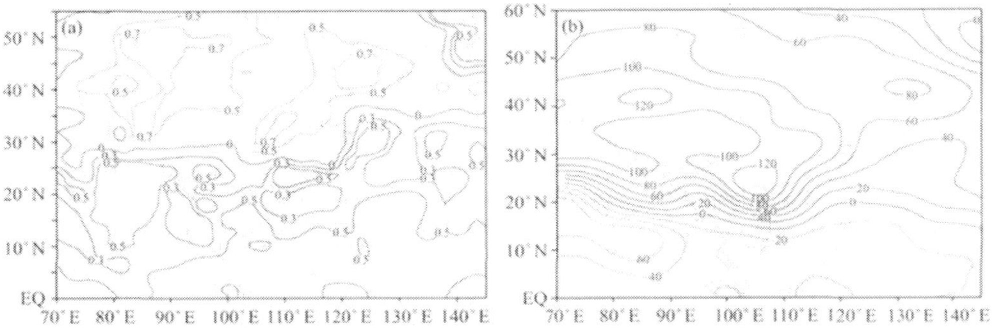

Fig. 1. Sulphate-induced changes in surface temperature and barometric pressure during summer (°C) [8]

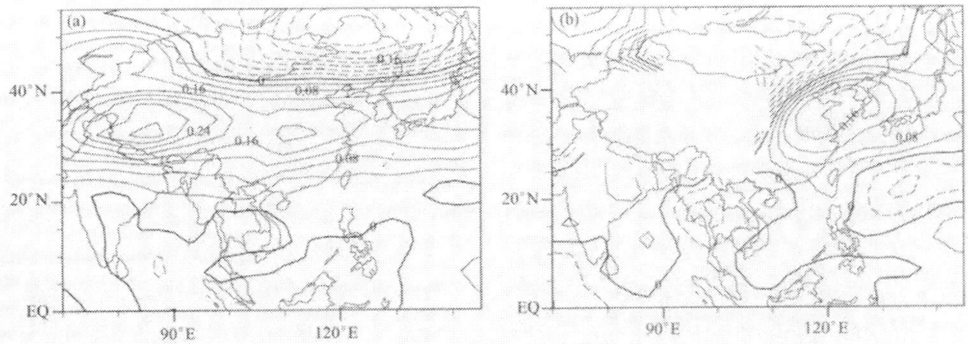

Fig. 2. Summer 500 hPa and 850 hPa temperature changes due to black carbon aerosols [9]

3.1.2 Impact on atmospheric transparency

The presence of aerosol particles reduces the transparency of the atmosphere, making it impossible for light to penetrate the atmosphere [10]. The primary reason for this is the scattering and absorption of light caused by aerosol particle as the concentration of aerosols increases, the transparency of the atmosphere is further reduced, leading to the formation of meteorological hazards such as haze.

3.1.3 Acting as condensation nuclei for water vapor

Aerosol particles can act as condensation nuclei for water vapor, facilitating cloud formation and precipitation processes. When water vapor molecules condense on the surface of aerosol particles, cloud droplets or raindrops are formed. The number, size, and distribution of these cloud droplets and raindrops are influenced by the aerosol particles. The higher the concentration of aerosol particles and the smaller the particle size, the greater the number of cloud droplets and the smaller the droplet size; conversely, a lower concentration of cloud droplets results in larger individual droplet sizes

3.2 Chemical effects

3.2.1 Provision of chemical reaction sites

Aerosol particles, with their large specific surface area and surface energy, are capable of adsorbing gas molecules and chemical reactants in the atmosphere, thereby providing a site for non-homogeneous chemical reactions. These chemical reactions may result in the production of new substances that influence the chemical composition and properties of the atmosphere.

3.2.2 Impact on stratospheric ozone climatic effects

The existence of aerosols in the stratosphere can affect the chemical balance of ozone. For example, volcanic aerosols from volcanic eruptions can reflect sunlight and reduce the intensity of solar radiation reaching the stratosphere, thus slowing down the rate of ozone photolysis [11]. At the same time, some components of volcanic aerosols may also participate in ozone chemistry, further affecting ozone concentrations.

3.3 Climatic effects

3.3.1 Direct climate effect

Aerosols can direct scatter and absorb solar radiation, which can impact the Earth's energy equilibrium and climate system. For instance, sulfate aerosols have a cooling effect by reflecting sunlight and reducing the total amount of solar radiation that reaches the Earth's surface. This effect is particularly pronounced in industrialized regions, where it has a significant impact on the local climate and ecosystem.

3.3.2 Indirect climate effect

Aerosols can also indirectly affect the climate system by influencing the microphysical structure of clouds and precipitation processes. For example, aerosols can increase the number of cloud droplets and reduce the size of individual droplets, making clouds more reflective of sunlight and reducing precipitation. This effect is known as the 'first indirect effect' of aerosols. In addition, aerosols may have more complex indirect climate effects by affecting the life cycle of clouds and the efficiency of precipitation.

4 Future prospects for aerosol research

4.1 Technology upgrading and innovation

Future research on aerosols will require some upgrading and innovation of existing technologies, such as the development of new aerosol preparation technologies. With the progress of nanotechnology and material science, the preparation technology of aerosols needs to be more refined and diversified. New preparation techniques such as microfluidic technology, laser induction technology, ultrasonic atomization technology, etc., will be able to prepare aerosol particles with smaller particle sizes, more uniform distribution and more stable performance. At the same time, the monitoring and control system will become more intelligent, and the development of intelligent technology will promote the upgrading of aerosol monitoring and control systems. In the future, there will be more aerosol monitoring equipment based on the Internet of Things, big data and artificial intelligence technology, which will be able to monitor the concentration, distribution and trend of aerosols in the atmosphere in real time and accurately, providing strong support for environmental protection and climate research.

4.2 Expansion of application areas

The application areas of aerosols are expanding with the improvement of related technologies. In the field of healthcare, aerosol inhalation therapy can be used as a new type of drug delivery. With the advantages of fast drug delivery and high bioavailability, this therapy plays an important role in the treatment of respiratory diseases such as asthma and chronic obstructive pulmonary disease. In addition, aerosols can be used for vaccine delivery and drug delivery. In the field of environmental protection, aerosol technology can be used in air treatment, water purification, soil remediation, etc., through adsorption, catalysis, decomposition and other mechanisms to remove pollutants in the atmosphere and water to improve environmental quality. In the field of industrial production, aerosol technology can be used in the preparation of nanomaterials, paint spraying, lubricant spraying and so on. In the future, as the degree of automation and intelligence of industrial production increases, it will play a greater role in improving production efficiency, reducing production costs and improving product quality.

4.3 Interdisciplinary collaboration expansion of application areas

Future research on aerosols requires further understanding of aerosol composition, physicochemical properties, climate system interactions, and aerosol health effects [12]. More accurate modelling of the direct and indirect effects of aerosols in regional and global climate models is an important direction for future research. In addition, interdisciplinary cooperation with each other is key to improving research on aerosols. If aerosols are cross-fertilized between physical chemistry and biology, more in-depth research can be carried out on the biochemical reactions of aerosols and the transport and dispersion of microbial aerosols. This will help to reveal the mechanism of aerosol impacts on ecosystems and human health. For example, cross-collaboration between environmental and atmospheric sciences can help to better understand the complexity and dynamics of the atmospheric environmental system through in-depth studies on the physical and chemical properties, sources and transport mechanisms, and climatic effects of aerosols. The development of environmental and atmospheric sciences will also be promoted.

4.4 International cooperation and policy promotion

Aerosol research requires international cooperation to jointly address global environmental issues. In the future, countries will strengthen cooperation and exchanges in aerosol monitoring, technology research and development, data sharing and other aspects to jointly promote the in-depth development of aerosol research. To strengthen international cooperation, the government can introduce more relevant policies to support aerosol research and technology application. For example, increasing investment in aerosol monitoring network construction, encouraging aerosol technology research and development and innovation, and promoting the application of aerosol technology in areas such as environmental protection and industrial production. These policy measures will provide a strong guarantee for the future development of aerosol research [13].

5 Conclusion

As a significant constituent of the atmosphere, aerosols exert considerable and pervasive influences on the Earth's climate and environment, largely due to their inherent diversity and complexity. First, aerosols have a significant effect on the climate system, both directly and indirectly. The direct effect is reflected in their ability to scatter and absorb solar radiation, thereby affecting the Earth's energy balance. This effect varies among different types of aerosols (e.g., sulfates, black carbon, etc.) and is influenced by some factors such as aerosol particle size, concentration and distribution. Indirect effects are even more complex, as aerosols act as cloud condensation nuclei and can change the microphysical properties of clouds, such as cloud thickness, brightness, and precipitation efficiency, which in turn have a long-term impact on the climate. Secondly, with the advancement of science and technology, aerosol research tools and techniques are also evolving. Breakthroughs in nanotechnology and materials science have made it possible to more precisely prepare and control the particle size, shape and surface properties of aerosol particles, thus enabling in-depth study of their physicochemical properties. Meanwhile, the development of intelligent monitoring technology enables us to monitor the concentration, distribution and trend of aerosols in the atmosphere in real time and accurately, providing important data support for environmental protection and climate research. In addition, the application fields of aerosol technology are expanding, such as environmental protection, medical health and industrial production. Advances in science and technology, such as nanotechnology and intelligent monitoring, have provided strong support for aerosol research and expanded its application areas. By strengthening international cooperation and policy promotion, the in-depth development of aerosol research can be better promoted, contributing more wisdom and strength to the fight against climate change and environmental protection.

References

1. Jietai Mao, Junhua Zhang, Meihua Wang, A review of atmospheric aerosol research in China. Acta Meteorologica Sinica. **60**, 625 (2002)
2. Shaodong Xie, Miao Yu, Ming Jiang. Current status of research on the sources and formation of organic aerosols. Acta Sci. Circumstantiae, **26**, 1933 (2006)
3. Chengcheng Liu. Mechanisms and climate effects of natural source aerosols: observations of marine aerosols and modelling of stratospheric wildfire smoke， Ph.D. thesis， China Univ. Sci. Technol, (2023)
4. Chengcheng Liu. Mechanisms and climate effects of natural source aerosols: observations of marine aerosols and modelling of stratospheric wildfire smoke， Ph.D. thesis， China Univ. Sci. Technol, (2023)
5. Li Hao, Wen Yang, Tongwen Wu, et al. Radiometric characterization of black carbon-sulphate mixed aerosol. Plateau Meteor, **29**, 1238 (2010)
6. Ping Tian. Physicochemical properties and source analysis of summer aerosols in Beijing， Ph.D. thesis， J. Beijing Norm. Univ., Nat. Sci, (2014)
7. Huaping Li, Ming Zhang, Lunche Wang, Yunbo Lu, Lan Yu, Yingying Ma, Wei Gong, Effects of aerosol on downward diffuse radiation under blowing dust and haze conditions, Atmos. Environ, **334**, 120682 (2024)
8. Jiaren Sun, Liu Yu, Possible impacts of regional aerosols on the East Asian summer winds in China(Ⅰ):Impacts of sulphate aerosols, ACCR, 4, 111(2008)

9. Jiaren Sun, Liu Yu, Possible impacts of regional aerosols on the East Asian summer winds in China(II): black carbon aerosols and their combined impacts with sulphate aerosols, ACCR, **4**, 161(2008)
10. Xin Zhang, Yue Wang, Chengduo Yuan, Zibo Zhuang, Study on visibility forecast optimization based on aerosol- meteorological feedbacks in wet conditions, Urban Clim, **55**, 101951 (2024)
11. Tao Deng, Shanshan Ouyang, Guowen He, Xue Zhang, Jeremy Cheuk-Hin Leung, Xiaoyang Chen, Qing Wang, Zebiao Zhang, Yu Zou, Boru Mai, Li Liu, Xuejiao Deng, Impact of aerosol actinic radiative effect on ozone during haze pollution in the Pearl River Delta region, Atmos. Environ, **332**, 120610 (2024)
12. Mingxing Wang, Renjian Zhang. Frontiers in atmospheric aerosol research. Climatic Environ Res, **6**, 119 (2001)
13. Yueyi Feng, Miao Ning, Yu Lei, Yamei Sun, Wei Liu, Jinnan Wang, Defending blue sky in China: Effectiveness of the "Air Pollution Prevention and Control Action Plan" on air quality improvements from 2013 to 2017, J. Environ. Manage, **252**, 109603 (2019)

Study on current climate change issues and response strategies

Jiayu Li[1 *], *Chongshuo Tao*[2], and *Yixian Zhang*[3]

[1]Nanjing University of Information Science & Technology, School of Ecology and Applied Meteorology, 210044 Nanjing, China
[2]Wuhan Experimental school, 430014 Wuhan, China
[3]Griggs International Academy, 215300 Suzhou, China

Abstract. Today, human activities are generating massive emissions of greenhouse gases, intensifying the issue of global climate change. This paper provides a comprehensive analysis of the current situation, impacts, and response strategies related to climate change. It also discusses the influences of climate change on agricultural production, socio-economic systems, and human health. Global warming, primarily caused by human-induced greenhouse gas emissions, has intensified due to feedback effects. Frequent extreme weather events and significant changes in ecosystems have profoundly affected agricultural production, socio-economic systems, and human health. To tackle climate change, mitigation and adaptation approaches are suggested, including the development of clean energy, ecosystem conservation, technological innovation, and policy support. It emphasizes the importance of international cooperation and points out that enhancing public education and participation is key to addressing climate change. In the future, global attention to and research on climate change must be continuously strengthened, and response strategies need to be constantly improved. Nations should further increase their investments in clean energy development, ecosystem protection, and technological innovation, faithfully fulfil international cooperation agreements, and jointly work to slow down the pace of climate change.

1 Introduction

In the current global environmental context, climate change has emerged as one of the paramount challenges confronting humanity. With the acceleration of industrialization and urbanization and the increasing frequency of human activities, the concentration of greenhouse gases in the atmosphere continues to rise, leading to significant changes in the global climate. Frequent extreme weather events, such as heavy rains, floods, droughts, and hurricanes, pose serious threats to human life and property safety. Concurrently, climate change has also caused huge impacts on ecosystems, with matters like biodiversity loss, sea-level rise, and glacier melting becoming increasingly prominent. These changes not only affect the lives of the current generation but will also have profound effects on the survival and development of future generations.

* Corresponding author: 202213020041@nuist.edu.cn

Currently, scholars have carried out research on climate change issues. In terms of the causes of climate change, human activity-induced greenhouse gas emissions are commonly regarded as the main driving factor. In terms of impact assessment, studies have covered the effects of climate change on a wide range of areas, including agriculture, water resources, ecosystems, and human health. In terms of response strategies, the international community has taken a series of measures, such as the signing of the Paris Agreement, with countries setting emission reduction targets and increasing the development and utilization of renewable energy. Simultaneously, technological innovation also plays an important role in addressing climate change, with technologies like carbon capture and storage and new energy technologies emerging continuously. However, despite certain research achievements and practical progress, the problem of climate change remains severe, and further in-depth research and exploration of effective response strategies are still needed, as analyzed by scholars like H Mohajan on the relationship between greenhouse gas emissions and temperature rise [1], and by Xiu Xin and Liu Jun on the economic effects of some emission reduction policies [2].

This paper will provide a comprehensive analysis of the current status of climate change, its impacts, and response strategies. The research content includes the scientific basis of global climate change, its extensive impacts at global and regional levels, and tactics for mitigating and adapting to climate change. This paper will also explore the roles of technological innovation and policy support in addressing climate change, and how to strengthen the global response to climate change through international cooperation and public participation. It aims to provide a comprehensive perspective on the issues of climate change and its response strategies, facilitating the implementation of global climate actions.

2 Status of climate change

2.1 Global warming

Global warming refers to the phenomenon of rising global temperatures resulting from the accumulation of the greenhouse effect due to the substantial emissions of greenhouse gases from human activities. Greenhouse gas emissions are the primary cause of global warming today. Because of their infrared activity, greenhouse gases are capable of absorbing and emitting infrared radiation within the same wavelength range emitted by the Earth's surface, clouds, and the atmosphere, thereby preventing heat from escaping into space [3]. This phenomenon leads to an increment in Earth's surface temperature, known as global warming. Additionally, greenhouse gas emissions trigger various feedback effects, such as reduced reflectivity from melting ice and snow which further absorbs heat, and warmer oceans releasing more methane, increasing greenhouse gas concentrations. These feedback mechanisms are also significant contributors to global warming caused by greenhouse gases. The IPCC's Sixth Assessment Report (AR6) explicitly states that human activities primarily cause global warming through greenhouse gas emissions. The warming caused by greenhouse gases is mainly attributed to carbon dioxide and methane. Although this warming is to some extent masked by the cooling effects of aerosols, the rise in global surface temperatures due to greenhouse gases is conspicuous. Future warming trends depend on future greenhouse gas emissions, with cumulative net carbon dioxide emissions playing a dominant role [4].

Volcanic activity has complex impacts on the global climate. It can cause short-term cooling and also potentially contribute to global warming to some extent. Volcanic eruptions are accompanied by the ejection of numerous aerosols into the atmosphere, like sulfur dioxide. These particles block and reflect sunlight, reducing solar radiation received at the surface and causing a short-term drop in near-surface temperatures. However, in the long term, volcanic activity accelerates global warming. On one hand, substances released during volcanic eruptions, such as hydrogen chloride, can react with atmospheric ozone, depleting the ozone layer and enhancing surface solar radiation. On the other hand, greenhouse gases such as carbon dioxide, due to their stable chemical properties, can remain in the atmosphere for long periods, absorbing surface radiation heat, and enhancing the greenhouse effect, thereby driving global warming.

In addition, solar activity, changes in land surface properties, and aerosol emissions are also important factors affecting global warming. Solar activities, such as sunspots and flares, cause periodic changes in solar radiation. Changes in land surface properties affect the climate by altering the absorption and reflection of solar radiation. The melting of ice caps and glaciers not only triggers sea level rise but also releases greenhouse gases, exacerbating global warming and creating a vicious cycle. It is the impact of aerosol emissions that is more complex: although some aerosols can reflect solar radiation and temporarily bring about a cooling effect, they can also absorb radiation or deposit on ice and snow surfaces, accelerating melting and indirectly promoting global warming. Furthermore, aerosols affect cloud formation and precipitation distribution, further exacerbating climate change.

2.2 Extreme weather events

In recent years, extreme weather events have become frequent, and these events are closely related to climate change. The occurrence rates of high temperatures and wildfires have notably increased. In 2023, various regions in the Northern Hemisphere experienced extremely high temperatures, especially in China, Europe, and the United States, where temperatures repeatedly set new records, with some areas in Europe even exceeding 40 degrees Celsius. The AR6 report also reveals that under the scenario of continuous global warming, extreme heat events will continue to increase in frequency and intensity [5]. Additionally, heatwaves can trigger droughts and fires, especially in regions with scarce water resources.

Simultaneously, another significant extreme weather event that cannot be ignored is precipitation and flood disasters. Climate change significantly increases the likelihood of extreme precipitation events. As the atmosphere warms, the content of water vapor in the air increases, leading to enhanced precipitation intensity and a rise in the frequency of extreme rainfall, thereby significantly increasing flood risks. Moreover, global warming also intensifies glacier melting and sea-level rise, further elevating the risk of flooding in coastal areas. A warming atmosphere can carry more moisture, releasing large amounts of water during precipitation processes, increasing the likelihood of local heavy rain and flooding.

2.3 Ecosystem changes

Ecosystems have undergone extensive and profound alterations due to climate change. To begin with, biodiversity is facing significant threats. Climate change has intensified the risk of species extinction, particularly for those species that are extremely sensitive to climatic conditions of their habitats, such as cold-water fish and Arctic wildlife. Changes in temperature and climatic patterns hinder these species' ability to swiftly adapt to their current environments, leading to potential extinctions. Climate change also alters the foraging, reproduction, and migration behaviors and patterns of wildlife, causing some species to struggle to survive in the rapidly changing new environments. A decline in key species can trigger a domino effect due to food chain disruptions and migration shifts, potentially leading to the collapse of entire ecosystems.

Furthermore, the changes induced by climate change vary across different ecosystems. In forests, global warming and resultant temperature increases lead to more frequent droughts and extreme heat events. This not only increases the frequency and intensity of forest fires but also affects the growth cycles and distribution of plant species, reducing forest biodiversity. For wetlands, the sea level rise caused by global warming results in some coastal wetlands being submerged, posing habitat reduction risks for many wetland species and reducing their carbon sequestration capabilities. In oceans, the increase in carbon dioxide levels due to climate warming causes ocean acidification, disrupting the pH balance in marine ecosystems and affecting the survival of coral and shellfish [6]. Additionally, rising temperatures lead to warmer sea waters, potentially altering the distribution of many fish species and marine mammals.

3 Problems caused by climate change

3.1 Impact on agricultural production

Climate change has a significant and complex impact on agricultural production, primarily through changes in key climatic variables such as temperature, precipitation, and sunlight. These changes directly affect the growing environment, growth cycles, yield and quality of crops.

Firstly, rising temperatures have increased heat resources in many regions. For crops in cooler regions, such as wheat and potatoes, excessively high temperatures may inhibit growth, leading to reduced yields or even crop failures. Secondly, frequent extreme heat events not only increase the risk of heart damage to crops but may also accelerate soil moisture evaporation, exacerbating drought conditions. Thirdly, changes in precipitation patterns bring challenges to agricultural production. In some areas, reduced rainfall exacerbates drought issues, increasing irrigation demands for crops. Moreover, increased rainfall in other areas, particularly through extreme weather events like heavy rain and floods, can directly destroy farmland and crops and may lead to soil erosion and fertility decline. Changes in sunlight conditions also impact agricultural production. Excessive sunlight can cause leaf burns, affecting photosynthesis efficiency. Changes in cloud cover and foggy conditions brought about by climate change may also affect the intensity and duration of sunlight, further influencing crop growth and yields.

The impacts of climate change on agriculture are multifaceted, featuring both positive stimulatory effects and negative inhibitory effects. It is necessary to enhance research and application of agricultural climate adaptability and to improve the stability and sustainability of agricultural production. Moreover, strengthening meteorological monitoring and early warning systems is crucial to providing timely and accurate meteorological information and services to agriculture.

3.2 Impact on socio-economic stability

Climate change also impacts socio-economic stability, including economic fluctuations triggered by issues like reduced agricultural production and water scarcity, and potentially exacerbating social conflicts and poverty. Climate change may lead to reduced agricultural yields and water shortages. In recent years, increased extreme weather events and more frequent meteorological disasters, such as flooding, severely damaged farmlands and agricultural infrastructure. Long-term flooding can cause oxygen deprivation in plant roots, while droughts lead to insufficient soil moisture and substantial reductions in food production. On one hand, reduced crop yields decrease farmers' incomes. On the other hand, reduced food production and water shortages may lead to higher food prices, raising living costs for the general populace and sparking socio-economic tensions. Some scholars have researched the relationship between climate change and regional conflicts, finding that climate change indeed correlates with an increased rate of military conflicts [7]. Frequent extreme weather events can severely damage arable land, destroy urban infrastructure, and even damage urban buildings, causing substantial economic losses to governments and people, and exacerbating poverty issues.

3.3 Impact on human health

The impact of climate change on human health is multi-layered and complex. Firstly, frequent extreme weather events caused by climate warming pose direct threats to human health. Heatwaves significantly increase the incidence and mortality rates of heat-related illnesses, such as heatstroke, particularly affecting vulnerable groups like the elderly, children, and those with chronic illnesses. Secondly, natural disasters like floods not only cause casualties but also lead to water pollution and outbreaks of infectious diseases, further exacerbating health risks. Additionally, climate change may also worsen air quality. With rising temperatures, pollutants in the atmosphere are more likely to form and accumulate, such as ozone pollution, which is particularly severe in the summer and can damage the respiratory and cardiovascular systems. Extreme events like forest fires also release a large number of harmful substances, affecting air quality and threatening human health. Furthermore, the indirect effects of climate change are even more profound, involving the instability of food supplies, water shortages, pollution, and the degradation of ecological environments. These changes not only weaken basic human survival conditions but also erode human health defenses through food chain contamination, the spread of waterborne diseases, and the emergence of new pathogens.

4 Coping strategies for climate change

4.1 Measures to mitigate climate change

In the face of increasingly severe climate change, a series of measures, such as promoting energy conservation and emission reduction in production and daily life, enhancing ecosystem protection, and formulating policies to promote international cooperation, can effectively mitigate the phenomenon of climate change. First, the development of clean energy sources such as solar, tidal, and wind energy can effectively curb the emissions of greenhouse gases like carbon dioxide and methane. According to a report by the International Energy Agency, global wind power capacity has significantly increased, with a projected additional growth of over 200% by 2030. In the first half of 2023, China's wind power construction rapidly developed, with cumulative power generation increasing by 20% compared to the previous year. By the end of June 2023, China's wind power capacity reached approximately 390 million kilowatts, a year-on-year increase of 13.7% [8]. Second, protecting ecosystems is crucial for mitigating climate change. Forests, oceans, and wetlands can store and absorb atmospheric carbon dioxide through photosynthesis and store it in plants and their soil. Thus, greenhouse gases accumulated in the atmosphere can be effectively absorbed. Local governments need to encourage forest and wetland conservation efforts, regularly clean up garbage pollutants in ecological reserves to prevent erosion and soil degradation and strengthen personnel supervision by enforcing strict forest conservation regulations to prevent rampant deforestation. Third, promoting global climate governance cooperation and strengthening the construction of international regulations help share the responsibility for climate among countries. For example, in the 2015 Paris Agreement, countries discussed and set their own energy conservation and emission reduction targets and implemented climate change response measures, working together to keep warming within 1.5 degrees Celsius. China promotes green and low-carbon infrastructure construction through the "Belt and Road" initiative, encouraging partner countries to implement environmentally friendly development policies through green investments and the development of renewable resources, achieving low-carbon development.

4.2 Measures to accommodate to climate change

Climate change has become one of the severe challenges facing the globe, with average global temperatures and extreme weather events continuously increasing, affecting all facets of society and human life. Therefore, adapting to climate change is especially crucial while efforts to mitigate it are ongoing.

First, the development of agricultural adaptability strategies such as crop planting structures, improved irrigation systems, and the development of drought-resistant crops are essential. These strategies enable crops to better adapt to a climate change environment characterized by global warming and increased extreme drought and precipitation. Since 2004, Shandong Province has actively implemented late harvesting of summer corn and late planting of wheat. At the beginning of September each year, the Shandong Provincial Climate Center produces a forecast for the suitable initial sowing period for winter wheat, guiding farmers to scientifically select the sowing period based on the current year's climate, laying a solid foundation for achieving a bountiful harvest the following year [9].

Second, in response to the increase in extreme precipitation events, local governments need to enhance urban drainage system capabilities. By constructing flood prevention facilities, cities can ensure normal operations during extreme precipitation. Further dredging of urban rivers and connecting urban rivers with natural rivers and lakes can achieve rapid rainwater discharge. For undeveloped urban low-lying areas, the construction of artificial lakes can enhance the city's water retention capacity. In the surrounding areas of key cities, the construction of flood storage and detention areas is strengthened to leave space and routes for floodwaters, enhancing the overall regional capacity to defend against floods [10].

Additionally, improving public health security is urgent. Governments can enhance the capacity to respond to extreme weather events and the spread of infectious diseases by strengthening public health systems, ensuring the safety of people's lives. For example, establishing a comprehensive infectious disease prevention and control system, and enhancing the construction of medical infrastructure, especially in remote rural areas where medical resources are scarce.

4.3 Technological innovation and policy support

In terms of technological innovation, technologies such as clean energy, carbon capture and storage, climate-smart agriculture, green building and smart city technologies are rapidly evolving. Continuous breakthroughs in these technologies not only improve resource utilization efficiency and reduce greenhouse gas emissions but also provide robust technical support for the sustainable development of human society. At the same time, policy support plays an indispensable role in guiding, incentivizing, and ensuring the research and application of these technologies. Governments establish clear emission reduction targets and timetables, creating a solid legal foundation for climate change response through relevant laws and regulations. Financial subsidies, tax incentives, a green financial system, and a carbon emission trading market effectively reduce the costs for businesses and individuals to engage in climate change responses, stimulate market vitality, and promote the widespread application of low-carbon technologies.

Moreover, public education and participation are also supported by policies. By disseminating knowledge about climate change, enhancing public awareness and participation, a favorable societal atmosphere for jointly addressing climate change is formed. This comprehensive effect not only helps mitigate the trend of climate change but also lays a solid foundation for achieving sustainable economic and social development. Therefore, strengthening the integration of technological innovation and policy support will be an important path for addressing climate change and promoting global sustainable development in the future.

5 Conclusion

Climate change is a complex and urgent global issue that not only affects the natural environment worldwide but also profoundly impacts all aspects of human society. This paper, through an analysis of the current status of climate change, reveals its extensive impacts on agricultural production, socio-economic stability, and human health. To tackle this challenge, a series of measures for mitigating and adapting to climate change have been proposed, including the development of clean energy, protection of ecosystems, technological innovation, and policy support. The paper emphasizes that international cooperation, public education, and participation are crucial for effectively responding to climate change. In the future, it is necessary to further enhance technological innovation, improve the policy support system, and heighten public awareness to achieve effective climate change responses and global sustainable development.

Authors contribution

All the authors contributed equally, and their names were listed in alphabetical order.

References

1. Mohajan H, Greenhouse gas emissions increase global warming. International Journal of Economic and Political Integration. 1(2), 21-34 (2011).
2. Xiu Xin, and Liu Jun, Economic Effects of the EU ETS Carbon Emission Reduction Policy on Emission-Controlled Enterprises: Implications for China's 'Dual Carbon' Strategy Implementation and National Carbon Trading Market Construction. Econ Inq. 05, 108-124 (2024)
3. Matthews, J.B.R., V. Möller, R. van Diemen, J.S. Fuglestvedt, V. Masson-Delmotte, C. Méndez, S. Semenov, and A. Reisinger. "Glossary." Climate Change 2021: The Physical Science Basis. Cambridge University Press, 2215-2256 (2021).
4. Core Writing Team, H. Lee, and J. Romero. "Sections." Climate Change 2023: Synthesis Report. Contribution of Working Groups I, II, and III to the Sixth Assessment Report of the Intergovernmental Panel on Climate Change. IPCC, Geneva, Switzerland, 35-115 (2023).
5. Zhou, Botao, and Qian Jin, Interpretation of IPCC AR6 Report: Changes in Extreme Weather and Climate Events. Adv Climate Change Res. 17, 713 (2021)
6. Seneviratne, S.I., X. Zhang, M. Adnan, W. Badi, C. Dereczynski, A. Di Luca, S. Ghosh, I. Iskandar, J. Kossin, S. Lewis, F. Otto, I. Pinto, M. Satoh, S.M. Vicente-Serrano, M. Wehner, and B. Zhou. "Weather and Climate Extreme Events in a Changing Climate." Climate Change 2021: The Physical Science Basis. Cambridge University Press, 1513–1766 (2021).
7. Ko J, Lee H F, Leung C K, War and warming: The effects of climate change on military conflicts in developing countries (1995–2020). Innovation and Green Development. 3(4), 100175 (2024)
8. Wang, Qian, Research on the Carbon Emission Reduction Effect of China's Clean Energy Development. Hebei University of Geosciences. 16 (2023).
9. Yan, Hong, Agricultural Production Strategies for Adapting to Climate Change. China Meteorological News. 2023-12-13(002). 2023.002272.
10. Wu, Hao, Enhancing Urban Resilience to Flood Disasters. Economic Daily. 2024-07-28(006). 2024.005149.

The contribution of artificial meat in reducing carbon emission and mitigating climate change

Qiongyue Hu*

Beijing 101 High school, Beijing, 100091, China

Abstract. The consumption of conventional livestock meat contributes significantly to greenhouse gas emissions, exacerbating climate change. As an alternative, artificial meat, or cultured meat, has emerged as a potential solution to reduce the environmental impact of meat production. In this paper, contribution of artificial meat in reducing carbon emission and mitigating climate change is analyzed. Comparison of the impact on carbon emission between artificial meat and conventional lifestock production system is conducted. The results indicated that artificial meat has advantage over natural meat. In terms of life cycle assessment in cultured meat, current LCA lack some practical data in commercial-scale, and the factor about transportation, usage, retail, and waste disposal should be included in the future assessment, which will give a more comprehensive and accurate assessment in whole life cycle of artificial meat production. In conclusion, while artificial meat holds promise for reducing carbon emissions and mitigating climate change, more research and development are needed to realize its full potential and address the associated challenges.

1 Introduction

Recently, with the raise of environmental awareness and continuous development in food industry, meat production gradually becomes a popular topic in reduction of carbon emission. According to the statistical report on Per Capita Major Food Consumption of National Residents in 2023 published by National Bureau of statistics of the People's Republic of China (NBSPRC), the consumption of meat production had been oscillating upward from 2016 to 2022, increasing from 26.1kg to 34.6 kg, demonstrating the rising demand in meat consuming [1]. On the other hand, the process of production require plenty of land use, which causes the loss of habitats for wildlife and irreversible damage of ecosystem, indirectly leading to extinction of some threatened species and imposing overwhelming pressure on environment. Recent research shows that the process of raising livestock contributes to 14.5% greenhouse gas emission, and the production of beef emit 44% anthropogenic methane emission and 53% of anthropogenic nitrogen dioxide emissions [2]. In order to follow the trend of sustainable development and carbon emission reduction, artificial meat is gradually emerging and become hotspot in food industry in recent decades. Cultured meat cultivated by stem cell of animals and plant-based meat shares indistinguishable flavor and appearance

* Corresponding author: eva.hu26@beijing101id.com

may become proper substitute of conventional meat. Some researches revealed that artificial meat, especially cultured meat includes "approximately 7-45% lower energy use (only poultry has lower energy use), 78-96% lower GHG emissions, 99% lower land use, and 82–96% lower water use depending on the product compared", which effectively relieve the pressure imposing on the environment [3]. This paper aims to holistically review the process of producing different categorization of artificial meat and evaluate whether artificial meat plays an important role in reducing carbon emission and slowing down the process of climate change. In the end, this paper will focus on challenges in future developing and give some practical advices towards these potential problems.

2 Manufacture of artificial meat

Artificial flesh can be primarily categorized into three varieties: those being meat substitutes derived from vegetal extracts and mycota, vitro-cultivated (or lab-generated or synthetic flesh) produced via tissue biotechnology or originating from genetically altered organisms and cloned fauna through genetic manipulation, as illustrated in fig. 1 [4].

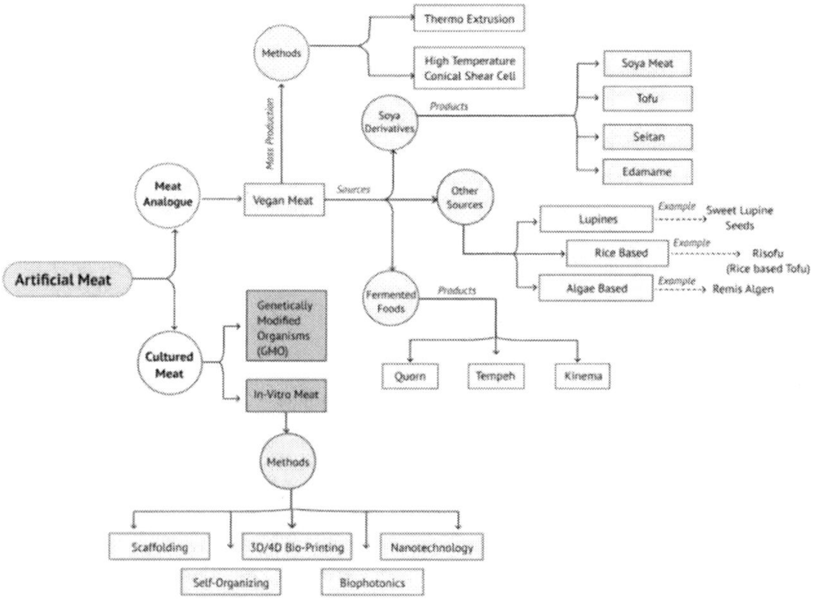

Fig. 1. Classification of Artificial meat [4]

2.1 Vegan meat

Vegan meat, which is also extracted from plant and fungi, use thermo-extrusion technology to texturize protein extracted from plant into fiber under relative high temperature as 140-180 Celsius. Formatting author names and author affiliations

2.2 Cultured meat in vitro environment

For cultured meat, there are typically 2 main technologies to cultivate the meat in vitro environment as scaffolding technique and self-organizing technique. Scaffolding technique is to use embryonic myoblast (skeletal stem cells) extracted from tissue of agricultural animals, and allow the stem cell to develop on scaffold within bioreactor, which is indispensable circulatory system that offers moderate conditions (temperature, pH, pressure, nutrient feeding) for cells to growth and differentiate while eliminating metabolic waste. Combined forces in bioreactor help improve diffusion and achieve a high mass transfer rate with a low shear stress level. In terms of scaffold, this technique uses a plant-derived growth medium or animal derived growth medium such as collagen in vitro meat production system to provide nutrient for cell growth and help them change into fiber when the cells are placed on scaffold. Another method to cultivate cultured meat is to explant tissue from animals to Petri plates with nutrient mixture, growth medium of fetal bovine serum to support cell proliferation. After it finishes growing and achieve meat size, it was marinated in olive oil, garlic and deep-fried, and then the meats are submitted to the sensory panel reporting that the explants and newly grown tissue looked and smell end edible [4].

3 Comparison of the impact on carbon emission between artificial meat and conventional lifestock production system

3.1 Cultured meat

Some earlier literatures use life cyle assessment conclude that cultured meat is generally environmentally friendly comparing to conventional meat, as demonstrated by Tuomisto and de Mattos in 2011[3]. They found that the cultured meat involves lower energy use, lower greenhouse gas emissions, lower land use, and lower water use depending on the product compared. However, this result is not definitive and certainty since it lacks data from practical production in commercial-scale and fail to consider the complex reaction between society and technology, such as consumer acceptance and economic factor [5]. Additionally, the feedstock used in the system evaluated by previous literature does not match current situation of cultured meat production, requiring further evaluation in energy usage.

In terms of carbon emission, although the prediction significantly varies from each other due to different system and different media cultivating cultured meat [6], the overall Global Warming Potential carbon dioxide equivalent is lower than conventionally produced meat, such as beef. Considering the long -term impact in mitigating climate change brought by cultured meat, however, the effect is not explicit. According to previous analysis, although carbon emission is far less than emission of other greenhouse gas (GHG) that has larger global warming potential (GWP), carbon dioxide causes more propounded effect since it stays longer than any other GHG. As a result, the impact brought by carbon dioxide is irreversible while other GHG is reversible. Unfortunately, the main gas emitted by production of cultured meat is carbon dioxide, because it needs fossil fuel to warm cultured cells, meaning the long-term effect is not obvious. According to the prediction from John Lynch and Raymond Pierrehumbert in 2019 shown in fig. 2 [6], in next 1000 year all the 4 cultured meat has relative advantages over conventional meat at first, but as the time proceed cultured-d even exceed the worst cattle system, suggesting the role cultured meat played in mitigating climate change is not absolute and conclusive [6, 7] (fig. 3). Moreover, recent study about cradle-to-gate life cycle assessment conveyed an opposed opinion in carbon emission, arguing that when adding energy usage in providing necessary growth medium for cell growth and energy required for constructing facility, they found that global warming potential (GWP) for all scenarios was greater than the minimum reported GWP for retail beef [8]. It states that cultured meat may have more severe impact in carbon emission than conventional meat, contradicting to previous assumption that cultured meat is environmental-friendly product.

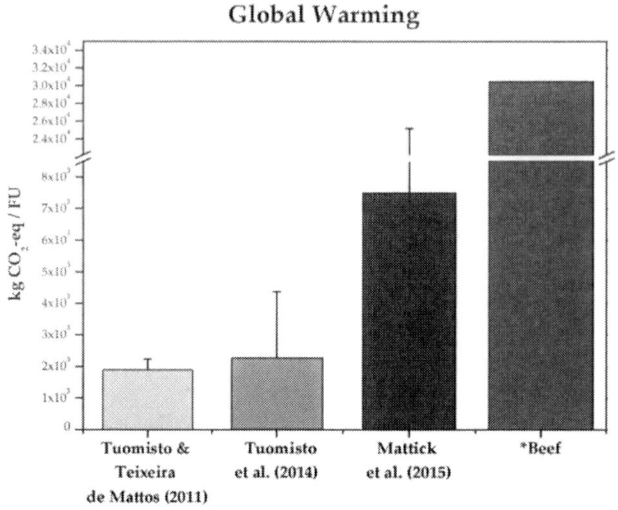

Fig. 2. CO_2 and GHG emission predicted by previous researches [7, 9]

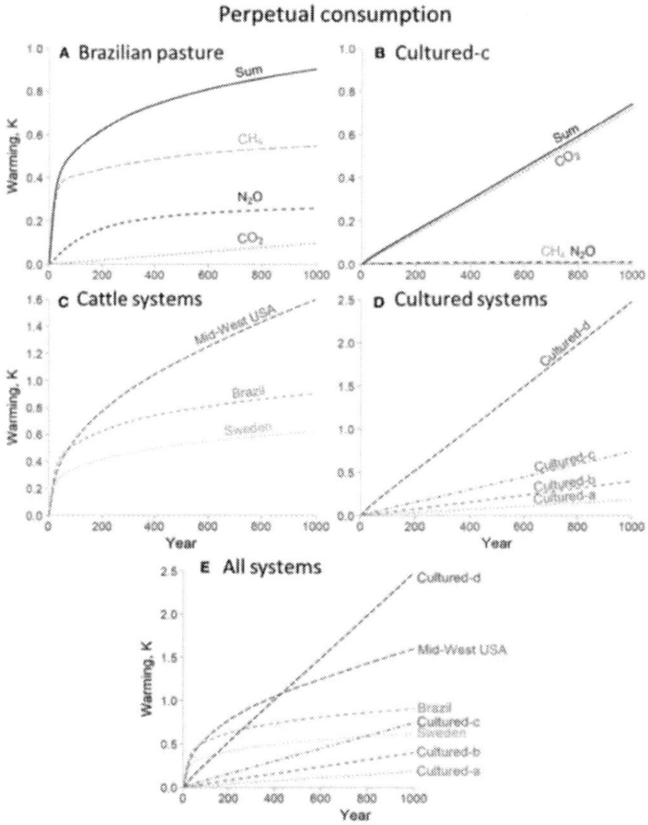

Fig. 3. Prediction of effect in climate change from 3 conventional meat production system and 4 cultured meat system in 1000 years [7]

3.2 Vegan meat

Life-cycle assessments show that products like the Beyond Burger and Impossible Burger, burgurs that are composed of plant-based meat, result in lower greenhouse gas emissions, approximately 3.2 to 3.5 kg CO_2-eq per kilogram of product, compared to the range of 10.2 to 48.5 kg CO_2-eq per kilogram for beef raised in feedlots [10]. However, this situation is not absolute as well since when considering amino acid and nutrient density, the differences between vegan meat and real meat is not clear because of high efficiency of animal meat to meat nutrient need. Furthermore, using some advanced land management, such as pastured beef system, involving rotating grazing and crop grazing, can counteract greenhouse gas emission (GHGE) and result in net negative carbon footprint that is less than carbon emission of vegan meat. Moreover, just like situation in cultured meat, plant-based meat mainly emits significant carbon dioxide when animal meat dominantly produces methane, suggesting long-lasting impact in climate change [10].

4 Conclusion

Based on previous literature, artificial meat, especially the vegan meat that has wide range of application, generally play an important role in reducing carbon emission and alleviating global warming in recent decades. Nevertheless, more detailed examination and evaluation should consider since the long-term effect of artificial meat is not significantly advantageous than conventional meat if we put situation in a larger time scale. In terms of life cycle assessment in cultured meat, current LCA lack some practical data in commercial-scale, and the factor about transportation, usage, retail, and waste disposal should be included in the future assessment, which will give a more comprehensive and accurate assessment in whole life cycle of artificial meat production. Although the specific effect in reducing carbon emission and softening global warming remains uncertain, the overall contribution, such as reducing land use and water use, reveals that artificial meat is more sustainable than conventional meat. Artificial meat is full of potential due to its environmental impact; however, it still needs some time to achieve commercialization and wide range of application, and social and economic factors such as different level of acceptance among different kinds of consumers need to further evaluate in the future.

References

1. Huxiu. Chinese Meat Consumption: How Far Are We from "Eating Well"? Retrieved on 8th November, 2024, retrieved from
2. Gerber, P.J., Steinfeld, H., Henderson, B., et al. 2013. Tackling climate change through livestock: a global assessment of emissions and mitigation opportunities. Food and Agriculture Organization of the United Nations (FAO).
3. Hanna L. Tuomisto, M.Joost Teixeira de Mattos. Environmental Impacts of Cultured Meat Production. Environmental Science & Technology. **45**, 6117-6123 (2011). https://pubs.acs.org/doi/10.1021/es200130u
4. Mateti, T., Laha, A. & Shenoy, P. Artificial Meat Industry: Production Methodology, Challenges, and Future. JOM. **74**, 3428-3444 (2022). https://doi.org/10.1007/s11837-022-05316-x
5. Carolyn S Mattick, Amy E Landis, Braden R Allenby. A case for systemic environmental analysis of cultured meat. Journal of Integrative Agriculture. **14,** 249-254 (2015). https://doi.org/10.1016/S2095-3119(14)60885-6.

6. Mattick CS, Landis AE, Allenby BR. Anticipatory, life cycle analysis of in vitro biomass cultivation for cultured meat, production in the United States. Environ Sci Technol. **49**, 11941-9 (2015) doi: 10.1021/acs.est.5b01614
7. John Michael Lynch, Raymond Pierrehumbert. Climate Impacts of Cultured Meat and Beef Cattle, Front. Sustain. Food Syst., Sustainable Food Processing. 3(2019). https://doi.org/10.3389/fsufs.2019.00005
8. Derrick Risner, Yoonbin Kim, Cuong Nguyen et al. Environmental impacts of cultured meat: A cradle-to-gate life cycle assessment. bioRxiv. **04**, 537778(2023) doi:https://doi.org/10.1101/2023.04.21.537778
9. Rodríguez Escobar MI, Cadena E, Nhu TT, Cooreman-Algoed M, et al. Analysis of the Cultured Meat Production System in Function of Its Environmental Footprint: Current Status, Gaps and Recommendations. Foods. **10**, 2941(2021). https://doi.org/10.3390/foods10122941
10. Stephan van Vliet, Scott L. Kronberg, Frederick D. Provenza. Front. Sustain. Food Syst. Agroecology and Ecosystem Services, **4**, 2020. https://doi.org/10.3389/fsufs.2020.00128

Comparative study of biosynthesis technology and traditional methods for organic solid waste

Yuzhou Wang*

Environment Science and Engineering, CHANGZHOU UNIVERSITY, Changzhou, 213164, China

Abstract. With the development of industrialization and trade, the amount of organic solid waste is increasing rapidly, which has a serious impact on the environment. Traditional methods of organic solid waste treatment method will have negative impacts during the process of the treating organic solid waste. The theme of the study is to find an efficient, environmentally friendly and economical organic solid waste treatment method. The best organic sold waste treatment method is determined by comparing landfill, incineration and biosynthesis technology in aspects of resource utilization, cost and environmental impact. Through research, in aspects of resource utilization and cost, biosynthesis technology has its own advantages and disadvantages compared with traditional landfill and incineration technologies. In aspects of environmental impact, biosynthesis technology has obvious advantages. This is not only because of the mild reaction conditions and no secondary pollutants of biosynthesis technology, but also because biosynthesis technology can improve the soil condition and reduce the emission of the greenhouse gas.

1 Introduction

As industrial production and trade growing continuously in many countries, the amount of solid organic waste is increasing rapidly, which has a serious impact on the environment. More than 2 billion tons of solid waste are generated worldwide every year, and organic solid waste account for a large proportion in the solid waste [1]. Traditional methods of organic solid waste treatment, such as landfill and incineration, have solved the problem to a certain extent. But at the same time, they have brought about a series of other problems, such as soil pollution and air pollution. The greenhouse gases from landfill and incineration, such as CO_2, CH_4 and N_2O, mix over the whole earth when they stay in the atmosphere long enough [2]. This will lead to global warming and do harm to the human health. The efforts to reduce greenhouse gas emissions becomes the key part to solve the problem. Therefore, carbon emissions should be increased when disposing the organic solid waste. Additionally, huge economic benefits are associated with it, exceeding $3 trillion per year [3]. Biosynthesis technology, an efficient, environmentally friendly and economical organic solid waste treatment technology, fits the advantages above and should be widely used.

* 2300380126@smail.cczu.edu.cn

Biosynthesis technology uses biological systems, such as microorganisms, to degrade pollutions in the environment. Organic solid waste degradation requires the action of microorganisms to break down complex polymers into simpler molecules [4]. Furthermore, biosynthesis technology can not only treat the amount of organic solid waste, but also turn waste into treasure, transforming waste into energy products such as organic fertilizer and biomass natural gas [5]. Improving the management efficiency of organic solid waste holds significant value for safeguarding the environment, conserving resources, fostering economic growth, and safeguarding public health [6]. The maturation of biosynthetic technologies necessitates a contemplation of challenges pertaining to biosecurity and the scalability of production processes. As a result, the widespread acknowledgment of biosynthetic technology is evident.

The goal of the study is filling the gap in the existing academic articles, which about the organic solid waste treatment. The study will compare biosynthesis technology with traditional technology in aspects of resource utilization, cost and environmental impact.

2 Landfill

The method of landfilling is accepted and used widely because of its economic advantages [7]. Currently, landfills are easy, adjustable and low-budget than other disposal methods. Virtually all waste can be disposed of through landfill, not just the organic solid waste [7]. But the organic solid waste in the landfill is not completely disposed harmlessly, and many bacteria, germs, heavy metal ions remain in it. Therefore, in the process of waste stacking and landfilling, leachate formed by mixing water and waste is highly toxic for the following reasons. Firstly, it will directly affect the quality of human drinking water. Secondly, these pollutants produced by organic solid waste will enter the biosphere and affect human health potentially by the pathway of bioaccumulation and biomagnification [8]. If a significant amount of financial resources and time is once again allocated to address the issue of landfill leachate, it will undoubtedly undermine its economic advantages. Additionally, a large amount of gas is emitted during the processes of landfilling. The typical composition of landfill gas is 60% methane and 40% carbon dioxide, respectively [9]. These gases act like a cover, absorbing Infra-Red radiation and preventing it from escaping to outer space [10]. Consequently, the temperature of the atmosphere and the surface increased significantly. The trend of the landfill gas over time is shown in the fig 1. As shown in this figure, in stage IV, landfill gas reaches its maximum volume and then decreases with time. The emission of the landfill gas will lead to global warming, which is not conform to the goal of environmentally friendliness.

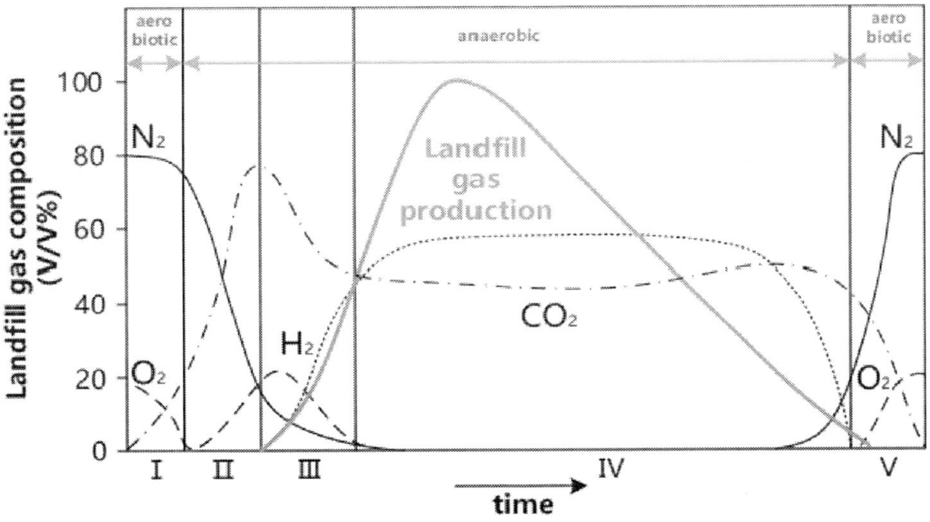

Fig. 1. Stages of landfill [8]

Landfill shows the certain feasibility in some aspects of the resource utilization. Humus formed after landfill is beneficial to the soil after removing the toxic substances in the organic solid waste. Application of humus-rich compost can increase soil dissolved organic carbon, microbial biomass carbon and fertility [11]. Furthermore, some useful materials can be formed by the screening of the organic solid waste in landfills. These organic solid materials have high calorific value, such as plastic [12]. Although these materials are not highly recycled or reused [13], they lay the foundation for waste derived fuels for heating and electric power generation. However, most of landfills are primarily built to solve the local pollution problems or to increase landfill capacities rather than to recycle the organic solid materials as secondary resources [14]. Therefore, landfill need to be further developed and widely used to solve the problem of the resource utilization of organic solid waste.

3 Incineration

Incineration is a traditional method of organic solid waste treatment. In the incinerator, the organic solid waste and air react to burn which releasing a large amount of energy. Meanwhile, the flue gas and ash produced by incineration are emitted into the atmosphere without any further treatment [15]. In aspects of economy, incineration has good prospects. Because the cost of the incineration is also relatively affordable and electricity generated from incineration can bring a lot of benefits. The net profit margin of the incineration power plant is attractive, reaching 25 percent [16]. In aspects of waste removal rate, the organic solid waste is directly turned into ash in the incinerator. After the process of incineration, the ash which falls to the bottom of the incinerator accounts for nearly 25 percent of the initial total mass of organic solid waste [17]. In addition, compared to the landfill, incineration power plant requires a smaller area, so that it is easier to select the site for the incineration plant [18]. But incineration has a huge impact on the human health and the environment. Organic solid waste is transformed into the secondary pollutants after the process of incineration. Major of them is dioxin, heavy metal element and greenhouse gas [19]. When people are exposed to a large amount of the dioxins, the incidence of cancer of the gastrointestinal sites and of the lymphatic and hematopoietic tissue increases [20]. Therefore, this may result in a big challenge for the global public health. Furthermore, the greenhouse gas, such as carbon dioxide, methane and nitrous oxide have a huge impact on the atmosphere

[21]. By statistically analyzing the concentrations of greenhouse gases generated by different types of incinerators, the concentration and emission factors of different gases are roughly with a certain range. The concentration of carbon dioxide is from 5.66% to 9.04%, the concentration of methane is from 2.35% to 5.88% and the concentration of nitrous oxide is from 1.49% to 42.64% [21]. Among these gas, nitrous oxide plays an important part in regulating stratospheric ozone levels, causing atmospheric greenhouse phenomena and participating in the generation of acid rain [22]. Therefore, while benefiting from the great energy advantages of incineration, the impact of secondary pollutants on the environment should be paid much attention.

4 Biosynthesis

At the microbial level, biosynthesis technology will synthesize some new substances to make microorganisms with better physical and chemical properties, which compared with the unsynthesized microorganisms. With the help of the synthetic micro-organisms, organic solid waste is transformed into the recyclable products. By comparison, the principles of traditional composting technology are similar with the biosynthesis technology in aspects of microbial treatment of organic solid waste [23]. But for the traditional composting technology, it has some disadvantages, such as long composting time, relatively low nutritional level and quality of compost product [24]. Biosynthesis technology can cleverly solve these problems. Firstly, supported by biosynthesis technology, effective microorganisms (EM) are generated by mixed culture of fermentation technology. It is composed of photosynthetic bacteria, lactic acid bacteria, yeast, ray fungi and other microorganisms [25]. During the process of the composting, the reactants with treatment of EM can reach a higher temperature than the reactants without treatment of EM [26]. As shown in the figure 2, the highest temperature of reactants with treatment reached more than 50 degrees Celsius, which was 5 degrees higher than the control group. The heat generated by the respiration of microorganisms, the breakdown of sugars, starches and proteins will lead to the increase in temperature [27]. The increase in temperature is beneficial to accelerate the degradation of organic solid waste.

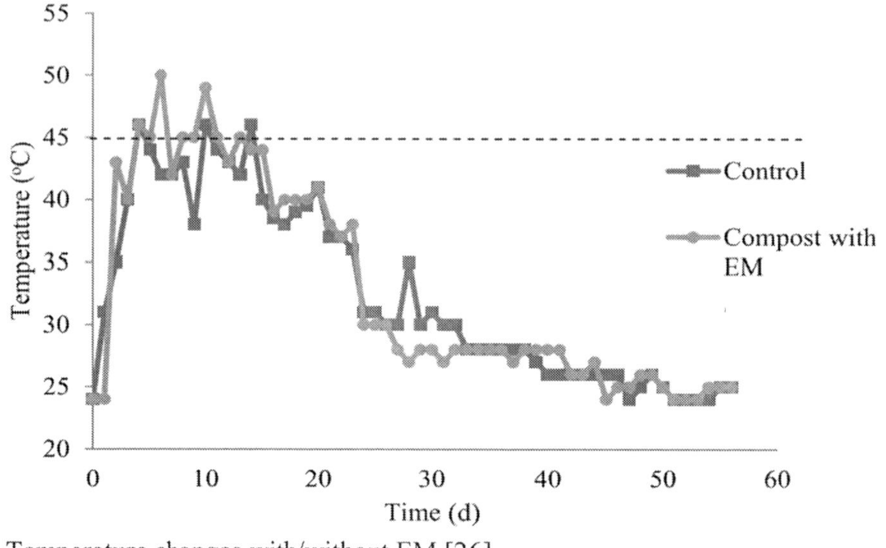

Fig. 2. Temperature changes with/without EM [26]

Secondly, combining biochar with nanomaterials through biosynthesis technology can synthesis a series of biochar nanocomposites. Due to the large surface area, porous structure and surface hydrophobicity, the biochar nanocomposites can treat organic solid waste [28].

During the composting process, adding with 10% biochar nanocomposites is beneficial to accelerate humification and generate the important nutrients [29]. Nitrogen, phosphorus and potassium elements in ionic form to enhance microorganisms' activity, to accelerate the conversion of organic solid waste. Compared with landfill, biosynthesis technology can improve the soil nutrient content instead of generating toxic substances. Additionally, biochar nanocomposites can reduce the emission of the greenhouse gas, such as carbon dioxide and nitrous oxide, which generate from the degradation of biomass [30]. Compared with incineration, biosynthesis technology decreases the amount of the greenhouse instead of increasing it.

Thirdly, the utilize of biosynthesis technology in the secondary fermentation of compost can improve the quality of compost products by adding bacteria, which can generate indole-3-acetic acid (IAA). After adding with these bacteria, the degree of humification and germination rate of seed increased [31]. The content of IAA in compost product increased between 2.9 and 5.2 times [31], showing that biosynthesis technology can improve the quality of compost product.

5 Comparation

In aspects of resource utilization, landfill, incineration and biosynthesis technology have all shown their respective advantages, which is transforming the organic solid waste into corresponding energy and materials. If the energy and materials can be effectively utilized, the value of them should not be judge whether is good or not. In aspects of cost, compared with landfill and incineration, although the initial research and technical support of biosynthesis technology is high, its resource recycling and emission reduction potential can bring benefits. In aspects of environment protection, biosynthesis technology has less impact on the environment due to its mild reaction conditions and no secondary pollution. In addition, biosynthesis technology can improve soil conditions and reduce greenhouse gas emissions, while landfill may cause soil and groundwater pollution, and incineration may produce harmful gas emissions. In summary, biosynthesis technology has the most obvious advantages in the environmental direction. In other directions, the three treatment all have their advantages and disadvantages.

6 Conclusion

The study compared biosynthesis technology with traditional landfilling and incineration for treating organic solid waste in aspects of resource utilization, cost and environment impact. It fills academic gap in the field of biosynthesis technology and provide suggestions for treating organic solid waste. But the study does not involve a large amount of data and algorithmic comparison of different treatment method. In the future experiments, each treatment method will be digitized and the specific organic solid waste will be track in different treatment method through life cycle assessment. Biosynthesis technology shows great potential in the treatment of organic solid waste, beginning a new era of environmental protection and resource recycling. By transforming organic solid waste into valuable chemicals, biomaterials and bioenergy, biosynthesis technology not only reduces the environmental pressure of waste treatment, but also promotes the environmental and economic integration. In the future, with the advancement of gene editing and microbial engineering, human expect to further improve the efficiency and economy of the conversion processes. In addition, the application of biosynthesis technology will also expand to the field of biorefining, improve the utilization efficiency of biomass, and develop new biomaterials, thereby reducing dependence on fossil fuels. With the development and innovation of technology, biosynthesis technology is expected to become a key tool for achieving the goals of economy and environmental protection.

References

1. UNEP, Global waste management outlook 2024. UN environment programme. (2024) https://www.unep.org/resources/global-waste-management-outlook-2024
2. Rabl, Ari, Joseph V. Spadaro, Assaad Zoughaib, Environmental impacts and costs of solid waste: a comparison of landfill and incineration. Waste Management & Research. **26**, 147-162 (2008)
3. Erickson, Larry E., Gary Brase, Reducing greenhouse gas emissions and improving air quality: Two interrelated global challenges. Taylor & Francis. (2020)
4. Sarkar, Payel, Rounak Chourasia, Bioconversion of organic solid wastes into biofortified compost using a microbial consortium. International Journal of Recycling of Organic Waste in Agriculture. **6**, 321-334 (2017)
5. Guo, Hao-nan, et al, Application of machine learning methods for the prediction of organic solid waste treatment and recycling processes: A review. Bioresource technology. **319**, 124114 (2021)
6. Xinhuanet. Spotlight: Xi wraps up LatAm tour as China's regional, global role growing bigger. Retrieved on 5 November 2024, retrieved from http://www.xinhuanet.com/energy/20220210/4db164eadcf44cbc953faa7d4360ef08/c.html
7. Vaverková, Magdalena Daria. Landfill impacts on the environment. Geosciences. **9.10**, 431 (2019)
8. Collin, M. Samuel, et al, Bioaccumulation of lead (Pb) and its effects on human: A review. Journal of Hazardous Materials Advances. **7**, 100094 (2022)
9. Zhao, C., Zhang, Y., Xie, D. The Multi-energy High precision Data Processor Based on AD7606. IOP Conf. Ser. Earth Environ. Sci. **94**, 012138 (2017)
10. Kweku, Darkwah Williams, et al, Greenhouse effect: greenhouse gases and their impact on global warming. Journal of Scientific research and reports. **17.6**, 1-9 (2018)
11. Solaiman, Zakaria M., et al, Humus-rich compost increases lettuce growth, nutrient uptake, mycorrhizal colonisation, and soil fertility. Pedosphere. **29.2**, 170-179 (2019)
12. Sonawane, Y. B., M. R. Shindikar, M. Y. Khaladkar, High calorific value fuel from household plastic waste by catalytic pyrolysis. Nature Environment and Pollution Technology. **16.3**, 897 (2017)
13. Shen, Li, Ernst Worrell, Plastic recycling. Handbook of recycling. 497-510 (2024)
14. Winterstetter, Andrea, et al, Framework for the evaluation of anthropogenic resources: A landfill mining case study–Resource or reserve?. Resources, Conservation and Recycling. **96**, 19-30 (2015)
15. Nidoni, Pooja G., Incineration process for solid waste management and effective utilization of by products. International Research Journal of Engineering and Technology. **4.12**, 378-382 (2017)
16. Xin-Gang, Zhao, et al, Technology, cost, a performance of waste-to-energy incineration industry in China. Renewable and Sustainable Energy Reviews. **55**, 115-130 (2016)

17. Silva, Rui Vasco, et al, Environmental impacts of the use of bottom ashes from municipal solid waste incineration: A review. Resources, Conservation and Recycling. **140**, 23-35 (2019)
18. Nabavi-Pelesaraei, Ashkan, et al, Modeling of energy consumption and environmental life cycle assessment for incineration and landfill systems of municipal solid waste management-A case study in Tehran Metropolis of Iran. Journal of cleaner production. **148**, 427-440 (2017)
19. Li, Juming, The key to avoiding secondary pollutants in the incineration of domestic waste lies in prevention. Procedia Environmental Sciences. **16**, 669-673 (2012)
20. Bertazzi, Pier A., et al, The Seveso studies on early and long-term effects of dioxin exposure: a review. Environmental health perspectives. **106**, 625-633 (1998)
21. Hwang, Kum-Lok, et al, Emission of greenhouse gases from waste incineration in Korea. Journal of environmental management. **196**, 710-718 (2017)
22. Badr, Ossama, S. D. Probert, Environmental impacts of atmospheric nitrous oxide. Applied energy. **44.3**, 197-231 (1993)
23. Lekasi, John K., Keziah W. Ndung'u, Mary N. Kifuko, A scientific perspective on composting. Organic resource management in Kenya: perspectives and guidelines, FORMAT, Nairobi. 65 (2003)
24. Ayilara, Modupe Stella, et al, Waste management through composting: Challenges and potentials. Sustainability. **12.11**, 4456 (2020)
25. Higa, Teruo, G. N. Wididana, The concept and theories of effective microorganisms. Proceedings of the first international conference on Kyusei nature farming. US Department of Agriculture, Washington, DC, USA. (1991)
26. Van Fan, Yee, et al, Evaluation of Effective Microorganisms on home scale organic waste composting. Journal of Environmental Management. **216**, 41-48 (2018)
27. Jusoh, Mohd Lokman Che, Latifah Abd Manaf, Puziah Abdul Latiff, Composting of rice straw with effective microorganisms (EM) and its influence on compost quality. Iranian journal of environmental health science & engineering. **10**, 1-9 (2013)
28. Chausali, Neha, Jyoti Saxena, Ram Prasad, Nanobiochar and biochar based nanocomposites: Advances and applications. Journal of Agriculture and Food Research. **5**, 100191 (2021)
29. Nguyen, Minh Ky, et al, Evaluate the role of biochar during the organic waste composting process: A critical review. Chemosphere. **299**, 134488 (2022)
30. Qambrani, Naveed Ahmed, et al, Biochar properties and eco-friendly applications for climate change mitigation, waste management, and wastewater treatment: A review. Renewable and Sustainable Energy Reviews. **79**, 255-273 (2017)
31. Cai, Guanjing, et al, Compost-derived indole-3-acetic-acid-producing bacteria and their effects on enhancing the secondary fermentation of a swine manure-corn stalk composting. Chemosphere. **291**, 132750 (2022)

Study on the Water Ecology Protection and Water Environment Remediation Technology, focusing on the heavy metal removal

Zihan Zhao*

College of Water Conservancy and Hydropower Engineering, Sichuan Agricultural University, Yaan, 625014, China

Abstract. Water environmental pollution has become a growing problem with the increase in pollutants released because of the expansion of human activities. Since water is peculiarly susceptible to pollution, water pollution control has garnered copious attention among the most crucial environmental conundrums. Therefore, water environment remediation is not only of great significance for river ecology, but also of theoretical and practical importance for sustainable human development. China's central government passed the Yangtze River Protection Law in response to the need for changes in governance arrangements and cooperation. The purpose of this study is to explore the technology of heavy metal treatment in water environment remediation, to study and solve a series of problems such as water ecological maintenance methods, river desilting, and water body renewal. This paper will analyze the benefits of water environment remediation technologies and explore their potential for sustainable development, focusing on the removal of heavy metals from water body.

1 Introduction

The Yangtze River, as a large water ecosystem, is home to a diversity of biological species, thus providing a huge ecosystem for organisms. As China's longest river, the Yangtze River's ecosystem is rich in biodiversity and contributes to the continuation of human survival and sustainable development. The water environment is of great significance to human survival. With industrial development river ecosystem health indices are deteriorating, and an increasing number of the 4,300 species are at highly depleted or endangered levels, there's no shortage of endangered animals and plants among those 4,000 species. The biological resources of the Yangtze River, especially the number of fish species, still show imperfections, while the fishing trend in the Yangtze River Basin has not declined significantly. the recovery of the Yangtze River's fishery resources has so far been insufficient in terms of catch trends, dolphin population status and fish biotic integrity indices. In order to respond positively to the integrated monitoring data implemented while advancing the process of improving ecosystem health, it is necessary to enhance the effectiveness of coordinated governance. There is a need to integrate Yangtze River governance with water

* Corresponding author: zhaozihan@stu.sicau.edu.cn

and environmental remediation techniques to manage the Yangtze River water environment and protect water resources and biodiversity.

In order to strengthen its efforts to protect the water ecology of the Yangtze River Basin, China has released a number of conservation policies. In 2018, a new policy document set phased goals: the initial phase focuses primarily on introducing a year-round fishing ban in parts of this region; while the long-term goal, set for 2035, aims to restore the river's ecology, ensure the effective protection of habitats, promote the growth of aquatic biodiversity, and restore the water body's ecological functions [1].

To improving the efficiency of river training, different river training methods should be used for different rivers, to avoid the inefficiency caused by the wrong methods. Aeration is one of the widely used techniques in physical technology because of its ability to increase the rate of degradation of organic pollutants in rivers and increase biological activity. Physical techniques are effective but costly and harmful to riverine water environment. Although chemical treatment can solve the problem in the short term, it should be improved because of its high cost and the risk of secondary pollution of the environment caused by chemical substances. Ecological engineering methodologies possess the merits of elevated economic, environmental, and ecological efficacy, coupled with straightforward sustainability and the absence of secondary contamination. There is some variability between the performance of artificial wetlands, microbial injections, eco-floating beds and biofilm technologies, but all are among the most widely used ecological technologies. Artificial wetlands have a good regulating effect when the water quality is less polluted and the hydraulic load is smaller [2].

Various water environment remediation technologies can well improve the water environment, improve the biodiversity and protect freshwater resources. The most suitable water environment remediation technology should be selected according to the specific analysis of different rivers in different situations. Among the many environmental problems, water pollution in rivers poses a very serious challenge, especially in developing countries and less economically developed regions. Rivers are not only the main source of drinking water for the population, but also support a wide range of needs, including domestic use, agricultural irrigation, commercial operations, industrial production and recreation. However, rivers in some countries have lost their use due to high levels of pollution. Worse still, this pollution contributes to the spread of numerous waterborne diseases in developing countries, releasing intolerable odors and further polluting air quality. Effective management of water quality relies on strict policy regulation of solid waste, wastewater discharges and stormwater runoff, as well as clear standards for the treatment of wastewater. Successfully addressing water pollution in rivers, however, requires both financial investment and time.

Heavy metals are very harmful to water bodies, not only to aquatic organisms but also to human health. Heavy metal pollutants can accumulate in water bodies and pose a threat to the growth, reproduction and survival of aquatic organisms. For example, heavy metals such as Cd and Pb can affect the reproductive and nervous systems of fish, leading to deformities and death. In addition, heavy metals are also enriched in living organisms through the food chain, which in turn can cause harm to higher-level organisms. Heavy metal pollutants will change the physical and chemical properties of water bodies, affecting their self-purification ability; at the same time, heavy metals will also accumulate in the soil, affecting the structure, fertility and microbial communities of the soil, which will in turn affect the growth and yield of crops. In the field of environmental pollution, several heavy metal elements and the metalloid arsenic are of great concern because they are difficult to degrade in the environment, are enriched, and may enter the human body through the food chain, posing a hazard to human health. Levels of heavy metals and other toxic substances in water are increasing as a result of increasing discharges from both natural and anthropogenic sources. Excessive intake of heavy metals by individuals can exacerbate cellular responses triggered by oxidative stress, which can increase the probability of developing tumors, neurological disorders, diabetes mellitus, infertility, developmental delays, renal impairment and cardiovascular diseases [3].

The purpose of this treatise is to elucidate the contamination of aquatic milieu, while also presenting rectification methodologies predominantly aimed at heavy metals within the aquatic environment, and elucidating their pivotal function in the rehabilitation of river systems.

2 Principles of water environment remediation technology

There are various methods of treating polluted river water, and more than one method is usually used to treat river water [4]. This is because using only one of these methods often fails to address multiple causes of polluted river water or to effectively clean up heavily polluted river water. Therefore, the use of hybrid technologies, which incorporate several different approaches, can address a wider range of problems and show greater effectiveness in treating polluted river water. Given this context, it is advisable to advocate for the employment of amalgamated methodologies for the purification of fluvial waters. The primary technologies presently utilized for addressing high concentrations of heavy metals encompass microfiltration, chemical precipitation, coagulation, and flocculation [5]. The process of wastewater purification can be delineated into three phases: primary, secondary, and advanced treatment. During the primary phase, both physical and chemical modalities are applied to eliminate pollutants from the wastewater. The secondary phase consists of anaerobic or aerobic biodegradation. The advanced treatment encompasses methodologies such as chemical oxidation, electrochemical deposition, crystallization, distillation, and photocatalysis, along with adsorption, membrane separation techniques, and ion exchange processes.

When tertiary treatment is fully integrated with primary and secondary treatment, removal efficiencies of up to 99% can be achieved [6]. Heavy metal removal methods can be categorized as adsorption, chemical precipitation, coagulation-flocculation, ion exchange, and membrane technology. In adsorption, because of the porous nature of the organic or inorganic material in the adsorbent, it is capable of adsorbing large amounts of heavy metals. Adsorption is considered to be one of the best choices for the heavy metal removal [7].

2.1 Adsorption

Adsorbents that can be used as adsorbents are generally inorganic materials from nature or biomass, such as fly ash, treated tree sap, zeolites and microorganisms.Activated carbon, which is made from a variety of different materials capable of providing microbial carbon, is the most common adsorbent used by the general public today. Only biosorbents with high adsorption capacity and low preparation cost are considered for heavy metal removal.
Nano adsorbent materials synthesized using minerals and organic matter can completely adsorb and remove heavy metal ions from wastewater [8].

2.1.1 Nanomaterials

Fig.1 shows the keywords of the literature related to nanomaterials and heavy metals in recent years, from which it can be seen that nanomaterials play a crucial role in this field, and they have a great contribution to sewage treatment, thus realizing the water environment remediation. This not only helps to reduce the heavy metal content of sewage, but also

 https://doi.org/10.1051/e3sconf/202560603010

provides new ideas and solutions for sustainable development and promotes the efficient recycling of resources.

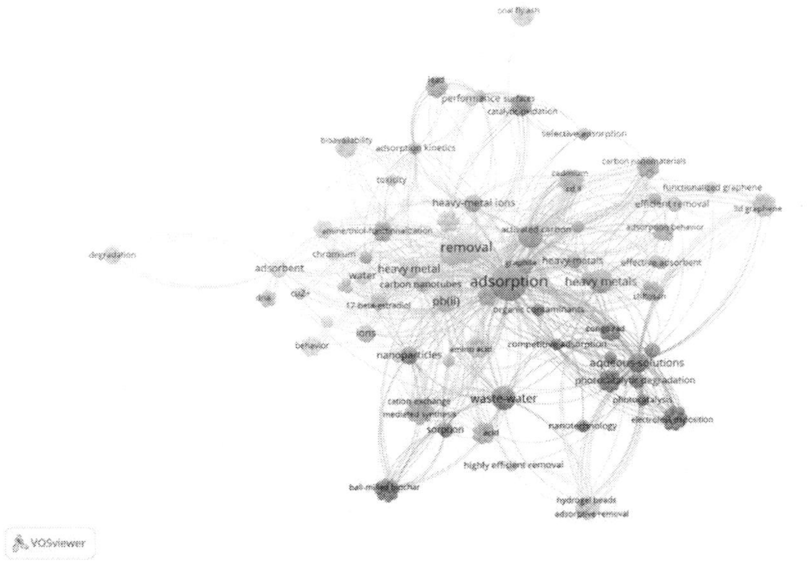

Fig. 1. Keyword map of nanomaterials synthesized with adsorbed heavy metal articles (Photo/Picture credit: Original)

2.1.2 The role of nanomaterials in adsorption

Because nanomaterials generally have a large total surface area per unit volume of material, they adsorb well. Nanoparticles, with their uniquely high specific surface area, have demonstrated a wide range of compelling and innovative applications in a variety of disciplines. The principle of adsorption of heavy metals in the aqueous environment by nanomaterials is to adsorb pollutants using different types of nanoparticles that compose themselves, such as bimetallic, metallic, carbonaceous, metal oxides and zeolites. Much effort has been invested in the use of nanomaterials to treat wastewater. Once the adsorbent has adsorbed the hazardous substances in the wastewater, the adsorbent can be reused after some treatment of the adsorbent, thus making the cost of water adsorption remediation lower and more sustainable. Tests have shown that after 8 rounds of reuse of the adsorbent, a removal rate of about 87% can be achieved[9]. The results of the study show that nanosorbents because of their high specific surface area and reusability, nanomaterials have a strong adsorption capacity for heavy metal ions. So far, a wide variety of nanomaterials such as nanorods, nanorings and nanotubes have been tested in the adsorption of heavy metal oxides [10]. Although nanomaterials are very good adsorbents, they have certain drawbacks, such as expensive material costs, limited reuse potential, and harm to human health.

2.2 Separation of heavy metals

Chemical precipitation (i.e., coagulation and sedimentation) is widely used in industry and is one of the most effective and proven methods. The principle is simply summarized as adding a precipitating agent to the wastewater, which is heavy metal ions precipitated and thus removed. The main methods are hydroxide precipitation, sulfide participation and carbonate precipitation. Among them, hydroxide precipitation is widely used because of its relative cheapness and simplicity. Sulfide precipitation removes more heavy metal ions and is more efficient in remediating wastewater. It has been shown that sulfide precipitation can remove toxic heavy metal ions, but this method produces bad odor and toxic hydrogen sulfide gas, so sulfide precipitation is recommended to be carried out under neutral conditions [11].

3 Conclusion

Various methods for removing heavy metal ions from wastewater, and in this paper nanomaterial adsorption and chemical precipitation is analyzed. In adsorption, nanomaterials have been widely adopted as adsorption media for pollutants in water and wastewater treatment for their excellent surface area-to-volume ratios and efficient reuse properties. In addition, their strong adsorption efficacy for heavy metal ions further establishes their preferred status as promising technological solutions for sewage remediation. Although nanomaterials have many advantages, their disadvantages are also more obvious. One is that nanomaterials are less commercially available, making them less applicable and underutilized. The second is the sustainability of nanomaterials, more green synthesis methods of nanomaterials should be explored in the future to synthesize environmentally friendly and economically sustainable materials. As an emerging field, nanomaterials have a broad application prospect in wastewater treatment.

References

1. Chen T G, Wang Y, Gardner C, Wu F. Threats and protection policies of the aquatic biodiversity in the Yangtze River. Journal for Nature Conservation, **58**, 125931 (2020) https://doi.org/10.1016/j.jnc.2020.125931
2. Anawar H M and Chowdhury R. Remediation of Polluted River Water by Biological, Chemical, Ecological and Engineering Processes. Sustainability. **12**, 7017(2020) doi:10.3390/su12177017
3. Paithankar J G, Saini S, Dwivedi S, et al. Heavy metal associated health hazards: An interplay of oxidative stress and signal transduction. Chemosphere. **262** ,128350 (2021) https://doi.org/10.1016/j.chemosphere.2020.128350
4. Bai X Y, Zhu X F, Jiang H B, et al. Purification Effect of Sequential Constructed Wetland for the Polluted Water in Urban River. Water. **12,** 1054 (2020) do i:10.3390/w12041054
5. Yenkie, K.M., Burnham, S., Dailey, J., et al. Generating efficient wastewater treatment networks: an integrated approach comprising of contaminant properties, technology suitability, plant design, and process optimization. Comput. Aided Chem. Eng. **46**, 1603-1608 (2019) http://dx.doi.org/10.1016/B978-0-12-818634-3.50268-X.
6. Ince Mand Olcay Ince O K, Heavy Metal Removal Techniques Using Response Surface Methodology: Water/Wastewater Treatment. Biochemical Toxicology-Heavy Metals and Nanomaterials. DOI: 10.5772/intechopen.88915
7. Zamora-Ledezma C, Negrete-Bolagay D, Figueroa F, et al. Heavy metal water pollution: A fresh look about hazards, novel and conventional remediation methods. Environmental Technology & Innovation, **22**, 101504 (2021) https://doi.org/10.1016/j.eti.2021.101504
8. Wu Y H, Pang H W, Liu Y, et al. Environmental remediation of heavy metal ions by novel-nanomaterials: A review. Environmental Pollution, **246**, 608-620 (2019) https://doi.org/10.1016/j.envpol.2018.12.076

9. Rajendran S, Priya A K, Kumar P S, et al. A critical and recent developments on adsorption technique for removal of heavy metals from wastewater-A review. Chemosphere, **303**, 135146 (2022) https://doi.org/10.1016/j.chemosphere.2022.135146
10. Shahrokhi-Shahraki R, Benally C, El-Din M G, et al. High efficiency removal of heavy metals using tire-derived activated carbon vs commercial activated carbon: Insights into the adsorption mechanisms. Chemosphere, **264**,128455 (2021) https://doi.org/10.1016/j.chemosphere.2020.128455
11. Qasem N A A, Mohammed R H, Lawal D U. Removal of heavy metal ions from wastewater: a comprehensive and critical review. npj Clean Water, **4**, 36 (2021). https://doi.org/10.1038/s41545-021-00127-0

Recent Advances in Electrocatalysts for Carbon Dioxide Reduction

Xiaojie Deng[1], *Ziying* Huang[2,*], and *Xuheng* Wu[3]

[1]Guangzhou No.1 High School, Guangzhou, 510610, China
[2]School of Energy and Power Engineering, Changsha University of Science and Technology, Changsha, 410114, China
[3]Shanghai Hongwen School, Shanghai, 201399, China

Abstract. With the progress of science and technology and the economy, realizing "sustainable development" has become the common development goal of many countries and regions in the world today. Carbon dioxide (CO_2) is widely used as an important carbon resource, which aggravates the problem of global warming. electrocatalytic Carbon Dioxide reduction reaction (ECO_2RR) is a key technology for converting CO_2 into useful carbon-based products (such as C_2H_4, CO, CH_3OH). Thereby increasing the economic and commercial value of clean energy worldwide. In this paper, different electrocatalysts commonly used in ECO_2RR reactions were studied. By summarizing the definition, characteristics, deficiencies, and modification strategies of different electrocatalysts, different effects of different electrocatalysts on ECO_2RR reaction were studied. This paper helps to make up for the shortage of different electrocatalysts in the literature review of ECO_2RR reaction, promote the development of ECO_2RR reaction electrocatalysts, and promote the progress of ECO_2RR reaction. Let the ECO_2RR reaction enable energy conservation and emission reduction, and provide more effective measures for the early realization of the "double carbon goal". Due to the limited literature on electrocatalysts, the current research is not comprehensive. Future studies should summarize the characteristics of different electrocatalysts in ECO_2RR reactions more comprehensively.

1 Introduction

With the rapid development of global science and technology and economy, "sustainable development" has become a common development trend in many countries and regions in the world today, and the rise of new energy vehicles with renewable energy as the main energy and the widespread use of various fuel cells are important contents of sustainable development strategy. In the development process, carbon dioxide (CO_2) is recognized as a key carbon resource [1]. CO_2, crucial for maintaining life and ecological balance, has historically remained stable. However, in the last century, human activities, particularly fossil

* Corresponding author: 201931331019@stu.qhnu.edu.cn

fuel use, have led to CO_2 levels surpassing natural absorption capacities, contributing to global warming [2].

In this context, the electrocatalytic conversion of CO_2 into useful carbon-based products (such as C_2H_4, CO, and CH_3OH) offers a promising approach to reducing carbon emissions and advancing sustainable development [1,3]. As a key technology for electrocatalytic CO_2 reduction, Electrocatalytic Carbon Dioxide Reduction Reaction (ECO2RR) technology holds potential for optimizing CO_2 and supporting clean energy growth. If ECO_2RR is widely adopted, it could increase the economic and commercial value of clean energy globally. Consequently, research on CO_2 reduction electrocatalysts has surged, with ongoing updates on various materials and structures [4].

Despite its promise, ECO_2RR still lacks the refined reaction mechanisms and conditions seen in other renewable technologies like solar, wind, geothermal, and tidal energy. How these gaps are closed is critical to advancing the technology. Multi-carbon products, due to their high market value and energy density, are particularly attractive but face significant barriers to large-scale application [5,6].

Numerous scientific studies have shown that ECO_2RR (electrochemical CO_2 reduction reaction) can efficiently couple protons and electrons with various catalysts. Given that this coupling process plays a crucial role in driving technological progress, further exploration in this area is essential. Several strategies have been proposed to enhance ECO2RR's efficiency. These include developing new acid-resistant and highly efficient catalysts and improving the catalyst's micro-environment through structural engineering and surface modification [7].

This essay aims to define ECO_2RR and elucidate its reaction mechanism. It will compare major electrocatalysts--such as alloys, metal oxides, metal complexes, and metal polymers--and suggest viable approaches for advancing ECO_2RR to some extent. By analyzing catalyst adoption and reaction prerequisites, this paper will offer insights into selecting appropriate carbon dioxide reduction electrocatalysts, thus promoting further progress in this field.

2 Reaction mechanism of ECO_2RR

Electrochemical carbon dioxide reduction (ECO_2RR) holds significant promise for achieving carbon neutrality by converting CO_2 into high-energy chemicals or fuels using renewable energy sources. Copper-based catalysts, in particular, are highly promising due to their cost advantages and excellent catalytic activity. Among the various products of ECO_2RR, formic acid stands out for its potential in industrial production, given its applications as a hydrogen storage material and a fuel for internal combustion engines. Typically, ECO_2RR is conducted with a liquid electrolyte, but this setup has limitations. The solubility of CO_2 in the electrolyte is often low, and the electrolyte can fill the pores of the catalyst, confining the reaction to the catalyst's surface and thus limiting catalytic efficiency [8].

Heterogeneous catalysis generally involves several key steps: diffusion of the reactant from the bulk phase to the catalyst surface, adsorption of the reactant on the surface, reaction of the reactant with protons and electrons to produce the product, desorption of the product from the catalyst surface, and diffusion of the product back into the bulk phase. The steps of adsorption, surface reaction, and desorption involve chemical or physico-chemical changes in the adsorption mode and strength of the reaction species on the catalyst surface. These changes reflect the reaction path and mechanism, including the formation of formic acid from CO_2.

Specifically, formic acid formation involves the intermediate step of *OCHO, which is formed after proton-electron transfer from CO_2. This intermediate is reduced to form formic acid. Alternatively, CO_2 can gain an electron to form $*CO^{2-}$, which then reacts with protons to form the OCHO or OCHO intermediates, eventually coupling with protons and electrons to produce formic acid. Recent studies emphasize the importance of optimizing the activation energy of CO_2 on the catalyst surface to minimize overreaction potential and enhance energy efficiency. Additionally, optimizing the adsorption of intermediates like *OCHO or OCHO can improve the electronic structure of the catalyst surface, reducing the formation of undesired *COOH intermediates and facilitating higher yields of formic acid. Given the critical role of ECO_2RR and the importance of electrocatalysts in this process, this study focuses on investigating the electrocatalysts used in ECO_2RR [9].

3 Electrocatalysts for carbon dioxide reduction

3.1 Alloy electrocatalyst

Alloy electrocatalysts consist of an active metal and an inert metal. Despite the term "alloy," these catalysts may not always be in a true alloy form under reaction conditions; they can exhibit different metal phases. The combination of active and inert metals in an alloy catalyst can create a "ligand" effect, where one metal influences the behavior of the other through electronic interactions or a collective effect.

For instance, adding iridium (Ir) to Pt facilitates naphtha reforming at lower pressures, leading to an increased production of heavier distillates. Similarly, alloying nickel (Ni) with copper (Cu) does not alter the dehydrogenation activity of cyclohexane but significantly reduces the hydrogenolysis activity of ethane.

Alloy catalysts can also improve the active site density of single atoms. While individual metal atoms can exhibit high intrinsic activity (as measured by the turnover frequency, TOF), their mass or volume activity is often limited, restricting their practical applications. Therefore, enhancing the active site density of single atoms is crucial for improving catalyst performance.

Single-atom catalysts offer notable advantages in selective reaction control. They excel in selective hydrogenation of nitro compounds, alkynes/diolefins, carbon dioxide reduction in thermal or electrocatalytic processes, and selective hydrogen peroxide formation in electrochemical oxygen reduction reactions. This selectivity arises from their unique MOx (MNx, MSx, etc.) structures and electronic properties, including partial positive charges on the single-atom centers [10].

Regarding stability, there is a common misconception that single-atom catalysts are less stable than nanocatalysts. However, recent research indicates that single atoms can be more stable than their nanocatalyst counterparts. This increased stability is due to the strong covalent bonds formed between the metal single atoms and the non-metal atoms on the support surface.

3.2 Metal oxide electrocatalyst

Metal oxide catalysts such as ZnO, CeO_2, WO_3, and In_2O_3 have demonstrated strong performance in photothermal catalytic CO_2 hydrogenation reactions. Among these, ZnO is particularly favored due to its high CO_2 adsorption capacity and catalytic efficiency. CeO_2 (Fig. 1) is notable for its excellent REDOX and hydrogen storage properties, allowing it to release additional oxygen without altering its structure. The oxygen vacancies on CeO_2's surface also enhance CO_2 catalytic reactions [11].

Researchers have focused on improving CO_2 reaction efficiency and product selectivity through the preparation and modification of copper-based materials. This includes surface modifications and nano-structural engineering. Additionally, complex oxides, which consist of multi-component oxides or transition metals like V_2O_5-MoO_3 and V_2O_5-MoO_3-Al_2O_3, are

https://doi.org/10.1051/e3sconf/202560603011

utilized. These alloys often feature interactions between their metallic components, where some act as primary catalysts while others serve as supports or carriers [12].

Due to their unique energy band structures, chemical properties, stability, and environmental benefits, metal oxides are widely used in photothermal catalytic hydrogenation. However, studies indicate that under electrocatalytic conditions, copper-based catalysts may exhibit instability in their subsurface, with minimal impact on the activity of oxide-derived copper catalysts.

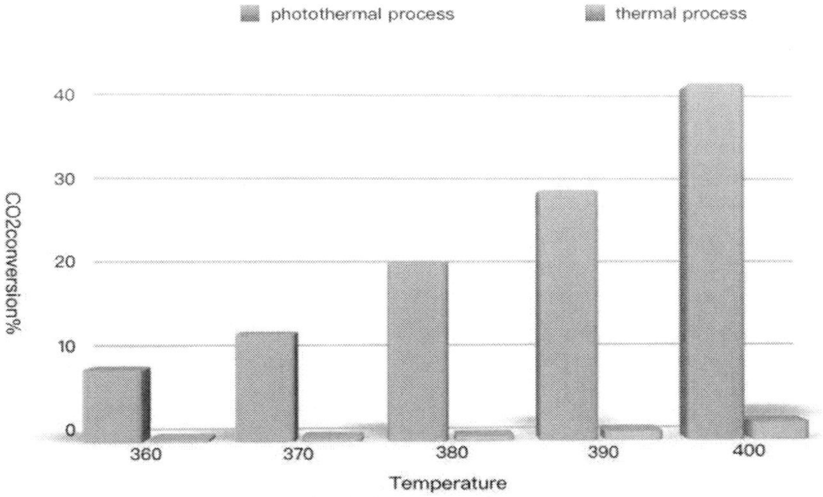

Fig. 1. Catalytic performance on Au/CeO2 during different situations [11].

These catalysts exhibit defects that significantly impact their effectiveness in CO_2 reduction reactions (ECO_2RR). Defects in metal oxides can be categorized as point defects, line defects, surface defects, and bulk defects. Specifically, "introduced defects," such as point and surface defects, directly influence catalytic performance [13].

Adding oxygen defects to metal oxides boosts CO_2 reduction performance. These defects expose metal cations on the surface, increasing the contact area with CO_2 and reducing the activation energy of C=O bonds, thereby improving reduction efficiency.

3.3 Metal complex electrocatalyst

Metal complex catalysts have a wide range of applications in the pharmaceutical, medical, and carbon dioxide fields. Gold catalysts are mainly used in cyclic reactions to synthesize various heterocycles and thick rings with high resistance. Metal N-heterocyclic carbene complexes are widely used in the fields of medicine, materials, and catalysis, and have high activity and stability. The antibody-nail catalyst conjugate can be used for tumor-targeted therapy, and the metal nail catalyst can be delivered to the tumor cell surface through the antibody fragment to achieve collaborative therapy.

Metal complex catalysts are crucial for carbon dioxide reduction. Advances in supporting ligands have significantly influenced the reactivity and stability of these catalysts within their coordination spheres. Common ligands include organic compounds such as ethers, amines, carbonyl groups, and phosphines, which stabilize metal complexes. Recently, metal-like catalysts have demonstrated superior catalytic activity and selectivity due to their unique electronic and spatial environments, opening new avenues in synthetic chemistry and CO_2 reduction [14]. These catalysts often face challenges related to electrochemical operating conditions, leading to difficulties in formation within aqueous electrolytes and instability in molecular structure. Weak electronic communication between the fixed metal complex and the substrate causes aggregation, which reduces electrical conductivity and overall reaction activity. Thus, metal complex catalysts are most effective when uniformly distributed on a substrate or electrically attached to an electrode surface, limiting their practical application. To enhance stability, multi-toothed ligands can act as scaffolds for stable metal complexes. Incorporating the main group of metal-like elements into transition metal complexes and optimizing their design improves stability. Additionally, integrating molecular catalysts with conductive carriers and using polymer-catalyst composites enhance stability, charge transfer, and active site exposure [15].

3.4 Colour illustrations

Metal polymers are essential for stabilizing metal nanoparticles in electrochemical CO_2 reduction reactions (ECO_2RR). They offer high catalytic performance and Faraday efficiency. The development of metal-polymer cathodes falls into two categories: polymer encapsulation and polymer binding. Polymer encapsulation enhances metal-catalyst interaction and activity but involves complex synthesis and stringent conditions. Polymer binding, on the other hand, simplifies material design and allows flexible configuration of ECO_2RR systems. Polymers can be tailored in terms of chemical function, molecular weight, and crystallinity, enabling precise control over metal-polymer interactions and catalytic properties [16]. Electropolymerized films can further improve catalyst stability and reactivity. The understanding of cathode activity with metal-polymer catalysts, the surface reaction process in ECO_2RR, and the role of surface charge and polymer interactions remains incomplete. The lack of detailed mass and charge transfer spectra at complex metal-polymer interfaces poses a significant challenge. Additionally, the significantly lower dielectric constant of polymer coatings on metal surfaces leads to inconsistent surface polarity compared to pure metal cathodes, an aspect that has been inadequately addressed in previous studies. Recent research shows that hybrid metal polymers improve ECO_2RR performance. A new multi-scale model (Fig. 2) offers insights into metal-polymer interactions. Increasing polytetrafluoroethylene (PTFE) content and adjusting the contact angle can alter the surface charge, with PTFE enhancing the availability of active sites in Cu nanoparticle catalysis.

Recent research has demonstrated that hybrid metal-polymers significantly enhance the catalytic performance of electrochemical CO_2 reduction reactions (ECO_2RR). A new multi-scale model (see Fig. 2) can understand the interaction of metal polymers. By analyzing this multi-scale model, it can show that by increasing the content of polytetrafluoroethylene (PTFE), and thus changing the contact Angle, it can change the surface charge of the cathode. In the case of copper nanoparticles catalyzing ECO_2RR, PTFE acts as a bridge to connect the less accessible active sites, which improves the overall catalytic efficiency [17].

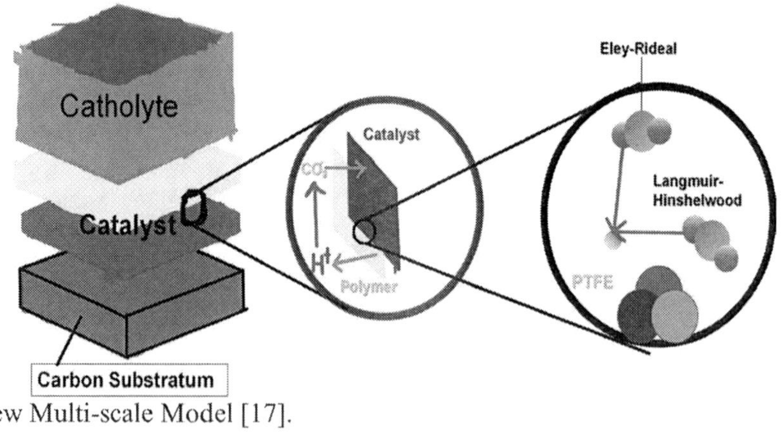

Fig. 2. New Multi-scale Model [17].

 https://doi.org/10.1051/e3sconf/202560603011

4 Conclusion

This paper provides a comprehensive analysis of electrocatalysts used in the electrochemical carbon dioxide reduction reaction (ECO_2RR). By reviewing their definitions, characteristics, limitations, and modification strategies, this article has highlighted the diverse effects these electrocatalysts have on ECO_2RR performance. Despite extensive research, literature summarizing the advantages and disadvantages of various electrocatalysts in ECO_2RR remains sparse. This gap is addressed by the study, which aims to simplify the selection process for researchers working in the field of carbon dioxide reduction. By offering a clearer overview of electrocatalysts, the paper supports the advancement of ECO_2RR technology and its role in energy conservation and emission reduction. This, in turn, contributes to the achievement of global "double carbon" goals. However, due to the limited existing literature, this review is not exhaustive.

Future studies should focus on a more detailed and comprehensive evaluation of the characteristics and performance of different electrocatalysts in ECO_2RR reaction, so as to highlight the importance of electrocatalysts in ECO_2RR reaction, further promote the development of electrocatalysts and carbon dioxide reduction industry, enable energy conservation and emission reduction, and promote the future process of "dual carbon plan".

Authors Contribution

All the authors contributed equally and their names were listed in alphabetical order.

References

1. J. Li, N. Kornienko, Electrocatalytic carbon dioxide reduction in acid. Chem. Catal. **2(1)**, 29-38 (2022).
2. X, Zhang, S.-X. Guo, K. A. Gandionco, A. M. Bond, J. Zhang, Electrocatalytic carbon dioxide reduction: From fundamental principles to catalyst design. Mater. Today Adv. **7**, 100074 (2020).
3. Q. Lu, J. Rosen, Y. Zhou, G.-S. Hutchings, Y.-C. Kimmel, J.-G. Chen, F.-A. Jiao, Selective and efficient electrocatalyst for carbon dioxide reduction. Nat. Commun. **5(1)**, 3242 (2014).
4. P. Prabhu, V. Jose, J.-M. Lee, Heterostructured catalysts for electrocatalytic and photocatalytic carbon dioxide reduction. Adv. Funct. Mater. **30(24)**, 1910768 (2020).
5. O.-S. Bushuyev, P. De Luna, C.-T. Dinh, L. Tao, G. Saur, J. van de Lagemaat, S.-O. Kelley, E.-H. Sargent, Construction of efficient and stable low-temperature reverse-bias bipolar membrane electrolyzer for CO2 reduction. Joule **2**, 825-832 (2018).
6. W. Ma, S. Xie, T. Liu, Q. Fan, J. Ye, F. Sun, Z. Jiang, Q. Zhang, J. Cheng, Y. Wang, Improving co-electrocatalytic carbon dioxide reduction by optimizing the relative potentials of the redox mediator and catalyst. Nat. Catal. **3**, 478-487 (2020).
7. R. Zhang, H. Wang, Y. Ji, Q. Jiang, T. Zheng, C. Xia, Electrifying the future: the advances and opportunities of electrocatalytic carbon dioxide reduction in acid. Sci. China Chem. **12**, 3426-3442 (2023).
8. B. Endrődi, et al., Continuous-flow electroreduction of carbon dioxide. Prog. Energy Combust. Sci. (2017).

9. E. Andrews, et al., Electrocatalytic reduction of CO2 at Au nanoparticle electrodes: effects of interfacial chemistry on reduction behavior. J. Electrochem. Soc. **162**, F1373-F1378 (2015).
10. T. Zhang, A.-G.Walsh, J. Yu, P. Zhang, Single-atom alloy catalysts: structural analysis, electronic properties and catalytic activities. Chem. Soc. Rev. **50(1)**, 569-588 (2021).
11. Y. Wan, F. Fang, R. Sun, J. Zhang, K. Chang, Metal oxide semiconductors for photothermal catalytic CO2 hydrogenation reactions: recent progress and perspectives. Acta. Phys. Chim. Sin. **39(11)** (2023).
12. L. Hao, Z. Sun, Metal oxide-based materials for electrochemical CO2 reduction. Acta. Phys. Chim. Sin. **37(7)** (2021).
13. S.-S.-A. Shah, M.-S. Javed, T. Najam, C. Molochas, N.-A. Khan, M.-A. Nazir, S.-J. Bao, Metal oxides for the electrocatalytic reduction of carbon dioxide: Mechanism of active sites and defect engineering strategies. Coord. Chem. Rev. 471214716 (2022).
14. Y. Liu, S. Lyu, F. Wen, W. Nie, S. Wang, Polymer-encapsulated metal complex catalysts: An emerging and efficient platform for electrochemical CO2 reduction. J. Mater. Sci. Techno. **172**, 33-50 (2024).
15. J. Takaya, Catalysis using transition metal complexes featuring main group metal and metalloid compounds as supporting ligands. Chem. Sci. **12(6)**, 1964-1981 (2021).
16. Y. Park, K. Hyun, S. Yun, M. Choi, M, Unique Possibilities of Dynamic Metal-Polymer Interactions in Selective Hydrogenation. Chem. Cat. Chem. **16(6)**, e202301378 (2024).
17. X. Wang, S. Sahoo, J. Gascon, M. Bragin, F. Liu, J. Olchowski, B. Li, Deciphering electrochemical interactions in metal-polymer catalysts for CO 2 reduction. Energy Environ. Sci. **16(10)**, 4388-4403 (2023).

E3S Web of Conferences **606**, 03012 (2025) https://doi.org/10.1051/e3sconf/202560603012
ICNAOE 2024

Research progress on platinum catalysts for hydrogen-oxygen fuel cells

Yifei Chen*

School of Chemistry and Chemical Engineering, Yantai University, 264005 Yantai, China

Abstract. In recent years, China's fuel cell technology and industry have developed rapidly. However, the high-cost problem caused by the extensive use of Pt-based catalysts remains one of the most important factors restricting its development. As an important component of hydrogen-oxygen fuel cells, platinum-based catalysts are crucial for promoting their application in this field. The development and research of Pt-based catalysts with higher catalytic efficiency are of great significance for promoting the development of fuel cell technology and industry. In this review, some feasible strategies for improving the performance of Pt-based catalysts were summarized based on reducing platinum usage. In addition, the methods and mechanisms of each strategy in enhancing catalytic activity and reducing the amount of platinum catalyst were discussed in detail. At the same time, the shortcomings and challenges faced by this type of catalyst were pointed out, and the future research and application of this type of catalyst were discussed.

1 Introduction

The global energy crisis and climate change have become pressing issues in recent decades. As our reliance on fossil fuels continues to grow, projections suggest that oil and gas reserves could be depleted by 2042 and coal by 2112. This situation highlights the urgent need to explore alternative energy sources to reduce our dependence on fossil fuels for energy and transportation. Among potential solutions, energy-efficient and environmentally friendly fuel cells are emerging as viable alternatives for portable, mobile, and stationary power applications [1]. Hydrogen fuel cells, in particular, are notable for their high theoretical energy density, zero pollution, and high efficiency, making them a focal point of research.

In hydrogen-oxygen fuel cells, catalysts are critical for driving the oxidation of hydrogen and the reduction of oxygen at the electrodes. These catalysts are crucial in influencing the activation polarization of hydrogen fuel cells and are therefore essential materials. Selecting suitable catalysts requires ensuring high resistance to temperature and corrosion under operational conditions. Currently, platinum (Pt) supported on carbon (Pt/C) is the most widely used catalyst. However, the high cost of platinum necessitates the development of more cost-effective alternatives.

This review examines recent advancements in platinum catalysts for hydrogen-oxygen fuel cells and explores strategies to enhance performance while reducing platinum usage.

* Corresponding author :chenyifei@s.ytu.edu.cn

2 Mechanism of ORR and HER

Currently, platinum (Pt) is the most effective electrocatalyst in fuel cells and promotes key reactions in hydrogen-oxygen fuel cells. In these batteries, the anodic reaction is hydrogen oxidation (HOR), while the cathodic reaction is oxygen reduction (ORR). The ORR process can be divided into two main paths. [2] The first method is to partially reduce oxygen to hydrogen peroxide (H_2O_2) or superoxide (HO_2) via a two-electron path. But this approach has its lower potential and stronger oxidation performance, which is not very beneficial and can also damage the catalyst. The second approach is to completely reduce oxygen to water (H_2O) or hydroxide (OH) through a four-electron process, which is usually preferred. [3] The four-electron ORR pathway differs in acidic and alkaline solvents. This difference is influenced by the fact that the protons required for oxygen reduction come from water and hydrogen ions in solvents. Furthermore, the mechanism of O-O bond cleavage can be divided into two categories: association and dissociation [4]. In acidic solvents, these mechanisms can be described by the following reactions: [5]

1. $O_2 + M^* + H^+ + e^- \rightarrow M\text{-}OOH$
2. $M\text{-}OOH + H_2O \rightarrow M\text{-}OH + H^+ + e^-$
3. $M\text{-}OH + H_2O \rightarrow M^* + H^+ + e^-$

Dissociative Pathway:

1. $O_2 + 2M^* \rightarrow 2M\text{-}O$
2. $2H^+ + 2e^- \rightarrow 2M\text{-}OH$
3. $2H_2O \rightarrow 2M^* + 2H^+ + 2e^-$

In basic solvents, the mechanisms are:

Associative Pathway:

1. $O_2 + M^* + H_2O + e^- \rightarrow M\text{-}OOH + OH^-$
2. $M\text{-}OOH + H_2O + e^- \rightarrow M\text{-}OH + OH^-$
3. $M\text{-}OH + OH^- \rightarrow M^* + H_2O$

Dissociative Pathway:

1. $O_2 + 2M^* \rightarrow 2M\text{-}O$
2. $2H_2O + 2e^- \rightarrow 2M\text{-}OH + 2OH^-$
3. $2OH^- \rightarrow M^* + H_2O$

Here, M* represents the catalytic active site.

The dual functional mechanism in the hydrogen oxidation reaction (HOR) is the reaction between adsorbed hydrogen (Had) and adsorbed hydroxide (OHad), which desorbs to form water. The activity of HOR is positively correlated with the gas affinity of the catalyst. Therefore, the HOR activity can be improved by adding materials that are prone to forming OHad on the catalyst surface.

3 Electrocatalyst for ORR and HOR

In hydrogen fuel cells, the oxidation of hydrogen and the reduction of oxygen at the electrodes are primarily controlled by catalysts. These catalysts significantly impact the activation polarization, overall performance, and fuel economy of hydrogen fuel cell vehicles. Selecting appropriate catalysts involves ensuring high temperature and corrosion resistance. The commonly used catalyst is Pt/C.

However, Pt/C faces issues such as Pt particle dissolution, migration, and agglomeration over time, which reduces its active surface area and fails to meet the strength requirements of carbon carriers. As a precious metal, Pt is often used in small particle sizes to enhance performance and minimize usage. Despite this, nano-sized Pt particles have high surface free energy and weak interactions with carbon carriers, leading to particle migration and agglomeration. Smaller Pt particles tend to oxidize, diffuse as Pt ions, and deposit on larger particles, which then grow and agglomerate.

3.1 Pt-based catalysts with special morphologies

Research has shown that the design of platinum-based catalysts with unique morphologies is a very effective strategy for controlling the amount of platinum used. In special forms of wave catalysts, the specific surface area and the atomic utilization efficiency of the catalyst have been improved. Research has shown that the active sites of catalysts lie on the surface of the catalyst and Pt atoms in nanoparticles usually do not participate in the reaction, resulting in low utilization of Pt. Therefore, in order to improve the utilization rate of Pt, we can design some special forms of catalysts that expose more active sites on the catalyst surface. To achieve this goal, special morphologies such as facets with high refractive index (HIF), nanowires (NW), and hollow structures were synthesized. In addition, three-dimensional structured nanomaterials typically have a large specific surface area, which can provide more active sites for reactant molecules. Therefore, the three-dimensional structure can effectively improve the mass transfer and gas diffusion on the alloy surface, thereby enhancing the electrocatalytic performance. The construction of nanomaterials with three-dimensional structures can usually form crystal planes with high refractive index with special structures and other crystal planes with special structural features on the material surface, thereby significantly improving catalyst activity.

In 2014, Stamenkovic et al. reported for the first time on a platinum-based catalyst with a three-dimensional skeletal structure. This catalyst has a high specific surface and is rich in crystal planes with a high refractive index. The mass-specific activity of Pt can reach 5.7A/mg-Pt, which is more than 20-times that of the widely used Pt/C catalyst, making research on three-dimensional skeleton Pt based catalysts in the field of electrocatalysis ever hotter [6]. Traditionally, the utilization efficiency of Pt can be improved by reducing the size of Pt-based catalyst particles. However, catalysts can clump strongly during use. The three-dimensional nanoframework has unique structural advantages. First, it has a highly open three-dimensional structure, a higher specific surface area, a higher density of active sites and a more complete contact with reactant molecules, which can accelerate the reaction. Secondly, its structure is more stable and can withstand harsh reaction environments. Therefore, three-dimensional metal nano scaffolding can effectively improve the activity and stability of catalysts and reduce the amount of precious metal Pt used.

The synthesis of noble metal Pt-based nano frameworks is mainly based on the following two strategies: (1) selective deposition and etching of active sites; (2) Dealloying of hollow or bi (poly) metal nanocrystals. In general, the synthesis mechanism includes an electric displacement reaction, an oxidation etching, diffusion of internal atoms, or a combination thereof.

Pt-based catalysts with three-dimensional special morphology are a kind of highly efficient electrocatalysts with great development in recent years, in terms of material preparation technology, activity and structure characterization, and mechanism exploration. A lot of research work has been carried out and many important results have been achieved. Several important preparation methods have been formed, and some basic understanding of its activity and mechanism has been obtained. However, this type of catalyst still has the following problems:

The preparation process is complex, which leads to the fact that the current research mainly stays at the stage of micro-preparation in the laboratory, and it is necessary to further explore a simple, repeatable and easy-to-scale preparation technology;

At present, the research on this kind of catalyst is still at the level of rotating disk electrodes, and there is almost no report on the application of this kind of catalyst in single fuel cell;

The current 3D Pt-based catalyst model can not well reflect the real reaction conditions of the catalyst, so it is necessary to establish a theoretical calculation model closer to the real environment as soon as possible and develop a new and efficient theoretical calculation method.

The understanding of the structure formation mechanism and catalytic mechanism of these catalysts is still insufficient.

Future research should focus on exploring and developing new preparation technologies; The formation mechanism of the special morphology of the catalyst and the catalytic mechanism were studied by means of advanced technologies such as in-situ technology. The performance and practical application prospects of these catalysts in fuel cells were studied[7].

3.2 Pt-based alloy catalysts

Pt-based alloy catalysts are gaining attention due to their reduced platinum (Pt) usage and enhanced properties. By incorporating foreign atoms into the Pt lattice, alloying modifies the catalyst's electronic structure, affecting the adsorption energy of intermediates and potentially creating dual-function mechanisms. This synergistic effect between surface and electronic structure often results in improved catalytic activity [8,9].When only Pt is used for catalysis, the significant decrease in Pt utilization efficiency is due to the electrocatalytic oxidation of intermediates (such as CO) on the Pt surface, which occupies the catalytic active sites. The oxidation of desorption and adsorption intermediates can be improved by adding a second metal. Adding a second metal to the catalyst will re expose the active sites, thus improving the reaction kinetics. In addition, the catalytic activity of Pt and the tolerance of CO can be improved through electronic or synergistic effects, thereby promoting interactions between other catalyst components.

In Pt based core-shell structures, common noble metal core elements are Pd, Au, Ag, Ru, and Ir, among which Au and Pd/Pt are more typical for intermetallic alloying. Au exhibits high catalytic activity in nanoscale reactions. Au@Pt Nanocatalysts can exhibit synergistic catalytic effects by changing the surface adsorption force to alter the structure of Pt electron bands. Platinum based catalysts can effectively improve their electrocatalytic activity and stability, as well as enhance their CO resistance through synergistic catalysis. In addition to precious metal alloys, extensive research has been conducted to prepare inexpensive and efficient PEMFC catalysts. By alloying Pt with other single metals, the use of Pt can be reduced and catalytic performance can be improved. Reducing the usage of Pt while ensuring good electrocatalytic activity is an urgent problem to be solved.

Alloying Pt with non precious metal elements such as Co, Ni, Cu, Fe, Sr, Gd, and Mg can improve catalytic efficiency and reduce costs. In addition, the atomic utilization efficiency of Pt can be improved by integrating with other metals. Co/Pt and Cu/Pt catalysts are mainly prepared by two methods: one is to first synthesize elemental nuclei, and then coat the surface of the nuclei with a Pt shell layer. According to the synthesis method of elemental nuclei, it can be further divided into hydrogen thermal reduction, sodium borohydride reduction, electrodeposition, and solvothermal methods; Another method is to form PtCo and PtCu alloys with internal Co or Cu enrichment and surface Pt enrichment through acid treatment, heat treatment, and electrochemical treatment Co@Pt and Cu@Pt Core shell structure [10].

3.3 Pt-based atomic-level-dispersed catalysts

In heterogeneous catalysis, nanoparticle size and shape significantly impact activity and selectivity. Reducing nanoparticles to sub-nanometer scales increases the number of uncoordinated metal atoms available as active sites, optimizing catalyst utilization. The comparison between experimental results and theoretical calculations can be made through a simplified model of an ideal atomic-level dispersed catalyst.

Cluster catalysts are a type of catalyst consisting of several metal atoms or clusters that are normally formed from several metal atoms connected by coordination bonds or direct bonds. The main carrier for the production of cluster catalysts is usually made of carbon-containing materials, such as graphene doped with N, S, P, B, F, porous carbon, soot, carbon nanotubes, etc. These heteroatoms form stable precious metal precursors during wet chemical synthesis due to anchoring effects, and then cluster/carbon composite catalysts are obtained through reduction methods. Cluster catalysts have high catalytic activity and selectivity, which can effectively catalyze molecular conversion reactions while avoiding problems such as excessive sintering of the catalyst. Compared to conventional single-atom catalysts, cluster catalysts have richer electronic structures and surface reaction sites and thus exhibit higher catalytic activity and selectivity. Researchers have developed a highly stable and highly dispersed surface catalyst that firmly anchors 1.8+/-0.6nm Pt nanoparticles to a porous 3D carbon carrier through a combination of electrochemical methods and extrusion. Changing the reaction temperature and ligand type during the experimental process can precisely regulate the composition, structure, and atomic number of noble metal clusters at the atomic level.

Metal nanoclusters are a new type of nanomaterial that can obtain atomic level precise structures. They have many advantages such as easy synthesis, strong designability, clear geometric/electronic structure information, and easy regulation of catalytic active sites. They can not only help us understand the structural transformation laws of metal materials, but also be applied to the precise design and construction of catalyst model molecules. Studying the fundamental correlation between catalytic kinetics and surface sealing of metal nanoparticles is crucial, but it remains challenging at present. Researchers have designed a unique 1.7nm Pt nanocluster catalytic system and found that the activity and selectivity of electrocatalytic ORR can be affected by surface caps [11]. In addition, Pt cluster catalysts with different atomic numbers (clusters Pt7, Pt10, and Pt11) have been discovered in the application of ORR. These Pt cluster catalysts not only exhibit similar catalytic activity on ORR, but also have catalytic activity comparable to traditional Pt nanoparticle catalysts. Thermodynamic instability is an inherent problem of metal nanoclusters in catalytic applications. In high-temperature catalytic reactions, metal nanoclusters are prone to rapid deactivation due to particle sintering and growth. Liang et al. found in their study that when sulfur is doped into a carbon matrix, it can effectively stabilize 1nm metal nanoclusters and prevent them from undergoing thermal sintering at high temperatures up to 700°C. Through spectral characterization and DFT calculations, it can be seen that due to the thermally stable and strong metal sulfur bond, the adhesion strength at the metal/S-C interface is enhanced, which greatly suppresses the particle sintering of metal nanoclusters by delaying the diffusion of metal atoms and the migration of nanoparticles. In addition, the prepared Pt nanocluster catalyst exhibits excellent propylene selectivity and stability in the high-temperature propane dehydrogenation to propylene reaction, supported on sulfur doped carbon (S-C) Pt nanocluster catalyst.

Efforts to reduce platinum-based catalysts, enhance power density (catalytic activity), and optimize membrane electrode assembly (MEA) preparation remain crucial for lowering the commercial cost of hydrogen fuel cell systems.

4 Challenge and future

Significant progress has been made in the research of precious metal platinum-based catalysts in the past few years. The several different structures of platinum catalysts mentioned in this article can significantly improve the activity and stability of half-reactions, but there are still some doubts about whether these catalysts can achieve stable performance in fuel cells. To further advance the development of platinum-based catalysts, it is still necessary to address many issues that arise when using platinum-based catalysts in fuel cells. Platinum-based catalysts are an important component of fuel cells, and improving the utilization rate and reducing costs of platinum catalysts are crucial for the commercialization of fuel cells. In the future, platinum catalysts can be studied by optimizing catalyst structures and simplifying synthesis routes to meet the requirements of fuel cells. In the future, the use of platinum can be reduced by exploring highly dispersed catalyst supports, further improving the catalytic performance and stability of the catalyst. In addition, further research can be conducted on the electrocatalytic mechanism of platinum-based catalysts through a combination of theoretical analysis and experimental analysis. Building efficient fuel cell catalysts with good CO tolerance is also essential. Finally, the development of innovative technologies plays an irreplaceable role in researching catalysts with high power, strong stability, and strong environmental adaptability.

5 Conclusion

Hydrogen-oxygen fuel cells represent a promising energy technology with vast application potential. As fossil fuel resources dwindle, hydrogen emerges as a sustainable and inexhaustible energy source. Notably, hydrogen can be produced from renewable energy sources, positioning hydrogen-oxygen fuel cells as a cornerstone for sustainable development.

One of the significant advantages of these fuel cells is their emission profile: they produce only water vapor, avoiding harmful pollutants such as carbon dioxide and nitrogen oxides. This characteristic is crucial for mitigating greenhouse gas emissions and protecting our atmosphere.

However, the widespread adoption of hydrogen-oxygen fuel cells faces challenges, particularly regarding catalysts. Most current catalysts rely on platinum, whose scarcity and high cost pose significant hurdles. This article explores several proposed strategies to enhance platinum-based catalysts by reducing platinum usage while improving overall performance.

Despite progress in the development of platinum-based catalysts, critical issues remain. Reducing platinum content and enhancing power density are vital for decreasing the production costs of hydrogen-oxygen fuel cells. With ongoing technological advancements, the high-performance, cost-effective catalysts will emerge that can significantly enhance the performance and economic viability of hydrogen-oxygen fuel cells across various applications.

References

1. L. An, T. Zhao, X. Yan, et al. The dual role of hydrogen peroxide in fuel cells. Science Bulletin, **60(1)**, 55-64 (2015)
2. B.-Q. Li, C.-X. Zhao, J.-N. Liu, et al. Electrosynthesis of Hydrogen Peroxide Synergistically Catalyzed by Atomic Co–N–C Sites and Oxygen Functional Groups in Noble-Metal-Free Electrocatalysts. Advanced Materials, **31(35)**:,1808173 (2019)

https://doi.org/10.1051/e3sconf/202560603012

1. K. Gao, B. Wang, L. Tao, et al. Efficient Metal-Free Electrocatalysts from N-Doped Carbon Nanomaterials: Mono-Doping and Co-Doping. Advanced Materials, **31(13),** 1805121 (2019)
2. C. Wan, X. Duan,Y. Huang Molecular Design of Single-Atom Catalysts for Oxygen Reduction Reaction. Advanced Energy Materials, **10(14),** 1903815 (2020)
3. X. Zhang, J. Wang,Y. Zhao Enhancement Mechanism of Pt/Pd-Based Catalysts for Oxygen Reduction Reaction. Nano.**13(7),** 1275 (2023)
4. C. Chen, Y. Kang, Z. Huo, et al. Highly Crystalline Multimetallic Nanoframes with Three-Dimensional Electrocatalytic Surfaces. Sci. **343(6177),** 1339-1343 (2014)
5. T. Xiangchun, C. Jiaxiang, L. Lina, et al. Pt-Based Electrocatalysts with Special Three-Dimensional Morphology or Nanostructure. Progre. in Chem. **33(7),** 1238-1248(2021)
6. T. Zhao, Y. Hu, M. Gong, et al. Electronic structure and oxophilicity optimization of mono-layer Pt for efficient electrocatalysis. Nano. Energy, **74**, 104877(2020)
7. F. Zhan, K.-S. Hu, J.-H. Mai, et al. Recent progress of Pt-based oxygen reduction reaction catalysts for proton exchange membrane fuel cells. Rare. Metals. **43(6),** 2444-2468(2024)
8. M.J.-j.L.H.-c.T.Z.-y.L. Jian-feng Progress in high-efficiency utilization of PEMFC cathode Pt-based alloy catalyst. Battery Bimonthly. **52(04),** 467-470(2022)
9. W. Liu,H. Wang Influence of surface capping on oxygen reduction catalysis: A case study of 1.7nm Pt nanoparticles. Sur. Sci.**648,** 120-125(2016)
10. L. Liu,A. Corma Metal Catalysts for Heterogeneous Catalysis: From Single Atoms to Nanoclusters and Nanoparticles. Chemical Reviews, **118(10),** 4981-5079 (2018)
11. P. Yin, X. Luo, Y. Ma, et al. Sulfur stabilizing metal nanoclusters on carbon at high temperatures. Nat. Commun. **12(1),** 3135 (2021).

Introduction to Mycelium Insulation as a Sustainable Insulation Material and Carbon Sink

Michael Junyou CHEN

Shenzhen College of Internation Education, Shenzhen, China

Abstract. Climate change is becoming an even bigger threat with the continued production of pollution in the atmosphere. Mycelium insulation offers a potential part of the solution by being a carbon sink and a sustainable material in an otherwise very unsustainable industry of construction and insulation. In this review paper, I present the various materials that can form mycelium insulation, the process of how mycelium insulation is made, and some examples of potential composites that have been studied. Furthermore, I discuss the environmental impact of the insulation industry, and how mycelium insulation can reduce such impacts. Additionally, mycelium insulation is then compared to current alternatives on the market and evaluated by outlining its capabilities across various metrics to determine its performance. I then introduce specific properties of a mycelium composite that can be achieved depending on the substrate make up. Finally, market viability is analysed to show the potential of mycelium insulation. Results show that the current iteration of mycelium insulation may appeal to select individuals that are environmentally conscious, or for specialised cases. Widespread us of mycelium will only be achieved if further innovation occurs or an increase in the need to become environmentally friendly.

1 Introduction

As the effects of climate change become increasingly apparent and extreme, and with no substitute available for many industries that produce greenhouse gasses, The global population is estimated to keep growing to 9 billion by 2050 [1]. This will in turn further increase our need for housing, waste disposal, and transport; the manufacturing of these materials will have to become sustainable. Carbon capture technology is a crucial field that may help the world achieve net zero emissions alongside innovation in manufacturing. One such technology is the growth and use of mycelium as insulation.

Earth is about 1.36 degrees Celsius warmer than in the 19th century, with the past 10 summers consistently breaking records for being the warmest summers to date [2]. While governments have committed to curbing carbon emissions, many heavy industry sectors are proving difficult to decarbonize, such as trucking, aviation, chemical production, and steel production. The manufacturing industry alone accounted for 12% of total carbon emissions in the US in 2021 [3]. This poses a huge threat to our ideal of a circular economy being implemented. A possible solution is to use mycelium as a carbon sink for insulation. Studies have shown that there is huge potential for abatement technologies in the construction industry with short-term reductions in greenhouse gas emissions [4]. From concrete roads to brink houses, many materials used during the construction of buildings are ripe for innovation and incredibly costly in terms of GHG emissions.

Currently, the materials used for insulation include fibreglass, cellulose, and mineral wool, among others. However, manufacturing these materials is either very labour-intensive or produces a high carbon footprint through the production of plastic. These insulators are becoming increasingly less attractive to our current world as society's aspirations to be eco-friendly increase. Mycelium Insulation is a steadily developing solution to this issue. Furthermore, mycelium has many different combinations of fungus species and substrates that boast different properties and uses [5]. Mycelium itself is already being used in many other industries around the world such as cosmetics, medicine, and food packaging [6].

Mainstream carbon capture is mostly focused on geological storage because of its speed and efficiency compared to biological storage. The initial capital investment is also high due to the need for machinery in carbon scrubbers and the setup of underground geological formations for carbon storage. This, however, presents further issues, such as the potential leakage of pipes and health concerns for personnel. Finally, a combination of a lack of public acknowledgement of this technology, market incentives, and an unclear legality of the transport and storage of carbon, makes geological storage slow to innovation [7]. Another point of concern about developing geological storage technology is its misuse in the oil industry. An example of this is the buyup of Ion Energy, to Chevron the carbon scrubber technology developed by Ion Energy is an air scrubber that captures carbon dioxide from ambient air and stores it underground [8]. However, it can also be used in oil wells, to pump out even more petroleum. This is counterintuitive as the net decrease in carbon captured is vastly topped by the carbon emissions produced from the petroleum that was pumped out. On the other hand, biological storage of carbon, such as in mycelium, may prove to have much fewer side effects while being cost-effective. This paper aims to discuss how Mycelium may replace traditional insulation in some places. Along with its effectiveness as a carbon sink.

2 Production and mechanism

2.1 Material

These species of fungus can then be further customized using different combinations of substrates to form new composites. The reason mycelium has so much potential as a material is its ability to grow from byproducts and waste, making it an affordable alternative to normal materials. It can also be recycled by compost and requires little energy to be produced [4]. The compressive strength of the mycelium ranges from 1.2 N/mm^2 to 1.8 N/mm^2 and is impacted by changes in porosity, pore size, and characteristics of the materials that it is grown from [9]. This variability also is a drawback of this technology, making it inconsistent and hard to upscale.

2.2 Studies of Potential Materials

Previous research revealed that the best mix proportion for a mycelium composite was 0.3:1:0.1 of Ganoderma resinaceum mushroom, Miscanthus x giganteus fibres, and potato starch (G0.1_M1_P0.1) Yielded a low-density (122 kg/m^3) material that was suitable as thermal insulation material [9]. It was found that the material had a high water absorption and increased in weight by 40-580 wt% when in contact with water for 48-192 hours [9]. This does not result in a disadvantage for mycelium insulation as you can just install a protective membrane or a render.

https://doi.org/10.1051/e3sconf/202560604001

On the other hand, another study used sawdust from birch and poplar for the creation of lightweight mycelium composites (100 kg/m3) and high porosity (>90%) which resulted in high insulation capabilities [10].

2.3 Environmental impact of mycelium insulation

As stated before, mycelium insulation is fully biodegradable while also providing nutrients to the soil. The production of mycelium also produces less emissions compared to other common insulators. However, if the amount of CO_2 absorbed during the growth of mycelium is taken into accounts, it suddenly becomes carbon negative at around 244 kg CO_2eq/m^3 with the possibility of reducing even more carbon as energy efficiency increases and renewable energy becomes more commonplace [11]. This can be seen in figure 1 where emissions from mycelium are contrasted with other common insulators.

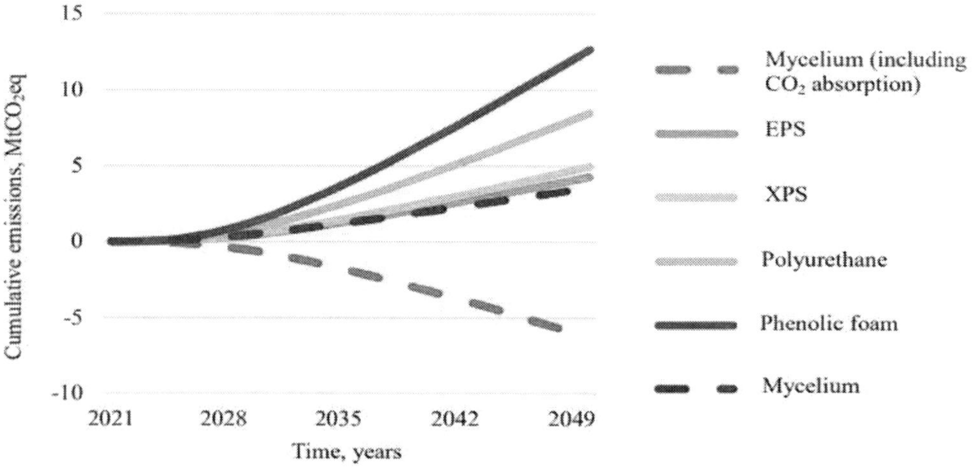

Fig. 1. Comparison of different materials [11]

2.4 Manufacturing Process of mycelium

The process of growing mycelium begins with finding a pure culture of the fungal species in the form of spores, liquid culture, agar plates, or other methods. The growth medium is then prepared by creating a block of substrate, the substrate is usually made of materials that are chosen to make up the composite of the final product, some of these materials include sawdust, grain, agricultural waste, or other organic material. The fungus is then spread evenly using various methods by layering the substrate and inoculum, or mixing the inoculum with the substrate, the inoculum will have to be thoroughly distributed throughout the substrate to achieve desired results. In a sterilized environment such as a laminar flow hood, the mycelium is incubated at around 20-30 degrees Celsius. Once the substrate is fully colonized, the hyphae will have permeated throughout the substrate and formed a dense network that is irregular and has insulation properties. The composite can then be moulded and further processed for specific applications.

3 Evaluation of general and specialized characteristics of mycelium insulation

3.1 Performance of mycelium insulation compared to other materials

Mycelium insulation presents an intriguing alternative to traditional insulation materials across various applications. Typically, mycelium insulation has a R-value of 3 per inch. This is comparable to other organic insulators such as wood fibre, straw, or wool. However, this still falls short compared to the performance of traditional insulators such as fibreglass insulation, which boasts an R-value of 3.7-4.3 per inch. This in turn means that mycelium insulation is less effective at resisting heat transfer, particularly in cold climates. Mycelium insulation demonstrates superior thermal performance– effectively reducing heat transfer and enhancing energy efficiency [11]. This translates to potential cost savings on heating and cooling bills while simultaneously increasing occupant comfort. Additionally, mycelium insulation exhibits sound-absorbing properties– contributing to a quieter indoor environment. However, long-term durability and comprehensive cost comparisons with traditional insulation warrant further investigation [12].

In industrial contexts, mycelium insulation's versatility is apparent. Its moldable nature allows for customization to fit specific insulation requirements– showcasing a promising option for various industries. Furthermore, its inherent fire resistance and ability to absorb pollutants position it as a potential game-changer in sectors such as construction and manufacturing. While research is ongoing, the initial findings are both encouraging and promising.

3.2 Mycelium composites

The true essence of the advantages of mycelium insulation comes from its ability to gain different attributes depending on the substrate chosen for the mycelium to grow in. One very prominent substrate is wood chips and sawdust. The main benefit that they provide is the fibrous structure which provides a high surface area that encourages mycelial growth. Furthermore, mycelium breaks down cellulose and lignin in wood which acts as a glue that forms a stronger and more complete composite [13]. Biochar is another material that is implemented in composite materials due to its inherent fire-resistive capabilities. Incorporation of biochar allows the result to become more resistant to ignition and combustion. It also has selective anti-fungal properties when optimized properly. These properties create an antifungal effect that can suppress mould without a sizable decrease in mycelium growth. Natural polymers are sometimes also added into substrates to act as a binding agent within the composite. It results in a stronger and more stable material along with increasing insulation capabilities. Some polymers such as starch can add flexibility to an otherwise rigid material, this can allow easier handling of the insulation or for when slight movement may occur throughout the insulation's lifespan. Other materials like agricultural waste and recycled cardboard or paper, which make up most of a composite, are mostly used due to it being readily accessible and cheap.

4 Evaluation of market viability

The viability of mycelium in the insulation market highly depends on the success of a few factors. Mycelium insulation has the potential to massively appeal to audiences that value and appreciate being eco-friendly - which is currently a market with huge potential for growth, or with companies that are seeking to reach goals for their green initiatives. However, for successful implementation of mycelium, it still has many hurdles it has to overcome. A notable issue is the slow production rate of mycelium, and subsequently uncertainties regarding the scalability of this technology. Currently, there have only been tests for industrial implementation at the pilot scale. Scaling up further from this will require significant capital investment [14]. The insulation market has historically been an industry with little innovation in terms of embracing new inventions. In an industry that focuses on being consistent and has huge consequences for failure, new material with no history will need time to thrive and garner interest. This problem may be combated by asking governments to help push the initiative of using mycelium insulation by pushing out government policies. Additionally, as mycelium insulation is still a very new technology, the scale and speed of production still is very small. Solutions to streamlining the production method, and the inability to utilize economies of scale result in the cost of mycelium insulation being higher compared to traditional insulation [14]. As the raw material used for mycelium production - agricultural waste, is extremely cheap and abundant, the possibility for vertical integration there is still much room to lower costs. Another risk that mycelium faces is the inconsistency of the supply chain. Yes, relying on organic waste material to produce the product indeed does facilitate circular economy and sustainability, but it also poses issues with how the process can be standardized and how quality control on the industrial scale can be maintained in a cost-effective manner. Finally, mycelium has problems with moisture resistance and in turn longevity of the material.

5 Conclusion

This paper discussed the potential of mycelium insulation as an organic carbon sink in combination with being a promising sustainable insulator. Current insulators are often one-time use and are made using unsustainable materials that cannot be recycled. Additionally, carbon capture technology is mainly focused on the underground storage of carbon. These practices have historically been implemented in a counterintuitive manner, leading to a negative public reputation and a decrease in funding and innovation in this sector. The production of mycelium is further illustrated to help readers understand the process that mycelium production goes through, along with introducing some notable composites and substrate materials that have been studied. The carbon footprint of the insulator was then compared with traditional insulation to conclude that mycelium insulation produces significantly less carbon dioxide compared to other technologies when carbon absorption is considered. The market potential for mycelium insulation was then analysed. While mycelium insulation has the advantage of being extremely customisable depending on the situation, as it is still a developing technology, it has a higher cost of production and currently does not perform as well as traditional insulators. On the other hand, as time passes more consumers may be inclined to choose mycelium to be more environmentally conscious and as the product gains a more sophisticated track history. the innovation potential for mycelium insulation is high, and given time, mycelium insulation may even become cheaper than traditional insulation. Mycelium's insulating ability will naturally increase as more efficient composites are discovered.

References

[1] Al-Qahtani, S., Koç, M., & Isaifan, R. J, Mycelium-Based thermal insulation for domestic cooling footprint reduction: A review. Sustainability, **15**, 13217 (2023) https://doi.org/10.3390/su151713217

[2] Alaux, N., Vašatko, H., Maierhofer, D., Saade, M. R. M., Stavric, M., & Passer, A, Environmental potential of fungal insulation: a prospective life cycle assessment of mycelium-based composites. The International Journal of Life Cycle Assessment, **29**, 255 (2023) https://doi.org/10.1007/s11367-023-02243-0

[3] Alemu, D., Tafesse, M., & Mondal, A. K, Mycelium-Based Composite: the future sustainable biomaterial. International Journal of Biomaterials, 1-12, (2022) https://doi.org/10.1155/2022/8401528

[4] Haneef, M., Ceseracciu, L., Canale, C., Bayer, I. S., Heredia-Guerrero, J. A., & Athanassiou, A, Advanced materials from fungal mycelium: Fabrication and tuning of physical properties. Scientific Reports, **7**, (2017) https://doi.org/10.1038/srep41292

[5] Change, N. G. C. (n.d.). Global Surface Temperature | NASA Global Climate Change. Climate Change: Vital Signs of the Planet. Retrieved from https://climate.nasa.gov/vital-signs/global-temperature/?intent=121#:~:text=Overall%2C%20Earth%20was%20about%202.45,change%20in%20global%20surface%20temperatures

[6] Dias, P. P., Jayasinghe, L. B., & Waldmann, D Investigation of Mycelium-Miscanthus composites as building insulation material. Results in Materials, **10**, 100189 (2021) https://doi.org/10.1016/j.rinma.2021.100189

[7] Congressional Budget Office. Emissions of greenhouse gases in the manufacturing sector. (2024) retrieved from https://www.cbo.gov/publication/60030#:~:text=CBO%20estimates%20that%20the%20manufacturing,that%20transform%20materials%20into%20products

[8] Gauvin, F., Tsao, V., Vette, J., & Brouwers, H. J. H., Physical properties and hygrothermal behavior of Mycelium-Based composites as Foam-Like wall insulation material. Construction Technologies and Architecture, **1**, (2022) https://doi.org/10.4028/www.scientific.net/cta.1.643

[9] Horst, R. K. Overview of plant diseases. Encyclopedia of microbiology (Fourth Edition). (2019) https://doi.org/10.1016/b978-0-12-809633-8.05025-1

[10] Johnson, L. Chevron leads $45M investment in carbon removal company Ion Clean Energy. Utility Dive. 2024, retrieved from https://www.utilitydive.com/news/chevron-leads-45m-investment-carbon-removal-ion-clean-energy-carbon-direct/712943/

[11] Jones, M., Mautner, A., Luenco, S., Bismarck, A., & John, S. Engineered mycelium composite construction materials from fungal biorefineries: A critical review. Materials & Design, **187**, 108397 (2020) https://doi.org/10.1016/j.matdes.2019.108397

[12] Zhang, M., Zhang, Z., Zhang, R., Peng, Y., Wang, M., & Cao, J. Lightweight, thermal insulation, hydrophobic mycelium composites with hierarchical porous structure: Design, manufacture and applications. Composites Part B Engineering, **266**, 111003 (2023) https://doi.org/10.1016/j.compositesb.2023.111003

[13] Shirmohammadi, R., Aslani, A., & Ghasempour, R. Challenges of carbon capture technologies deployment in developing countries. Sustainable Energy Technologies and Assessments, **42**, 100837 (2020) https://doi.org/10.1016/j.seta.2020.100837

[14] Lutz, W., & Kc, S. Dimensions of global population projections: what do we know about future population trends and structures? Philosophical Transactions of the Royal Society B Biological Sciences, **365**, 2779 (2010) https://doi.org/10.1098/rstb.2010.0133

Planning Concept for the Circular Economy Industrial Park in Heyuan City

Hengran Liu[1*]

[1]University of Glasgow, Adam Smith Business School, Glasgow G12 8QQ, The United Kingdom

Abstract. The circular economy industrial park built based on the 3R principle is not only for pursuing greater economic benefits and less resource consumption, but also an important measure for building zero-waste cities. The article first studied the circular economy industrial parks in Hong Kong and Japan, summarizing their advanced experience in pursuing lower pollution and sustainable use of materials and energy. Then the article selected Heyuan City in Guangdong Province as the object, and evaluate the necessity and feasibility of building a circular economy industrial park by analyzing the policy environment, location conditions, and current situation of general sanitation facilities. Preliminary planning will be carried out for the functional zoning, energy system, water resources system of the circular economy industrial park, and discuss the environmental protection technologies that can be applied in the proposed scenario. Finally, it is recommended that the park and enterprises minimize the impact of facilities on the surrounding environment to better integrate into the surrounding urban space.

1 Introduction

In 1994, Germany promulgated the Circular Economy and Waste Management Act [1]. This law takes the 3R principle, namely Reduce, Reuse, and Recycle, as its core idea and fundamental [1]. On the basis of the 3R principle, people have been seeking an advanced economic model that pursues greater economic returns, less resource consumption, and less pollution, so that both material and energy can be reasonably and sustainably utilized in an economic cycle. The circular economy industrial park is a disposal and recycling site developed based on the 3R principle.

The embryonic form of circular economy industrial parks originated from the exchange of industrial waste between sectors within industrial parks. Costs and pollution are reduced by using industrial waste from some sectors as production materials for others. At the end of the last century, circular economy industrial parks began to build on the basis of the solid waste treatment industry. Compared to the material exchange between production departments within industrial parks, the types of upstream and downstream industries of solid waste treatment are more diverse, because it includes the treatment of urban household waste, construction waste, hazardous waste, and sludge. In addition to production and

* Corresponding author: hengran.liu@dukekunshan.edu.cn

manufacturing functions, there are also productive services in the park, such as warehousing, logistics, accommodation and technical training. In the past two decades, circular economy industrial parks have entered a development stage led by technological innovation. The construction of circular economy industrial parks at this stage is generally jointly managed by the local governments, with active participation from enterprises and research institutions, forming an integrated operation mode of government, industry, and academia. In addition to maintaining an environmentally friendly, clean, and livable city image, the park has also begun to provide public welfare for the city residents, gradually integrating with the city in terms of functionality.

Heyuan City is an important node connecting the Pearl River Delta, southern Jiangxi, and western Fujian in Guangdong Province. With the economic development of Heyuan City, urban solid waste continues to grow. The demand for proper disposal and resource utilization of solid waste is becoming increasingly prominent. To promote the construction of a green development demonstration zone in Guangdong Province, the government plans to build a circular economy industrial park with the goal of realizing the resource utilization, harmless treatment, and reduction of urban solid waste [2]. This article will discuss advanced cases, project implementation necessity, regional condition analysis, conceptual layout plan, and proposals for the future.

2 Advanced experience

2.1 Hong Kong circular economy infrastructure

T·PARK is the world's largest sludge incineration treatment facility built at a cost of 5 billion yuan (T·PARK Home Page, 2024), which includes multiple sectors such as sludge incineration, power generation, and seawater desalination, providing heat, electricity, and fresh water to the city. There are also scenic spots, educational bases, cinemas, tea rooms and other facilities in the park to promote environmental protection concepts. T · PARK can process up to 2000 tons/day of sludge, and the current actual processing capacity is 1200 tons/day [3]. Over 90% of the sludge weight can be reduced [3], producing only ash and a small amount of residue. At the same time, the incineration process converts heat energy into electricity, which can be connected to the public power grid to supply power to 4000 households [3]. The adjacent bay provides abundant water sources for T. PARK. The seawater desalination facilities can produce 600 cubic meters of fresh water per day [3], which is sufficient to supply the daily operation of the facilities. Purified fresh water will be used for incinerator cooling, boiler heating, and domestic water. The sewage generated by the entire T · PARK will be treated and used for cleaning and irrigation. T·PARK is a representative circular economy infrastructure that provides comprehensive ecological services, which has functions such as energy self-sufficiency, resource recycling, and public education while disposing of pollutants for cities.

The WEEE·PARK waste electronic equipment treatment facility covers an area of 3 hectares and has the processing capacity of 30000 tons per year [4]. Its collection network for waste electrical appliances covers the entire Hong Kong, with 5 regional collection centers that track and collect discarded household appliances from citizens through online tracking and waste collectors. Waste electronic products generally contain metals or chemicals that are harmful to human health. WEEE·PARK converts these toxic substances into valuable secondary materials through processes such as detoxification, dismantling, and recycling. Chlorofluorocarbon (CFC) or hydrofluorocarbon (HFC) refrigerants in electrical appliances will be extracted for proper disposal. The lead containing cone glass in the cathode ray tube screen will be separated, and the fluorescent powder or mercury inside will be properly disposed of. The electrical casing will be disassembled in a safe environment with inert protective gas, with the removal of all the flammable materials [4]. The remaining metal and plastic components will be classified and stored. Many buildings in WEEE·PARK utilize natural light and ventilation, and their rainwater collection systems can provide irrigation and reclaimed water for the park. There is also a public exhibition hall inside, where the public can gain a deeper understanding of the process of processing and recycling waste electronic products.

O'PARK 1 is the first organic resource recycling center in Hong Kong, with a daily processing capacity of 200 tons of kitchen waste [5]. O'PARK 1 uses anaerobic decomposition technology to convert kitchen waste into biogas (methane and carbon dioxide) and use it for power generation. The generated electricity is not only used internally by O'PARK 1, but is expected to output approximately 3000 households' electricity consumption per year. The collected kitchen waste will be crushed into kitchen waste slurry. The slurry will first pass through a drum screen to remove large impurities of 16 millimeters or more, and then enter a sand removal well to remove heavier impurities such as sand, stones, glass, and metal. After removing impurities, the slurry will be injected into the anaerobic digestion tank, where microorganisms are used to convert the kitchen waste into biogas. Then the collected biogas will be transported to the desulfurization tower for purification. The purified biogas are stored in a steel double membrane gas storage tank with a capacity of 1500 cubic meters , ready for use in power generation. The residue generated by the anaerobic digestion tank will be first dehydrated using a centrifuge and mixed with a leavening agent, these mixtures will be stored in tunnels at 55 degrees Celsius for about 20 days to reduce residual organic matter and moisture content, ultimately forming compost that can be used for agricultural production.

2.2 Kitakyushu Ecological Industrial Park

Kitakyushu is located in the northern part of Kyushu Island, and is one of the largest port city in Japan. Its core industries include steel, chemicals, machinery, kiln industry, and information industry [6]. The industrial pollution in Kitakyushu City was once very serious, with the occurrence of multiple pollution incidents. To alleviate industrial pollution, the Kitakyushu Ecological Industrial Park was established in 2001 [7]. The land within the park is purchased by the government and leased to enterprises, with investment compensation provided to encourage the development of the environmental protection industry.

The Kitakyushu Ecological Industrial Park mainly consists of four areas: academic research area, experimental verification area, environmental protection enterprise cluster, and recycling factory cluster. Academic research area promotes technological innovation in the park, technology cooperation relationships with more than ten universities had been established [7]. The function of the experimental verification area is to verify and transform academic research results. Research achievements such as food residue regeneration, foam polystyrene regeneration and other technologies have been widely used in the industrial production links in the park. The environmental protection enterprise cluster and the recycling factory cluster have established upstream and downstream links related to the utilization of waste disposal resources. These two sectors are centered around waste incineration power plants for the treatment of industrial waste discharged, and the melted material after incineration is recycled into building materials such as concrete bricks. At the same time, the heat generated by incineration is used for power generation. The park also relies on the surrounding mountain and sea landscape conditions to build beautiful ecological parks, science education squares and other public welfare facilities, providing a series of free social welfare services for surrounding residents.

3 Demand forecast

3.1 Policy interpretation

On June 7, 2018, Guangdong Province held a working conference on solid waste management [8]. The meeting emphasized the need to comprehensively accelerate the planning and construction of solid waste disposal facilities in Guangdong Province, actively use information technology to implement the full process management of solid waste, and ensure that solid waste is safely and properly disposed of. In March 2021, the General Office of the People's Government of Guangdong Province issued the Pilot Work Plan for Promoting the Construction of waste free cities in Guangdong Province [9]. This plan proposes to solve the problems of large solid waste generation, low reuse rate, illegal dumping, and difficulty in selecting sites for disposal facilities. The work plan also encourages pilot cities to construct ecological industrial parks and circular economy industrial parks according to their industry development characteristics, and actively promote advanced industrial solid waste utilization technology and equipment. Against the backdrop of actively promoting ecological civilization construction in Guangdong Province, in order to build a green development demonstration zone throughout the province, Heyuan City has identified the ecological and environmental protection industry as a key development industry. According to the Industrial Development Plan of Heyuan City (2021-2030) [10], constructing circular economy industrial parks is an important measure to meet the requirements of zero-waste city construction. This measure can not only meet the needs of urban solid waste disposal, but also lead the development of the city's environmental protection industry.

3.2 Prediction of solid waste generation

The formula for predicting the amount of household waste generated is $Q=R*q*e*10$. Q is the daily production of household waste (tons/day), R is the planned population size of the planning area (10000 people), q is the daily per capita production of household waste, and e is the coefficient of variation of waste production (related to the local economic level). If the population in the service area of the industrial park is 1.2 million, the per capita production of household waste is taken as 1.2kg per day. Based on the local economic level, the variation coefficient of garbage production is set as 1.2, resulting in a daily production of 1728 tons of household waste, estimated to be 1800 tons per day. The per capita production of kitchen waste is about 0.15kg per day, so the predicted scale of kitchen waste is 216 tons/day. The amount of large waste generated usually accounts for 10% of the amount of household waste, so the predicted amount of large waste generated is $1800\times10\%=180$ tons/day.

According to the Announcement on the prevention and control of solid waste pollution in Heyuan City for the year 2022, the sludge production reached 33577 tons, an increase of about 10% compared to last year [11]. Based on this average annual growth rate, the sludge production in Heyuan City is about 400 tons per day. The amount of hazardous waste generated by industrial enterprises in Heyuan City is 44670 tons. Considering the reserved processing scale of 0.3 times for long-term increment, the industrial hazardous waste disposal gap will reach 421 tons per day by 2035. Therefore, it is predicted that in the long term, $421\times60\%=250$ tons/day of industrial hazardous waste will need to be incinerated for harmless disposal. The annual disposal capacity of medical waste is 2478 tons, with a rated scale of 3800 tons per year, approximately 10 tons per day. According to the experience of similar cities, the amount of construction waste generated is about 2000 tons per day.

4 Regional condition analysis

4.1 Geographical condition analysis

Heyuan City is mainly composed of mountains and hills, with mountains accounting for 53%, hills accounting for 36%, valleys and plains accounting for 11% [12]. The city belongs to the subtropical monsoon climate zone, with mild climate, abundant rainfall, and short winter time. Dongjiang River is the largest river in this city, originating from Jiangxi, with a large number of tributaries and abundant water resources. Xinfengjiang Reservoir is a drinking water source for cities such as Shenzhen and Hong Kong in the lower reaches of Dongjiang River. The geological structure in the upper reaches of the main stream and some tributaries of Dongjiang River is complex and severely weathered, which may lead to landslides in high precipitation weather. The planning and site selection of industrial parks can be located between hills or in valleys, mountains and hills may form natural barriers, reducing the impact of the project on the surrounding area. At the same time, by relying on the surrounding mountain terrain, multi-layer transportation system and pipe gallery can be constructed. Additionally, adequate protection distance needs to be reserved. It is advisable to choose an area that is far away from sensitive targets such as ecological resources, surface water systems, transportation hubs, cultural relics and scenic areas. Site selection should avoid areas prone to earthquake faults, landslides, mudslides, and subsidence. To cope with the threat of floods, tides, or water-logging, reliable flood control and drainage facilities should be constructed.

4.2 Economic condition analysis

According to the calculation results of Guangdong Province's regional gross domestic product, the proportion of the first, second, and third industries in Heyuan City in 2023 is 12.8:37.0:50.2 [13]. At the end of year 2023, there were a total of 649 industrial enterprises above designated size in the city, including 203 with an annual output value exceeding 100 million yuan. The added value of industrial enterprises above designated size reached 37.268 billion yuan. The secondary industry is the economic pillar of Heyuan City, among which the production of electronic information equipment is the leading industry. More than 100 electronic information enterprises and the national communication terminal quality inspection center have settled in Heyuan National High tech Zone [14], forming an electronic information industry cluster with mobile phones and electronic communication as core business. The total output value of Heyuan's electronic information industry has exceeded 50 billion yuan, and the overall scale accounts for more than one-third of the city's industrial output. Under this industrial orientation, the urban development of Heyuan City is positioned as an emerging industrial base in Guangdong Province and a logistics center in the northern Guangdong and southern Jiangxi regions.

The circular economy industrial park is an important platform for technological progress, and an attempt to alleviate regional development inequality. Guangdong Province has long faced the problem of uneven regional development. The eastern region of Guangdong is mountainous and hilly, with inconvenient transportation and single industrial structure, mainly consisting of traditional manufacturing industries. Circular economy industrial parks are important carriers for alleviating resource scarcity and promoting technological progress. At the same time, they can also integrate resources on a larger scale, bringing more opportunities and possibilities for the development of surrounding areas.

4.3 Current situation of environmental sanitation facilities

There are currently six operational domestic waste landfills in Heyuan City, with a disposal capacity of up to 2000 tons per day. The Qizhai Municipal Solid Waste Sanitary Landfill is the first landfill to be built strictly in accordance with national standards. The Qizhai landfill site is located in a hilly area with a lower central terrain and surrounded by mountains on three sides. During the operation period of the landfill, the garbage treatment scale is 300-900 tons/day, with an average treatment scale of 500 tons/day, covering an area of 335000 square meters [15]. The total volume of the landfill area is 4.1 million cubic meters, which can accommodate 6.4 million tons of garbage [15]. The Qizhai Municipal Solid Waste Sanitary Landfill consists of an entrance area, two landfill areas, a sewage regulation facility, a sewage treatment plant, flood control facilities, a landfill gas treatment station, and a comprehensive management area. The first phase of the Qizhai municipal solid waste landfill has been in operation for nearly 10 years, with over 1.3 million tons of landfill waste, and was closed in February 2021 [15]. At present, all urban domestic waste is transported to Dongyuan County landfill for burial, with no facilities for incineration of household waste, construction waste treatment, harmless treatment of kitchen waste, and harmless treatment of sludge. There are 10 units in the city holding hazardous waste operation licenses, including 4 hazardous waste disposal units and 6 hazardous waste collection units. The uncontaminated infusion bottles and bags used by medical institutions in the city are transported to the Medical Waste Treatment Center and other qualified enterprises for recycling. Sludge generated by urban sewage treatment plants is recycled and disposed of by sludge disposal enterprises. Large scale breeding farms are basically equipped with manure treatment facilities.

At present, the solid waste treatment system in Heyuan City has the following problems: 1. Lack of transfer facility. The main manifestation is that cleaning and shoveling are all done manually, manual trucks are still used as the main means of transporting garbage. 2. The harmless treatment rate and resource utilization rate of garbage are low. The existing landfill sites only carry out simple land-filling of domestic waste without taking effective environmental protection measures, which has caused certain pollution to the environment and affected the lives of surrounding residents. It is mainly manifested in exposed garbage, paper, and small particles fluttering in the wind, as well as the leachate from garbage has polluted the surrounding water bodies. 3. Poor safety. The main gases produced by the garbage are methane and carbon dioxide. Methane is prone to spontaneous combustion and explosion, posing serious safety hazards. 4. A small number of citizens have weak hygiene awareness, and some bagged garbage is collected haphazardly along the streets. There is no implementation of garbage classification and collection, and the facilities of private waste collection points do not meet hygiene standards.

The construction of a circular economy industrial park can meet the urgent demand for solid waste treatment in Heyuan City. Under the driving force of the Greater Bay Area, the rapid development of Heyuan City has led to increasing pressure on urban solid waste treatment. At present, simple landfill technology occupies a large amount of land and may cause irreversible pollution. Building circular economy industrial parks may improve solid waste treatment capacity while reducing construction and operational costs.

5 Initial planning

5.1 Functional partitioning

The entire circular economy industrial park can be divided into seven functional sectors: waste incineration and power generation sector, landfill sector, comprehensive utilization area of construction waste, water resource purification sector, sludge incineration sector, hazardous waste resource utilization sector, and environmental education sector.

The core area of the entire park is the garbage incineration and power generation sector (figure 1). Due to the low level of dry and wet garbage classification in China, the transported garbage has a high moisture content, making it difficult to ignite. Therefore, after weighing and measuring, the garbage needs to be unloaded into a closed storage room, stacked in the garbage storage room for fermentation, and the moisture in the garbage will eventually precipitate in the form of leachate. Afterwards, the dry garbage in the storage room is sent to the incinerator for combustion, and the fermentation gas in the garbage pool is pumped into the incinerator to assist combustion. During the process, it is necessary to maintain a slight negative pressure in the garbage pool to ensure that the odor does not leak out. Domestic waste undergoes three stages of re drying, combustion, and burnout in the incinerator, and the flue gas temperature during combustion can reach over 1000 °C. High temperature flue gas undergoes heat recovery through a waste heat boiler, generating superheated steam that drives the steam turbine generator set to generate electricity, and directly delivers the produced electrical energy to the power grid. The SNCR denitrification system inside the furnace can initially reduce the NO_x content at the boiler outlet [16]. After passing through the generator set, the flue gas needs to undergo SDA desulfurization process combined with activated carbon, bag filter and SCR reactor to meet the emission standards of Standard for pollution control on the municipal solid waste incineration(GB18485-2014) [17].

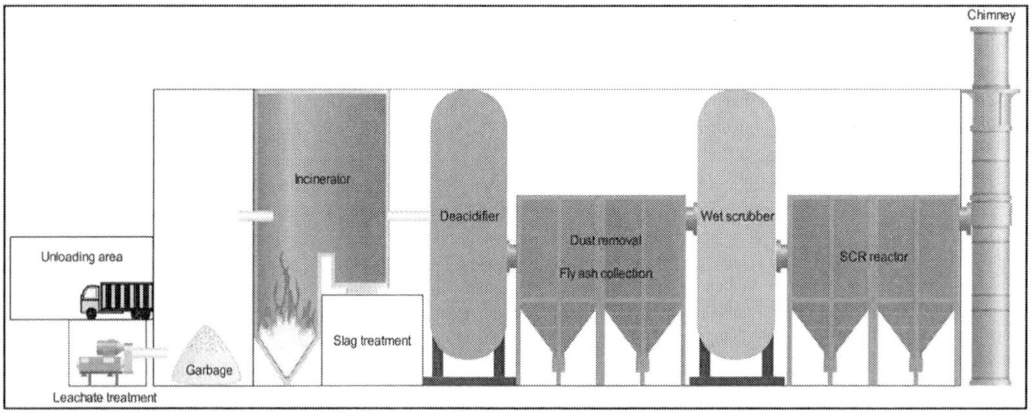

Fig. 1. Waste incineration and power generation sector.

The comprehensive utilization area of construction waste will jointly utilize the construction waste, slag and fly ash. Slag belongs to general solid waste, and after recycling the metals, ceramics, and glass, the remaining part is used to produce building materials. The fly ash generated during the flue gas treatment process will also be transported to this area after stabilization. The remaining slag and fly ash that have not been resourceful will be buried in landfills.

The comprehensive sewage treatment plant constructed in the water resources purification sector concentrates and treats various types of wastewater such as leachate, surface water, rainwater, and industrial production wastewater. Due to the high concentration of pollutants, garbage leachate and industrial wastewater generally require a multi-step complex recovery process to meet the Industrial Water Quality Standards for Urban Wastewater Reuse (GB119923-2005) standard [18]. The pretreated wastewater enters the anaerobic reactor, organic substances in wastewater can be converted into biogas through anaerobic reactions, which can be used as fuel or as a combustion aid for incinerators. Afterwards, the wastewater enters the MBR reactor, where organic pollutants are further degraded through the action of microorganisms. Finally, the wastewater is processed through reverse osmosis technology, utilizing the separation characteristics of reverse osmosis membranes to effectively remove dissolved salts, colloids, and other impurities. The concentrated water treated by reverse osmosis technology contains a large amount of salt substances, which can be used for the preparation of lime. The wastewater generated after acid washing of flue gas can recover metal salts such as sodium chloride and sodium sulfate. The entire water resource purification process has achieved harmless treatment and resource utilization of wastewater.

Due to the fact that most cities have not yet implemented separate treatment of industrial and domestic wastewater, incineration is the most thorough method for sludge disposal. The sludge incineration sector is similar to the garbage incineration and power generation sector, adopting a process of incineration, waste heat utilization, and flue gas treatment. There are also supporting processes such as fly ash collection, slag collection, wastewater pretreatment, and deodorization. The incinerated fly ash and slag can be used for phosphorus recovery or as raw materials for building materials.

The hazardous waste resource utilization sector will serve as a supplement for surrounding enterprises to utilize and dispose of hazardous wastes. The environmental education sector is mainly responsible for receiving visitors and providing leisure area for citizens. By laying a tour line that runs through the main areas of the park, visitors can have a more intuitive and in-depth understanding of the solid waste treatment process. The indoor area of the environmental education sector can be equipped with small cinema, activity room, reading area and cafe, providing an area for visitors to communicate, learn and relax.

5.2 Energy system planning

The common utilization methods of solar energy can be divided into passive utilization and active utilization. Passive utilization refers to placing the main space or thermal storage wall of a building on the sunny side to directly obtain solar heat, while active utilization utilizes solar energy collection devices installed on the roof and exterior walls of the building to convert solar energy into thermal and electrical energy. The annual sunshine hours in Heyuan City are about 1700 hours [19], with ample sunshine. The buildings in the park can adopt a combination of passive and active solar energy collection methods. Some sun rooms can be equipped with amorphous silicon solar panels with good transparency on the top. Production, office, and living buildings can obtain thermal and electrical energy by installing solar collectors and solar photovoltaic panels on the roof to meet the needs of hot water, lighting, air conditioning, and cooling. Solar powered street lamps can be installed on various levels of roads in the park.

5.3 Water resources system planning

The main sources of sewage in the park are leachate from garbage, industrial wastewater, and domestic sewage generated in the production and living areas. The reclaimed water generated from the comprehensive sewage treatment plant in the park can continue to be used as cooling water, irrigation, fire protection, and domestic water. Many facilities in the park generate a large amount of high-temperature steam after production operations, which can be used for sludge drying treatment before incineration. The steam undergoes heat exchange through the shell of the heat exchanger, evaporating the moisture in the sludge to dry it. The remaining steam can be supplied to the refrigeration station.

5.4 Systematic application of environmental protection technology

The low impact development(LID) method is an environmentally friendly rainstorm management technology, which can minimize the adverse impact of construction and development activities on the ecological environment. Traditional rainwater treatment usually only alleviates peak flow , while using LID helps maintain the natural runoff cluster layout [20]. Grass ditches, rain gardens,biological culverts, and permeable sidewalks, all of which may reduce the effective impermeable area of watersheds.

Under various climatic conditions, green roofs can retain a significant proportion of precipitation (an average of 63%) [21]. The rainwater collected on the roof can be guided to the water storage facility through the ecological curtain wall. A catchment area can be set up on the ground to collect rainwater, and the remaining rainwater can seep into the ground through permeable or perforated roads. Rainwater can be reused as reclaimed water after filtration. The ground paving of the park will present alternating changes in soft and hard surfaces according to different usage rates. Areas frequently traversed by vehicles or pedestrians will be paved with hard surfaces, while areas with low usage rates will be paved with vegetation rich soft surfaces. Biological interception on soft surfaces can effectively intercept a large amount of runoff and pollutants, and continuously reduce the concentration of certain pollutants, such as suspended solids, oily substances, and toxic substances [6].

6 Implementation suggestions

6.1 Not-in-my-backyard attitude

The Not-in-my-backyard attitude is a manifestation of social failure, which reflects the concerns and opposition of surrounding residents towards facilities that may have negative impacts. The appearance of circular economy industrial park buildings should adopt deindustrialization design to avoid the exposure of a large number of pipelines and equipment. The by-products such as slag and fly ash generated during the production process can be sold at low prices or gifted to surrounding township enterprises. The industrial park can provide employment for surrounding villages and towns, and hire representatives of surrounding residents as supervisors to regularly inspect whether the industrial park's emissions comply with regulations. The non core areas of the industrial park can be opened to the public and regular visits, exhibitions, and other activities can be organized. The above measures to address the neighbor avoidance effect not only reduce the potential negative impact of the project, but also ensure that the benefits brought by the project are shared by all members of society.

7 Implementation suggestions

The construction and operation of circular economy industrial parks may face many challenges. For Heyuan City, building a circular economy industrial park is undoubtedly a daunting task, therefore the overall planning of the park still requires a lot of related work. Further planning involves collecting and analyzing data from various local sanitation departments.The circular economy industrial park not only has a high initial construction cost, but also has a high operating cost. The economic return of the park directly affects its sustainability. The manufacturing industry in Heyuan City is mainly focused on the production and processing of electronic products. This single industrial structure leads to the concentration of related industries in a certain link in the production chain. Homogenized competition and low product added value are obstacles to creating greater profit. High risk operations such as landfill, leachate treatment, and hazardous waste treatment are concentrated in the industrial park, accompanied by the transportation and use of a large number of hazardous chemical products, which pose significant risks to employees and surrounding residents. Overall, the construction of circular economy industrial parks presents both opportunities and challenges. Through continuous exploration and practice, it can provide experience and lessons for the construction of similar projects.

References

1. Q. Han, J. Wang. Theoretical exploration on the priority order of the 3R principles of circular economy. Environmental Protection Science. **32**, 59-62 (2006)
2. Our city is carrying out the special research on revision work of the action plan to comprehensively promote the construction of Heyuan City into a green development demonstration zone in the province and integrate it into the ecological forefront of the Greater Bay Area. The People's Government of Heyuan City, (2019)
3. T·PARK, Home Page, https://www.tpark.hk/tc/
4. WEEE@HK, Home Page, https://weee.gov.hk/sc/background/weee-park-weee-treatment-and-recycling-facility/
5. O·PARK 1, Home Page, https://www.opark.gov.hk/tc/index.php/
6. L. Dong, N. Li. The Model and Experience of Developing Ecological Industrial Parks in Japan. Contemporary Economy of Japan. **6**, 11-16 (2009)
7. H. Li, X. Cui. Japan Kitakyushu Ecological Industrial Park embark on the Shinkansen train to build a circular society. Circular economy. **22**, 11-13 (2006)
8. Ma Xingrui presided over a special work meeting on solid waste management, comprehensively consolidating the responsibility for solid waste pollution prevention and control, and continuously improving the level of solid waste management capabilities. The People's Government of Guangdong Province (2018)
9. Notice from the General Office of the People's Government of Guangdong Province on Issuing the Pilot Work Plan for Promoting the Construction of "Waste Free Cities" in Guangdong Province. The People's Government of Guangdong Province (2021)
10. Industrial Development Plan of Heyuan City (2021-2030). The People's Government of Heyuan City, (2022)
11. Announcement on the Release of Solid Waste Pollution Prevention and Control Information in Heyuan City for the Year 2022，The People's Government of Heyuan City
12. YangHeyuan. County Annals, (Guangdong People's Publishing House, 2000)
13. Statistical Bulletin on National Economic and Social Development of Heyuan City. The People's Government of Heyuan City (2023)
14. Secretary of the Heyuan Municipal Party Committee: Strive to be a pioneer in ecological and green development to integrate into the Greater Bay Area. Nanfang Daily, (2022)
15. Ministry of Ecology and Environment of the People's Republic of China. Qizhai Landfill Site in Heyuan City.
16. B. Lian, W. Zhou, J. Long. Research and application of seven stage flue gas purification technology for extra large garbage power plants. Installation. **43**, 160 (2023)
17. Ministry of Ecology and Environment of the People's Republic of China, Standard for pollution control on the municipal solid waste incineration Standard number: GB 18485-2014 replaces GB 18485-2001

18. National Digital Library of Standards, The reuse of urban recycling water Water quality standard for industrial uses. https://www.nssi.org.cn/nssi/front/62516200.html
19. The People's Government of Heyuan City, Administrative Divisions and Natural Resources (2021 Statistical Yearbook)
20. M. Dietz. Low Impact Development Practices: A Review of Current Research and Recommendations for Future Directions. Water, Air, and Soil Pollution. **186**, 351-363 (2007)
21. B. Derek, C. Jackson. Urbanization of aquatic systems: degradation thresholds, stormwater detection, and the limits of mitigation 1. JAWRA Journal of the American Water Resources Association. **33**, n. pag (1997)

Innovative Integration and Optimization of RE and Smart Energy Management Systems in Skyscrapers: Enhancing Sustainability and Energy Self-Sufficiency in Vertical Cities

Bingzhuo Yan[1,*]

[1]Nanjing No.1 high School, Nanjing, Jiangsu, 210000, China

Abstract. Skyscrapers have become a vital part of modern urban development, offering efficient space utilization in increasingly dense cities. However, their huge input of energy as well as environmental impact necessitate integration of RE with intelligent energy management systems. This paper reviews key strategies for incorporating renewable energy (RE) technologies such as solar, wind, geothermal, and hybrid systems into skyscrapers. It also explores the role of Smart Energy Management Systems (SEMS) in optimizing energy use through real-time monitoring, predictive analytics, and automated control. Case studies like Masdar City and Shanghai Tower illustrate successful applications of these technologies. Despite challenges such as limited space, technological constraints, and high costs, emerging advancements in solid-state batteries, artificial intelligence, and Internet of Things (IoT)-based energy systems present significant opportunities. The review concludes that integrating RE and SEMS is critical for reducing the environmental impact of skyscrapers and creating more sustainable urban environments.

1 Introduction

Over the past few decades, skyscrapers have become integral to the development of urban landscapes. Unlike in the last century, there is a growing enthusiasm for constructing skyscrapers in urban areas, driven by the need for efficient space utilization in increasingly dense cities. According to data from the Council on Tall Buildings and Urban Habitat (CTBUH) [1], skyscrapers are classified into three categories based on height: Skyscrapers: Buildings over 100 meters; Supertall: Buildings over 300 meters; Mega tall: Buildings over 600 meters.

The construction of skyscrapers offers numerous advantages, particularly in cities with limited land availability, such as New York. Skyscrapers provide highly efficient use of space for commercial, residential, and entertainment purposes, thereby optimizing urban development. Additionally, these tall buildings often serve as iconic symbols of cities, contributing to economic growth by attracting tourism and investment. As urbanization

* Corresponding author: niyinsong@ldy.edu.rs

continues to expand, skyscrapers are playing an increasingly prominent role in shaping the economic, social, and cultural fabric of cities.

Despite their benefits, skyscrapers pose significant environmental challenges, particularly in terms of energy consumption and sustainability. Super-tall buildings, due to their scale, require vast amounts of building materials such as steel and concrete, which are associated with high carbon emissions during production. Furthermore, skyscrapers contribute to the exacerbation of urban heat islands, light pollution, and the generation of greenhouse gases. The operational energy requirements for heating, cooling, and maintaining these structures are significantly higher than those of typical buildings, presenting a challenge to the sustainability of urban environments. Addressing these challenges is crucial for ensuring that the development of skyscrapers aligns with the growing emphasis on environmental responsibility and sustainable urban planning.

Due to the significant energy demands of skyscrapers, which include air conditioning, heating, elevator operation, lighting, and various facilities such as office equipment, hot water systems, and kitchen utilities, integrating RE sources has become a critical strategy to mitigate overall urban energy consumption. According to the International Journal of High-Rise Buildings [2], skyscrapers are major consumers of energy, and without intervention, their impact on the energy grid could be unsustainable. Incorporating RE sources such as solar, wind, and geothermal power into skyscrapers can significantly reduce their reliance on non-RE, contributing to the overall sustainability of cities. In addition to RE integration, Smart Energy Management Systems (SEMS) are essential for improving energy efficiency in skyscrapers. These systems automate the regulation of energy use across various building operations, optimizing energy consumption patterns, minimizing waste, and reducing overall energy costs [3,4]. SEMS enhance operational efficiency by adjusting heating, cooling, lighting, and equipment usage in real-time based on demand, ensuring that energy is used where and when it is needed most, thereby minimizing unnecessary consumption.

This paper aims to explore the integration and optimization of RE sources with smart energy management systems in skyscrapers. The review will examine various forms of RE—solar, wind, geothermal, and hybrid systems—and how they can be effectively incorporated into super-tall buildings. Additionally, the role of SEMS in enhancing energy efficiency will be discussed, focusing on optimization techniques that can reduce energy waste and improve the sustainability of skyscrapers. Through this review, the paper will offer insights into the most effective strategies for integrating RE and smart systems to create more energy-efficient vertical cities.

2 RE integration in skyscrapers

2.1 Solar energy

Building-Integrated Photovoltaics (BIPV) systems offer a seamless solution for generating solar energy within the architectural fabric of skyscrapers. These systems convert solar energy directly into electricity, contributing to reduced greenhouse gas emissions and lowering overall energy consumption. BIPV can be integrated into various building components, including roofs, exterior walls, and windows, eliminating the need for additional space dedicated to photovoltaic panels. Furthermore, BIPV systems can be designed to align with the aesthetic and architectural style of the building, ensuring that their functionality does not compromise the visual integrity of the structure.

Despite the advantages of BIPV systems, several challenges arise when integrating large-scale solar panels into high-rise buildings. These include structural concerns, energy storage limitations, and maintenance complexities. Reinforcing the structural integrity of skyscrapers to accommodate solar panels is essential to ensure safety and stability. Additionally, improving the efficiency of solar panels, particularly under varying weather conditions, is crucial for maximizing energy output. Advanced energy storage solutions are needed to store excess energy generated during peak sunlight hours. Maintenance of BIPV systems, particularly at extreme heights, poses further challenges. Innovative approaches, such as using drones or robotic systems for inspection, repair, and replacement of faulty components, offer promising solutions for maintaining these systems efficiently and safely [5].

By meeting these issues, skyscrapers can significantly benefit from the integration of solar energy, contributing to urban sustainability and energy independence.

2.2 Wind energy

Wind energy, traditionally harvested in remote locations, presents substantial engineering and economic challenges when transporting electricity to urban centers [6]. However, skyscrapers offer a unique opportunity to harness wind energy directly in urban environments due to the natural increase in wind speed with height. The elevated structure of skyscrapers allows them to capture stronger wind currents compared to lower buildings. Nevertheless, the integration of wind energy in urban environments is complex, as wind patterns around tall buildings are often turbulent and highly variable. The height, shape, and layout of surrounding structures can significantly influence wind behavior, causing disturbances and vortices that complicate wind energy capture.

Many wind energy technologies, like Vertical Axis Wind Turbines (VAWT), Horizontal Axis Wind Turbines (HAWT), and micro wind turbines, have been designed to operate in urban settings. Among these, VAWTs are particularly suited for skyscrapers due to their ability to capture wind from all directions without needing reorientation, making them more adaptable to the unpredictable airflow in urban environments. However, the location and positioning of these turbines must be carefully optimized. Computational Fluid Dynamics (CFD) simulations are essential for understanding airflow patterns and determining the ideal placement of VAWTs on a skyscraper to maximize energy capture.

In addition to airflow considerations, the impact of noise and vibrations generated by wind turbines must also be addressed, as these factors could affect the comfort of building occupants. Proper vibration dampening mechanisms and noise reduction technologies are critical for ensuring that wind turbines do not negatively impact the living and working environments within skyscrapers.

2.3 Geothermal energy

Geothermal energy systems in urban environments harness the Earth's internal heat to provide heating and cooling solutions for skyscrapers, offering substantial environmental and energy-saving benefits. These systems typically operate using a closed-loop mechanism, where heat is transferred to and from the ground, making them suitable for cities with limited available space. This system minimizes environmental impact as it utilizes naturally occurring heat from the Earth's crust without requiring extensive external infrastructure. Notably, Reykjavik, Iceland, is an exemplary case of successful large-scale geothermal energy use in urban settings.

Ground-source heat pumps (GSHPs) are a particularly effective application of geothermal energy in skyscrapers. A GSHP system consists of underground piping (either horizontally or vertically installed), a heat pump unit, and heat exchangers integrated into the building's internal cooling and heating systems. In winter, GSHPs extract heat through the underground and transfer it to the skyscrapers for heating. Conversely, during the summer months, the system operates in reverse, transferring excess heat from the building into the ground for cooling. GSHPs are highly energy-efficient compared to traditional HVAC systems because they rely on the relatively stable and renewable heat stored underground, resulting in reduced energy consumption and operational costs [7]. This makes GSHPs an ideal solution for sustainable skyscraper development, contributing to lower emissions and operational efficiency in urban environments.

2.4 Hybrid systems

Hybrid energy systems in skyscrapers integrate multiple energy sources—conventional energy, RE, energy storage technologies—to optimize energy efficiency as well as reduce carbon emissions. By combining different sources like wind, geothermal, and solar energy, these systems create a more resilient and reliable energy infrastructure.

One notable example of a hybrid RE system is Masdar City in Abu Dhabi. The city's core buildings feature an integration of solar photovoltaic panels, a centralized solar water heating system, and wind energy turbines, demonstrating a balanced mix of RE technologies [8]. Similarly, Shanghai Tower in China is another example, where approximately 1,000 solar photovoltaic panels are installed on the building's façade, complemented by four vertical wind turbines. Additionally, the tower utilizes a geothermal energy system for air conditioning and heating, reflecting an efficient combination of solar, wind, and geothermal energy [9]. These case studies showcase how skyscrapers can incorporate diverse RE technologies to enhance sustainability while reducing energy consumption and emissions.

3 Smart energy management systems (EMS)

3.1 Overview of smart energy management systems (SEMS)

A SEMS means a comprehensive platform that integrates sensors, data analytics, and automated controls to supervisory control and optimize energy use in real-time. The primary goal of EMS is to enhance energy efficiency by analyzing energy data and automatically adjusting various systems within buildings, such as lighting, HVAC, and other operational functions. By doing so, EMS not only reduces energy input and the building's carbon footprint but also improves occupant comfort and overall management efficiency, making it an essential component in modern, sustainable buildings.

A smart SEMS typically consists of several key components designed to monitor, analyze, and optimize energy usage. These include sensors and meters, data collectors, data processing units, communication modules, user interfaces, and control systems. Sensors and meters measure key parameters such as current, voltage, and environmental conditions. Data collectors convert sensor information into digital signals, while data processing units analyze and store this information for further use. Communication modules transmit data across the network, and user interfaces offer visual dashboards and reports for real-time monitoring. Finally, control systems execute optimization strategies based on data analysis, ensuring continuous improvements in energy efficiency. Through this comprehensive setup, EMS enables real-time management of energy resources, significantly reducing unnecessary energy consumption while enhancing operational effectiveness.

3.2 Integration of RE with smart grids

Smart grids play a crucial role in managing and optimizing RE within modern energy systems, especially in the context of skyscrapers and urban environments. By integrating advanced technologies, smart grids enhance the energy network's capacity for better monitoring and control, enabling efficient use of electricity. Through real-time data analysis, smart grids are able to effectively control the supply and demand of RE sources like solar panels as well as wind turbines, ensuring optimized resource allocation and minimizing energy waste [10]. Moreover, smart grids support decentralized RE systems, including small-scale solar and wind installations, by integrating them seamlessly into the broader energy network.

In skyscrapers, smart grids enable real-time monitoring and control of energy flows, providing visibility into the consumption of critical systems such as air conditioning, heating, and lighting. This real-time data allows for automatic adjustments to energy supply based on current demands, significantly reducing waste and improving overall energy efficiency. Through continuous monitoring, skyscrapers can better manage their energy consumption, aligning energy use with operational needs and minimizing unnecessary energy expenditure.

Smart grids employ various techniques to balance energy supply and demand efficiently. Demands response programs incentivize users to shift their energy usage during off-peak and peak times, optimizing the load distribution across the grid. Additionally, smart grids utilize advanced load forecasting and energy management systems to forecasting future energy requirements as well as adjust distribution strategies accordingly [11].

3.3 Energy storage solutions in skyscrapers

Energy storage systems play a vital role in maintaining a stable energy supply for skyscrapers and large urban areas. By capturing surplus energy from sources like solar or wind during low-demand periods, these systems can release it when demand rises, ensuring a consistent power supply. This process not only balances energy supply and demand but also reduces dependence on the grid during peak times, promoting overall energy efficiency and enhancing grid stability.

Integrating storage technologies such as batteries, pumped hydro storage, and flywheel systems with RE sources is key to the reliability of skyscraper power systems. Batteries, like lithium-ion and hydrogen fuel cells, are frequently used to store excess energy generated by solar panels or wind turbines, providing power during times of high demand or inconsistent RE generation. Additionally, other storage methods like pumped hydro and flywheel systems help manage energy by converting and storing it in different forms, ensuring it can be efficiently used when needed.

4 Optimization techniques for energy efficiency

4.1 Predictive analytics and machine learning

The use of predictive analytics and machine learning plays a pivotal role in forecasting energy demand and optimizing energy usage in skyscrapers. AI-driven models can predict future energy consumption trends with high accuracy. These predictions enable buildings to adjust energy usage dynamically through smart control systems and demand response management, leading to increased energy efficiency. Machine learning algorithms also help in optimizing HVAC systems, lighting, and other energy-intensive operations by learning from past usage patterns and suggesting real-time adjustments.

The iconic building implemented an Energy Management System (EMS) that leverages sensors, predictive algorithms, and machine learning to analyze energy usage patterns. The system optimizes HVAC operations, lighting, and temperature management in real time, leading to significant improvements in energy efficiency. As the tallest building in the world, Burj Khalifa employs predictive models to manage its extensive HVAC systems, lighting, other energy-demanding operations. A centralized system monitors and adjusts energy consumption in real-times, using predictive analytics to optimize energy usage and reduce costs.

4.2 Demand response strategies

Demand response strategies are essential for managing energy consumption based on grid demand and the availability of RE. Techniques such as Time-of-Use Pricing (TOU), Real-Time Pricing (RTP), and Critical Peak Pricing (CPP) help balance supply and demand by encouraging energy consumption during off-peak periods and reducing it during peak times. Automated demand response systems in smart buildings enable seamless adjustments to energy use, responding to signals from the grid or based on RE production levels. These strategies ensure efficient energy utilization and reduce stress on the grid, particularly during high-demand periods.

4.3 Dynamic load balancing and energy distribution

Dynamic load balancing is a crucial technique for optimizing energy distribution among various building systems. By intelligently allocating energy loads based on real-time electricity demand and system conditions, dynamic load balancing enhances both energy efficiency and system reliability. In skyscrapers, this involves the distribution of energy among HVAC, lighting, energy storage systems, and RE sources, ensuring that power is efficiently allocated where it is most needed. Integrating RE sources and smart grids with dynamic load balancing allows for the seamless adjustment of energy flows, further optimizing energy distribution and reducing energy waste.

5 Challenges and opportunities

The integration of RE and smart management systems in skyscrapers faces several technological challenges. One of the primary limitations is the current efficiency of RE technologies. For example, solar panels are often constrained by limited surface area on skyscrapers, while wind energy systems must overcome challenges related to inconsistent wind patterns in urban environments. Additionally, energy storage technologies, such as batteries, still face issues related to capacity, longevity, and cost-effectiveness, limiting their widespread use in high-rise buildings. Smart EMS also need further development to handle the complexity and scale of skyscraper energy demands, requiring advancements in data processing, real-time monitoring, and automation technologies.

Urban environments present unique barriers to the large-scale integration of RE systems. Dense city layouts, limited available space, and structural constraints of tall buildings make it difficult to install large-scale RE technologies like wind turbines or extensive photovoltaic systems. Additionally, urban wind flows can be highly turbulent due to surrounding buildings, making wind energy generation less efficient. Moreover, retrofitting existing skyscrapers with renewable systems and smart grids can be logistically challenging and costly, further complicating efforts to modernize energy infrastructure in cities.

The economic feasibility of implementing advanced energy systems in skyscrapers remains a significant consideration. While RE technologies and smart energy management systems offer long-term savings in terms of reduced energy costs and improved efficiency, the upfront capital investment required for installation, maintenance, and integration into existing buildings can be prohibitive. Cost-benefit analyses often show that retrofitting skyscrapers with renewable systems may not provide immediate financial returns, posing a challenge for stakeholders.

Emerging technologies in RE and smart systems offer exciting opportunities for enhancing skyscraper energy management. Innovations such as solid-state batteries, which promise higher energy density and longer lifespans, could revolutionize energy storage in urban environments. Additionally, advancements in artificial intelligence and machine learning will further enhance the predictive capabilities of EMS, allowing for more accurate energy forecasting and dynamic load balancing. The integration of IoT devices into energy systems will enable even more granular control and monitoring of energy usage within buildings, improving efficiency and reducing waste.

6 Conclusion

The integration of RE sources and SEMS in skyscrapers is essential for sustainable urban development. Skyscrapers face significant environmental challenges due to their high energy consumption and carbon emissions. This review has explored solutions like BIPV, VAWT, GSHP, and hybrid systems, which can significantly reduce energy reliance and improve efficiency in high-rise buildings.

SEMS play a vital role in optimizing energy use by enabling real-time monitoring, predictive analytics, and automated control. These systems help balance energy supply and demand while minimizing waste, improving energy efficiency in skyscrapers.

Despite the potential, challenges such as limited space for RE integration, technological limitations, and high costs remain barriers. However, emerging innovations in solid-state batteries, AI-driven energy management, and IoT-based systems present promising opportunities for improving the energy efficiency of skyscrapers. As these technologies advance, skyscrapers can contribute to more sustainable and energy-efficient cities in the future.

Reference

1. A. Nicolas, J.-L. Barrat, J. Rottler, Effects of inertia on the steady-shear rheology of disordered solids. Phys. Rev. Lett. **116**, 058303 (2016)
2. Ctbuh, C. Council on tall buildings and urban habitat. Chicago: CTBUH. (1969)
3. I. Shakir, M. A. Jasim, & S. S. Weli. High Rise Buildings: Design, Analysis, and Safety: An Overview. International Journal of Architectural Engineering Technology, **8**, 1-13 (2021).
4. S. L. Arun, & M. P. Selvan. Smart residential energy management system for demand response in buildings with energy storage devices. Frontiers in Energy, **13**, 715-730 (2019).
5. S. Mischos, E. Dalagdi, & D. Vrakas. Intelligent energy management systems: a review. Artificial Intelligence Review, **56**, 11635-11674, (2023).
6. A. Riaz, R. Liang, C. Zhou, & J. Zhang. A review on the application of photovoltaic thermal systems for building façades. Building Services Engineering Research and Technology, **41**, 86-107 (2020).
7. P. Škvorc, & H. Kozmar. Wind energy harnessing on tall buildings in urban environments. Renewable and Sustainable Energy Reviews, **152**, 111662 (2021).
8. Omer, A. M. Ground-source heat pumps systems and applications. Renewable and sustainable energy reviews, **12**, 344-371 (2008).
9. Lau, A. Masdar City: A model of urban environmental sustainability. Social Sciences, **9**, 77-82 (2012).

10. Gong, J. Shanghai tower. Frontiers of Engineering Management, **4**, 106-109 (2017).
11. J. J. Moreno Escobar, O. Morales Matamoros, R. Tejeida Padilla, I. Lina Reyes, & H. Quintana Espinosa. A comprehensive review on smart grids: Challenges and opportunities. Sensors, **21**, 6978 (2021).

A Review of Exterior Waterproofing Technologies in Prefabricated Buildings

Baiheng Liao[1], *Meng* Zhou[2*], *Yutong* Wang[3], and *Zhiduo* Wu[4]

[1]Civil Engineering, Hubei University of Technology Engineering and Technology College, Wuhan, China, 430068
[2]Civil Engineering, Tianjin Chengjian University, Tianjin, China, 300192
[3]Port, Waterway and Coastal Engineering, Harbin Engineering University, Heilongjiang Province, China, 150006
[4]Civil Engineering, Southwest Jiaotong University, Sichuan Province, China, 611756

Abstract. This paper analyses the status quo of exterior wall waterproofing technology of prefabricated building, introduces the application of waterproofing coil laying, sealant waterproofing, grouting method, exterior wall waterproofing coating, waterproof concrete and the integration technology of heat preservation and drainage, and analyses their advantages and disadvantages. In the future, the research and application of new materials should be strengthened, and the effects of different waterproof technologies should be quantitatively compared through experiments to further improve the waterproof performance of prefabricated building external walls.

1 Introduction

With the development of society, prefabricated buildings, as a representative of modern industrial production, have made great progress in the field of construction, and have gradually become one of the important ways of modern building construction. The waterproofing technology of prefabricated building exterior wall is a special technology to ensure that the construction structure is not corroded, and it is indispensable.

As an important part of the waterproofing of prefabricated buildings, exterior wall waterproofing technology is an important means to ensure the performance of the building throughout its life cycle. Among them, there are physical shielding waterproofing technologies such as waterproofing membranes, chemical reaction waterproofing such as waterproof coatings, and composite waterproofing such as thermal insulation and drainage integrated systems [1]. Since prefabricated buildings often have joint leaks and hidden dangers, effective exterior wall waterproofing can prevent a series of problems caused by moisture penetrating into the interior of the building, thus Significantly extend the life of buildings [2]. Effectively improve economic, social, environmental protection, safety and other benefits [3].

* Corresponding author: 85849@student.pb.edu.pl

Since the development in the middle of the last century, the prefabricated building waterproofing technology has experienced a transformation from manual to modern assembly, and has gradually developed into waterproofing membrane technology, joint waterproofing technology, etc. In recent years, with the development of science and technology, new building waterproofing materials and supporting construction methods have shown a blowout, which has promoted the development of building waterproofing.

This paper analyzes the current research status of prefabricated building exterior wall waterproofing technology, focusing on the laying of waterproof membranes, sealant waterproofing, grouting, exterior wall waterproofing coatings, exterior wall waterproofing concrete, and integrated insulation and drainage technologies, and explores future research directions to provide ideas for the future development of prefabricated building exterior wall waterproofing technology.

2 Overview of prefabricated building facades

2.1 Prefabricated building craft

The wallboard edge at the horizontal joint of the upper and lower edges of the component is designed as the shape of outer low and inner high. In the joint, the outdoor side of the lower end of the weather-resistant sealant for sealing, indoor side of the higher end of the waterproof adhesive for sealing. Thus, a cavity is formed in the middle and a gradient for water conduction is arranged in the length direction as shown in Figure 1.

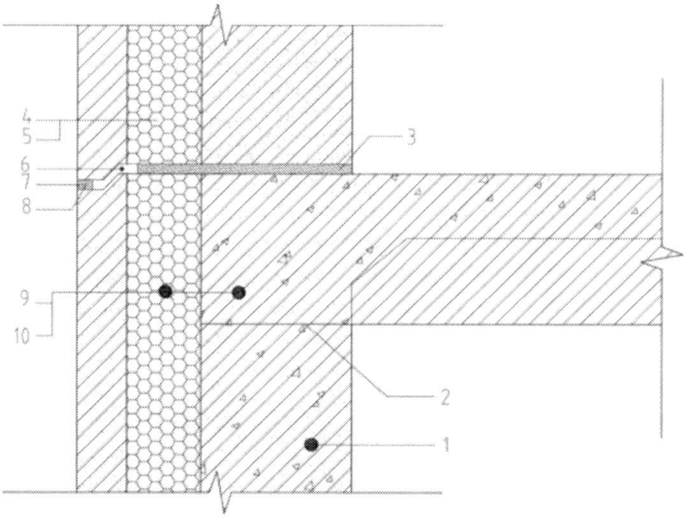

Fig. 1 Composition of waterproof concrete (Picture credit : Original)

(1- Inner panel wallboard 2- Rough surface 3- Fine aggregate concrete 4- Waterproof tape 5- Foam insulation strip or polyethylene insulation strip of the same material 6- Horizontal cavity 7- Sealant 8- Backing material 9- Cast-in-place concrete 10- Sandwich insulation layer)

2.2 Outline of waterproofing technology for external walls,

For waterproofing roll laying technology, the base is treated first, then the waterproofing roll is laid. Next, the joints are sealed with bitumen paste, followed by being fixed with galvanized sheet metal and steel nails. In this way, the sealing work of the waterproof layer on the facade is completed. Finally, the steel wire mesh is hung, and the mortar is put on as the transitional zone. When laying waterproofing membrane, it should pay attention to the standard operation, including fixing and sealing, to avoid operating in an inappropriate situation.

For sealant waterproof technology, it needs to combine the building characteristics of flexible choice. Due to the performance of hot expansion and cold shrinkage, it needs to match with concrete materials in the right environment together. In the choice of sealant material should be considered when the fit with concrete.

As for the grouting method, it needs to check whether the members are clean, and then in accordance with the proportion of the deployment of slurry to ensure that the fluidity and uniformity. Next, when using professional grouting equipment, it is necessary to ensure that there is no blockage occurring in the grouting pipes. Care should be taken to ensure that the strength of the slurry is sufficient after the completion of irrigation.

3 Significance of prefabricated building exterior wall waterproofing technology application

From the aspect of material selection, new waterproofing materials are based on nanotechnology, which can improve the waterproof performance and durability of the material. Furthermore, the biological waterproof material based on renewable resources can reduce the environmental pollution and the impact on human health [4].

For safety, due to the characteristics of the prefabricated building assembly, there will be some inevitable gaps between the various building parts during the assembly process. Therefore, there are certain defects in waterproof [5]. The outer wall is the first barrier to prevent water from infiltrating into the building, and the service life of the building will be greatly shortened under the condition of poor waterproof performance of the outer wall [6] . Consequently, the waterproof consideration of exterior wall is particularly important.

4 Exterior wall waterproofing technology of prefabricated building

Prefabricated exterior wall waterproofing is an important part of prefabricated building waterproofing system. According to different waterproof requirements and exterior wall design, different waterproof technical measures are adopted.

4.1 Waterproofing membrane laying technology

Waterproofing membrane laying is a physical waterproofing technology in building construction, which can use its physical properties to prevent water penetration building. It is mainly used for exterior walls, roofs and basements and other parts of the building. Correct waterproofing membrane laying technology for building waterproofing is essential to reduce the damage caused by moisture infiltration on the building. The construction sequence is shown in Figure 2. First, the base layer is treated, and then the waterproof coil is laid. The laying effect is shown in Figure 3. Next, the galvanized iron strip and steel nail are used to fix it. Finally, the waterproof layer of the facade is sealed with asphalt paste.

When choosing the joining technology for waterproofing membranes, adhesive fixing technology is more reliable in the repair of APP-modified membranes than thermally welded joints, due to its higher deformation capacity and superior peeling force. Besides, the adhesive joints can be separated by sliding in shear, which enables them to cope with thermal ageing more flexibly. Although heat welded joints are strong, aging may lead to increased brittleness of the membrane and affect its overall performance. Therefore, for APP membranes, the adhesive fixation technique is a more appropriate choice [7]. When sealing the joints, asphalt putty is utilized to fill the joints, effectively preventing moisture penetration from damaging the building. The termination of the waterproofing roll-roofing closure using galvanized iron skin closure compression strip, and in the transition surface hanging steel wire mesh wiping mortar to enhance the overall effect [8]. At present, many buildings have made use of waterproofing membrane laying technology to improve the waterproof performance of the building, such as Huawei Tianjin regional headquarters in the construction of the exterior wall with a hot joint welded head of waterproofing membrane, and waterproofing coatings to strengthen its waterproofing [9].

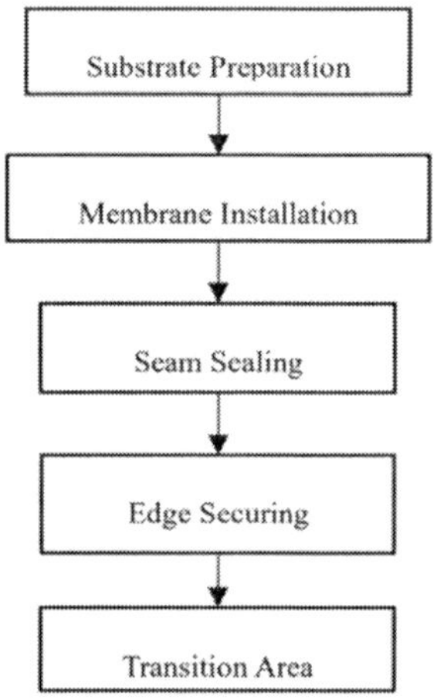

Fig. 2 Sequence of technical processes for laying waterproofing membranes (Picture credit : Original)

Fig. 3 Installation of waterproof materials [10]

4.2 Sealant waterproofing technology

The sealant waterproofing technology uses sealant to fill the gaps between the joints of prefabricated exterior wall panels, forming an uninterrupted sealing layer to isolate moisture and achieve waterproofing [4]. The construction principle of sealant is shown in Figure 4.

Firstly, on the basis of cleaning the gaps between the prefabricated exterior wall panels, add the backing material and paste the decorative paper onto the outside of the wall, preventing the avoidable pollution during construction. Then sealant is added and smoothed on the outside of the wall. Finally, remove the decorative paper and cure the sealant to achieve the predetermined waterproof effect [11].

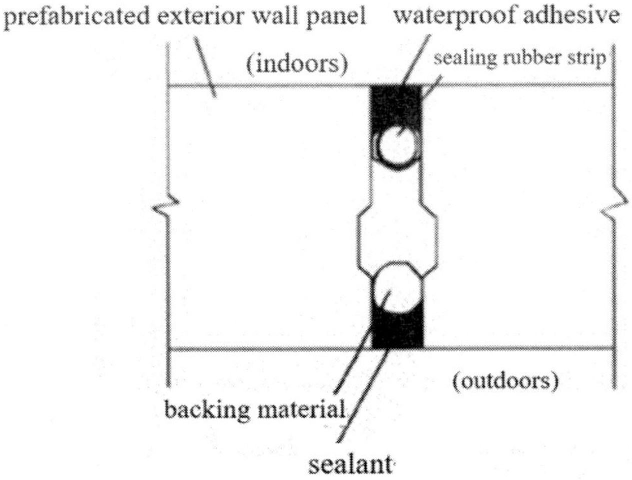

Fig. 4 Principle of sealant construction technology [11].

At present, Silane Modified Polyethylene Adhesive (MS sealant) is widely used in industry, which has reliable adhesion and anti-aging properties. When Branched Polyethylene Epoxy Resin (FBEPO) and Aminosilane Coupling Agent are added into the MS sealant at the same time, the mechanical properties are optimized [12]. In 1980s, Hilton hotel in Japan had already used MS sealant during construction. After 30 years of use, it still exhibits good durability and stability, without any quality problems such as water leakage.

4.3 Grouting method

For grouting method, it fills the gaps between the prefabricated exterior wall panels with grouting materials including cement mortar through grouting holes set in the walls, thereby improving the waterproof performance and impermeability of the exterior wall joints [4]. In engineering, grouting methods are often used at the horizontal gap between the prefabricated exterior wall and the floor slab, which is shown in Figure 5.

During construction, it should first determine the designated location, hoist the wall panel to the location and fix it with slant support. Then use the grouting hole for grouting after the preparation is completed until the grout outlet starts to discharge grout. Finally, when inspecting the waterproof quality of the exterior wall, it is necessary to conduct a water pouring or storage test to ensure that no leakage occurs [13].

Fig. 5 Grouting method construction [13].

When using grouting methods for construction, the following two points should be given special consideration. Firstly, the quality of grouting and compatibility of grouting materials should be ensured. Unqualified grouting quality and unsuitable grouting materials can cause cracks in the gaps of the exterior wall, thereby reducing the waterproof effect. Secondly, the cost of using grouting method to improve the waterproof performance of prefabricated exterior walls in practical engineering is relatively high, as it requires specific waterproof materials to prepare grouting aggregates and special tools and equipment for grouting operations [4].

4.4 Waterproof coatings for external walls

Exterior waterproof coating is a composite waterproofing technology for building exterior walls. It uses different physical material properties and chemical mechanisms to effectively prevent moisture penetration, not only to enhance its adhesion and durability, but also to extend the service life of the building.

In the construction, the first grass-roots level treatment, to ensure that the wall is clean and flat and free of debris. Then apply waterproof coating evenly.

Wang Guojian through the preparation and performance testing of silicone modified elastic emulsion exterior waterproof coating and obtained that this coating has good waterproof performance, stain resistance and aging resistance, suitable for exterior wall decoration and waterproofing applications [14]. Wu Runping et al. further illustrated through the preparation and performance study of fly ash inorganic exterior waterproof coating that the inorganic coatings using fly ash as filler can reduce the pollution of fly ash to the environment, and also meet the requirements of people for coatings and the environment, and at the same time greatly reduce the cost of coatings [15]. In addition, the use of hydrophobic protective materials can significantly improve the waterproofing performance of external wall materials, especially in the application of stone and rendering mortar, which can effectively improve the moisture resistance and contact angle, and play a role in enhancing the waterproofing effect [16].

4.5 External wall waterproof concrete

Fabricated concrete structural components waterproof including structure self-waterproofing and additional waterproofing layer waterproof. The structural self-waterproofing material is mainly waterproof concrete, but the concrete is a porous medium. Besides, the water will gradually penetrate through the micro-pores to cause a series of physical and chemical reactions in the concrete, especially in the presence of corrosive ions, it will seriously affect the durability of concrete. Therefore, waterproof concrete must meet certain impervious grade requirements [17]. In the concrete application process, the precast concrete component shows the characteristics of good production quality and high stability. Compared with the traditional cast-in-place concrete form, the precast concrete component waterproof and stability is better.

The technology of structure self-waterproofing, that is, using the impermeability of reinforced concrete to achieve the goal of waterproofing. Through controlling the concrete proportion and construction quality, the building can achieve the waterproofing effect. The self-waterproofing of the structure needs three necessary conditions, which are a complete closed structure at least, flexible sealing waterproofing at the position of displacement deformation, non-leakage of structural concrete.

The advantages of structural self-waterproofing are as follows. Firstly, the working life of structural self-waterproofing is longer. Secondly, the construction time of external waterproofing is greatly shortened. Thirdly, the maintenance after structural leakage is more convenient. Fourthly, the cost of structure self-waterproofing is lower in later maintenance [18].

To make concrete to achieve self-waterproof effect, usually need to do the following. First, it needs to strengthen the heat preservation and moisture conservation. Next, the structure design of post-cast concrete strip should be improved. Then the quality of coarse aggregate and cement is controlled. Besides, the thickness of cement should be controlled to prevent the cracking caused by hydration heat [19][20].

4.6 External wall insulation and waterproofing integrated system`

The integrated exterior wall insulation and waterproofing system is composed of waterproof mortar and thermal insulation materials. The system can achieve the functions of waterproofing and thermal insulation at the same time. It is usually composed of a multi-layer structure including an external insulation board, anti-cracking mortar, and waterproof coating, through which the waterproofing function is achieved. It also has the advantages of saving energy, improving the durability of the external wall, and being environmentally friendly [21]. By optimizing prefabricated parts, setting waterproof measures for cast-in-place concrete, sealing gaps, and setting diversion pipes at the joints to achieve water discharge in the exterior wall. These measures are combined to achieve the purpose of waterproofing [20]. The construction technical steps of the extruded board exterior wall insulation layer are shown in the Figure 6. The above construction steps should be avoided in rainy weather [22].

https://doi.org/10.1051/e3sconf/202560604004

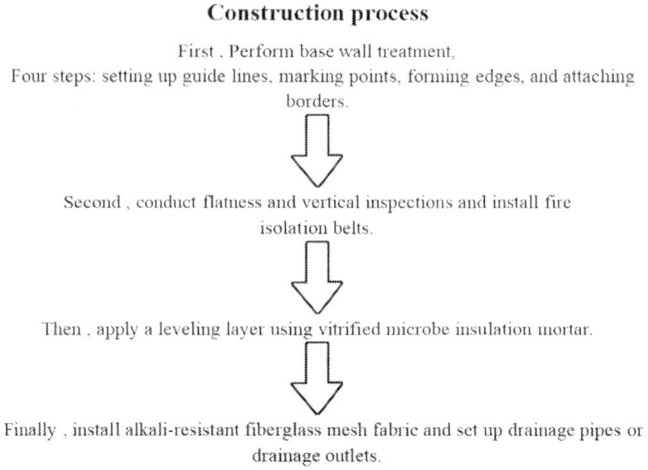

Fig. 6 Construction technical steps of extruded board exterior wall insulation layer (Picture credit : Original)

In order to make full use of exterior wall insulation materials, expand exterior wall functions and meet demand, it is necessary to adjust the materials according to the differences in their functions. Figure 7 is a schematic diagram of the commonly used rock wool exterior insulation system.

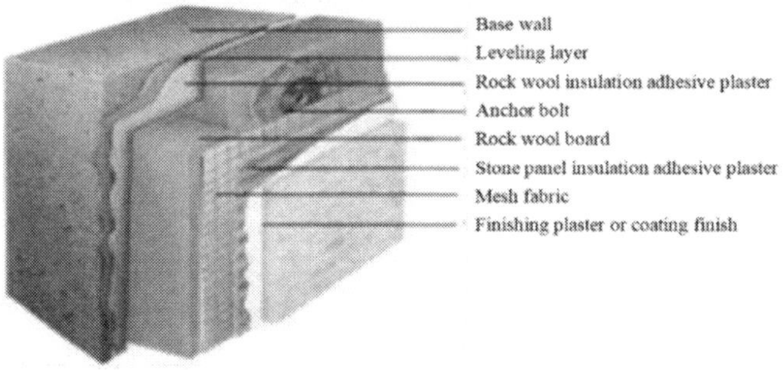

Fig. 7 Commonly used rock wool external insulation system components [22]

It is worth noting that the essence of the integrated external wall insulation and waterproofing system in terms of waterproofing is a combination of blocking and drainage, which is a combination of structural self-waterproofing technology and drainage technology. Drainage technology is the key to the exterior wall waterproofing system of prefabricated buildings and it can enhance the waterproofing ability of buildings. As shown in Figure 8, the gutter external wall drainage system effectively channels rainwater and prevents water accumulation at the facade. Furthermore, Figure 9 illustrates the external wall diversion pipe drainage system, which ensures that any water infiltrating the waterproof layer is promptly discharged through an integrated structure [20]. In prefabricated buildings, the application of waterproof materials and reasonable drainage design complement each other. The key lies in layered waterproofing and drainage design, water-conducting structure, and integrated drainage system. These play a role in each link of waterproofing, ensuring that even if water penetrates into the waterproof layer, it can be quickly discharged through the drainage system to avoid erosion [23].

https://doi.org/10.1051/e3sconf/202560604004

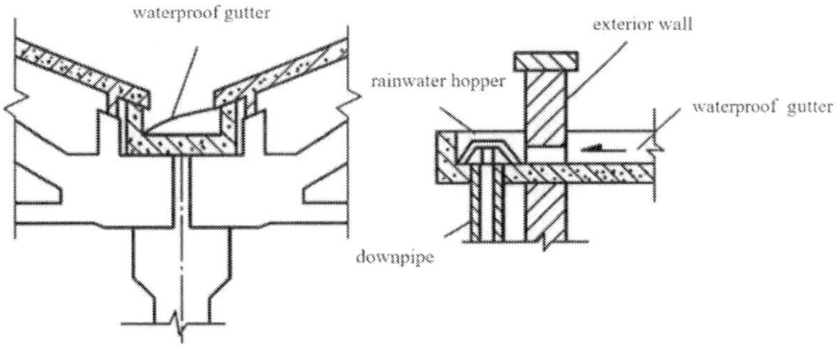

Fig. 8 Schematic diagram of gutter external wall drainage system (Picture credit : Original)

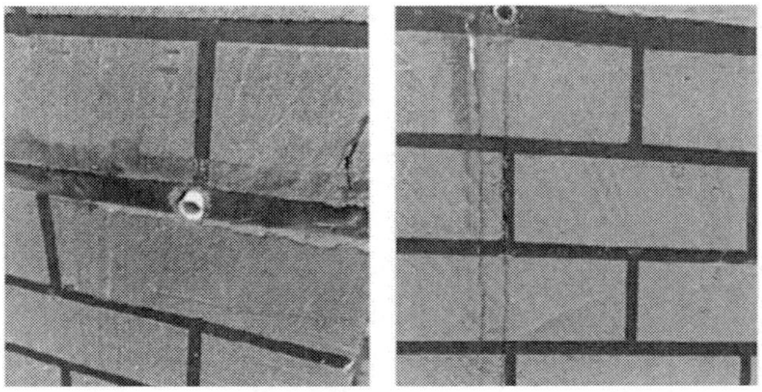

Fig. 9 Schematic diagram of drainage of external wall diversion pipe [20]

5 Conclusion

Prefabricated building exterior wall waterproofing technology is an important part of building performance, directly affecting the durability and service life of the building. Using waterproofing membrane laying, sealant technology and grouting methods, as well as developments in construction technology and material selection, the current prefabricated building exterior wall waterproofing technology has become mature, effectively solving the pain points of prefabricated buildings and improving the overall quality and waterproofing capabilities of prefabricated buildings.

The various waterproofing technologies introduced in this article have their own characteristics. Waterproofing roll laying is suitable for large-scale waterproofing, but it has the disadvantage of complex construction. Sealant waterproofing and grouting methods are suitable for small-scale repairs. Among them, the grouting method has good effects but complex construction, and sealant waterproofing is prone to aging. Exterior wall waterproofing paint is easy to construct but has the problem of easy damage. Exterior wall waterproof concrete has a long-lasting waterproofing effect, but the subsequent maintenance is complicated. The integrated insulation and drainage system has a good comprehensive effect, but the cost is high. In general, the above technologies are suitable for different environments and conditions.

Although this paper has systematically analyzed the above waterproof technologies, there are still some shortcomings. Most of the waterproof technologies analyzed in this paper are based on physical shielding and lack chemical material related technologies. In addition, the impact of different technologies on waterproof results is often qualitative analysis and lacks experimental quantitative analysis. In future research, we should focus on breakthroughs in waterproof technology in chemical materials and achieve quantitative comparisons between different waterproof technologies through experiments.

Authors Contribution

All the authors contributed equally and their names were listed inalphabetical order.

References

1. Q.M. He, et al., Analysis of the key points of testing waterproof coatings for chemical building materials. Chemical engineering management. J. (01), 60-63 (2023) https://doi.org/10.19900/j.cnki.ISSN1008-4800.2023.01.018
2. Z.M. Zhe, Analysis of prefabricated building waterproofing technologies in construction. Ceramics. J. **(09)**, 115-117 (2023) http://doi.org/10.19397/j.cnki.ceramics.2023.09.027
3. D.K. Feng, et al., Waterproofing technologies and measures of prefabricated concrete buildings. Construction technology. J. **53(17)**, 7-14+66 (2024)
4. J.S. Nie, et al., New materials and construction technology innovation of prefabricated exterior wall waterproof sealing of prefabricated buildings. Adhesion. J. **51(05)**, 37-40 (2024)
5. G. Gao, M. Du, X.R. Ran. Research on waterproofing construction technology for exterior walls of prefabricated buildings. Brick-Tile. J. (12), 76-77 (2020)
6. J.R. Yang. Preliminary analysis of waterproofing technology for prefabricated exterior walls of concrete buildings. Construction materials and decoration. J. **(28)**, 15-16 (2020) https://doi.org/10.3969/j.issn.1673-0038.2020.28.008
7. M. Garrido, et al., Performance of different jointing techniques used in the repair of bituminous waterproofing membranes. J. **158**, 346-358 (2018) https://doi.org/10.1016/j.conbuildmat.2017.09.180
8. M.H. Lu. Key control technology for waterproofing construction of basement in high-rise buildings. Low carbon world. J. **9(09)**, 249-252 (2019)
9. S.W. WEI, W. Li, W. Yan, et al., Research on Waterproofing and impermeability Construction Technology of prefabricated building Exterior Wall / Construction Technology Magazine, Proceedings of National Civil Engineering Construction Technology Exchange Conference (Part2). C. 751-753; (2022) https://doi.org/10.26914/c.cnkihy.2022.076755.
10. J.L. Liu, et al., Research on optimization of underground wall waterproof roll process. Henan science and technology. J. **40(30)**, 43-45 (2021)
11. S.G. Zhang. Construction technology of joint sealant for prefabricated concrete building exterior walls. Science and technological innovation. J. **10**, 191-194 (2024)
12. Y. Chen, et al., Synthesis and application of tacklifying resin in silane modified polyether adhesive for prefabricated building, in Proceedings of the 2022 international conference on green building, civil engineering and smart city, Singapore, C. 176-184 (2022)
13. J.Y. Wang. Analysis and application of waterproof system for prefabricated exterior wall joints of prefabricated buildings. Guangzhou architecture. J. **50(02)**, 63-68 (2022)
14. G.J. Wang. Development of exterior wall waterproof coating based on siloxane modified elastic acrylic emulsion. Green building. J. **(06)**, 31-35 (2000)
15. R.P. Wu, et al., Preparation and performance study of inorganic exterior wall waterproof coating with fly ash. J. **45(02)**, 1-2 (2016)

16. C. Esteves, et al., The influence of hydrophobic protect on building exterior claddings. J. **16**, 1379-1388 (2019)
17. S.L. Lv. Study on waterproof technology of prefabricated concrete building exterior wall. Fujian construction science and technology. J. (03), 69-73 (2017)
18. J. Hu, et al., Self-waterproofing system and flexible waterproofing system of underground building structure. China building waterproofing. J. **(06)**, 1-5 (2021) https://doi.org/10.15901/j.cnki.1007-497x.2021.06.001
19. G.P. Qiao, et al., Research on key technologies for self-waterproofing of ultra long structures in underground space engineering. Construction machinery. J. **(12)**, 32-34 (2021) https://doi.org/10.14189/j.cnki.cm1981.2021.12.016
20. L. X, et al., Water technology of prefabricated exterior wall with assembled building thermal insulation integration. Building structure. J. **52(S2)**, 1636-1639 (2022) https://doi.org/10.19701/j.jzjg.22S2073
21. Z. H. Wang, A brief discussion on the construction technology of building exterior wall insulation, energy saving and waterproofing system, 2022 Engineering Technology Innovation and Management Seminar (ETIMS 2022), C. 523-525 (2022-12-18) https://doi.org/10.26914/c.cnkihy.2022.075543
22. L.L. Qiu, et al., Construction technology of building exterior wall insulation, energy saving and waterproofing system. Sichuan building materials. J. **50(08)**, 116-118 (2024)
23. P. Guo, D. Z. Xie, S. H. Wei, X. Li. Comprehensive technology for anti-leakage of deep foundation pit basement engineering. 2024 National Engineering Construction Industry Construction Technology Exchange Conference. C. 328-330 (2024-06-28) https://doi.org/10.26914/c.cnkihy.2024.027950.

A Critical Review of Structural Topology Optimization Algorithms

Yuhang Jiao[1,*]

[1]School of Civil Engineering, Chongqing University, 174 Shazheng Street, Shapingba District, Chongqing, China

Abstract. In the past 30 years , the field of structural topology optimization has developed rapidly, and many representative algorithms such as homogenization algorithm, solid isotropic material with penalization algorithms and evolutionary structural optimization algorithms have emerged. In this paper, the above three representative algorithms' principles and development history are briefly introduced. Secondly, with the classic example of long cantilever beams, the three methods are compared in all aspects of the optimisation process, and their similarities and differences in terms of optimisation objectives, constraints, and characteristics of the results are analysed. Finally, the above algorithms are summarized, and their advantages, disadvantages and applicable scenarios are listed, which will provide reference for future designers to a certain extent.

Keywords: Topology optimisation; homogenisation; SIMP method; ESO method

1 Introduction

With continuous developments in the realm of architecture, the complexity and innovativeness of structures have reached an unprecedented level, and the requirements for material utilisation efficiency and structural safety are also increasing. In this context, the traditional experience-based optimisation methods have been difficult to meet the needs of today's structural optimisation, and many complex structures are difficult to be optimised using traditional methods. The emergence of topological optimisation methods is a good solution to this problem. Compared with traditional structural optimisation methods, topological optimisation methods show higher optimisation efficiency and greater versatility, and thus have gained the attention of researchers.

In the past 30 years, the methods of topology optimisation have made great development, and many mature algorithms have appeared both domestically and internationally [1].Bendsoe et al proposed the homogenisation algorithm, which transforms the topology optimisation problem from a problem of considering existence or non-existence of the materials at each individual point in the domain to a continuum material distribution problem,

[*] Corresponding author: 07107850@alu.cqu.edu.cn

and provides a computational framework for the topology optimisation of a continuous structure [2]. In order to reduce the complexity of the homogenisation method, Bendsoe proposed the solid isotropic material with penalization(SIMP) method on its basis, which takes the artificially assumed 0-1 density value as the design variable, and completes the material redistribution through the change of the density value, so as to realise the topology optimisation of the structure [3].Xie et al proposed and perfected the evolutionary structural optimization(ESO) algorithm, which can be optimised for a wide range of stress, stiffness, frequency and other constraints [4]. Currently, these topology optimisation algorithms have been extensively used in many realms such as building structure design, aircraft design, 3D printing, etc., showing high potential and practical value [5].

However, at present, domestic and international research on structural topology optimisation algorithms mainly focuses on the improvement and validation of the algorithms, while less attention is paid to the differences between the various algorithms and the applicable scenarios. Based on the above three algorithms, this paper analyses the advantages and disadvantages of each algorithm and the applicable scenarios with the example of long cantilever beam, which will help future researchers to choose the appropriate optimization algorithm and drive topology optimisation to more areas of application.

2 Concept of structural topology optimisation

2.1 Definition of structural topology optimisation

Structural topology optimisation refers to the design approach to achieves goal of using less material and better performance without affecting the design function of the structure by optimising the material distribution and structural form with the help of technical means such as mathematical algorithms and computer programs under the given material properties, boundary conditions and loads [6]. The three elements of topology optimisation include design variables, objective function and constraints.

Topology optimisation is significantly different from traditional size optimisation and shape optimisation. Traditional optimisation methods can be solved step-by-step using mathematical planning methods with parameters like the cross-sectional area of the beam and the slab thickness as design values. However, In the topology optimisation issue, due to the existence of an infinite number of potential topologies at each point within the structure, nodes or beams may exist at any position within the computational domain under the complex structure, and it is not possible to build up a system of equations with a finite number of parameters to accurately simulate the mechanical analysis of a real structure, but only to use a finite number of parameters and combine with the algorithm to approximate the expression, which is exactly the difficulty of the current structural optimisation of the topology [7].

2.2 Process of structural topology optimisation

As is shown in the Fig. 1, a systematic structural topology optimisation process should consist of three phases: modelling, analysis and optimisation. In the initial mathematical modelling stage, the researcher needs to construct a mathematical model that can approximately characterise the topological properties of the structure using finite parameters in the design domain and determine the optimisation objectives. The parameters used to construct the mathematical model can be the volume, displacement, solid frequency, etc. of the structure. Considering the practical needs of engineering, researchers often use the maximum structural stiffness as the optimisation target [8]. The parameters chosen for constructing the mathematical model greatly affect the efficiency and accuracy of structural topology optimisation, as well as being the fundamental difference between different topology optimisation algorithms.

In the structural analysis stage, the model is analysed with the help of finite element method to determine the target function and sensitivity function of the structure. The target function is generally a flexibility function, which is used to determine when the structure converges. The sensitivity function is used to reflect the influence of parameter changes on the overall structure index.

Finally, in the topology optimisation stage, the researcher needs to make the parameters of the mathematical model gradually approach the optimal values through the optimisation algorithm. When the objective function converges to a certain accuracy under the manufacturing constraints , it can be considered that a topologically optimised structure that meets the requirements has been obtained [9].

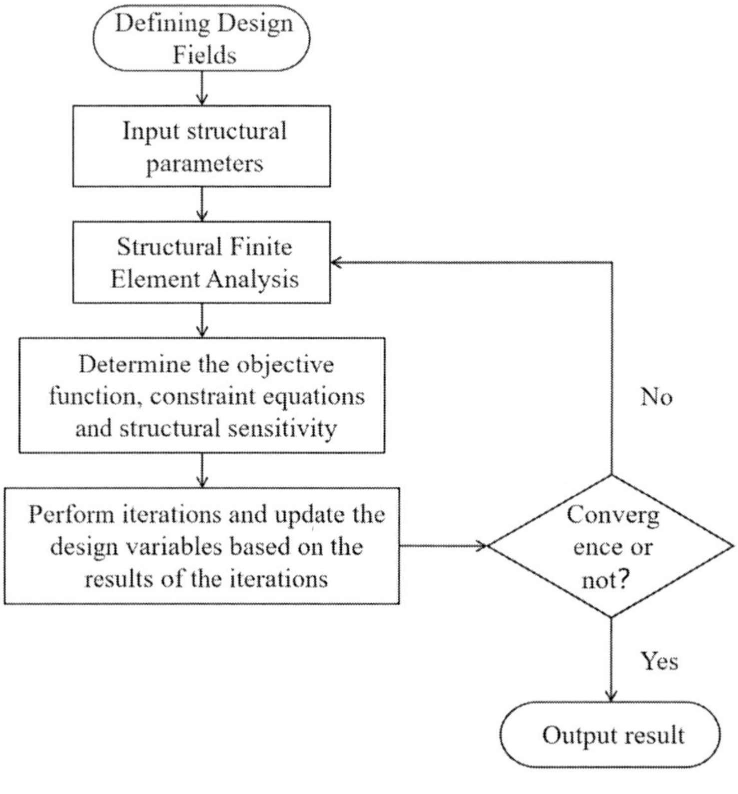

Fig. 1. Topology optimisation process. (Picture credit: Original)

2.3 Classification of structural topology optimisation

The two main categories of structural topology optimization are illustrated in Fig. 2, respectively continuous structure topology optimization and discrete structure topology optimization, and this classification is due to the characteristics of the original structure. Continuum structure topology optimisation usually assumes that the material distribution is continuous and the design variables are continuous values. This approach is suitable for most engineering applications, especially when the distribution of structural materials can vary arbitrarily. Therefore, most current topology optimisation algorithms use the continuum assumption. Discrete structure topology optimisation treats the design variables as discrete, discontinuous values, and the main representative methods include genetic algorithms and particle swarm optimisation algorithms. Discrete structure topology optimisation methods are difficult to describe more complex structures, with a narrow range of applications and poor engineering practicability, so current research mainly focuses on continuum topology optimisation.

Structural topology optimisation is also categorized into material-based and geometry-based according to the parameters used in building the mathematical model [5]. Material-based topology optimisation methods mainly include homogenisation and SIMP methods. These methods start from describing certain properties of the material, such as the density and elastic modulus of the material, and achieves structural optimisation by improving the materia distribution of the structure. Geometry based topology optimisation methods include ESO method, level set method. These methods focuses on defining the boundary condition of the structure and constrain the structure by describing it.

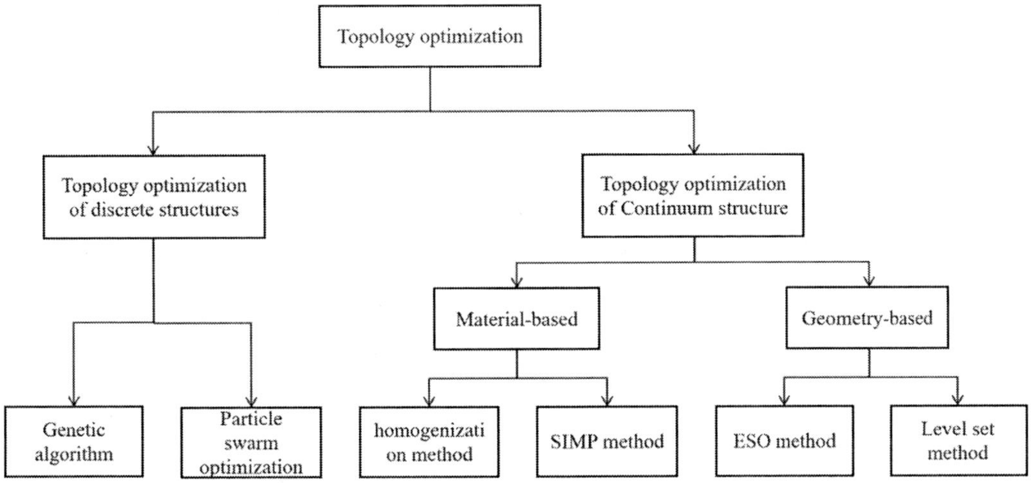

Fig. 2. Topology optimisation classification. (Picture credit: Original)

3 Representative algorithms for structural topology optimisation

3.1 Homogenisation method

In 1988, Bendsoe et al proposed the homogenisation theory, which establishes a new computational framework for topological optimisation of continuum structures [2]. As is shown in the Fig. 3, The principle of homogenisation is to define the structure by the material distribution in the design region, and the conventional method is to discretise the design region into a rectangular or square grid cell of equal size, and in the optimisation process, the material density coefficient of each grid cell is adjusted according to the stress magnitude in every grid cell by increasing the material density coefficient in the area with high stress and decreasing the material density coefficient in the area with low stress. With the goal of minimising the flexibility of the structure, after several iterations of calculation, a reasonable material density value is selected, and the cells below this value are deleted to obtain the final structure after topology optimisation. Compared with the original structure, the more inefficient parts are discarded, and the efficiency of material use is improved.Groen introduced a projection method for topology optimisation of orthogonal heterogeneous filler materials at larger sizes, which allows the materials to be modelled on coarser meshes and reduces the computational cost considerably [10].

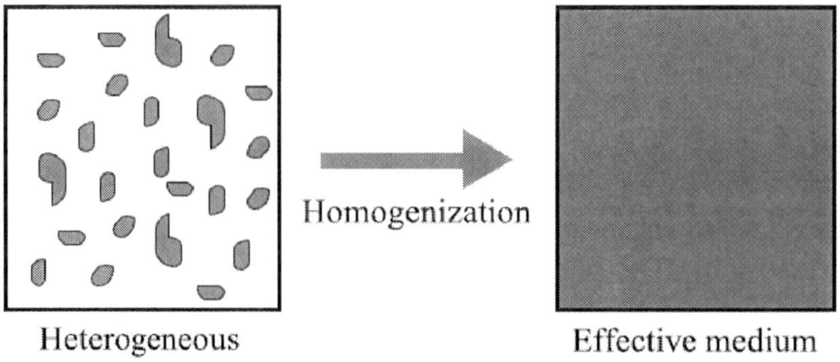

Fig. 3. The principle of homogenisation. (Picture credit: Original)

3.2 Solid isotropic material with penalization method

The solid isotropic material with penalization method introduces the concept of penalty factor based on the homogenisation method, where instead of deleting the material directly during the iterative calculations, the low-density region of the material is penalised by a penalty factor, which reduces the effective elastic modulus of the region. In the SIMP method, Bendsoe gives the relationship between the density design variables and the material properties by a power function [3]:

$$\mathrm{E}(\rho_\mathrm{i}) = \mathrm{g}(\rho_\mathrm{i})\mathrm{E}_0 = \rho_\mathrm{i}^{\mathrm{P}}\mathrm{E}_0 \qquad (1)$$

$$\mathrm{g}(\rho_\mathrm{i}) = \rho_\mathrm{i}^{\mathrm{P}} \qquad (2)$$

In the equation (1), p is the penalty parameter, the value range is p>1, commonly used value is p=3; E_0 is the Young's modulus of the solid material; $\square_i$ is the i-th unit density.

This method significantly reduces the complexity of the homogenisation method, but at the same time introduces a new problem, i.e., it is often difficult to make a decision on whether to remove structural units at intermediate densities or not [5]. Yarlagadda improved the SIMP method by proposing the solid isotropic material with thickness penalization (SIMTP) method, which uses 2.5D planar units (2D for the state of stress and 3D for the state of volume) instead of the 2D units in the SIMP method [11]. The SIMTP method is more direct, cost-effective and efficient in generating high-resolution images of structural boundaries.

3.3 Evolutionary structural optimization method

Xie introduced the evolutionary structural optimization(ESO) method in 1993, the basic principle of which is to remove parts of low inefficiency from the structure gradually and eventually obtain an optimised structure with better material utilisation and performance [12]. In Fig. 4, the structural shaping process of the column by the ESO method is shown when the column is subjected to a vertical compressive force on its upper surface. The ESO method aims at minimising the weight of the structure, and finite element analysis is carried out on the model at each step of the iteration to eliminate structural elements that are in a low-stress state.The definition of the low stress state unit is given by the following equation:

$$\frac{\sigma_{\mathrm{ij}}}{\sigma_{\mathrm{maxj}}} \le \mathrm{RR}_\mathrm{n} \qquad (3)$$

In the equation (3), σ_{maxj} is the maximum stress of the unit; RR_n is the deletion rate of each iteration step, the value range is generally 0.05-0.2; σ_{ij} is the *Von Mises* stress of the unit, calculated as follows:

$$\sigma_{ij} = \sqrt{\frac{1}{2}[(\sigma_1 - \sigma_2)^2 + (\sigma_2 - \sigma_3)^2 + (\sigma_3 - \sigma_1)^2]} \quad (4)$$

In the equation (4), σ_1,σ_2 and σ_3 are the first, second and third principal stresses, respectively.

When the maximum stress in a structural cell reaches the permissible stress at a certain operating condition, the iteration is stopped and the optimised structure is obtained. This method was refined by Huang as a bi-directional evolutionary structural optimization (BESO) method [13]. It allows the removal of cells in the part of the structure where the material is utilised less efficiently while adding cells in the part of the structure where it is utilised more efficiently. On this basis, Rut proposed an automated image-based SBFE-BESO method using a quadtree/octree decomposition algorithm within the SBFE framework, which greatly reduces the number of degrees of freedom and the memory required for computation, while the topology results maintain a good accuracy compared to traditional grid methods [14].

Fig. 4. Structural Formation Process of ESO method [4].

4 Comparative analysis of various optimisation algorithms

4.1 Introduction to long cantilever case.

In the field of topological optimisation, cantilever whose length to height ratio is greater than 1 (L>H) are called long cantilever. The topology optimisation of long cantilever is a classical example in the realm of structural topology optimisation. In this paper, a long cantilever with L=2H is selected as an example to analyse the similarities and differences among the topology optimization methods. This cantilever is subjected to a concentrated force in the vertical direction at the midpoint of the right end face, as shown in Fig. 5. The modulus of elasticity of the cantilever E=1,Poisson's ratio ν=0.33 and the optimised material occupancy are all 50%. The topology optimisation results of long cantilever for different algorithms are shown in Fig. 6.The number of iterations for the three methods are 50, 150 and 50 respectively.

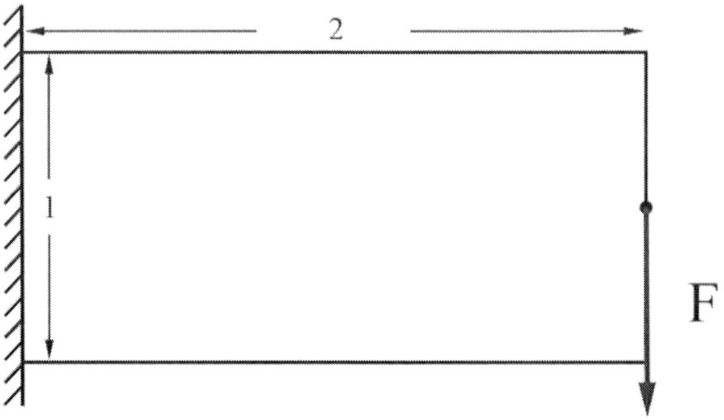

Fig. 5. long cantilever case. (Picture credit: Original)

(a) Homogenisation method [15] (b) SIMTP method [11] (c) BESO method [16]

Fig. 6. Optimisation results for long cantilever [11,15,16].

As is shown in Fig. 6, optimisation results obtained by various optimisation methods are largely the same, although the optimisation results of the homogenisation method show a more severe checkerboard phenomenon. There is an obvious jagged phenomenon on the boundary of the optimisation results of homogenisation algorithms and the BESO algorithms. This is a numerical instability problem due to the discrete nature of the finite cell lattice [3].The optimisation results of the SIMTP method have very smooth boundaries, which is due to the fact that the planar cell of 2.5D avoids the instability problem of the boundaries in the SIMP method, and stable, high-resolution structure boundaries can be obtained without the need for a smoothing algorithm [6]. Comparison with the optimisation results of the homogenisation method shows that the BESO method converges slower, which is due to the nature of its evolutionary method.

4.2 Analysis and discussion

In this paper, we summarise the four aspects of each topology optimisation algorithm, namely the basic idea, optimisation objective, constraints and result characteristics, and the results are presented in Table 1.

Table 1. Summary of different optimisation methods. (Picture credit: Original)

Methods	Basic idea	Optimisation targets	Constraints	Result characteristics

Homogenisation	Uniform meshing of the structure	Minimum structural flexibility	Volume	Large number of holes in the result
SIMTP	Cells with densities below a certain threshold are deleted	Minimum structural flexibility	Volume	Resulting boundaries are smooth, stable and high resolution.
BESO	Gradual removal of inefficient materials from the structure	Multiple objectives	Stress , displacement	Severe jaggedness in the optimised boundary is evident

From the above table, it can be seen that the three representative methods have their own characteristics and each has its own advantages and disadvantages when performing structural optimisation. The homogenisation method can handle materials with complex microstructures, especially composites and non-homogeneous materials, because it is based on the principle of homogenising the structure into a uniform mesh, but its topology optimisation results contain a large number of holes and the phenomenon of tessellated lattice. SIMTP method is more accurate in analysing the structure under the influence of static loads, but it is not applicable to the optimisation of the structure under dynamic loads, and the BESO method is suitable for structures affected by dynamic loads and stability, especially when considering the limit states and failure modes of structures.

The homogenisation method and SIMTP method are both based on the optimisation objective of minimising structural flexibility, which is inconsistent with the objective of achieving the minimum amount of material usage in construction projects, and thus are often impractical.The BESO method is capable of being adapted to a wide variety of optimisation objectives, and has a wider range of applications. However, the BESO method is an evolutionary method that requires a higher number of iterations, leading to its lower computational efficiency. The homogenisation method and SIMTP method are only applicable to volumetric constraints, while the BESO method is able to apply to a wide variety of constraints like stress constraints and displacement constraints.

In addition, in order to satisfy the manufacturability of the structure, equal-thickness uniform isotropic structures with holes are often used as the optimisation objective in engineering. This isotropic material of equal thickness can also be interpreted as a constraint on the optimised structure, which should be taken into account when selecting the optimisation method.

5 Conclusions

This paper compares three representative topology optimisation algorithms, namely, the homogenisation method, SIMTP method and BESO method, describes the characteristics of each method in terms of optimisation results, optimisation efficiency and applicability, and compares the differences between the three methods in terms of optimisation objectives, constraints and optimisation results in conjunction with the example of a long cantilever beam, and obtains the following conclusions:

(1) The homogenisation method performs well in dealing with complex boundary conditions and composite material problems, and can effectively optimise the material distribution. However, its calculation is relatively complex, and the optimisation efficiency is relatively low.

(2) The SIMTP method has a simple optimisation process, fast convergence speed, and performs well in terms of the accuracy of the optimisation results, but it is unable to deal with non-uniform and composite materials, and has certain limitations.

(3) The BESO method, although inferior to the above two methods in terms of optimisation efficiency, has wide applicability as it is capable of being adapted to a wide variety of optimisation objectives and constraints.

Although this study provides a comprehensive comparison of the homogenisation method, SIMTP method and BESO method, there are still some shortcomings. This paper only analyses the simplified cantilever beam example, which is not universally representative and cannot characterise the variation of different algorithms under more complex structures and loads. In future research, more complex structural forms and diversified loading conditions should be considered, and combined with actual engineering cases, the various algorithms should be studied in greater depth.

References

1. Y M Xie, Y Y Zhou, J Wang, Application of Topological Optimization Technique to the Integration of Architectural and Structural Design. J. Architecture Technique. **04**, 32-34 (2016) https://doi.org/10.3969/j.issn.1674-6635.2016.04.006
2. P. Martin, K. Noboru, Generating optimal topologies in structural design using a homogenization method. J. Computermethods in applied mechanics and engineering. **71(2)**, 197-224 (1988) https://doi.org/10.1016/0045-7825(88)90086-2
3. P. Martin, O. Sigmund, Material interpolation schemes in topology optimization. J. Archive of applied mechanics. **69**, 635-654 (1999) https://doi.org/10.1007/s004190050248
4. Y M Xie, Z H ZUO, J C Lu, Architectural Design through Bi-directional Evolutionary Structural Optimization. J. Time Architecture. **05**, 20-25 (2014) https://doi.org/10.13717/j.cnki.ta.2014.05.004
5. S Y Shi, X Z Guo, K M Zhou, Research Progress in Numerical Methods of Structural Topology Optimization. J. Journal of Huaqiao University(Natural Science). **45(02)**, 150-157 (2024) https://doi.org/10.11830/ISSN.1000-5013.202312025
6. P Du, W F Du, B Q Gao, Research and Application of Free Topological Building Structure. J. Journal of Henan University(Natural Science). **54 (02)**, 245-252 (2024) https://doi.org/10.15991/j.cnki.411100.2024.02.011
7. K M Zhou, J F Li, X Li, A Review on Topology Optimization of Structures. J. Advances in Mechanics. **35(01)**, 69-76 (2005) https://doi.org/10.3321/j.issn:1000-0992.2005.01.007
8. M C Cheng, Y Zhao, Y M Xie, Topology Optimization and Additive Manufacturing of Nudes in Spatial Structures. J. China Civil Engineering Journal. **52 (02)**, 1-10 (2019) https://doi.org/10.15951/j.tmgcxb.2019.02.001
9. W W Wang, Research onTopology Optimization of Shell Structure with Parameterized Level Set Method forArchitectural Form Innovation Design, MA thesis, South China University of Technology. (2019) https://doi.org/10.27151/d.cnki.ghnlu.2019.000620

10. P. Jeroen, J Wu, O. Sigmund, Homogenization-based stiffness optimization and projection of 2D coated structures with orthotropic infill. J. Computer Methods in Applied Mechanics and Engineering. **349**, 722-742 (2019) https://doi.org/10.1016/j.cma.2019.02.031
11. Y. Tejeswar , Z X Zhang, L M Jiang, B. Pradeep, U. Asif, Solid isotropic material with thickness penalization – A 2.5D method for structural topology optimization. J. Computers & Structures. **270**, 106857 (2022) https://doi.org/10.1016/j.compstruc.2022.106857
12. Y M Xie, G. Steven, A simple evolutionary procedure for structural optimization. J. Computers and Structures. **49**, 885-896 (1993) https://doi.org/10.1016/0045-7949(93)90035-C
13. X Huang, Y M Xie, Convergent and mesh-independent solutions for the bi-directional evolutionary structural optimization method. J. Finite Elements in Analysis & Design, **43(14)**, 1039-1049 (2007) https://doi.org/10.1016/j.finel.2007.06.006
14. S. Rut, T. Sawekchai, C M Song, Automatic image-based SBFE-BESO approach for topology structural optimization. J. International Journal of Mechanical Sciences. **263**, 108773 (2024) https://doi.org/10.1016/j.ijmecsci.2023.108773
15. A. Grégoire, C. Lorenzo, et al, The homogenization method for topology optimization of structures: old and new. J. Interdisciplinary information sciences. **25(02)**, 75-146 (2019) https://doi.org/10.4036/iis.2019.B.01
16. X Huang, Y M Xie, Bi-direetional evolutionary topology optimization of continuum structures with one or multiple materials. J. Computational Mechanics. **43**, 393-401 (2009) https://doi.org/10.1007/s00466-008-0312-0

Research on Green Construction Technology and Management Systems in Engineering Projects

Zhongye Lyu*

Xiamen University Tan Kah Kee College, Civil Engineering Department, Zhangzhou city, China

Abstract. In the context of increasingly strained global environmental resources, green building construction can achieve a win-win scenario by reducing resource consumption and environmental pollution during the construction process while simultaneously realizing economic and ecological benefits. This study primarily explores the current research on green construction technologies and management systems and their similarities and differences compared to traditional construction practices. It reflects the issues present during the construction process and assesses the current level of green construction in China, highlighting future development trends and objectives for green construction. By validating the effectiveness and significance of Building Information Modeling (BIM) technology and organizational management structures, this research provides relevant insights for the green transformation of the construction industry. It offers valuable references for the continuous improvement and promotion of green construction concepts in China.

1 Introduction

With the continuous development of China's construction industry, the environmental pollution caused by carbon emissions has become increasingly severe. In the context of a global trend toward low-carbon energy and industrial development, China has set strategic goals to achieve "peak carbon emissions" by 2030 and to strive for "carbon neutrality" by 2060. The "Green Building Evaluation Standards" introduced by China emphasize that green buildings are essential for promoting a new green development concept, aiming to achieve high-quality green development in construction. Among these, green construction is a critical component of green building development, widely promoted and applied due to its advantages in pursuing high efficiency, low energy consumption, and environmental protection.

In the current state of research on green construction, China is still in its early stages. Scholars such as Wang Zhen-Shuang, Tong Lin, and Tang Li-li argue that traditional construction methods can no longer meet the development demands of the modern construction industry [1-3]. Introducing scientifically sound and reasonable green construction practices can effectively address the conflicts between construction projects and

* Corresponding author: zhongyelyu@ldy.edu.rs

the ecological environment. Construction enterprises should eliminate outdated construction models and integrate information technology, green energy-saving construction techniques, and management innovations as future research themes in green construction. However, there needs to be more comprehensive regulatory frameworks and practical standards within the industry. The utilization of construction waste and the construction and innovation of enterprise management also require greater attention. International scholars Suman Paneru and Bahaa Chammoutcontend that green construction is a critical driver in addressing environmental degradation, resource depletion, and climate change issues caused by the construction industry [4-5]. Among the various factors, decarbonization is essential in determining.

whether green construction standards are met. This necessitates the establishment of labor-management models during the construction process and reforms in carbon emission technologies to address challenges in green construction. Therefore, further in-depth research is needed to promote the establishment of comprehensive green construction technologies and management systems to achieve green construction.

This paper comprehensively analyzes recent research on green construction technologies and management systems. By integrating various green construction measures, such as environmental protection, water and energy conservation, and the preservation of construction sites, the study examines the advanced application of Building Information Modeling (BIM) technology in green construction. Additionally, it reviews organizational management, planning management, and material management initiatives within the green construction management system, using the organizational structure of the Volcano Hospital as a case study. The analysis concludes that it is essential to implement the principles of green, low-carbon, energy-efficient, and environmentally friendly practices throughout the construction and management processes, advocating for a green and low-carbon transformation in the construction industry. This is of significant importance for reducing resource and energy consumption in construction, enhancing the quality of green building development, and achieving sustainable development within the construction sector in China.

2 Green construction technology

2.1 Summary of green construction technology

Green construction technology operates to ensure project quality and safety while adhering to a people-centered and context-specific approach. It aims to maximize environmental, economic, and social benefits, balancing and coordinating these three aspects. At its core, it emphasizes technological innovation and progress, which can reduce resource waste and minimize the negative impacts on the environment during the construction process. Green construction technology is not merely a reformative innovation; it is not entirely separate from traditional construction techniques but rather represents the practical application of sustainable development principles in the construction process. To meet the technical innovation evaluation criteria outlined in the "Green Construction Evaluation Standards for Building and Municipal Engineering" and to achieve the goals of resource conservation and environmental protection, recent research on green construction technologies can be categorized into several specific types based on their functions, as illustrated in Figure 1.

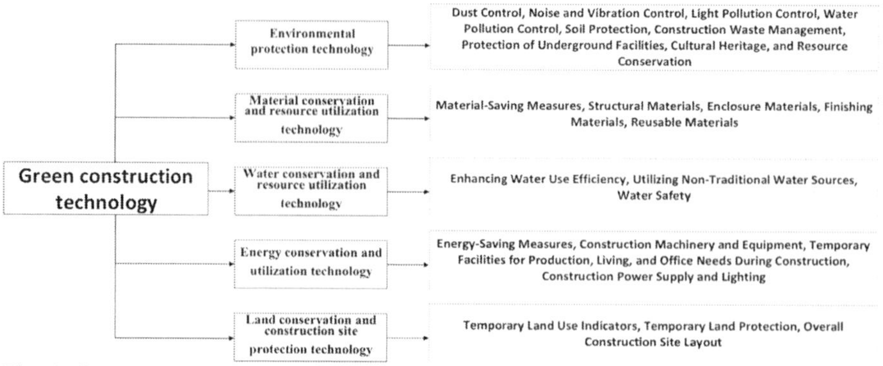

Fig. 1. Green construction technology classification chart [6]

Environmental protection technology refers to a series of measures to reduce or eliminate negative impacts on the natural environment, ensuring a harmonious coexistence between humans and nature. By controlling factors such as dust, noise, water resource pollution, and construction waste, CO2 emissions can be reduced, air quality improved, and the living conditions of both construction workers and residents enhanced [7]. This approach promotes harmonious relationships between construction enterprises and society while maintaining ecological stability in the natural environment.

Material conservation and resource utilization technology involves optimized design, efficient use, and recycling to reduce material consumption and maximize utilization efficiency [8]. This technology encompasses two main aspects: control of material usage and the application of environmentally friendly materials. During construction, dynamic management of material resources is conducted based on construction schedules, material inventories, and spatial layouts. Using insulation, waterproofing, and high-performance building materials, such as high-strength steel and concrete, aims to minimize resource consumption and improve construction methodologies.

Water conservation and resource utilization technology are vital to addressing global water scarcity. Advanced water-saving techniques enhance water resource utilization efficiency during construction by minimizing losses during water delivery and reducing waste [9]. Additionally, utilizing unconventional water resources, such as rainwater, reclaimed water, and desalinated seawater, alleviates the pressure on traditional water resources. The government, enterprises and the public must collaborate, employing various water-saving technologies and management practices to ensure the safety of construction water use and promote the sustainable utilization of water resources.

Energy conservation and utilization technology refers to measures taken during construction to achieve energy savings, environmental protection, and sustainable development by enhancing energy efficiency, reducing energy losses, and improving energy structure [10]. In architectural design and construction, adopting energy-efficient materials, green roofs, natural lighting, and high-efficiency air conditioning systems reduces energy consumption and environmental pressure. It promotes the upgrading and improvement of industrial economic structures. This approach is a significant pathway for advancing sustainability in the building industry.

Land conservation and construction site protection technology involves adopting scientific management and technical measures during the construction phase to minimize land occupation and degradation, enhance land utilization efficiency, and protect and improve the ecological environment. This technology achieves a harmonious integration of construction land and ecological space through optimized planning and construction, standardization of construction site layouts, and promotion of vegetation restoration on construction sites [11]. Promoting this technology and related policies to revitalize abandoned land contributes to the harmonious advancement of engineering construction and environmental protection, thereb y facilitating the sustainable utilization of land resources.

2.2 Comparative similarities and differences between traditional construction technology and green construction technology

Traditional and green construction technology involves costs related to materials, labor, equipment, and management, with effective cost control crucial for project profitability and corporate competitiveness. Although green construction typically incurs higher initial investments due to the use of advanced equipment and materials, it can significantly reduce resource waste and operating costs over the long term through measures aimed at water conservation, energy savings, and material efficiency [12]. Additionally, green construction considers the impact of environmental pollution on surrounding communities, safeguarding residents' quality of life and enhancing social benefits. Using strategies such as green supply chains and prefabricated components, green construction technology optimizes building materials and energy use while effectively managing construction waste. The construction phase must adhere to green principles, encompassing raw materials production, transportation, and storage, to minimize the environmental costs associated with transportation and production activities.

In construction, cost reduction and environmental impact considerations must be balanc ed to enhance efficiency. Green construction technology represents an innovation over tradi tional methods, which rely heavily on manual operations and often lack efficiency and preci sion, as well as a strong awareness of resource conservation. In contrast, green construction technology employs advanced techniques to address challenges associated with delays and quality defects typical of traditional construction while simultaneously reducing environme ntal impacts. This technology enhances resource utilization efficiency through local materia ls, optimized material selection, and advanced water-saving technologies. By integrating int elligent construction technologies, real-time monitoring and analysis of the construction site can be achieved, effectively resolving construction issues and improving overall work effic iency [13].

Green construction technology has been widely adopted globally, demonstrating superio r structural, aesthetic, and innovative characteristics compared to traditional construction m ethods. Amid rapid economic growth, conventional techniques have led to a homogenizatio n of architectural styles. The focus in the construction industry has shifted towards standard ization and financial efficiency, while green construction emphasizes environmentally frien dly design. This includes using natural lighting, ventilation, modular structures, rainwater h arvesting, and green roofs to reduce environmental impact. In terms of material innovation, the incorporation of prefabricated components effectively reduces construction waste. Low-carbon and renewable materials such as bamboo, wood, and stone are increasingly utilized. Additionally, using low volatile organic compound (VOC) paints and sealants minimizes ai r pollution and improves the quality of the construction environment.

Traditional and green construction technology fulfill engineering construction tasks, aim ing to meet contractual requirements for quality, safety, schedule, and cost. However, their f ocal points differ. Traditional construction technology emphasizes cost reduction while ensu ring compliance with quality, schedule, and safety standards, often neglecting consideration s for energy conservation, emissions reduction, and environmental friendliness. In contrast, green construction technology prioritizes resource conservation and environmental protectio n during the construction process, striving for the sustainable development of project constr uction.

2.3 BIM technology analysis and case application

Building Information Modeling (BIM) technology represents a transformative advancement in the construction industry. As shown in Figure 2. It provides visual support for project pla nning, design, construction, and operation by integrating three-dimensional model informati on with shared data. This integration fosters team collaboration and significantly improves d esign efficiency, operational effectiveness, and cost control [14]. For instance, in the Arnhe m region of the Netherlands, BIM technology enabled designers to revitalize a World War I I German military base by incorporating natural elements and preserving historical features, thereby enhancing the architectural cultural expressiveness. Furthermore, BIM technology can assess the potential for material recycling, optimize material selection, reduce dependen ce on traditional building resources, improve energy systems, and encourage using renewab le energy sources. During the construction design phase, it simulates interactions between b uildings and their environments, including lighting, air quality, and microclimates. This proj ect utilized local materials, recycled plastics, and solar panels to achieve resource circularity , showcasing an innovative application of green construction technology.

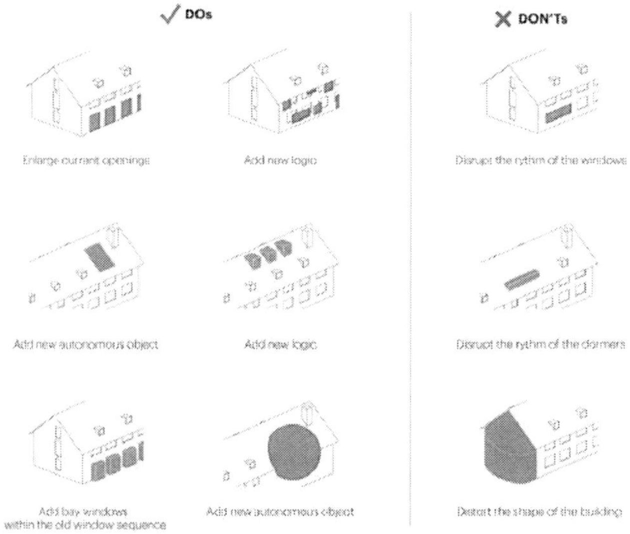

Fig. 2. Kamp Koning Noord building adjustment analysis diagram [15]

In the practical application of cross-sea bridges, objective factors such as extreme weather conditions, including typhoons, uncertainties in the hydrogeological environment surrounding the bridge, and the instability of imposed loads must be addressed. To effectively manage these challenges, cross-sea bridges must employ Building Information Modeling (BIM) technology, which differs from conventional construction methods. This approach should be progressive and integrated with other advanced technologies to achieve a more efficient, safer, and cost-effective construction process [16].

BIM technology in building project management needs to adequately consider the impact of the surrounding geographical environment, limiting its accuracy in spatial information management [17]. The integration of Geographic Information System (GIS) technology addresses this shortcoming by enhancing the efficiency of integrated management between architectural and geographical information, effectively reducing the negative environmental impacts of construction [18]. For instance, using Revit software for BIM modeling with GIS technology in bridge engineering enables refined three-dimensional modeling and visual management for the entire bridge construction process, optimizing resource allocation, logistics pathways, and cost control [19]. In the case of the Yongzhou Cross-Sea Bridge project, the application of BIM-GIS technology established a construction supervision platform, improving the precision of construction activities, shortening project timelines, addressing the complexities of bridge structures, and promoting the practice of green construction technology [20].

3 Green construction management

3.1 Summary of green construction management system

Implementing green construction in any engineering project necessitates establishing a robust green construction management system and regulatory requirements. The green construction management system is crucial for achieving resource conservation, environmental protection, pollution reduction, and ensuring the sustainability of the construction process within building projects. Traditional construction management focuses on achieving fundamental goals such as short construction durations, high quality, low costs, and efficient outcomes. In contrast, green construction management introduces new concepts that augment these original goals with sustainable development, resource conservation, and environmental protection, promoting high integration among social, economic, and ecological benefits [21,22]. As illustrated in Figure 3, the green construction management system primarily encompasses the following five categories.

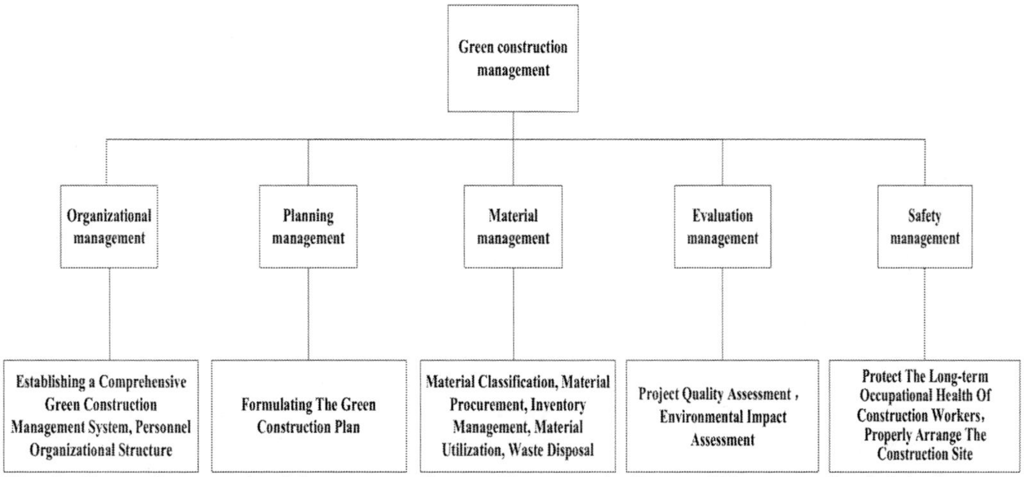

Fig. 3. Green Construction Management System Framework diagram(Picture credit: Original)

Organizational management refers to the systematic planning, organization, guidance, control, and coordination activities undertaken within construction projects to achieve green construction objectives. Its core components include establishing a green construction management system and allocating responsibilities among construction units. A management system founded on ecological friendliness and low-carbon environmental protection should be developed, along with establishing management goals and standards. An appropriate personnel structure aids in clarifying responsibilities, enhancing decision-making efficiency, and promoting communication and team collaboration. The project manager is responsible for allocating construction personnel to ensure the effective implementation of green construction practices [23].

Planning management refers to the detailed design and planning of the entire construction process in the early stages to ensure that construction activities maximize resource conservation, minimize environmental impacts, and promote the greening of the construction process. This requires developing a green construction plan that includes effective measures for environmental protection, energy conservation, and water savings. The plan must be reviewed and approved by relevant regulations and standards, reflecting the green construction objectives established during the design phase.

Material management primarily focuses on environmentally friendly, economical, and efficient use of materials during construction. It prioritizes selecting green building materials that are recyclable, low toxicity, and have low carbon emissions. A well-planned procurement strategy is essential for choosing reliable suppliers to ensure material quality. Appropriate storage methods should be employed to minimize material loss during storage. Information technology can track material usage in real time, enhancing the transparency of material management. Additionally, controlling the generation of construction waste, along with its classification, recycling, and reuse, is vital to prevent environmental pollution and degradation.

Evaluation management refers to the assessment and oversight of whether construction activities meet green construction standards during the project development process. Construction units must establish evaluation criteria within the green construction management system and form an expert evaluation team to conduct thorough assessments at various stages, from initiating the green construction plan to project completion. This includes comprehensive benefit assessment, engineering quality evaluation, and environmental impact assessment, ensuring the construction project complies with green construction requirements.

Safety management refers to a series of management measures and activities implemented during the green construction process to ensure the safety and health of construction personnel, the regular operation of construction equipment, and the protection of the environment and surrounding areas. This typically includes arranging appropriate construction sites, providing necessary medical emergency measures, improving the daily living conditions for construction workers, and safeguarding living and office areas from harmful impacts caused by construction activities.

3.2 Similarities and differences between traditional construction management systems and green construction management systems

Green construction managers advocate for dynamic monitoring of costs to identify and control factors associated with green construction expenses, thereby constraining project expenditures. Through an evaluation mechanism, construction activities are supervised, enabling timely correction of actions that violate green construction regulations, thus reducing resource waste and rework. While controlling costs, ensuring safety, quality, and environmental protection standards is essential. Unlike traditional construction, which primarily pursues profit, green construction emphasizes respect for people and the environment, avoiding disturbances to residents during nighttime work and protecting regional vegetation and wildlife to maintain biodiversity. Although initial costs may increase, this approach enhances corporate image and improves social and environmental benefits [24].

Traditional construction management focuses on reducing costs to maximize profits while ensuring progress, quality, and safety, often overlooking energy efficiency and construction waste management. Addressing these issues requires a shift in management perspectives to enhance environmental awareness and foster a green construction atmosphere through training and education. Project managers should establish green construction objectives, impose limitations on carbon emissions over time and space, and create evaluation and incentive mechanisms to reward environmentally friendly practices [25]. Collaboration with government, clients, and other stakeholders is essential to promote green construction, integrating these principles into corporate culture and continuously optimizing management processes to facilitate the long-term development of green construction practices.

In summary, both require construction enterprises to possess qualifications following national standards and utilize management methods for resource allocation to achieve the contractual objectives of safety, cost, schedule, and quality while conducting construction at the project site until the building product meets established standards [26]. The green construction management system represents an evolution of traditional management systems, emphasizing environmental protection, resource conservation, and sustainable development. This shift is of significant practical importance in addressing ecological pollution issues within the construction industry.

3.3 Organizational structure management analysis and case application

Optimizing the management of organizational structures in green construction is vital for the sustainable development of building projects. Traditional hierarchical management frequently leads to problem-solving and information flow inefficiencies, causing prolonged project timelines and higher labor costs. To tackle this challenge, it is essential to adopt a unified planning approach that involves all project stakeholders. This collaboration will help establish effective decision-making, supervision, and evaluation mechanisms within the construction leadership and operational organization.

In optimizing organizational structure management, enterprises should form a core leadership team comprising representatives from the government, suppliers, and other stakeholders tasked with overseeing project environmental protection. The project manager should be designated as the primary responsible for environmental protection. The core leadership team implements green construction policies and legal regulations, establishes rules and guidelines, and addresses significant ecological protection issues. The project manager's responsibilities include formulating a green construction plan, clearly defining environmental protection tasks, minimizing overlapping responsibilities, supervising and guiding ecological protection efforts in construction, exemplifying the execution of environmental policies, encouraging innovations in environmentally friendly technologies, and ensuring the advancement of green construction practices [27].

In the construction case of Wuhan Huoshenshan Hospital, faced with the dual challenges of the pandemic and the Spring Festival, there was an urgent demand for hospital construction with high-quality standards. An efficient organizational structure was established, comprising a construction command group formed by government agencies, a general contracting management organization led by China State Construction Engineering Corporation, and project subcontracting by various participating units. As illustrated in Figure 4, the project manager was comprehensively responsible for managing designated areas, integrating resources, and collaborating with the team to ensure task completion. In contrast, two deputy managers assisted in resource allocation and personnel coordination.

Through specialized teams and subcontractors, construction wastewater was collected and treated in a fully enclosed system, with disinfection processes implemented before being discharged into the municipal sewage network. The ventilation specialist team established independent fresh air and exhaust systems for the patient wards to ensure fresh air circulation and proper discharge of disinfected air. Suppliers provided HDPE impermeable membranes to protect underground foundations, while waterproof subcontractors prevented contamination. The medical specialist team designated a specific area to dispose of medical waste safely within the hospital. These measures exemplify the advantages of optimizing organizational structure; through collaborative division of labor and effective communication, management decision-making capabilities and employees' efficiency in environmental protection efforts were significantly enhanced.

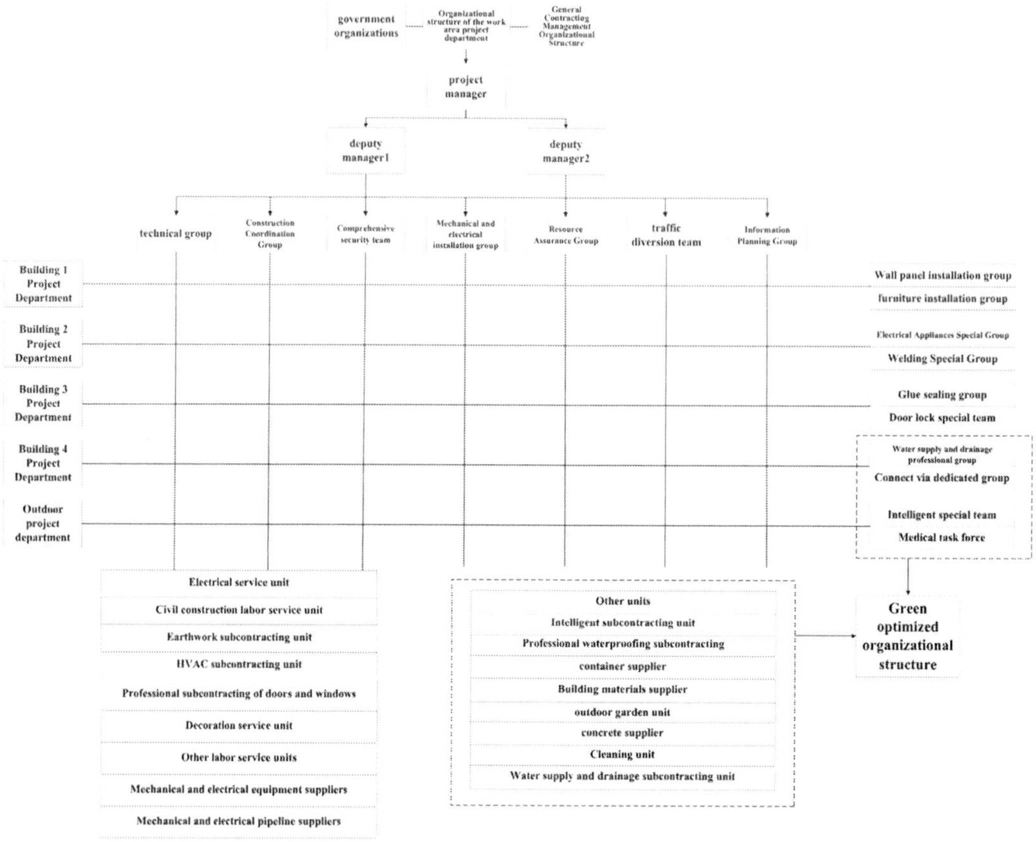

Fig. 4. Organizational chart of the Huoshenshan Hospital Project Department [28]

4 Conclusions

Currently, China has made significant research achievements in promoting realizing its "dual carbon" goals through building energy efficiency and the development of green building planning. This paper explores integrating sustainable building concepts into traditional construction processes from green construction technology and management systems perspectives. It examines the entire lifecycle of engineering projects during the construction phase. The following conclusions have been drawn:

1. This paper compares the similarities and differences between green and traditional construction technology, identifying factors contributing to environmental pollution arising from conventional construction, such as lifecycle operating costs, the lack of environmental protection concepts, and outdated technologies reliant on personnel experience. Solutions are also proposed. Additionally, BIM technology is analyzed as a case study, highlighting its advantages in information processing, construction planning, and problem-solving. This demonstrates that BIM technology better embodies sustainable development principles than traditional construction methods, showcasing the possibility of harmonious coexistence between modern engineering projects and the natural environment, thus representing a more rational and advanced level of construction technology.

2. The research identifies the similarities and differences between traditional and green construction management systems, reflecting the deficiencies in efficacy and concepts within traditional management frameworks. Through an empirical case study of the organizational structure during the construction of Huoshenshan Hospital, the remarkable advantages of the green construction management system are thoroughly demonstrated in optimizing personnel allocation, providing specific measures, and distributing construction tasks. This underscores the profound care that green construction demonstrates for the environment and the efficient use of resources.

This paper acknowledges that specific research achievements have been made in green construction technology and management. However, due to considerations regarding the feasibility of implementation costs and the challenges of market promotion, improvements in the study of green construction technologies for specific scenarios still need to be made. Future research is recommended to explore further integrating BIM technology with the Internet of Things, finite element analysis, artificial intelligence, and other technologies applied to green construction to enhance efficiency and standards. Currently, the green construction management system is discussed solely from an organizational management perspective; it is suggested that future studies incorporate quantitative evaluation mechanisms for a more comprehensive analysis.

References

1. Z S WANG, J S SUN, Investment Engineering College, Dongbei University of Finance and Economics.Current Status and Trends of Green Construction Research in China: Based on Analysis of Domestic Literature Knowledge Graph from 2005 to 2020. J. Journal of Shenyang University(Natural Science).33(03) 267-271+287 (2021) https://link.cnki.net/doi/10.16103/j.cnki.21-1583/n.2021.03.012
2. L TONG, Green Construction Studies. J. Real Estate World.(02) 127-129 (2023)
3. L Tang, M Liao , Y F Wang, et al, Research on the new era's construction technology and status quo of green and energy-saving buildings. J. Chinese building metal structure,36(03) 54-55 (2022)
4. S Paneru, et al, "Evaluating the Decarbonization Potential of Industrialized Construction: A Review of the Current State, Opportunities, and Challenges.". J. Journal of Construction Engineering and Management 150.9 03124005 (2024) https://doi.org/10.1061/JCEMD4.COENG-14609
5. C.Bahaa, and I.El-adaway, "A Contractual Approach to Resolving Disputes Arising from Failure to Achieve Green Certification." (*Construction Research Congress,*2024) https://doi.org/10.1061/9780784485286.003
6. Green Construction Guidelines. J. Construction Technology,(11) 1-5 (2007)
7. Chen, Talking about environmental protection measures in construction projects. J. Energy Conservation,38(04) 111-112 (2019) https://link.cnki.net/urlid/21.1115.TK.20190426.1526.088
8. Y Niu, Research on green building material saving and material resource utilization technology. J. Construction Materials & Decoration,(33) 48-49 (2019)
9. K QIN, Research on the application of water-saving technology in green housing construction. J. Building safety 36(03) 43-45 (2021)
10. X F LIU, Analysis of the Application of Modern Green Energy-Saving Technology in Building Construction. J. Foshan Ceramics,33(05) 55-57 (2023)
11. Y Wang, Research on the Application of Green Construction Techniques in Buildings. D. Southeastern University,004489 (2019)

https://link.cnki.net/doi/10.27014/d.cnki.gdnau.2019.004489
12. Y N Cheng, A Study on Environmental Cost Management in Construction Companies from the Perspective of Green Supply Chain. D. Dongbei University of Finance and Economics,000741 (2023) https://link.cnki.net/doi/10.27006/d.cnki.gdbcu.2023.000741
13. H J Zhu, B G Hwang, et al, "Applications of smart technologies in construction project management.". J. Journal of Construction Engineering and Management 148.4 04022010 (2022) https://doi.org/10.1061/(ASCE)CO.1943-7862.0002260
14. Cao, Yu, Syahrul Nizam Kamaruzzaman, and Nur Mardhiyah Aziz. "Green building construction: a systematic review of BIM utilization.". J. Buildings 12.8,1205 (2022) https://doi.org/10.3390/buildings12081205
15. "Buitenplaats Koningsweg / MVRDV" [Buitenplaats Koningsweg / MVRDV] 26.9. 2024. ArchDaily. Accesed 23.10. 2024. <https://www.archdaily.cn/cn/1021323/buitenplaats-koningsweg-mvrdv>
16. X Y LI, J LIU, W XU, et al, The application of BIM technology in constructing a sea-crossing bridge project. J. Information Technology for Civil Engineering and Architecture,15(06) 20-25. 2023.06.04 (2023) https://link.cnki.net/doi/10.16670/j.cnki.cn11-5823/tu.2023.06.04
17. N F Liu, et al. "BIM-based digital platform and risk management system for mountain tunnel construction." Scientific Reports 13.1 7585 (2023) https://doi.org/10.1038/s41598-023-34525-w
18. V. K. Bansal, "Integrated framework of BIM and GIS applications to support building lifecycle: A move toward nD modeling.". J. Journal of Architectural Engineering 27 405021009 (2021) https://doi.org/10.1061/（ASCE）AE.1943-5568.0000490
19. Z Y Chen, et al. "Review of Research and Application of BIM-GIS Integration in the Civil Engineering Industry.". J. ICCREM 2023 658-669 (2023) https://doi.org/10.1061/9780784485217.064
20. Y A ZHOU, The Microservice Architecture Design of the Cross-Sea Bridge Construction Supervision Platform Based on BIM+GIS. J/OL. Railway construction technology,1-5 (2024) https://link.cnki.net/urlid/11.3368.tu.20240914.1103.026
21. J K Han, Research on Green Construction Management of Building Engineering. D. Shijiazhuang Tiedao University,2019.000236 (2019) https://link.cnki.net/doi/10.27334/d.cnki.gstdy.2019.000236
22. T S Zhang, Research on Standardized Management of Green Construction in Construction Enterprises. D. Qingdao University of Technology,2021.000244 (2021) https://link.cnki.net/doi/10.27263/d.cnki.gqudc.2021.000244
23. W Z Fu, How should construction enterprises adjust their management, control mode, and organizational structure? J. Construction Enterprise Management,(06) 87-90 (2022)
24. Z M Zhu, Research on Green Construction Management System Based on Cooperative Theory . D. Shenyang Jianzhu University,2021. 000309 (2021) https://link.cnki.net/doi/10.27809/d.cnki.gsjgc.2021.000309
25. B Liu, Research on Project Manager Competency Model Design from the Perspective of Green Building Construction Management. D. Tianjin University of Technology, 000086 (2022） https://link.cnki.net/doi/10.27360/d.cnki.gtlgy.2022.000086
26. T Jin, Research on the application of green construction management [D]. Xi'an University of Architecture and Technology (2020)

https://link.cnki.net/doi/10.27393/d.cnki.gxazu.2020.001965
27. X R Sun, Evaluation and optimization of green construction organization design of Tibet highway based on combinatorial weighing cloud model . D. Tibet University,000460 (2023) https://link.cnki.net/doi/10.27735/d.cnki.gxzdx.2023.000460
28. W H DENG, FENG Huajun; ZHOU Jiegang; PU Yong; General Construction Co. CCTEB Group Co., Ltd.; Organization Structure and Working Mechanism Practice of Huoshenshan Hospital. J. Construction Technology,49(12) 11-14 (2020)

Research on construction management based on BIM-AR/VR technology

Songke Ma[1]

[1]School of Engineering, The University of Birmingham, Birmingham B15 2TT, UK

Abstract. With the digital transformation of the construction industry, construction management built by BIM and AR/VR technologies has gradually become a crucial mean to enhance efficiency, reduce costs and promote environmental protection. This paper discusses the application of BIM combined with AR/VR in construction management, and demonstrates its advantages in project visualisation, collaborative communication and safety management through literature analysis. The combined technology can significantly improve the transparency and precision of project management, help managers better control construction progress and costs, optimise resource allocation, and reduce rework. At the same time, the introduction of AR/VR improves worker training, decreases the risk of accidents and promotes the intelligent and green development of construction management. Although the technology still faces problems such as data compatibility, applying BIM and AR/VR in construction management is promising with the integration of new technologies such as artificial intelligence.

1 Introduction

With the rapid development of information technology, construction industry is undergoing a digital transformation. In traditional construction management, cost, efficiency and environmental issues are the main challenges faced: works often lead to budget overruns and wasted resources; many projects still rely on manual operations and paper documents, which result in slow information transfer and inefficient decision-making; and waste from construction poses a threat to the environment and the health of construction workers. The combination of BIM and AR/VR is driving construction management in the direction of intelligence. This integration can better support the visualisation, informatisation and intelligent management of projects, helping managers to more accurately grasp the progress, cost and quality of the project, while also optimising the allocation of resources, reducing errors and rework during the construction process [1].

Alizadehsalehi et al. created a model which can enhance abilities to learn engineering and provided recommendations and a future roadmap for AEC educators to integrate BIM with VR [2]; similarly, Lozano-Galant et al. developed a low-cost AR tool for visualising BIM models applied to civil engineering education, and the results showed that the AR tool integration can enhance students' learning experience and improve their understanding of complex projects [3]; in terms of research, Gan et al. proposed an integrated BIM and VR-based approach to optimise the aerodynamic design and wind comfort of modular buildings

* Corresponding author: SXM2035@student.bham.ac.uk

by visualising Computational Fluid Dynamics simulation (CFD) simulation results, thus helping designers to make wiser decisions at an early stage [4].

The aim of this paper is to discuss the application of technology combining BIM and AR/VR in construction management. Assila et al. proposed a model combining technology that can achieve the goal of cost saving [5]; Panya et al. strengthened collaboration and communication in the process of information transfer through the combination of technology to increase the efficiency of construction management [6]; Chen et al. demonstrated the combination of technology attenuates the fire effects, thus achieving the purpose of protecting the environment and the building [7]. By analysing these cases, this paper focuses primarily on the construction management phase and explores the potential advantages and development prospects of integrating BIM into AR/VR technologies in this scenario.

1.1 Overview of BIM and AR/VR

Building Information Modelling (BIM), also called n-D modelling or virtual prototyping, is a revolutionary development which is constantly changing the architecture, engineering and construction (AEC) industry [8]. BIM advances the construction process and improves the integrated efficiency of the construction process by creating three-dimensional digital models that integrate the information in a construction project and provide a systematic management platform for the construction phase.

Augmented Reality (AR) is defined as an interactive experience in a real-world environment, linked to the real world via computer-generated data information; Virtual Reality (VR) is different in that it takes place in an artificial data environment, where the users are a part of this world in an immersive or non-immersive way [9].

BIM improves the accuracy and efficiency of the construction management by creating digital building models. However, some technical bottlenecks have gradually emerged in the long-term use of this technology: the lack of rendering accuracy of BIM models leads to the reuse of project resources, thus increasing costs; construction conflicts due to data incompatibility and model information inconsistency; and a lack of competence in environmental protection, where the lack of fluid data leads to wasted materials and casualties, thus increasing the burden on the environment. The combination with AR/VR effectively improves these aspects.

According to Seyman et al., there is an increasing preference for utilising VR/AR in BIM-based construction management [10]. Integrating BIM into emerging technologies like AR/VR is currently considered an effective solution to enhance construction management [11]. This technology not only plays a crucial role in communication efficiency and safety training, but also improves collaborative construction site management. AR/VR integrates the BIM model with the real environment, which improves the standard of visualisation of the building model and helps construction personnel to understand and collaborate more intuitively.

1.2 Comparative analysis of BIM-AR/VR combination

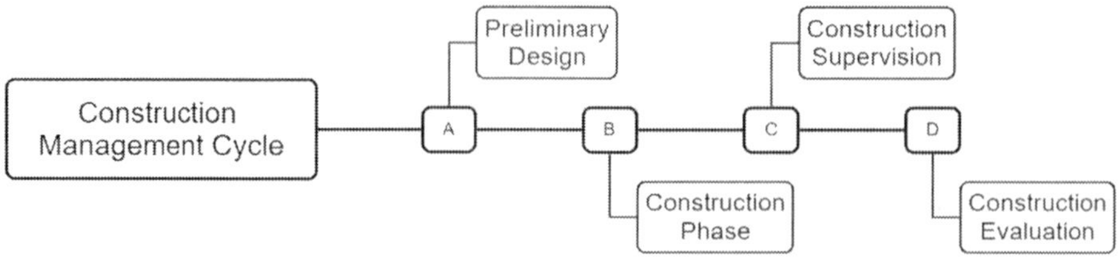

Figure 1. Construction management cycle (Picture credit: Original).

The integration of BIM and AR/VR not only improves construction efficiency, but also in an important position in project management, site supervision, and construction evaluation. The following is a comparative analysis of the changes before and after the combination of this technology during the construction management cycle in Figure 1.

In the design phase, BIM is able to generate 3D models that help designers and engineers obtain a comprehensive understanding of the project. However, with the gradual complexity of the building model, BIM gradually becomes difficult to fully demonstrate the construction situation in the real environment. With the introduction of AR/VR, the construction team can conduct simulation demonstrations in a virtual environment and anticipate construction problems in advance. This interactivity makes the design intent clearer and reduces rework due to misinterpretation.

On the construction site, binding BIM and AR/VR significantly improves the efficiency of teamwork. Project participants in traditional construction usually communicate with each other through paper drawings or 2D models, which can easily transmit error messages; AR can further enhance the visualisation of BIM models so that construction personnel can view construction progress and design models through the device and revise deviations in time; VR can also be used to train employees to familiarise themselves with the construction process and precautions in virtual environments, which can reduce the on-site incident Incidence rate.

In terms of construction supervision, BIM can track construction progress and resource usage through real-time data updates. Combined with AR/VR, project managers are able to visualise the status of the project through an immersive virtual experience, analyse the challenges that may be encountered during the construction process and make decisions accordingly. This data and real-time feedback-based decision-making enables managers to react more quickly, optimise resource allocation and ensure that projects are completed on time.

The combination of BIM and AR/VR also enriches the customer experience. By improving every aspect of the learning process, such as Ghanem's comparative analysis of AR/VR headsets, clients learning the technology are offered a more comprehensive range of options [12]. Technical details and construction processes that are difficult to understand can be displayed in AR/VR, allowing for an immersive 'tour' of a project that has yet to be built. This experience not only improves the client's confidence about the project, but also establishes the foundation for subsequent project negotiations and promotion.

In summary, this combination not only improves the efficiency and accuracy of buildings and structures, but also brings significant impacts in a variety of aspects such as on-site communication, project supervision and construction feedback and experience.

1.3 BIM-AR/VR technology applications

There are technical problems such as incompatible data formats and backward real-time rendering performance in the process of combining BIM and AR/VR, so it is very crucial to modify the quality of model data combining BIM and AR/VR devices.

Aung et al. proposed a mathematical algorithm on the basis of triangle mesh reconstruction [13]. This method optimises the overall model by reducing the number of vertices and triangles in the geometric model thereby reducing the transmission time and effort. The researchers conducted two tests on this algorithm as shown in the analysis of data in Tables 1, 2. The architectural model using this algorism shows significant improvement and optimisation in terms of vertices, triangles and file memory by 56.7%, 67.9% and 73.4% respectively.

Table 1. Steel bridge model before and after optimisation (Aung et al., 2022).

Model	Type	Initial	Optimised	Improvement (%)
Steel bridge	Number of vertices	590942	255625	↓56.7
	Number of triangles	335564	107573	↓67.9
	File size (KB)	50535	13451	↓73.4

Table 2. Engineering building model before and after optimisation (Aung et al., 2022).

Model	Type	Initial	Optimised	Improvement (%)
Engineering building	Number of vertices	4183198	2235879	↓46.6
	Number of triangles	3231090	1336165	↓58.6
	File size (KB)	364761	124173	↓66.0

Through the application of this mathematical algorithm, the time and quality of combining techniques for on-site modelling and construction is significantly improved. A significant reduction in modelling time and improvement in the quality of the model leads to improved design and construction efficiency.

Keyu et al. also proposed a data transmission mechanism aimed at simplification using different polygon reduction methods [14]. It was validated that the mechanism can notably decrease the number of triangles in the model thus increasing the frame rate of the model transfer while improving the overall shape consistency.

2 Advantages of BIM-AR/VR in construction management

As AR/VR continues to evolve, their integration with BIM brings new possibilities for construction management. Integrating BIM with AR/VR demonstrates significant advantages in these three areas shown in Figure 2, which not only improves the granularity and transparency of project management but promotes the sustainable development of the

https://doi.org/10.1051/e3sconf/202560604007

construction industry. This technological innovation lays a solid foundation for the future of construction management.

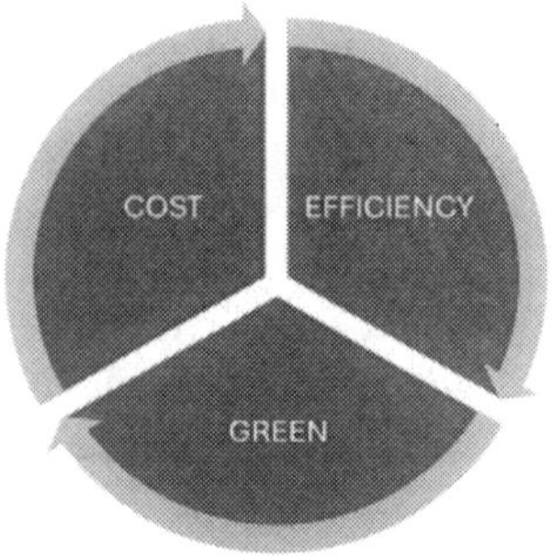

Figure 2. Three technical advantages (Picture credit: Original).

2.1 Cost

During the construction management phase, BIM provides comprehensive building information, including bill of materials, component specifications and construction steps. By creating accurate 3D models, construction teams can better understand the scale and complexity of a project; BIM also automatically generates cost budgets based on the model, avoiding cost overruns due to inaccurate budgets in later stages of construction. AR/VR can superimpose the BIM model on the real environment, allowing the construction team to identify potential problems in advance. Such foresight can significantly reduce cost losses caused by construction errors and material wastage.

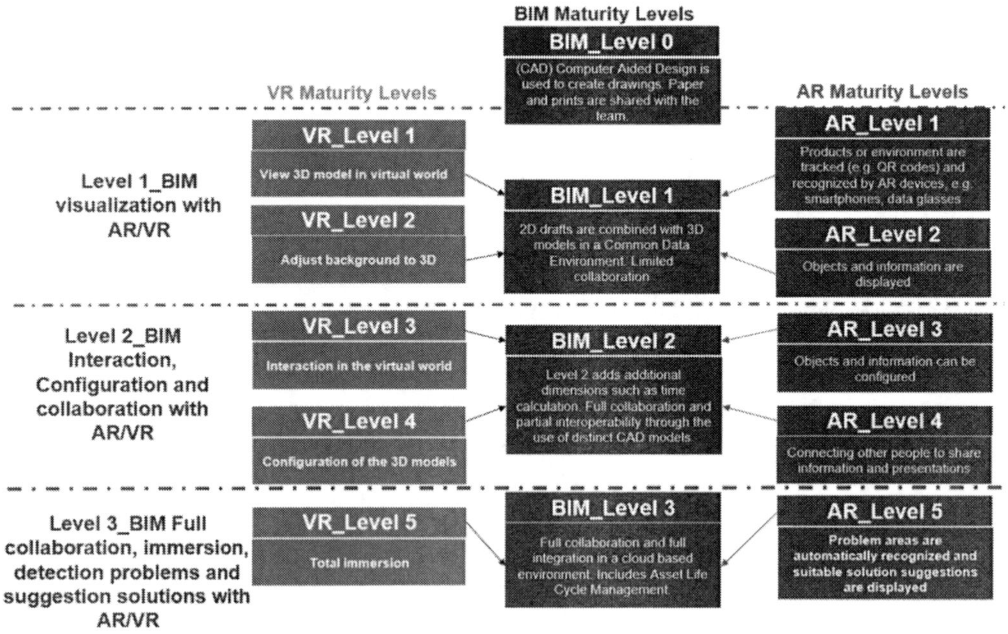

Figure 3. The recommended AR/VR maturity model for BIM [5].

Combination technology can also provide real-time project monitoring and data feedback mechanism, managers can quickly understand the construction progress and cost consumption, Assila et al. proposed a model of the combination of technology as shown in Figure 3, aimed at assessing the maturity level of the existing progress, reducing the misrepresentation of progress and cutting corners [5]. This transparency allows the project manager to make timely adjustments and optimise resource allocation, thus reducing the cost of capital employed and the cost of delays; the technique also enhances the identification of hazards, and according to Afzal et al. on a series of case studies, the VR visualisation of 4-D BIM can be an effective solution to the challenge of workers working together in different language environments [15]; and Ankit et al. used this technology to create safety training scenarios to enhance the perception of machinery for people operating tower crane machinery in the field [16]; combined with the fact that the technology also plays a vital role in real-time interaction and communication during the training of the workers, Bao et al. researchers suggested a cross-platform framework with which the workers can access VR safety training courses to improve construction safety knowledge [17].

2.2 Efficiency

The coordination function of BIM can effectively motivate the industry to work together. This technique has shown significant improvements in the modelling design phase and field modification phase [18]. However, some communication problems between design, construction, operation and maintenance still need to be overcome in the management aspect [19].

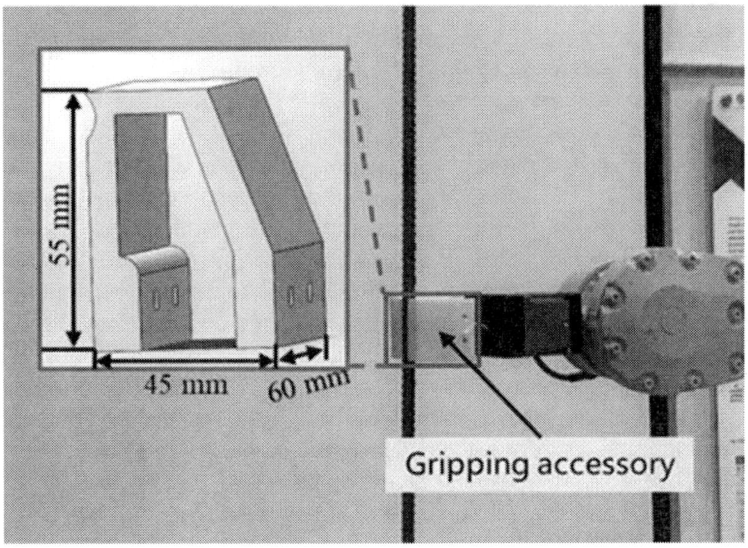

Figure 4. A designed gripping accessory [21].

Combined with AR/VR technology, it can enhance the interactive experience during the construction phase, enabling employees in different positions such as builders and technicians to obtain more detailed information about the BIM model when interfacing and confirming the information, and reducing conflicts of interest due to out-of-sync or incorrect information during the project. Panya et al. have confirmed the feasibility of this technology by combining the technology to improve collaboration and communication among construction project stakeholders by redesigning, maintaining the flow of information, and delivering the design in an effective way [6]. improved collaboration and communication between construction project stakeholders in rework, confirming the feasibility of the technique. However, the technology is still in the developmental stage, so there are many potential problems, Sidani et al. concluded in their study that: combining technology is far from the expected state, such as connection and positioning problems, lack of non-geometric information and practical obstacles of applying AR technology in construction sites; for the connection and positioning problems [20]. Du et al. invented a clamping attachment as shown in Figure 4 that the technology is one of the research advances for this problem through the extension of BIM real-time technology to improve the accuracy of real-time operation of existing robotic arms through new algorithms [21]. This technology has a clear potential for health and safety in on-site construction, especially with the progression of assembly technology, the quality of materials has been uniformly guaranteed, so the reliability of on-site installation has become an assessment criterion for the building, and the efficiency of on-site management will be greatly improved if this technology continues to be developed instead of the errors of manual installation. Amin et al. also emphasised the importance of combining the technology research with a view to future The technology can be better automated with the assistance of AI to improve efficiency [22].

2.3 Environmental protection

BIM can be modelled to provide detailed building information and resource requirements for projects to achieve the modern goals of green building and sustainability [23]. AR/VR can help BIM to update project information in real time and analyse which construction materials to use without compromising structural stability with less negative impact on the environment.

Incorporating technology can improve the safety of construction sites and thus indirectly protect the environment. Workers at construction sites are usually exposed to various dangers, such as collapses, equipment failures, and fires, which not only jeopardise the safety of the workers, but also cause damage to the surrounding environment. Incorporating technology can reduce the occurrence of accidents by obtaining real-time information about the hazards at the construction site. For example, Chen et al. proposed the impact of this technology on fire prevention, which efficiently extinguishing fires, also implies a reduction in environmental pollution, thus achieving the goal of sustainable development [7]; combining technology can also promote the design and implementation of green buildings, by comprehensively simulating the life cycle of a construction, and meticulously improving the design concepts and programs; designers could use this to observe the buildings' structure in details, while engineers and workers apply the combined technology to immerse themselves in the building's performance, understand the building's energy consumption and emissions, and select energy-efficient construction methods. The combined technology not only reduces environmental pollution during the operation of the building, but also provides data support for environmental practices during the construction management phase.

3 Conclusion

In an era of mainstream AR/VR development, combination of BIM and AR/VR has brought unprecedented changes to construction management in the construction industry. By offering real-time feedback on data, this technology enhances the transparency and visualisation of all aspects of the process, enabling project managers to quickly identify potential risks and optimise the decision-making process. Especially in terms of safety management, AR/VR technology greatly enhances employees' ability to perceive dangerous environments, and at the same time reduces the likelihood of accidents by providing safety training through realistic virtual scenarios; in addition, the combination of technology greatly enhances the interactive experience during the construction phase, enabling construction personnel to simulate operations in virtual environments, anticipate and rectify potential problems in the construction process beforehand, decline the rework rate. In addition, the combined technology greatly enhances the interactive experience in construction phase, enabling the construction personnel to foresee and correct potential problems in the construction in advance through the virtual environment simulation operation, reducing the rework rate and improving the accuracy of the mechanical work so as to improve the efficiency; what's more, the combined technology also plays a pivotal role in the design and implementation of the green building, and promotes the achievement of the goals of energy saving, emission reduction and environmental protection by simulating and analysing the energy consumption pattern of the building, and better controlling the use and emission of the resources in the construction process.

Current research focuses on the case study of BIM and AR/VR, while in-depth research on the related algorithms is still limited, especially in improving the accuracy of data processing and optimising construction simulation decision-making there is a great potential. With the advance of Artificial Intelligence (AI), the combination of BIM and AR/VR will gain new breakthroughs. AI can analyse construction data, optimise energy consumption and material usage through machine learning algorithms, thus promoting the realisation of intelligent construction management, further improving efficiency, reducing costs and facilitating the sustainable development of green buildings.

References

1 E. Yigitbas, A. Nowosad, and G. Engels, Supporting construction and architectural visualization through bim and ar/vr: a systematic literature review. in Human-Computer Interaction – INTERACT 2023, J. Abdelnour Nocera, M. Kristín Lárusdóttir, H. Petrie, A. Piccinno, and M. Winckler, Eds., Cham: Springer Nature Switzerland, pp. 145–166 (2023)
https://doi.org/10.1007/978-3-031-42283-6_8

2 S. Alizadehsalehi, A. Hadavi, and J. C. Huang, Assessment of AEC students' performance using BIM-into-VR. Applied Sciences, **vol. 11**, no. 7 (2021)
https://doi.org/10.3390/app11073225

3 F. Lozano-Galant, R. Porras, B. Mobaraki, F. Calderón, J. Gonzalez-Arteaga, and J. A. Lozano-Galant, Enhancing civil engineering education through affordable AR tools for visualizing BIM models. Journal of Civil Engineering Education, **vol. 150**, no. 3, p. 05024003 (2024)
https://doi.org/10.1061/JCEECD.EIENG-2007

4 V. J. L. Gan, T. Liu, and K. Li, Integrated BIM and VR for interactive aerodynamic design and wind comfort analysis of modular buildings. Buildings, **vol. 12**, no. 3 (2022)
https://doi.org/10.3390/buildings12030333

5 A. Assila, D. Beladjine, and M. Messaadia, Towards AR/VR maturity model adapted to the Building Information Modeling. Product Lifecycle Management Enabling Smart X, F. Nyffenegger, J. Ríos, L. Rivest, and A. Bouras, Eds., Cham: Springer International Publishing, pp. 753–765 (2020)
https://doi.org/10.1007/978-3-030-62807-9_59

6 D. S. Panya, T. Kim, and S. Choo, An interactive design change methodology using a BIM-based Virtual Reality and Augmented Reality. Journal of Building Engineering, **vol. 68**, p. 106030 (2023)
https://doi.org/10.1016/j.jobe.2023.106030

7 H. Chen, L. Hou, G. (Kevin) Zhang, and S. Moon, Development of BIM, IoT and AR/VR technologies for fire safety and upskilling. Autom Constr, **vol. 125**, p. 103631 (2021)
https://doi.org/10.1016/j.autcon.2021.103631

8 S. Azhar, M. Khalfan, and T. Maqsood, Building Information Modeling (BIM): now and beyond. The Australasian Journal of Construction Economics and Building, **vol. 12**, no. 4, p. [15]-28 (2012)
https://search.informit.org/doi/10.3316/informit.013120167780649

9 H. Ardiny and E. Khanmirza, The role of AR and VR technologies in education developments: opportunities and challenges. RSI International Conference on Robotics and Mechatronics (IcRoM) pp. 482–487 (2018)
https://ieeexplore.ieee.org/document/8657615

10 T. Seyman Guray and B. Kismet, VR and AR in construction management research: bibliometric and descriptive analyses. Smart and Sustainable Built Environment, **vol. 12**, no. 3, pp. 635–659 (2023)
https://doi.org/10.1108/SASBE-01-2022-0015

11 Z. Monla, A. Assila, D. Beladjine, and M. Zghal, Maturity evaluation methods for BIM-based AR/VR in construction industry: a literature review," IEEE Access, **vol. 11**, pp. 101134–101154 (2023)
https://ieeexplore.ieee.org/document/10138393

12 S. Y. Ghanem, Implementing virtual reality-building information modeling in the construction management curriculum. Journal of Information Technology in Construction, **vol. 27** (2022)
https://www.itcon.org/paper/2022/3

13 P. P. W. Aung, W. Choi, A. S. Kulinan, G. Cha, and S. Park, Three-dimensional engine-based geometric model optimization algorithm for BIM visualization with Augmented Reality. Sensors, **vol. 22**, no. 19 (2022)
https://doi.org/10.3390/s22197622

14 C. Keyu, C. Weiwei, C. J. C. P, and W. Qian, Developing efficient mechanisms for BIM-to-AR/VR data transfer. Journal of Computing in Civil Engineering, **vol. 34**, no. 5, p. 04020037 (2020)
https://doi.org/10.1061/(ASCE)CP.1943-5487.0000914

15 M. Afzal and M. T. Shafiq, Evaluating 4D‐BIM and VR for effective safety communication and training: A case study of multilingual construction job‐site crew. Buildings, **vol. 11**, no. 8 (2021)
https://doi.org/10.3390/buildings11080319

16 S. Ankit, A. Mehrdad, D. Tim, P. Arnaud, and L. Heng, Safety in off-site construction: simulation of crane-lifting operations using VR and BIM. Journal of Architectural Engineering, **vol. 29**, no. 1, p. 04022035 (2023)
https://doi.org/10.1061/(ASCE)AE.1943-5568.0000570

17 L. Bao, S. V.-T. Tran, T. L. Nguyen, H. C. Pham, D. Lee, and C. Park, Cross-platform virtual reality for real-time construction safety training using immersive web and industry foundation classes. Autom Constr, **vol. 143**, p. 104565 (2022)
https://doi.org/10.1016/j.autcon.2022.104565

18 Y. Liu, S. van Nederveen, and M. Hertogh, Understanding effects of BIM on collaborative design and construction: an empirical study in China. International

Journal of Project Management, **vol. 35**, no. 4, pp. 686–698 (2017) https://doi.org/10.1016/j.ijproman.2016.06.007

19 S. Safikhani, S. Keller, G. Schweiger, and J. Pirker, Immersive virtual reality for extending the potential of building information modeling in architecture, engineering, and construction sector: systematic review. Int J Digit Earth, **vol. 15**, no. 1, pp. 503–526 (2022)
https://doi.org/10.1080/17538947.2022.2038291

20 A. Sidani et al., Recent tools and techniques of BIM-Based Augmented Reality: a systematic review. Journal of Building Engineering, **vol. 42**, p. 102500 (2021) https://doi.org/10.1016/j.jobe.2021.102500

21 S. Du et al., A BIM-enabled robot control system for automated integration between rebar reinforcement and 3D concrete printing. Virtual Phys Prototyp, **vol. 19**, no. 1 (2024)
https://doi.org/10.1080/17452759.2024.2332423

22 K. Amin, G. Mills, and D. Wilson, Key functions in BIM-based AR platforms. Autom Constr, **vol. 150**, p. 104816 (2023)
https://doi.org/10.1016/j.autcon.2023.104816

23 A. Patel, A. Shelake, and A. Yadhav, Sustainable construction by using novel frameworks using BIM, LEED, and Lean methods. Mater Today Proc (2023) https://doi.org/10.1016/j.matpr.2023.02.238

Application of ultra-high performance concrete in the marine environment

Jingtong Zhang[1] *

[1]Department of Civil and Environmental Engineering, Hong Kong Polytechnic University, Hong Kong Special Administrative Region, China

Abstract. The development and utilization of marine resources are becoming the focus of all countries. However, the special characteristics of the marine environment pose severe durability challenges to traditional concrete materials. Ultra-high performance concrete (UHPC) has gradually been widely used in marine engineering due to its excellent durability and corrosion resistance. This paper discusses the basic concept of UHPC and its mechanism of action, evaluates its current application in marine engineering, and analyzes the challenges of its application in the marine environment. On this basis, this paper looks forward to the future development direction of UHPC in marine engineering, pointing out that with the progress of material science and construction technology, UHPC will play a greater role in the research and development of new types of structures and the improvement of construction technology, and promote its wide application in marine engineering.

1 Introduction

With the growth of the world economy and the advancement of science and technology nowadays, the development and utilization of marine resources have become the focus of attention of all countries. Port terminals and bridges across the sea and submarine tunnels in coastal cities have become an important part of the development of infrastructure. However, due to the special characteristics of the marine environment with high chloride ion concentration, alternating humidity and heat, dry-wet cycle, strong corrosivity, and wind and wave impact, which makes traditional concrete easily eroded and impact in the marine environment, the marine engineering with concrete as the main material is facing serious durability problems, for example, the structure of the piers of cross-sea bridges in the role of the harsh marine environment, the efficiency of the damage to the concrete is significantly accelerated, which will affect the bridge The service life of the bridge structure will be affected.

Ultra-High Performance Concrete (UHPC) is widely used in marine engineering for its durability and corrosion resistance as well as excellent mechanical properties compared to traditional cement-based materials. UHPC has a dense internal structure and ultra-high

* Corresponding author: jingtong.zhang@connect.polyu.hk

 https://doi.org/10.1051/e3sconf/202560604008

mechanical properties, which effectively prevents structural collapse caused by external erosion and corrosion of reinforcing steel [1].

The purpose of this paper is to discuss the present state and anticipated future developments in the use of ultra-high performance concrete in marine environments. Firstly, this paper will describe the basic concept of UHPC, and analyze of its mechanism of action. This paper will also assess the current application of UHPC in marine engineering and discuss in depth the challenges faced by its application in the marine environment. On this basis, the paper will look forward to the development direction of UHPC in marine engineering. With the current advances in material science and construction technology, the innovation of UHPC materials in the research and development of new structures and the improvement of construction processes will further promote its wide application in marine engineering.

2 Basic concepts and characteristics of UHPC

Ultra-high performance concrete, whose basic components include high-strength steel fibers, superplasticizer, quartz sand, cement, and silica fume. It is a novel kind of fiber-reinforced cementitious material with high strength, high durability, high impact toughness, and has compressive strength greater than 120 Megapascals (MPa) and sustained postcracking tensile strength greater than 5.2 MPa [2]. It possesses excellent mechanical properties and superior durability. The ultra-high performance of UHPC is mainly obtained by improving the structure and properties of cement stone and the contact surface between cement stone and aggregate. Compared with traditional concrete materials, UHPC adopts the optimal granular grading of the constituent materials, the water-cement ratio is typically less than 0.25, and a high proportion of steel fiber materials are added, which can adapt to the freezing and thawing cycle, the marine environment, and the weak acid attack, and other harsh environments [2].

One of the properties of UHPC is its extremely low porosity, due to its optimized material mix and dense microstructure. The introduction of silica fume not only improves the densification of the material but also enhances the strength of the matrix through the volcanic ash reaction. In addition, the use of superplasticizers significantly improves the fluidity of the concrete, allowing UHPC to achieve a self-compacting effect during construction.

Secondly, its extremely low water-cement ratio and dense microstructure significantly reduce the permeability of moisture and harmful ions, making it excellent in resistance to chloride attack, sulfate attack, and carbonation. This property is particularly suitable for engineering applications in harsh conditions such as marine environments and chemical plants.

The excellent durability and crack resistance of UHPC is due to the incorporation of steel fibers. The three-dimensional network structure formed by the steel fibers in the concrete matrix effectively prevents crack propagation and enhances its ability to perform under dynamic loading. This gives UHPC a significant advantage in projects such as bridges, tunnels, and protective structures where high durability and safety are required. This property allows UHPC to excel when subjected to dynamic loading and extreme environmental conditions [3].

3 Mechanisms of action of UHPC in the marine environment

3.1 Specificity of the marine environment

In marine engineering, concrete materials face severe environmental challenges. The form of concrete erosion by the marine environment varies according to the location of the exposure zone, which can be divided into corrosion zones with different characteristics, such as marine atmospheric zones, marine splash zones, marine tidal zones, and marine underwater zones. Chemically, the marine splash zone and tidal zone are in the inorganic salt chemical corrosion environment, which is often the most serious area of erosion deterioration of concrete structures [4]. Firstly, chloride ions, which are abundant in seawater, can penetrate concrete and lead to corrosion of reinforcing steel; therefore, concrete materials must have very low permeability to resist chloride ion attack. In addition, the sulfates in seawater can react chemically with the cementitious components of concrete, causing the material to swell and crack, so concrete is required to have excellent sulfate resistance.

Physically, structures in marine engineering are often subjected to the impact of waves, tides, and floating objects, thus requiring high durability properties of concrete to extend the service life of the structure. Concrete also needs to have good crack resistance due to the dynamic loading and temperature variations in the marine environment. In cold marine environments at high latitudes, concrete structures may experience frequent freeze-thaw cycles, which are highly susceptible to material fatigue and damage. Concrete materials are therefore required to have high freeze-thaw resistance [5].

In summary, the marine environment places more stringent requirements on concrete, emphasizing the durability and erosion resistance of the material to ensure the long-term stability and safety of the structure.

3.2 Mechanism of action of UHPC in marine engineering

In the marine environment, the mechanism of action of ultra-high performance concrete is mainly reflected in its unique microstructure and material properties. UHPC forms a highly dense microstructure through optimized particle gradation and a low water-cement ratio. This structure significantly reduces porosity, which prevents the penetration of chloride ions and other hazardous substances and enhances the material's resistance to erosion [5]. The compactness and the UHPC matrix's strength enhance its chemical stability, which makes it perform well in sulfate and other chemically aggressive environments.

The steel fibers added to UHPC form a network structure in the concrete that significantly improves the toughness and crack resistance of the material. This enhancement enables UHPC to effectively prevent cracks from forming and expanding when subjected to complex physical loads, maintaining structural integrity. The low porosity and high densification of UHPC reduce the accumulation of moisture in the material, thereby reducing the damage caused by freeze-thaw cycles [6].

In summary, the mechanism of UHPC in the marine environment mainly relies on its dense microstructure, chemical stability, and enhanced mechanical properties, which together ensure its excellent durability and stability in harsh marine environments.

 https://doi.org/10.1051/e3sconf/202560604008

4 Current status of UHPC applications in the marine environment

In recent years, the application of ultra-high performance concrete in marine engineering has gradually increased, and domestic and international research and engineering practice have shown that UHPC has significant advantages in marine environmental engineering.

4.1 Status of domestic applications

The specific application of UHPC in cross-sea bridges mainly focuses on key parts such as bridge deck slabs, piers and connection nodes. China's Hong Kong-Zhuhai-Macao Bridge (HZMB) project is located in the Lingdingyang Sea, with high temperature, high humidity, high salinity of seawater, etc., impacted by the tides, dry-wet cycle, freeze-thaw cycle, salt spray, seawater, sea breezes and many other factors, the concrete structure of the corrosion of the harsh environment, the durability of the problem is very prominent [7]. Hong Kong-Zhuhai-Macao Bridge immersed tube tunnel along the basal soft soil thickness of about 30 meters, the bottom of the tube section of the longitudinal geological conditions are complex and uneven; top of the tube back to the silt and shipwrecks load, section joints force and waterproofing risk are high; The project is located in the outer sea, with complicated meteorological and hydrological circumstances, the project area of the daily average of more than 4,000 ships,intricate trajectory line, the safety of marine construction risk; giant immersed tube manufacturing crack-resistant impermeability risk is high (As shown in Figure 1); designed for 120 years of service life, the structural resistance of the structure is very high [8]. The service life is 120 years, which requires high structural durability. Instead, UHPC is widely used in bridge piers, bridge decks and key connecting parts, effectively resisting seawater and sea breeze environmental erosion and dry and wet cycles, effectively responding to structural durability problems (As shown in Figure 2), extending the service life of the bridge, and effectively lowering maintenance expenses [8][9].

Fig. 1. HZMB Segmental Bridge Box Girder Construction [8].

Fig. 2. HZMB Immersed Pipe Launching [9].

The application aspect of UHPC in offshore wind facilities improves the durability performance and wind load resistance of the structure. Unlike terrestrial installations, the construction of offshore wind farms is more susceptible to the effects of wind and waves brought about by climate change. Offshore wind foundations are subject to complex environmental loads, including wave impacts, tidal changes and wind loads, and UHPC's high strength and toughness enable it to effectively resist these loads. The Southern Power Grid Guishan Offshore Wind Farm is responsible for delivering green and clean wind energy to the Zhuhai power grid. It is the first offshore wind turbine in China to use an interpolated four-pile conduit frame foundation type. Due to the high number of typhoons in the sea area where the Southern Power Grid Guishan Wind Farm is located, high surge waves, ultra-high temperatures in the summer, and more cloudy and rainy weather, it puts forward ultra-high performance and technical requirements for the selection of concrete materials, grouting construction and management [10]. The use of UHPC as the foundation material for offshore wind power facilities significantly extends the service life of the wind power foundation (As shown in Figure 3).

Fig. 3. Southern Battery Guishan Offshore Wind Farm [11].

4.2 Status of foreign applications

The application of UHPC in harbor terminals has also achieved remarkable results. Seaport terminals need to withstand the erosion of seawater and mechanical loads. The Port of Hamburg is situated at the mouth of the Elbe River in Germany, which is significantly affected by tides and has frequent water level changes [10]. The harbor needs to withstand strong wave impacts from the North Sea and has high requirements for structural stability. Extreme weather events such as storms and sea level rise pose a threat to harbor facilities. And the salt in seawater is corrosive to concrete and steel structures, affecting structural durability (As shown in Figure 4). The Port of Hamburg utilized UHPC for the terminal upgrade. The low permeability and high-density internal structure of UHPC provide excellent resistance to corrosion and abrasion, as well as resistance to ocean current impact loads, freeze-thaw cycles, and extreme weather conditions faced by port terminals at high latitudes.

Fig. 4. Ship Transit and Land Transportation in the Port of Hamburg [12].

UHPC excels in breakwater applications. Breakwaters are constructed to protect port facilities and surrounding communities from marine hazards. Breakwaters need to withstand strong wave impacts, and UHPC's high strength and toughness enable it to effectively resist wave erosion. Japan's coastal areas are frequently affected by typhoons and tsunamis, and the strength and durability of breakwaters are required to be extremely high. After the 311 tsunami disaster in Japan, the Tohoku region was severely damaged and a breakwater rehabilitation project was carried out in Kamaishi Harbor, where UHPC was used in key structural parts of the breakwater, such as wave dissipation blocks and revetment panels [13]. Its high strength and mechanical properties ensure structural stability under extreme conditions, and UHPC is able to withstand strong wave impacts and extreme weather conditions. Its application ensures the reliability of breakwaters in all kinds of harsh environments (As shown in Figure 5).

Fig. 5. Condition of the top of the south dike at Kamaishi Harbor [13].

Offshore platforms are subject to complex environmental loads and chemical corrosion, and UHPC's high durability, low permeability, and excellent mechanical properties make it an ideal material choice. Norway's Troll A platform is one of the world's largest natural gas production platforms, located in the Norwegian North Sea, facing strong wave impacts, and complex environmental loads, and need to resist seawater corrosion and high-latitude offshore projects face relatively frequent freeze-thaw cycles and extreme weather conditions [14]. The use of UHPC in key structural joints to enhance the durability and stability of the platform has significantly improved its safety and service life (As shown in Figure 6).

Fig. 6. Troll A offshore gas production platform in Norway [15].

The application of UHPC in marine engineering is promising, and its outstanding mechanical properties and durability make it an ideal material choice for marine engineering. Future research should further optimize the mix ratio design and construction process of UHPC to give full play to its potential in ocean engineering.

5 Challenges to the use of UHPC in marine engineering applications

The primary challenge facing the application of UHPC in marine engineering is cost-effective. The high cost of raw materials for UHPC, especially the use of high-strength cement, silica fume and steel fibers, makes its manufacturing cost significantly higher than the conventional concrete materials. In addition, the construction process of ultra-high performance concrete is relatively complex and requires high levels of construction equipment and technicians, which may lead to lower construction efficiency and thus increase overall project costs [16].

The production and application of UHPC also involves issues of environmental impact. The production process of high-strength cement generates large amounts of carbon dioxide emissions, which can place additional burdens on the environment. Therefore, when promoting the application of UHPC, it is necessary to consider how to reduce its carbon footprint and develop more environmentally friendly production processes and material substitutions to minimize the adverse effects on the environment [17].

For now, the application of UHPC in marine engineering also faces the problem of a lack of management norms and standards. Since UHPC is a relatively new material, many countries have not yet established perfect technical standards and construction norms. This reveals that in practical application, the design and construction may lack uniform guidance, affecting the quality and safety of the project. Therefore, the development and improvement of UHPC-related management norms and standards is the key to promoting its wide application in marine engineering [18].

It can be seen that although UHPC has significant application potential in offshore engineering, its promotion and application still need to overcome various challenges such as cost, productivity, environmental impact and, management standardization.

6 Conclusion

As material science and construction technologies continue to advance and grow, the application of UHPC in the marine environment is promising. First of all, UHPC shows excellent mechanical properties, durability and impermeability in the marine environment, which makes it an ideal cementitious composite material suitable for harsh marine conditions. The application of UHPC can significantly enhance the safety and durability of offshore constructions., and its excellent performance can help extend the service life of the structure and lower the cost of long-term maintenance. UHPC is also particularly suitable for the protection of UHPC is also particularly well suited to protect the functionality of critical parts of offshore structures, such as breakwaters, offshore platforms and port facilities, which require high material performance due to their functional importance, high utilization and cost.

Through in-depth research and application promotion of UHPC in the marine environment, it can not only improve the safety and durability of marine engineering but also provide support for sustainable development and green construction. Since this paper is mainly based on a literature review and lacks specific data and experimental verification, further research and exploration are still needed in practical application. How to stimulate the potential of this material to make it more suitable for different complex marine environments requires deeper and more specific research.

References

1. R. Ullah, Y. Qiang, J. Ahmad, N.-I. Vatin, M.-A. El-Shorbagy, Ultra-High-Performance Concrete (UHPC): A State-of-the-Art Review. Materials 2022, 15, 4131. https://doi.org/10.3390/ma15124131
2. J.-R. Lliso-Ferrando, J.-M.Gandía-Romero, J. Soto, M. Valcuende, OC, HPC, UHPC and UHPFRC Corrosion Performance in the Marine Environment. Buildings 2023, 13, 2439. https://doi.org/10.3390/buildings13102439
3. M. Xie, H. Li, B.-T. Yuan, M. Li, W.-L. Hou, Preparation Technology and Experimental Study of Fast-Hardening UHPC Repair Material. Buildings 2024, 14, 2124. https://doi.org/10.3390/buildings14072124
4. Z.-B.Haber, B.-A. Graybeal, and B. Nakashoji. "Ultimate behavior of deck-to-girder composite connection details using UHPC." Journal of Bridge Engineering 25.7 (2020): 04020038. https://doi.org/10.1061/(ASCE)BE.1943-5592.0001574
5. A. Mugahed, S. Huang, O. Ali M, et al. Recent trends in ultra-high performance concrete (UHPC): Current status, challenges, and future prospects[J]. Construction and Building Materials, 2022, 352: 129029. https://doi.org/10.1016/j.conbuildmat.2022.129029
6. L. Dong, Y. Yang, Z. Liu, T. Yang, C. Xue, R. Shao, & C. Wu, (2024). Effect of chloride ion migration behaviour on the microstructure and mechanical properties of ultra-high performance concrete: A review. Journal of Building Engineering, 82, 108233- https://doi.org/10.1016/j.jobe.2023.108233
7. L.-S. Lv, J.-Y. Wang, R.-C. Xiao, et al. Chloride ion transport properties in microcracked ultra-high performance concrete in the marine environment[J]. Construction and Building Materials, 2021, 291: 123310. https://doi.org/10.1016/j.conbuildmat.2021.123310
8. M. Li, et al. "Physical model study for effects of the Hong Kong - Zhuhai - Macao Bridge on harbors and channels in Lingdingyang Bay of the Pearl River

Estuary." Ocean & Coastal Management 177 (2019): 76-86. https://doi.org/10.1016/j.ocecoaman.2019.04.010

9. S.-Y. He, et al. "Efficient Bayesian model updating for settlement prediction of the immersed tunnel of HZMB." Transportation Geotechnics 44 (2024): 101179. https://doi.org/10.1016/j.trgeo.2023.101179
10. J. Li, Z. Wu, C. Shi, et al. Durability of ultra-high performance concrete - A review[J]. Construction and Building Materials, 2020, 255: 119296. https://doi.org/10.1016/j.conbuildmat.2020.119296
11. A.-C. Pillai, et al. "Application of an offshore wind farm layout optimization methodology at Middelgrunden wind farm." Ocean Engineering 139 (2017): 287-297. https://doi.org/10.1016/j.oceaneng.2017.04.049
12. Hein, J., Hilder, N. Estuarine territorialization and the port of Hamburg. Maritime Studies 22, 39 (2023). https://doi.org/10.1007/s40152-023-00329-x
13. Kasama, Kiyonobu, et al. "Instability evaluation of rubble mound for breakwaters subjected to tsunami-induced overflow and seepage flow." Soils and Foundations 60.6 (2020): 1532-1545. https://doi.org/10.1016/j.sandf.2020.10.006
14. X. Guo, H. Li, & S. Wang, Microstructural characterization of ultra-high performance concrete further hydrating in the marine environment. Mater Struct 55, 207 (2022). https://doi.org/10.1617/s11527-022-02042-3
15. Gaidai, Oleg, et al. "Multivariate ocean waves dynamics in North Sea and Norwegian Sea by Gaidai reliability method." Energy Reports 12 (2024): 2346-2355. https://doi.org/10.1016/j.egyr.2024.08.040
16. D. Choi, K. Hong, O. Munkhtuvshin, M. Didar, & S. Piti, (2023). Mechanical Properties of Ultra-High Performance Concrete (UHPC) and Ultra-High Performance Fiber-Reinforced Concrete (UHPFRC) with Recycled Sand. International Journal of Concrete Structures and Materials, 17(1), 67 - 20. https://doi.org/10.1016/j.oceaneng.2017.04.049
17. X. Li, Z. Shui, R. Yu, et al. Magnesium induced hydration kinetics of ultra-high performance concrete (UHPC) served in marine environment: Experiments and modelling[J]. Construction and Building Materials, 2019, 224: 1056-1068. https://doi.org/10.1016/j.conbuildmat.2019.07.273
18. D. Anjaneya, V. Abhishek, S.-D. Pang, Dual waste utilization in ultra-high performance concrete using biochar and marine clay[J]. Cement and Concrete Composites, 2021, 120: 104049. https://doi.org/10.1016/j.cemconcomp.2021.104049

Research on Intelligent Construction Technology Based on Green Concepts

Kai Wang*
School of Architecture and Urban Planning, Yunnan University, 650504, Kunming, China

Abstract. Compared with other industries, the construction industry has significant energy consumption characteristics. Therefore, actively promoting the green ecological construction in the construction industry is a crucial aspect to follow the current global low-carbon transformation call. This paper introduces intelligent building technology and its embodiment of green building concept. From the analysis of Building Information Modelling (BIM), intelligent building robot, Internet of Things (IoT) and other key technology fields, the significance of the combining of green and intelligent building is summarized. By conserving energy and reducing emissions, efficient use of resources, pollution reduction and other means. The construction industry is enriched by the birth of the construction industry Internet, construction robots, and other new industrial forms, which not only promotes resource conservation and environmental protection, but also promotes technological innovation and industrial upgrading. Meanwhile, these technologies improve building quality and user experience, create a safe, comfortable and convenient living and working environment, and extend the service life of buildings. Therefore, the active promotion and application of smart building technologies based on green concepts is of great significance for building a better and more sustainable future.

1 Introduction

The traditional construction industry faces issues such as low energy consumption efficiency, outdated management methods, severe environmental pollution, lack of energy-saving awareness and technological innovation, and short building lifespans lead to energy waste. The construction industry's direct energy consumption contribution to total societal energy consumption is up to 30% [1]. As people's attention to building energy-saving design concepts continues to increase, the National Development and Reform Commission has clarified in the Medium- and Long-Term Special Plan for Energy Conservation that building energy-saving design should meet a 50% standard [2]. Although China has issued a series of regulations and policies related to building energy efficiency, some still lack specific operational details and standards, leading to difficulties in implementation. However, the introduction of intelligent building technologies such as building information modelling (BIM), Internet of Things (IoT), and big data has made intelligence run through the whole

* Corresponding author: mudae@ldy.edu.rs

life cycle of buildings, thereby improving efficiency, reducing costs, and providing an effective way for the green transformation of the construction industry.

Starting from the growth background and characteristics of green buildings, Liu et al. [3] expounded how intelligent technology can serve the "four sections and one environmental protection" of green buildings, namely, efforts to save energy, land, water, materials and protect the environment. It points out that green buildings and building energy efficiency are two inseparable parts. The construction industry will adopt intelligent construction technology that uses the green concept as a future trend, which will have a positive impact on the harmonious coexistence of buildings and the natural environment. In theory, it constructs a theoretical system of intelligent green buildings to support industry transformation. In practice, it integrates and enhances building performance, such as energy efficiency, indoor environment, and construction efficiency, reduces carbon emissions and energy usage, and promotes the sustainability of the construction industry. Jiang [4] studied the integration instances of intelligent and green buildings. The study mentioned that green intelligent buildings integrate multiple technological concepts, including energy-saving ideas and renewable concepts. From the perspective of contemporary architecture, it is imperative to abandon the traditional energy-intensive development model and transition towards a modern, green, and intelligent development approach, with smart buildings and green buildings being the primary manifestations of this transformation [4].

Based on research on BIM technology, IoT technology, intelligent construction robots, and other technologies in building construction, this paper analyses the intelligent application of green building concept to guide the intelligent construction of civil engineering and foster the green development of buildings.

2 Overview of intelligent construction and green buildings

2.1 Summary of intelligent construction technology

Intelligent buildings are widely regarded as the inevitable path for the construction industry will undergo transformation and upgrading in the future [5]. However, at the current stage, intelligent construction is still in the nascent phase of its product life cycle. The industry's understanding of this field is relatively general, tending to broadly classify all construction processes involving automation and digital technology as intelligent construction, lacking a clear and widely accepted definition to precisely delineate its scope [6]. Figure. 1 shows some specific definitions and elaborations given by scholars on intelligent construction. During the planning phase, big data analysis aids decision-making, and BIM technology lays the foundation for design. During the design phase, BIM, Virtual Reality/Augmented Reality (VR/AR), and Artificial Intelligence (AI) technologies enhance design accuracy and efficiency. During the construction phase, robots, IoT, automation, and unmanned aerial vehicle (UAV) technologies improve construction efficiency and quality. During the operation and maintenance phase, BIM is combined with IoT to achieve intelligent operation and maintenance, sensor and AI technology to optimize building performance and energy consumption management, and comprehensively foster the intelligent upgrading and digital transformation of the construction industry.

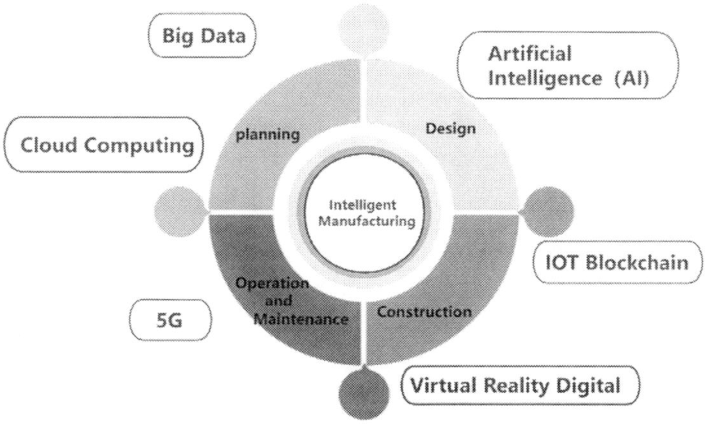

Fig. 1. Intelligent Construction Technology System [7].

2.2 The green building concept reflected in technology

In the process of constructing and developing, green buildings strive to save resources, protect the environment and reduce pollution to the maximum extent through scientific design, reasonable material selection, efficient construction and operation management. This approach aims to promote harmonious coexistence between man and nature and contribute to the building of a sustainable society [8]. The comprehensive integration of intelligent construction technology in various stages of building projects deeply reflects the concept of green buildings. During the planning and design phase, BIM technology is used to optimize building layout and orientation, ensuring that buildings can fully utilize natural light and ventilation. Meanwhile, green building materials and energy-saving designs are adopted to reduce environmental impact. During the construction stage, IoT and automation technologies precisely control the construction process, reducing energy consumption and material waste. Renewable materials and environmentally friendly construction techniques are also adopted to lower carbon emissions. In the operation and maintenance stage, intelligent systems and AI technologies monitor building energy consumption and equipment status in real-time, optimizing energy use efficiency, identifying potential issues in advance, and reducing maintenance costs and environmental impact. The deep integration of intelligent construction and green buildings not only improves the efficiency and quality of building projects but also promotes sustainable development and environmental protection in the construction industry.

3 Key technologies and applications

3.1 Application of BIM technology

BIM integrates diversified information, including building material types, building structure types, construction progress records, construction cost control, and material durability assessments. This platform ensures the integrity and consistency of information during transmission through interoperability and shareability, meaning the information content remains identical from transmission to reception. This feature significantly reduces project costs, saves time, and enhances overall productivity, aligning with the green building concept [9]. BIM technology optimizes building structures and systems (such as HVAC and plumbing) through precise modelling, improving energy efficiency and space utilization. It also supports multidimensional design analysis, including sunshine, wind energy, energy consumption, and acoustic environment, promoting green buildings and sustainable development. Figure 2 depicts the flowchart for using BIM technology in architectural design and construction management. It involves establishing building models using software such as Revit, conducting structural calculations with PKPM, performing environmental analysis with software like Ecotect and Phoenics, and ultimately integrating models and optimizing them in Navisworks to generate construction drawings. BIM technology integrates all information related to building projects, providing a platform for multi-party collaboration and driving digital transformation in the construction industry.

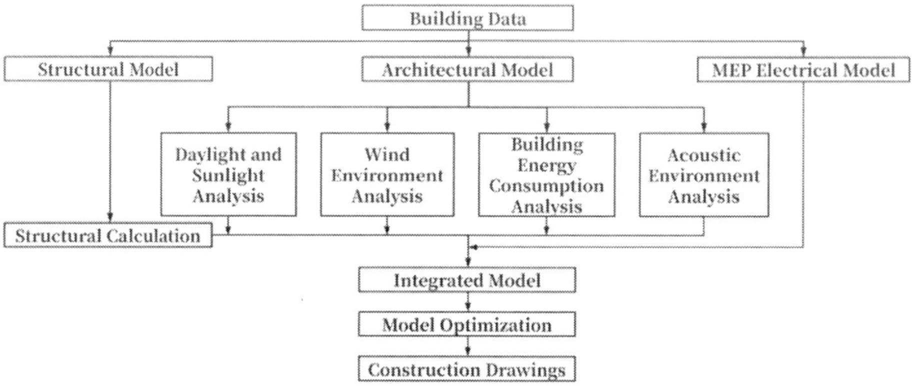

Fig. 2. Green Building Design Process Based on BIM Technology (Data from: [10])

3.2 Applications of intelligent construction robots

The construction industry is moving towards unmanned and labour-reduced operations, leveraging automation tools and intelligent robots. These devices can operate autonomously by following preset programs or immediate instructions, significantly reducing the burden on human labour and enhancing construction efficiency [11]. The application of intelligent equipment and construction robots during the construction phase is illustrated in Figure 3. Intelligent construction robots are high-tech products in the construction industry, designed to improve construction efficiency and safety. They are mainly categorized into several types: Surveying and Mapping Engineering Robots: such as setting-out and measurement robots, which provide precise construction data. Main Works Engineering Robots: performing arduous tasks like welding, steel bar handling, and reinforcement arrangement. Highway Engineering and Steel Structure Robots: covering bulldozing, scraping, concrete placement, and steel plate processing. Precast Concrete and Decoration Robots: automating the handling of precast components and decoration materials. Construction Engineering and Railway Engineering Robots: suitable for high-rise buildings, floor laying, and railway levelling. Underground Engineering Robots: such as shield tunnelling tool-changing robots, designed for complex underground construction. Through automation and precision, intelligent construction robots have significantly improved efficiency and quality in the construction industry.

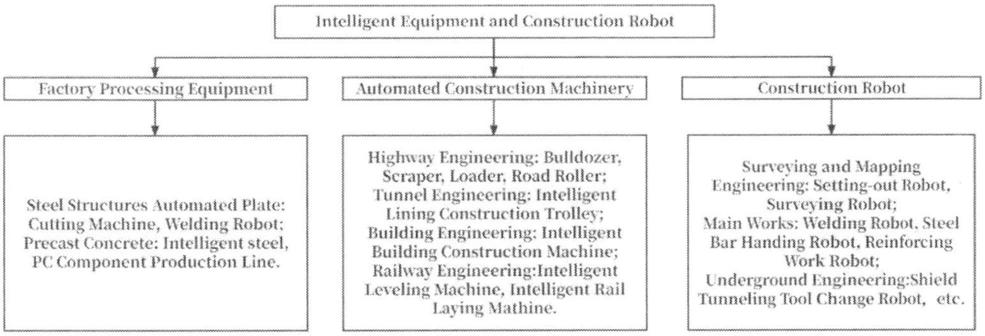

Fig. 3. Intelligent equipment and construction robots (Data from: [12]).

3.3 Applications of IoT technology

IoT technology is defined as the connection of any item to the internet through information-sensing devices including RFID, infrared sensors, GPS, laser scanners, and other equipment, which enables information exchange and communication for intelligent identification, location tracking, monitoring, and management. In Figure 4, IoT technology is juxtaposed with technologies such as two-dimensional barcodes, RFID, and laser scanners, suggesting their close collaboration and mutual support in fields like modern logistics, warehouse management, and automated production. Specifically, IoT technology collects vast amounts of data through various sensors and smart devices, which are then analysed and processed using cloud computing, big data, and other technologies to achieve efficient management and optimization of items, equipment, personnel, and more. These technologies can be applied in multiple domains such as smart homes, smart cities, intelligent transportation, industrial automation, bringing great convenience and benefits to people's lives and work. In summary, IoT technology represents an important direction in modern technological development, connecting various items to the internet and enabling intelligent processing and management of information. As shown in Figure 4, IoT technology stands alongside other technologies, collectively forming a crucial part of the modern technological system.

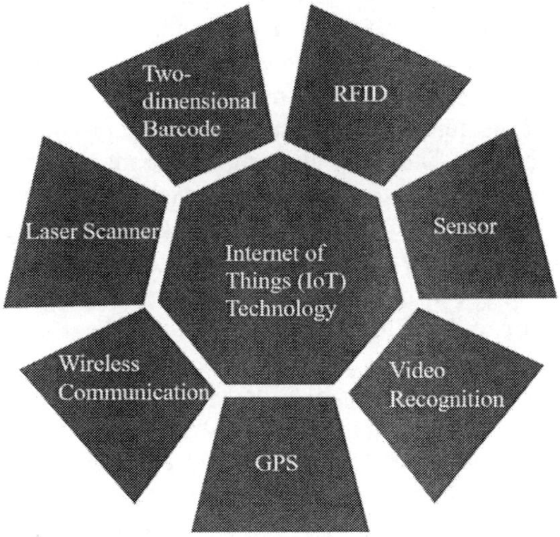

Fig. 4. Schematic Diagram of Key Technologies in the IoT (Data from: [13]).

4 The significance of the integration of intelligent construction and green building

The integration of intelligent construction and green building has significantly enhanced the efficiency of building construction. Intelligent technologies such as automated construction, sensor networks, and simulation modelling not only shorten the construction cycle but also optimize building operations through precise control and management systems. Meanwhile, green buildings reduce energy consumption through energy-saving designs. The combination of these two approaches ensures efficiency and intelligence throughout the entire process of building construction and operation, providing a more comfortable and convenient living environment for residents.

Using environmentally friendly materials, renewable energy and optimized building layout, green buildings greatly reduce the negative impact of buildings on the environment. On the other hand, intelligent building technology realizes accurate regulation of building energy consumption and efficient use of resources through intelligent control systems, further promoting energy conservation and emission reduction. The combination of the two not only protects the ecological environment, but also promotes the sustainable development of the construction industry.

Although the initial investment in intelligent construction and green buildings may be higher, in the long run, they bring significant cost control and return enhancements to investors. Green buildings reduce long-term operating costs by lowering energy consumption and maintenance costs. Intelligent construction technologies further reduce total costs by improving construction and operational efficiency. Moreover, as society's demand for green and intelligent buildings increases, these building projects often have higher market value and competitiveness, bringing higher investment returns to investors. Therefore, the integration of intelligent construction and green building is an inevitable choice for the future development of the construction industry.

5 Conclusion

This paper focuses on the significant trends of intelligent construction and green buildings in the construction industry's future development, proposing a theoretical framework for their integrated development. Firstly, the basic concepts of intelligent construction and green buildings are outlined. Subsequently, this paper comprehensively reviews the widespread applications of BIM technology, intelligent construction robots, and IoT technology in building construction. These technologies not only support the digital transformation and intelligent upgrading of the construction industry but also promote the deep implementation of the green building concept. The integration of intelligent construction and green buildings brings notable advantages in efficiency enhancement, environmental benefits, and cost control. By means such as precise modelling, optimizing building structures and systems, and improving energy efficiency and space utilization, it significantly boosts the design efficiency and quality of green buildings. At the same time, this integration shortens the construction cycle, optimizes building operations, reduces the negative environmental impact of buildings, provides inhabitants with a more comfortable and convenient living environment, and promotes sustainable development in the construction industry. Furthermore, the application of intelligent construction technologies can provide more precise and efficient support for various stages of green building design, construction, and operation, facilitating the in-depth implementation of the green building concept. However, it is necessary to intensify technology research and development efforts, improve technical standards, and address issues such as incomplete technical systems, technical barriers, and inconsistent standards. In the future, with the continuous advancement of technology and the

https://doi.org/10.1051/e3sconf/202560604009

expansion of application scenarios, the integration of intelligent construction and green buildings will become even closer, providing humanity with more efficient, environmentally friendly, and comfortable living environments.

References

1. J. T. Wang, Y. Liu, Changes and Countermeasures on Evaluation Standard for Green Building Version 2014. J. Const. Eco. **36(12)**, 75-78 (2015)
2. J. Feng, Research on Key Elements of Integrated Development of Green Building and Intelligent Building. J. Eng. Const. Design. **(06)**, 20-21+27 (2021)
3. D. J. Liu, Han, Y. Q. Liu, M. Wu, F. Pan, L. Cheng, Intelligent Technologies for Green Buildings. China Architecture & Building Press, 2023.
4. M.X. Jiang, Research on Integrated development of Green Building and Intelligent Building. J. Urban Arch. Space. **31(S1)**, 159-160 (2019)
5. Q. Kong, W. Diao, L. Yan, Application Analysis of Steel Bar Refinement Management Based on BIM and Intelligent Processing Technology. J. Jiangxi Build. Mat. **(6)**, 238-239 (2022)
6. C. Chen, X. Li, W. Yao, et al, Exploration on the Application of BIM Technology in the Intelligent Construction. J. Const. Tech. **51(20)**, 104-111 (2022)
7. Z. S. Liu, X. T. Sun, G. L. Shi, Summary of Application of Intelligent Construction in Civil Engineering Construction. J. Const. Tech. **50(13)**, 40-53 (2021)
8. Z. Ji, F. Xu, Application Analysis of BIM Technology in Green Public Building Design. J. Const. Design Project. **(7)**, 173-174+177 (2020)
9. L. Chen, The Design Method of Green Intelligent Building Based on BIM Technology. J. Intel. Buildi. Green Cities. **(08)**, 100-101 (2021)
10. S. J. Hou, R. Chen, Z. Wang, Research on the Development Trend of the Integration of Green Buildings and Prefabricated Buildings under the Background of BIM Technology. J. Bulk Cement. **(03)**, 62-64 (2021)
11. L. Han, The Application of Robot in Construction Industry under Intelligent Manufacturing Era. J. Const. Eco. **39(3)**, 23-27 (2018)
12. T. Hai, Q. C. Wang, J. P. Yang, et al, Monitoring system for PM2.5 based on LPWAN internet of things. J. Guilin Univ. Tech. **42(04)**, 1002-1008 (2022)
13. C. Wang, Application of Intelligent Construction in Civil Engineering Construction. J. Juye. **(08)**, 74-76 (2023)

Advances in the catalytic conversion of CO_2

Xiyang Gu[1]*

[1] Jinsha Middle School, 226300 Nantong, China

Abstract. The catalytic conversion of carbon dioxide (CO_2) offers significant potential for reducing greenhouse gas emissions and producing valuable chemicals, driven by advancements in sustainable technology and environmental regulations. Current catalytic methods for CO_2 conversion primarily include photocatalytic, thermocatalytic, and electrocatalytic processes, each using distinct energy sources—light, heat, and electricity. Among these, photocatalytic conversion stands out for its environmental friendliness, as it utilizes solar energy to reduce CO_2 into useful products. However, challenges such as low photon absorption efficiency and catalyst degradation have limited its large-scale application. Recent research has focused on enhancing photocatalyst performance by introducing nanomaterials and multifunctional catalysts. Innovations, such as noble metal-doped semiconductors and two-dimensional materials like graphene, have shown promise in improving light absorption, electron transfer, and stability. This paper reviews the latest advancements, explores the advantages and limitations of each method, and suggests optimization directions for effective CO_2 catalytic conversion, with an emphasis on achieving industrial viability.

1 Introduction

With the rapid development of industrialization and urbanization, the burning of fossil fuels has caused a large amount of CO_2 emissions and caused a series of ecological and environmental problems, such as the intensification of the greenhouse effect, ocean acidification, and sea level rise [1]. To alleviate the harm caused by CO_2, A lot of research has been conducted. For example, CO_2 could be directly captured and stored, but this method cannot fundamentally solve the problem. Another promising approach is to convert CO_2 into other high-value chemicals through chemical reactions, such as methanol, olefin, even glucose, and starch, which can not only realize the efficient use of CO_2, but also truly realize the recycling of carbon resources [2]. Therefore, chemical utilization plays a pivotal role in the management of CO_2. With the intensification of global climate change and rising CO_2 emissions, efforts to mitigate these impacts have become increasingly urgent. Converting CO_2 into valuable chemicals not only addresses carbon emissions but also enables the efficient recycling of carbon resources. While several methods have been developed, challenges such as high energy consumption, low efficiency, and product selectivity remain significant obstacles. So, this article will talk about catalytic conversation methods for

* Corresponding author: teaching@xjge.cn

CO_2.So that, it would make the knowledge of these more detailed. This paper reviews the primary catalytic methods for CO_2 conversion, discussing their mechanisms, advantages, and limitations, as well as the potential for future improvements in this critical area.

2 Catalytic conversion methods for CO_2

The catalytic conversion of carbon dioxide (CO_2) presents significant research opportunities, driven by advances in science and technology and by increasing environmental regulations aimed at mitigating greenhouse gas emissions. Moreover, growing awareness of sustainable and green chemistry has further encouraged the exploration of CO_2 conversion as a method for creating high-value chemicals and fuels from CO_2, contributing to carbon resource recycling and a circular economy. Currently, CO_2 catalytic conversion methods are primarily categorized into photocatalytic, thermocatalytic, and electrocatalytic conversion, each employing different energy sources—light, heat, and electricity, respectively. Each method has its own set of advantages and limitations that impact its efficiency, scalability, and industrial feasibility.

Photocatalytic technology, for instance, is considered environmentally friendly due to its ability to use sunlight as an energy source, making it a promising method for sustainable CO_2 reduction. However, practical application remains limited by several challenges. Photocatalysts typically have low photon absorption efficiency, with many only utilizing the ultraviolet (UV) spectrum, which constitutes less than 5% of sunlight. This narrow absorption range greatly restricts overall conversion efficiency. To address this, researchers have developed semiconductor catalysts doped with noble metals, such as platinum and gold, which have shown enhanced photon absorption and electron transfer capabilities. For example, platinum-doped titanium dioxide (TiO_2) has demonstrated significantly improved CO_2 conversion efficiency, as the noble metals expand the catalyst's light absorption to include the visible spectrum. Such doped catalysts also facilitate better charge separation, thereby reducing electron-hole recombination rates and enhancing reaction efficiency.
In addition, multifunctional catalysts, designed to simultaneously act as light absorbers and active sites for CO_2 reduction, have proven effective in increasing conversion rates. Recent studies indicate that these catalysts, often structured with multiple active components, can achieve improved CO_2 reduction efficiency by accelerating reaction kinetics and stabilizing reaction intermediates. The integration of two-dimensional materials, such as graphene and molybdenum disulfide (MoS_2), has further boosted photocatalytic activity. These materials provide a large surface area and unique electronic properties, which allow for more active reaction sites and enhanced electron mobility. The layered structure of these materials also contributes to their durability, making them suitable for long-term application in CO_2 reduction.

CO_2 electrochemical reduction technology is expected to become an important means of CO_2 conversion and utilization, the use of renewable energy to drive CO_2 electrochemical conversion, to achieve an effective artificial photosynthesis process, open new renewable energy storage and CO_2 resource conversion and utilization of sustainable development model. Researchers are developing new catalysts, such as nanostructured metals, ion-modified metals, and organic and inorganic hybrid materials, to improve the efficiency of the electrochemical CO_2 reduction process. Hsieh et al. prepared a coral-like nanostructured Ag electrocatalyst and found that a stable chemical bond was formed on the surface of Cl^- and Ag nanocoral [3]. Compared with polycrystalline Ag, the electrochemical surface area of this material increased by 20 times, showing relatively high catalytic activity.

2.1 Photocatalytic conversion

Photocatalytic technology is a highly efficient and environment-friendly purification technology that uses semiconductors as catalysts and light as activation energy. Research on photocatalytic solar energy conversion and utilization is of great significance for optimizing our country's energy structure and achieving the goals of "emission peak" in 2030 and "carbon neutrality" in 2060. Photocatalytic reduction of CO_2 involves absorbing light energy with photocatalysts, which then convert this energy into chemicals like CO, CH_4, and CH_3OH. Catalysts are crucial in this process, serving both as adsorbents for CO_2 and as agents for light energy conversion. However, finding efficient catalysts remains a challenge, and scaling up these processes for large-scale applications is still difficult. Similarly, photocatalytic conversion of nitric oxide is also a hot topic, such as photocatalytic oxidation, photocatalytic reduction, and photocatalytic direct decomposition.

2.2 Thermocatalytic conversion

Thermocatalytic conversion is a significant method for utilizing CO2. It involves providing thermal energy to drive chemical reactions that transform CO2. Key processes include CO2 hydrogenation, catalytic reforming, and transesterification. Among these, CO_2 hydrogenation is notable for producing various chemical raw materials, such as carbon monoxide, methane, methanol, and olefins, thereby offering efficient CO_2 utilization [4].Thermocatalytic conversion typically requires high temperatures and pressures, which pose challenges for catalyst durability. Common thermocatalytic reactions include the Fischer-Tropsch process, which converts CO_2 into liquid fuels and chemicals, and CO_2 reaction with epoxides to produce cyclic carbonates, an eco-friendly chemical. The choice of catalyst greatly impacts reaction pathways and product selectivity. Recently, dual-metal catalysts and composite catalysts based on porous materials have shown potential for enhancing both efficiency and selectivity in thermocatalysis, thereby reducing energy consumption while achieving higher yields.

2.3 Electrocatalytic conversion

Among all strategies for the conversion of CO_2, electro reduction represents a green way for the fixation of CO_2 via the use of clean and renewable electricity. As one of carbonaceous products from CO_2 electro reduction, methane exhibits the highest combustion heat per unit mass (a calorific value of about 56 KJ/g), which serves as a promising energy carrier to generate heat and electricity, as well as gas power for vehicles. In addition to combustion,CH4 could be adopted to prepare syngas via the steam reforming process. Accordingly, the synthesis of CH4 from CO2 electro-reduction opens up a perspective to reach a carbon-neutral economy. Electrocatalytic reduction of CO2 uses electricity to convert CO2 into carbon-based chemicals or liquid fuels3. This process is complex due to the multi-step electron-proton transfer required, leading to a range of products. The high thermodynamic stability and chemical inertness of CO2 necessitate significant energy input for its direct conversion5. Developing highly efficient electrocatalysts is crucial for scaling up these processes and producing valuable chemicals. The key to efficient CO_2 electrocatalysis lies in selecting optimal catalysts. Ideal electrocatalysts exhibit high conductivity, stability, and product selectivity. Currently, copper-based catalysts show high selectivity for CO_2 reduction; however, product diversity complicates the separation process. Studies indicate that modifying catalyst nanostructures or surfaces significantly enhances catalytic performance. Additionally, emerging catalysts like metal-organic frameworks

(MOFs) and single-atom catalysts are gaining attention for their excellent stability and efficiency, opening new possibilities for scalable, cost-effective CO_2 electrocatalysis.

2.4 Comparison of catalytic methods

Photocatalysis, thermocatalysis, and electrocatalysis are distinct catalytic methods that facilitate chemical reactions, each utilizing different energy sources: light, thermal, and electrical energy, respectively. Selecting the most suitable method for specific applications can enhance reaction efficiency while reducing energy consumption and environmental impact. Each catalytic method—photocatalysis, thermocatalysis, and electrocatalysis—has distinct advantages for specific applications. Photocatalysis is advantageous in low-energy, solar-driven applications but is limited by low efficiency and dependency on sunlight. Thermocatalysis is suitable for industrial CO_2 conversion, especially for synthesizing fuels and chemicals, as it achieves high reaction rates but at high energy costs. Electrocatalysis, when powered by renewable electricity, provides a clean and efficient pathway for CO_2 conversion, making it ideal for low-carbon applications. However, its scalability is restricted by current equipment costs and product selectivity challenges. In practice, selecting the most appropriate method requires careful consideration of process conditions, cost, and environmental benefits to maximize both economic and ecological impact. Catalytic processes offer a promising route to convert CO2 into green, high-value products, thus playing a strategic role in addressing environmental and energy challenges. While promising, photocatalytic CO_2 conversion still requires significant advancements to overcome its current limitations. Future research could focus on developing catalysts that not only absorb a broader range of light but also exhibit high stability under prolonged light exposure, addressing issues of catalyst degradation. Additionally, optimization of reaction conditions, such as controlling reaction environments to minimize energy losses and integrating co-catalysts to improve electron dynamics, may enable more efficient and scalable CO_2 photocatalytic conversion. In conclusion, although challenges remain, the combination of innovative materials, multifunctional catalyst design, and reaction engineering has substantially advanced the field, bringing photocatalytic CO_2 conversion closer to feasible industrial application.

3 Major challenges and limitations in CO_2 conversion methods

The catalytic conversion of CO_2 is a promising field, offering potential solutions for carbon emissions reduction and resource recycling. However, each method—photocatalytic, thermocatalytic, and electrocatalytic conversion—faces unique challenges that limit its efficiency, scalability, and economic viability in industrial applications. This section will discuss the primary limitations of each method and suggest potential strategies for improvement [5].CO_2 conversion methods, including photocatalytic, thermocatalytic, and electrocatalytic processes, offer promising solutions for reducing carbon emissions and recycling CO_2 into valuable products. These methods are central to efforts aimed at mitigating the impacts of climate change by transforming CO_2 into chemicals, fuels, and other resources. However, each approach faces significant challenges that hinder its efficiency, scalability, and economic viability. Overcoming these challenges is essential for achieving large-scale, sustainable CO_2 conversion processes that can contribute meaningfully to global efforts to reduce carbon emissions.

3.1 Low efficiency and catalyst stability issues

Photocatalytic conversion is an environmentally friendly approach that leverages solar energy to reduce CO_2, mimicking the natural photosynthesis process. Despite its potential, this method faces several major limitations. One of the primary challenges is low photon absorption efficiency. Most photocatalysts currently in use are optimized to absorb ultraviolet (UV) light, which constitutes less than 5% of the solar spectrum. As a result, the overall efficiency of photocatalytic CO_2 reduction is severely limited by the inability to capture a large portion of the sunlight. To address this, researchers are exploring catalysts that can absorb a broader range of the light spectrum, especially visible light. Some promising strategies include doping traditional semiconductor materials with transition metals or incorporating plasmonic nanostructures to enhance light absorption.

Another significant hurdle in photocatalytic processes is the fast recombination of electron-hole pairs. In photocatalysis, the separation of electrons and holes is essential for initiating chemical reactions. However, the rapid recombination of these charge carriers significantly reduces the reaction efficiency and, as a result, the CO_2 conversion rate. To mitigate this, the development of heterojunctions or multi-layered structures has been proposed. These structures can help create more efficient separation of charge carriers, which prolongs the lifetime of the electrons and holes involved in the reaction, thereby improving overall efficiency.

Catalyst stability is also a critical issue in photocatalytic CO_2 conversion. Photocatalysts are prone to deactivation due to photochemical corrosion, which limits their long-term performance. To improve catalyst stability, researchers are focusing on more robust materials like graphitic carbon nitride ($g\text{-}C_3N_4$) and titanium dioxide (TiO_2), which show greater resistance to degradation. Surface modifications and hybrid materials are also being explored as means to enhance catalyst durability, ensuring that they remain functional for longer periods in real-world applications [6].

Despite ongoing research into these areas, achieving high efficiency and stable performance in photocatalytic CO_2 conversion remains a significant challenge. Overcoming these limitations will require continued innovation in catalyst design, reaction engineering, and system integration to enable scalable and cost-effective applications.

3.2 High energy consumption and selectivity challenges

Thermocatalytic conversion of CO_2 has proven effective in producing valuable chemicals and fuels, such as methanol and synthetic hydrocarbons. However, it faces several challenges that hinder its commercial viability. One of the major concerns is the high energy demand of the process. Thermocatalytic conversion typically requires temperatures above 200°C and pressures exceeding 10 MPa. These extreme conditions lead to significant energy consumption, which makes thermocatalysis both economically and environmentally challenging, particularly when using fossil fuels as the energy source. Research efforts are focused on developing catalysts that remain active at lower temperatures, thereby reducing the energy burden. For example, metal-organic frameworks (MOFs) and nanostructured bimetallic catalysts have shown enhanced activity at reduced temperatures and are being actively explored as alternatives.

Another challenge in thermocatalytic CO_2 conversion is the issue of product selectivity. CO_2 hydrogenation can produce a wide range of products, including methanol, methane, carbon monoxide, and higher hydrocarbons, depending on the reaction conditions and the properties of the catalyst. Achieving high selectivity for a desired product is difficult due to the complexity of the reaction pathways. Recent advancements in catalyst design, such as the development of catalysts with controlled pore structures and tailored active sites, aim to improve selectivity. Additionally, advanced preparation techniques like atomic layer deposition and surface modification allow for more precise control over reaction pathways, which could enhance product selectivity.

Catalyst deactivation due to prolonged exposure to high temperatures and pressures is another major issue in thermocatalytic CO_2 conversion. High temperatures can cause sintering (the process where catalyst particles coalesce), while carbon deposits, known as coking, can accumulate on the catalyst surface. Both of these factors lead to a reduction in catalyst efficiency over time. To address this, researchers are focusing on developing catalysts with higher thermal stability and greater resistance to coking. Some promising solutions involve using metal oxide supports or incorporating alkaline elements to prevent carbon deposition and enhance the catalyst's longevity.

Addressing these challenges will require a multifaceted approach that combines advanced catalyst design, energy-efficient reaction conditions, and robust system engineering. Developing thermocatalytic conversion processes that are economically and environmentally feasible will be key to realizing its potential on an industrial scale.

3.3 Cost and product selectivity limitations

Electrocatalytic CO_2 reduction is a clean and efficient way to convert CO_2 into valuable chemicals and fuels, especially when powered by renewable electricity. Despite its promise, electrocatalysis faces notable challenges that limit its widespread adoption. A primary issue is the high cost of catalysts, which often rely on noble metals such as gold, silver, and platinum. These materials are costly and rare, making them unsuitable for large-scale applications. To make electrocatalytic CO_2 reduction more economically viable, researchers are seeking alternatives made from earth-abundant materials, such as copper, nickel, and iron. Metal-free catalysts, like nitrogen-doped carbon materials, are also being explored as potential substitutes. Additionally, single-atom catalysts, where individual metal atoms are dispersed on a support material, have shown promise in maximizing catalytic efficiency while minimizing the amount of precious metal used.

Low selectivity for the desired product is another major challenge in electrocatalytic CO_2 reduction. The process involves multi-step electron-proton transfer reactions that can yield a mixture of products, including carbon monoxide, methane, ethylene, and formic acid. Achieving selectivity for a specific product is challenging due to competing reaction pathways. Researchers are working to enhance selectivity by designing catalysts with tailored electronic structures and surface properties. Modifying the electronic environment of the catalyst through doping or alloying can help steer the reaction toward a single product. Additionally, improvements in reactor design, such as adjusting electrolyte composition or applying membrane separators, can help create more controlled reaction conditions, further improving selectivity.

Electrode stability and degradation also pose significant challenges for electrocatalysis. During the conversion process, electrodes can undergo corrosion, dissolution, or poisoning by reaction intermediates, which leads to a decline in performance over time. To address this, strategies such as protective coatings, alloying metals for enhanced durability, and the development of novel supports that maintain structural stability while providing high conductivity are being explored. The use of robust materials like titanium or graphene-based supports has shown promise in extending the lifetime of electrodes and improving the overall economic feasibility of the process.

4 Conclusion

With the continuous acceleration of China's industrialization process, environmental problems, especially air pollution, have seriously threatened people's normal lives. It is urgent to practice the concept of sustainable development and effectively control air pollution. Catalytic technology provides an effective approach to the treatment of atmospheric pollutants and has been widely applied. Carbon dioxide is not only the main component of greenhouse gases, but also an abundant and renewable carbon resource. This paper introduces three main CO_2 conversion catalytic methods, including photocatalysis, thermal catalysis and electrocatalysis. In particular, it is mentioned that photocatalytic conversion is environmentally friendly because of its use of solar energy, but it also faces the challenge of low photon absorption efficiency and catalyst degradation, which limits its large-scale application. This study highlights recent research on enhancing photocatalyst performance through the introduction of nanomaterials and multifunctional catalysts, such as precious metal-doped semiconductors and two-dimensional materials (such as graphene), which show potential for improving light absorption, electron transfer, and stability, discusses the advantages and limitations of each catalytic method, and suggests directions for optimization of efficient CO_2 catalytic conversion. Finally, the importance of achieving industrial feasibility is emphasized, and optimization directions are proposed to promote the practical application of CO_2 catalytic conversion technology.

Looking forward, research on CO_2 catalytic conversion must focus on developing highly efficient, low-cost processes for large-scale industrial applications. In photocatalysis, the development of high-efficiency catalysts and optimized photon absorption pathways will remain crucial. In thermocatalysis, lowering reaction temperatures and enhancing catalyst selectivity are essential to minimize energy consumption and improve product purity. For electrocatalysis, future research should focus on enhancing electrode durability and product selectivity to meet scalability requirements. Furthermore, combining different catalytic methods, such as hybrid photocatalytic-electrocatalytic systems, may leverage the strengths of each approach to achieve higher CO_2 conversion efficiencies, contributing significantly toward carbon neutrality.

References

1. X.Jiang, X. Nie, X. Guo, C. Song, J.G. Chen, Recent Advances in Carbon Dioxide Hydrogenation to Methanol via Heterogeneous Catalysis. Chem. Rev. **120**, 7984-8034 (2020).
2. C. Yang, S. Liu, Y. Wang, J. Song, G. Wang, S. Wang, Z.J. Zhao, R. Mu, J. Gong, The Interplay between Structure and Product Selectivity of CO2 Hydrogenation. Angew. Chem. Int. Ed. **58**, 11242-11247 (2019).
3. S. Das, J. Perez-Ramirez, J. Gong, N. Dewangan, K. Hidajat, B.C. Gates, S. Kawi, Core-Shell Structured Catalysts for Thermocatalytic, Photocatalytic, and Electrocatalytic Conversion of CO2. Chem. Soc. Rev. **49**, 2937-3004 (2020).
4. A. Bavykina, I. Yarulina, A.J. Al Abdulghani, L. Gevers, M.N. Hedhili, X. Miao, A.R. Galilea, A. Pustovarenko, A. Dikhtiarenko, A. Cadiau, A. Aguilar-Tapia, J.-L. Hazemann, S.M. Kozlov, S. Oud-Chikh, L. Cavallo, J. Gascon, Turning a Methanation Co Catalyst into an In–Co Methanol Producer. ACS Catal. **9**, 6910-6918 (2019).
5. J. Zhong, X. Yang, Z. Wu, B. Liang, Y. Huang, T. Zhang, State of the Art and Perspectives in Heterogeneous Catalysis of CO2 Hydrogenation to Methanol. Chem. Soc. Rev. **49**, 1385-1413 (2020).
6. X.-F. Bai, W. Chen, B.-Y. Wang, G.-H. Feng, W. Wei, Z. Jiao, Y.-H. Sun, Recent Progress on Electrochemical Reduction of Carbon Dioxide. Acta Phys.-Chim. Sin. **33**, 2388-2403 (2017).

E3S Web of Conferences **606**, 05001 (2025) https://doi.org/10.1051/e3sconf/202560605001
ICNAOE 2024

Comparative analysis of ternary lithium batteries and lithium iron phosphate

Ruoxuan Liu[1*]

[1]Environmental science and green management, Hong Kong Metropolitan University, Hong Kong, China

Abstract. Ternary lithium batteries (TLB) and lithium iron phosphate batteries (LIPB) are two popular battery types in the current battery market. They have their own advantages and disadvantages in performance and application areas. Through the analysis of the structure, performance and application of the two types of batteries, it can be seen that the anode of TLB is an octahedral structure with high energy density, strong fast charging ability and excellent low temperature discharge performance. The different ratios of nickel, cobalt and manganese in the anode material are suitable for a variety of unused occasions. However, the high-temperature stability of TLB is poor, and thermal runaway is easy to occur at high temperatures, and their cycle life is relatively short. LIPB are known for their high safety, long cycle life and relatively low cost. Its unique olivine crystal structure and stable P-O covalent bonds give it excellent thermal stability, even at high temperatures, the battery will not easily decompose. The disadvantages of LIPB are mainly reflected in their lower energy density and low-temperature discharge performance. Combining the advantages of the two materials to develop a new battery material with both high energy density and high safety will be an important research direction in the future.

1 Introduction

In today's society, batteries have been deeply used in all aspects of our lives, such as mobile devices, electric vehicles, energy storage systems and medical devices. This fully reflects the importance of batteries to our life. With the gradual depletion of fossil resources, energy transition is imminent, and the development of clean and sustainable energy has become a priority. Among them, solar and wind energy have been proven to be effective alternatives to fossil energy. However, due to the unstable supply of these energy sources, the stability of the power grid is seriously affected. To mitigate and solve this problem, the combination of solar or wind power generators with energy storage devices has become a viable option [1]. As a result, the battery market has been expanding year on year in recent years. The Asia-Pacific consumer battery market size is expected to reach USD 8.78 billion by 2023 and is projected to continue growing until 2032 [2]. Among the many battery types, Li-ion batteries are favored for their superior performance, making them an important energy option for electric vehicles and other sectors, and they stand out as the mainstream of the industry in terms of applications. As far as electric vehicles are concerned, 67.3% of electric vehicles loaded with TLB, 32.6% of electric vehicles loaded with LIPB, and 0.1% of electric vehicles loaded with other batteries will be used in 2023 [3]. This paper focuses on the structural and performance characteristics of TLB and LIPB, and compares and analyses their advantages and disadvantages.

2 Ternary lithium batteries

2.1 Structural characteristics of ternary lithium batteries

Ternary lithium batteries are chemical batteries in which the positive electrode is made of a mixture of nickel, cobalt and manganese, and the negative electrode is made up of graphite in which lithium ions are inlaid and released. As for the anode structure, the anode material of TLB has a layered structure, with nickel atoms, manganese atoms, cobalt atoms, and oxygen atoms forming an octahedron, as in Figure 1 [4]. Thus, the TLB has the advantages of lithium cobaltate, lithium nickel, and lithium manganate batteries at the same time. Where lithium ions are located between the layers of the octahedron and can be released by inserting them back and forth between the layers. The presence of nickel atoms increases the capacity of the battery, the presence of cobalt atoms can help to maintain structural stability and improve electronic conductivity, and the presence of manganese atoms reduces costs and improves structural safety [5]. Thus the ratio of nickel, cobalt and manganese in the ternary material can vary according to the demand. The structure of nickel-cobalt-manganese ternary material can be expressed as $LiNi_xCo_yMn_zO_2$ where x+y+z=1. Based on the elemental molar ratio structural types such as type 333, type 523 and type 811 can be obtained [6]. Graphite, the negative electrode material, has good electrical conductivity and stability, and is better for the storage and release of electrical energy. Electrolytes are usually found in the form of gel polymers, often formed from a mixture of polymers, organic solvents and lithium salts [4]. It has the advantages of both solid and liquid electrolytes, with good ionic conductivity and some thermal stability. For the diaphragm, it is often made of materials such as polypropylene or vinyl membranes [7].

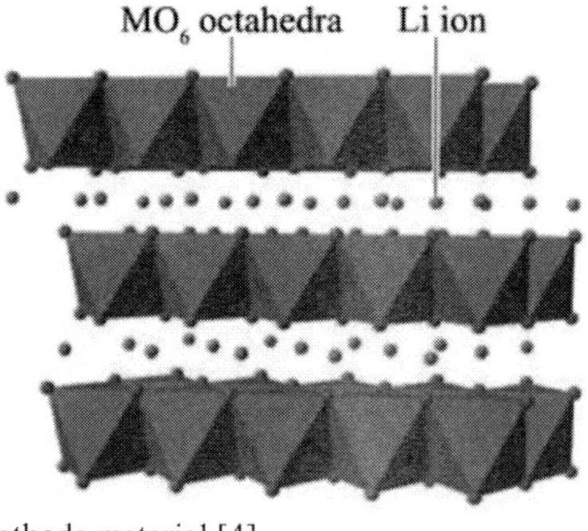

Fig. 1. Structure of TLB cathode material [4].

2.2 Performance Characteristics of Ternary Lithium Batteries

TLB has a high energy density usually 200 Wh/kg, and with in-depth research, it possibly up to 300 Watt-hours/kg in the future, which is able to provide a long time of endurance and higher power output, a characteristic that allows them to be used in lightweight or devices that require a long time of power supply [4]. It also has the ability of fast charging, which can significantly improve the convenience of the product. Moreover, lithium ternary batteries are environmentally friendly. Its constituent materials do not contain heavy metals and toxic substances, so it is less polluting to the environment and can strictly meet the environmental requirements. In addition, ternary lithium batteries have better discharge performance at low temperatures. At -20 °C, the battery capacity only decreased by 70.14% of 25 °C. (TLB battery capacity at 25 °C for 100%) [7]. But the cathode of the TLB is not resistant to high temperatures, it will be at 250 °C to 300 °C is to carry out internal decomposition, thermal stability is poor. The decomposed cathode encounters the combustible electrolyte and carbon materials in the battery, after which it will react rapidly to generate a large amount of heat. The safety of its positive electrode is poor, in the event of external impact, the positive electrode and the positive electrode in the TLB separator is easy to be damaged, resulting in a short-circuit battery. Short-circuit batteries are prone to thermal runaway and then rapidly warm up and finally spontaneous combustion. Secondly, the theoretical service life of lithium ternary battery is 2000 times of charging and discharging. After 3900 cycles, the battery capacity will be reduced to 66%, which shows that its cycle life is short [4].

3 Lithium iron phosphate batteries

3.1 Structural characteristics of lithium iron phosphate batteries

The negative electrode of LIPB is usually composed of graphite, which can be effectively embedded and embedded off lithium ions, and its positive electrode is usually composed of $LiNiO_2$, $LiMn_2O_4$, $LiCoO_2$, $LiFePO_4$ and other materials. Among them, LIPB material ($LiFePO_4$) is widely studied for its abundant resources, low toxicity and high stability [4]. This material is usually a stable olivine crystal structure with orthorhombic homology to the space group Pnma, as shown in Figure 2. $LiFePO_4$ consists of a polyoxo anionic framework composed of FeO-octahedral, LiO-octahedral and PO-tetrahedral sites. The oxygen arrays are hexagonally close packed with octahedra sharing edges and faces as shown in Figure 3 [8]. Lithium ions are located in the edge-octahedral (LiO) chains, divalent iron ions occupy the corner-sharing octahedra (FeO_6), and phosphorus ions are located in the tetrahedral sites (PO_4). Among them, the P-O covalent bond has an important chemical bonding energy, which avoids the release of oxygen in the high charge state, and it is its anode that keeps its structure stable even in high temperature environments [9]. The three-dimensional structure presented by LIP crystals forms a one-dimensional ion transport channel, thus limiting the diffusion of lithium ions [4]. The electrolyte for LIPB is usually a mixture of organic solvents and lithium salts with good ionic conductivity and thermal stability. Diaphragm is usually one or two of polyethylene and polypropylene, and is often used as a single-layer or multi-layer composite film. There are also an inorganic ceramic film on the surface of the membrane to enhance the treatment, to a certain extent, to improve the safety of the battery [10].

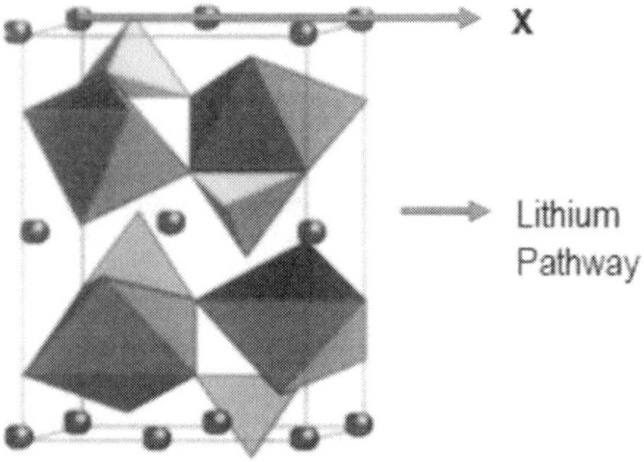

Fig. 2. LIPB cathode material structure [4].

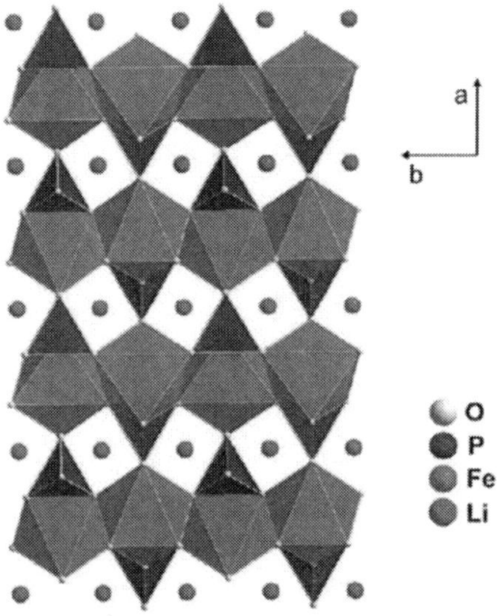

Fig. 3. Olivine structure of LIPB [9].

3.2 Performance Characteristics of Lithium Iron Phosphate Batteries

LIPB has high safety, due to its stable olivine structure, the battery cathode has excellent chemical properties and high temperature resistance. Due to its P-O covalent bond, the battery will not release oxygen in the face of impact, pinprick and short-circuit, which makes him even if the damage will not burn, only to reach 700 °C to 800 °C between the internal decomposition. LIPB has a high cyclability, after 5000 cycles of use still has 84% of the battery capacity. This is due to the high stability of the LIP lattice, which makes lithium ion embedding and embedding out of the lattice have less effect on the lattice and good reversibility [4]. And because the raw materials of the battery are iron and phosphorus this kind of rich materials, make the LIPB has a low cost, is conducive to large-scale production and application. In addition, LIPB does not produce toxic substances, in line with the requirements of green environmental protection. However, under low temperature conditions, the discharge performance of LIPB is poor. At -20 °C, the battery capacity drops to only 54.94% at 25 °C. (The capacity of $LiFePO_4$ battery is 100% at 25°C) and so far, the energy density of $LiFePO_4$ battery is usually 105Wh/kg, and a few of them can reach 130～150 Wh/kg. It is also challenging to break through 200Wh/kg in the future [11].

4 Comparison between lithium ternary batteries and lithium iron phosphate batteries

Nowadays, both LTB and LIPB are widely used in the field of electric vehicles and have become the two most popular types of electric vehicle batteries. This section will compare them in seven aspects: environmental friendliness, energy density, safety, cost and price, charging efficiency, low temperature discharge and cycle life.

Lithium ternary batteries and LIPB are environmentally friendly, and none of the constituent materials are toxic, and they do not produce toxic substances.

In terms of energy density, TLB has a higher energy density, usually 200 Wh/kg, and there is room for improvement in the future. In contrast, only a few LIPB can reach 130-150 Wh/kg, with less room for future improvement. This makes TLB smaller and more convenient to use while providing the same amount of energy [11].

In terms of safety, LIPB has high safety. It will not spontaneously combust after being subjected to external forces, and will only decompose internally when it reaches 700 °C to 800 °C. In contrast, lithium ternary batteries will decompose at 250 °C to 300 °C, which is less thermally stable. In contrast, lithium ternary batteries will decompose at 250 °C to 300 °C, with poor thermal stability. And it is easy to cause a short-circuit under the action of external force, and the temperature rises to reach the spontaneous combustion temperature and spontaneous combustion [4].

In terms of cost price, LIPB has a lower cost. Because the current iron and phosphorus resource reserves are sufficient, the price is low. In contrast, nickel, cobalt and manganese cost price is relatively high.

In terms of charging efficiency, the TLB is much more efficient. When charging at ten times the constant charging current, there is little difference in the charging efficiency of the two batteries. When charging at 20 times the constant charging current, the charging efficiency of TLB is significantly better than that of LIPB [12].

In terms of low-temperature discharge, TLB has better low-temperature discharge. At 25 °C, both TLB and LIPB have 100% battery capacity. However, at -20 °C, the battery capacity of Li-ion ternary battery drops to 70.14% of that at 25 °C. The LIPB decreased even more, down to 54.94% [7].

In terms of cycle life, LIPB has better cyclability. Lithium ternary batteries in the cycle after 3900 times the battery capacity will be reduced to 66%. And LIPB in the cycle after 5000 times and 84% of the battery capacity [4].

From the comparison of seven aspects, Li-ion ternary battery has environmental friendliness, while Li-$FePO_4$ battery has environmental friendliness, lower cost price, higher safety and longer cycle life. The two battery structures are different, both have their own respective advantages and disadvantages, and are widely used by customers in different needs and scenarios. But for now, neither battery is the optimal choice. In the future, our research on these two types of batteries should be continued and deepened. Continuously improve their performance, or try to combine their superior performance in the direction of research, to create a new combination of the advantages of the two new cathode materials to replace ternary materials and LIPB. At the same time, the research on new alternative energy sources is also developing rapidly. It is believed that a perfect solution will be applied to the automotive industry in the near future.

5 Conclusion

Lithium ternary batteries and LIP batteries have their own advantages and disadvantages in terms of performance and application areas. By analyzing the structure, performance and application of the two batteries, the octahedral structure of the anode of Li-ion ternary battery, with its high energy density, fast charging ability and excellent low temperature discharge performance, has been widely used in the occasions requiring high energy output, such as electric vehicles. The different ratios of nickel, cobalt and manganese in the cathode material allow it to be applied to different needs. However, TLB has poor high-temperature stability, are prone to thermal runaway at high temperatures, and have a relatively short cycle life. Therefore, the applicability of TLB is limited in high temperature and long life applications. LIPB are known for their high safety, long cycle life and relatively low cost. Its unique olivine crystal structure and stable P-O covalent bonds give it excellent thermal stability, even at high temperatures, the battery will not easily decompose. The disadvantages of LIPB are mainly reflected in their low energy density and low temperature discharge performance, which limits their application in some high energy demand and low temperature environments. However, due to its low cost and environmental friendliness, LIPB has been widely used in price-sensitive markets such as energy storage devices and economical electric vehicles.

Lithium ternary batteries and LIPB have their own advantages and limitations, future research should continue to focus on improving the energy density, safety and cycle life of the two, in order to achieve a comprehensive improvement in performance.

References

1. T. Li, X. Bai, U. Gulzar, et al. A Comprehensive Understanding of Lithium - Sulfur Battery Technology ,Advanced Functional Materials, **29**, 32 (2019) https://doi.org/10.1002/adfm.201901730
2. Fortune business insights Consumer Battery Market Size, Growth: Global Report 105526, (2032)
3. China Energy Storage Network, Analysis of the Domestic Power Battery Market in 2023, https://www.escn.com.cn/20240130/fca53014ab284b2cae6bb9fff6df27b8/c.
4. Y. Su. Comparative Analysis of Lithium Iron Phosphate Battery and Ternary Lithium have Battery, Journal of Physics: Conference Series, **2152**, 1 (2022) https://doi.org/10.1088/1742-6596/2152/1/012056
5. K. Zuo. Classification of ternary lithium battery, (Linkedin, K.Zuo, 2020)
6. Z. Tong, Z. Li, L.Tan, et al. Recent advances in synthesis and modification strategies for lithium-ion battery ternary cathodes, Journal of Energy, **98**, 113085-, (2024)
7. Sun and moon morning, Structure and Basic Principles of Lithium-Ion Batteries, http://www.szryc.com/?p=10574
8. B. Wu, Y. Ren, N. Li. $LiFePO_4$ cathode material. Electric Vehicles-The Benefits and Barriers, **3**, 199-216, (2011)
9. J. Yang. Development of nanostructured $LIMPO_4$ (M= Fe, MN) as Cathodes for high performance lithium-ion batteries, Doctoral dissertation, The University of Western Ontario (Canada), (2023)
10. W. K. Baidu Wenku, attery characteristics, https://wk.baidu.com/view/73d7370da02d7375a417866fb84ae45c3a35c263?pcf=2&bfetype=new&_wkts_=1723121272615

11. N. Omar, M. A. Monem, Y. Firouz, et al. Lithium iron phosphate based battery – Assessment of the aging parameters and development of cycle life model, Applied Energy, **113**, 1575 – 1585. (2014) https://doi.org/10.1016/j.apenergy.2013.09.003,
12. X. Su, B. Sun, J. Wang, et al. Experimental study on charging energy efficiency of lithium-ion battery under different charging stress. Journal of Energy Storage, **68**, 107793-. (2023) https://doi.org/10.1016/j.est.2023.107793,

From Coagulation to Disinfection: Key Steps and Challenges in Wastewater Treatment

Shengyuan Wang[1]*

[1]University of Southern California, Los Angeles, USA

Abstract. Wastewater treatment is a critical process to ensure the safe management of water for domestic, industrial and municipal use. The importance of wastewater treatment has become more pronounced as water consumption increases and environmental regulations become more stringent. Wastewater treatment plants remove harmful contaminants from wastewater and produce clean discharges that meet regulatory standards through a series of mechanical, biological, and chemical processes such as coagulation, sedimentation, filtration, and disinfection. This paper explores the advantages and disadvantages of each wastewater treatment stage and evaluates the wastewater treatment infrastructure in the United States and China. The U.S. has better wastewater treatment systems, while China still has inadequate wastewater treatment facilities in rural areas. Wastewater reuse is an important part of achieving a sustainable future, and further development and implementation of these systems can help improve global water management.

1 Introduction

Wastewater treatment is a critical process that ensures the safe and efficient management of water used in households, industries, and municipalities. As water consumption increases and environmental regulations tighten, the importance of treating wastewater before releasing it back into natural bodies of water becomes paramount. Wastewater contains a wide range of contaminants, including organic matter, pathogens, heavy metals, and nutrients, all of which can harm ecosystems and public health if left untreated. Through a series of mechanical, biological, and chemical processes, wastewater treatment plants (WWTPs) are designed to remove these pollutants and produce clean effluent that meets regulatory standards. This paper explores the various stages of the wastewater treatment process—coagulation, sedimentation, filtration, and disinfection—and evaluates each stage's benefits and detriments. Further, this paper also examines the wastewater treatment infrastructure in the United States and China, two of the biggest and most influential countries in the world.

2 Wastewater treatment method

* Corresponding author: twang417@usc.edu,

2.1 Coagulation

The first step in treating wastewater is coagulation, which seeks to clump fine, negatively charged suspended solids in the wastewater so that they can be easily removed later in the wastewater treatment process. This is achieved through the addition of positively charged coagulants, usually metal ions or pre-polymerized inorganic compounds, to the wastewater. The coagulants then neutralize the negatively charged solids, which allows for them to bind together to form floccules—aggregates of material in suspension. After the reaction concludes, the wastewater will become less turbid and clearer—one step closer to becoming potable.

Commonly used metal ion coagulants include aluminum sulfate, ferric sulfate, and ferric chloride. Aluminum sulfate is the preferred coagulant by wastewater treatment plants worldwide due to its availability, affordability, and efficiency; ferric sulfate and ferric chloride work well but have downsides in the heavier floccules formed and the chloride's potential to increase the treated water's corrosivity, respectively. Metal ion coagulants, while effective, can also be extremely sensitive to the pH and the temperature of the water, which detracts from their overall versatility. As such, scientists have developed pre-polymerized inorganic compounds, such as aluminum chlorohydrate and polyaluminum chloride, which can function efficiently over wider ranges of pH and water temperatures. Further, testing has shown that pre-polymerized inorganic compounds produce less residual material overall through the coagulation process than metal ions, which makes the treatment and disposal of the residuals easier and reduces the possibility of toxic material leaking into the environment. Natural biopolymer coagulants also exist as a versatile option, but overall, the coagulant used for treating wastewater depends on the resources and needs of the wastewater treatment plant.

2.2 Sedimentation

After the wastewater undergoes coagulation, it will undergo the process of sedimentation, where gravity will separate the suspended floccules from the wastewater. During sedimentation, the floccules sink to the bottom of the sedimentation tanks, after which they are removed from the solution by a trough or a vacuum and properly disposed. The remaining water is then transported to the next stage, filtration.

Though sedimentation appears to be a simple process, there are still some obstacles to it, namely costs and time. Industrial sedimentation tanks, often circular, are large—they must be able to accommodate tens of thousands of gallons of wastewater and there are often several sedimentation tanks at a wastewater treatment plant [1]. As such, the tanks require sufficient funding, resources, and land to construct and maintain smooth operation. The second factor is time. First, less dense floccules take more time to settle to the bottom of the container because of fundamentals of flotation [2]. Second, according to sedimentation theory and Stoke's Law, warm water, due to its decreased viscosity, allows for the floccules to settle faster than cold water [3]. Therefore, the temperature of the wastewater must be carefully regulated, which can further increase the operational costs of sedimentation. While some of sedimentation's obstacles have clear workarounds and others are unavoidable, sedimentation remains a crucial part of the wastewater treatment process.

2.2.1 Disposal of sediments

External to the main wastewater treatment process is the treatment and disposal of the sediments generated by sedimentation. Such sediments often come in high amounts due to the volume of wastewater treated by a wastewater treatment plant. For example, after coagulation and sedimentation are applied, 1000 gallons of water can produce 5 gallons of sediment [4]. Scaling the amount of wastewater by 40 or 50 times to account for industrial levels of wastewater treatment, the amount of sediment produced increases greatly. The method of treating and disposing the sediment depends on the type of coagulant used.

Metal ion coagulants result in the most complex treatment and disposal process for the sediments, as they often contain hazardous metal hydroxides and can sometimes be corrosive. As such, wastewater treatment companies must acquire the necessary equipment to ensure the proper treatment of the sediments, which can cost more than coagulation and sedimentation process. On the other hand, the sediments created by natural coagulants can be recycled and reused as soil conditioners and fertilizers, among other applications [4].

2.3 Filtration

2.3.1 Slow sand filtration

Wastewater then undergoes filtration to remove any solids and particles that remain after coagulation and sedimentation. The filters have varying pore sizes and are made of materials such as sand, gravel, and charcoal to ensure that all the particles are removed from the water, no matter the size. A common wastewater filtration method is slow sand filtration, which features layers of biofilm, sand, and gravel. The biofilm is made of a combination of particulate and organic matter and dense biomass growth, which further breaks down the particles in the water. Next, the water travels through the permeable sand and gravel layers. As the water seeps through the layer, the particles become trapped within the sand and gravel; the water becomes cleaner as a result.

Overall, slow sand filtration is a simple step in the wastewater treatment process. It is relatively inexpensive, as the materials used are mostly natural, and does not require specialized equipment or chemicals. Slow sand filtration is also energy efficient due to its reliance on gravity for filtering the wastewater. While industrial sand filters require regular maintenance and cleaning every four weeks, it still is one of the most efficient filtration methods.

2.3.2 Activated carbon filtration

Another filtration method is activated carbon, typically made from carbon-rich materials such as coal, wood, or coconut shells that are treated at high temperatures to create a porous structure. This takes advantage of the adsorption process, during which contaminants stick to the surface of the carbon. The porous structure also increases the surface area of the activated carbon, thereby increasing the effectiveness of adsorption and in turn, filtration.

There are two primary types of activated carbon: granular activated carbon (GAC) and powdered activated carbon (PAC). GAC is commonly used in fixed-bed filter systems where water flows through a bed of carbon granules, whereas PAC is often used as a dosing agent in water treatment plants, where it is mixed with water and later filtered out. Both types of activated carbon are highly effective at removing organic and inorganic contaminants through adsorption.

The main advantage of activated carbon filtration is its ability to remove a wide range of contaminants, including those that are not easily eliminated by other treatment methods. It is especially effective at targeting organic compounds, chlorine, volatile organic compounds, and toxic heavy metals such as arsenic, copper, and lead in the wastewater [5]. Just like slow sand filtration, activated carbon filtration is also energy efficient as it relies solely on gravity for filtering the wastewater. However, over time, the activated carbon becomes saturated with contaminants and must be replaced or regenerated, which increases operational costs, especially with carbon regeneration, as it relies on chemical and heating methods.

2.3.3 Reverse osmosis

Reverse osmosis (RO), widespread and effective, is another method of filtration in the wastewater treatment process, though it has certain niches to it—mainly in desalination. RO relies on the principles of osmosis and the use of a semi-permeable membrane to filter out contaminants from water, resulting in clean, purified water. It is especially valuable in regions where fresh water is scarce and the reuse of water or desalination of seawater is necessary to meet demand.

Osmosis is a natural process in which water molecules pass through a semi-permeable membrane from a region of low solute concentration (pure water) to a region of high solute concentration (contaminated water). The goal of this process is to balance the concentrations of solutes on both sides of the membrane. Reverse osmosis works by reversing this natural flow. By applying pressure to the contaminated water, it forces water molecules to move from the side with higher solute concentration to the side with lower solute concentration. This pressure needs to exceed the osmotic pressure of the water for the process to occur effectively. In addition, the semi-permeable membrane used in reverse osmosis is highly specialized. Its pores are incredibly small, typically around 0.0001 microns, which allows only water molecules and some miniscule molecules to pass through.

During reverse osmosis in a specialized unit, high pressure is applied to the water on the feed side of the membrane, forcing the water through. As the water is forced through the membrane, the water molecules and other small substances pass through the semi-permeable membrane. Remaining large contaminants from coagulation and sedimentation are left behind on the feed side of the membrane and are subsequently removed from the RO unit for disposal. The clean water that passes through is called permeate, while the concentrated contaminants that remain on the feed side are known as concentrate or reject. Afterwards, the permeate can proceed to disinfection.

Reverse osmosis has several significant advantages, the most notable being its ability to remove up to 99% of dissolved salts, organic matter, bacteria, and other contaminants from water [6]. This makes it an extremely effective method for producing clean water for drinking, industrial processes, and environmental discharge. Furthermore, reverse osmosis systems can be scaled to meet different needs of each community, from small towns to industrial wastewater treatment plants.

However, reverse osmosis also presents some challenges. One of the primary drawbacks is the high energy consumption required to generate the pressure needed to force water through the membrane, which makes RO unfeasible in areas that lack the necessary infrastructure and resources. Additionally, membrane fouling, which occurs when contaminants accumulate on the surface of the membrane, can reduce the system's efficiency and increase maintenance costs. To mitigate this, regular cleaning and maintenance are essential. Lastly, the disposal of concentrate or reject water, which contains high concentrations of contaminants, can be a challenge, especially in areas with strict environmental regulations. There are still many hurdles to overcome with reverse osmosis, but it shows promise as a filtration method.

2.4 Disinfection

The wastewater is now mostly contaminant-free, but it is crucial to disinfect the water to remove any remaining bacteria, as well as reduce the possibility of future bacteria growth in the water and the containers that store it. If disinfection does not occur, bacteria such as E.

coli and salmonella can enter the water systems of a city and cause mass disease outbreaks. Annually, waterborne pathogens in the United States cause approximately 7.15 million illnesses and 6630 deaths [7]; this number rises dramatically in third-world countries. There are three common methods of wastewater disinfection: ozone, ultraviolet (UV) radiation, and chlorine.

2.4.1 Ozone disinfection

Ozone disinfection uses ozone's chemical properties to clean the water. Ozone, generated from air or pure oxygen in an ozone generator, is pumped into a chamber with the filtered wastewater. Next, the ozone decomposes in the water and forms the free radicals hydroperoxyl and hydroxyl, both of which are powerful oxidizing agents. The free radicals then make contact with the cell walls of the bacteria in the wastewater, which induces protoplasmic oxidation and in turn, cell lysis, destroying the bacteria. The water is then discharged from the chamber to be converted into drinking water or released into the environment. Ozone is one of the most effective wastewater disinfection methods; however, it requires significant monetary and technological investments to ensure its effectiveness and safety [8].

2.4.2 Ultraviolet disinfection

Ultraviolet radiation disinfection relies on a system of high-intensity UV lamps, which are powered by a reaction of mercury and electricity. When shone into the wastewater, the UV rays penetrate through the cell walls of the bacteria in the water and destroy their essential biological functions, rendering them harmless. The water is then discharged from the container. UV disinfection is advantageous due to its avoidance of chemicals and large machinery. As such, it is a safer and more space-efficient method of wastewater disinfection compared to ozone or chlorine. UV disinfection becomes ineffective when there are low intensity UV rays or highly turbid wastewater, but these are minor issues.

2.4.3 Chlorine disinfection

Chlorine is the most widely used water disinfectant due to its oxidizing properties. Like ozone disinfection, chlorine gas is pumped into a chamber with the wastewater. The chlorine then damages the cell membranes of the bacteria, destroys their essential biological functions, and renders the bacteria harmless. However, while the water is now free from the bacteria, it is still full of chlorine residuals, which can last in the water for many hours. As such, the water is unsafe for release and must be dechlorinated. To dechlorinate the water, sulfur dioxide is added to the water, where it reacts with the chlorine residuals and renders them harmless. Overall, chlorine is an effective method of wastewater disinfection, but it also requires extensive safety measures due to the explosive nature of chlorine. Additionally, while chlorination itself is cost-effective, dechlorination is not.

3 Applications and usage in the United States and China

In the United States, many states, especially those in the Sun Belt region, routinely suffer from high temperatures and drought conditions. Therefore, the recycling and reuse of water resources is essential. There are more than 16000 wastewater treatment plants in the US, with 80 percent of the U.S. population receiving their potable water from the wastewater treatment plants and approximately 75 percent of the U.S. population having their sanitary sewerage treated by them [9]. In New York City, with a population of 8.336 million, there are 14 municipal wastewater treatment plants, which treat a total of 1.3 billion gallons of wastewater daily [10]. The wastewater treatment plants follow the wastewater treatment processes outlined above, with some minor variances between plants. For example, New York City, due to its large population, does not filter wastewater to reduce operational costs [11].

While wastewater treatment infrastructure is readily available in the United States, China does not have this luxury. As of 2021, 98.1% of municipalities and 28% of rural areas in China have access to wastewater treatment facilities [12]. The low percentage for rural areas is concerning considering that China's rural areas are severely underdeveloped relative to its cities. As such, it is crucial for China to invest in and implement more wastewater treatment facilities in its rural regions to ensure the health and safety of all. China has improved its wastewater treatment infrastructure rapidly in recent years, but more action is necessary.

4 Conclusion

The wastewater treatment process is a vital multi-stage system designed to remove contaminants from water used in residential, industrial, and municipal settings. By employing mechanical, biological, and chemical methods, treatment plants effectively eliminate pollutants such as organic matter, pathogens, heavy metals, and excess nutrients, ensuring that the treated water, or effluent, is safe for discharge or reuse. Developed countries such as the United States have implemented effective wastewater treatment systems that span nearly the entire country, while developing countries, such as China, still lack wastewater treatment systems in their rural areas. Reusable water is a key to a more sustainable future, and its further development and implementation will help make the world a better place.

References

1. Sedimentation tank standards for construction sites. King County. (2024) https://kingcounty.gov/en/dept/dnrp/
2. How Sedimentation Water Treatment Works and How to Make it Efficient. Clearwater Industries. (2024) https://clearwaterind.com
3. Tiwari, R. K. Sedimentation Tanks: Design and Working. (2024) https://www.slideshare.net/slideshow
4. Common Pitfalls When Choosing A Chemical Coagulation Water Treatment System To Treat Your Water. Water Online. (2019) https://www.wateronline.com
5. Frequently Asked Questions About Activated Charcoal. Wisewell, (2024) https://www.wisewell.com/blogs/news/carbon-water-filters
6. The Truth Behind Reverse Osmosis Water Systems: Filtering Out Fact vs. Fiction. Hawaiian Coolwater, (2023). https://hawaiiancoolwater.com
7. J. M. Kunz, H. Lawinger, S. Miko, et al. Surveillance of waterborne disease outbreaks associated with drinking water-united states. Surveillance Summaries, **73(1),** 1-23 (2024) https://doi.org/10.15585/mmwr.ss7301a1
8. Wastewater Technology Fact Sheet: Ozone Disinfection. United States Environmental Protection Agency. (2024) https://www3.epa.gov/npdes/pubs/ozon.pdf
9. Water and Wastewater Systems. Cybersecurity and Infrastructure Security Agency. (2024) https://www.cisa.gov/topics/
10. China - Environmental Technology. International Trade Administration. (2024) https://www.trade.gov

11. How to Choose the Best Water Filtration System for New York Drinking Water. High Water Standard, (2024) https://www.highwaterstandard.com/blog/drinking-tap-water-in-new-york
12. Wastewater Treatment System. NYC Environmental Protection. (2024) https://www.nyc.gov/site/dep/water/ wastewater-treatment-system.

E3S Web of Conferences **606**, 05003 (2025) https://doi.org/10.1051/e3sconf/202560605003
ICNAOE 2024

Waste Heat Recovery and Energy Reutilization Technologies in Industrial Processes: Research on Improving Energy Efficiency and Reducing Emissions

Yian Wang [1]

[1]Qingdao Academy, Qingdao, Shandong, 266000, China

Abstract. Industries are significant contributors to global energy consumption and greenhouse gas emissions, making energy efficiency improvements essential. WHR offers a valuable solution by capturing and reusing heat that would otherwise be lost, thus enhancing energy efficiency and reducing environmental impacts. This paper explores the application of WHR technologies across various industries, including steel, cement, and petrochemicals, where waste heat accounts for a large portion of energy losses. Technologies such as heat exchangers, ORC, and TEGs are discussed for their roles in converting waste heat into usable energy forms, such as electricity or heat for industrial processes. The review also covers advancements in materials and hybrid systems to optimize efficiency, as well as the potential of WHR in stabilizing renewable energy sources like solar and wind power. The integration of WHR systems into smart grids is highlighted as a future trend for improving energy management.

1 Introduction

To address environmental concerns, industries, as significant contributors to climate change, must take responsibility for reducing their environmental impact. In 2000, global energy consumption reached 10.3 million tons of oil equivalent (Mtoe) per year [1]. Research conducted in 2004 estimated that the technical potential for energy savings could be as high as 25–37 exajoules (EJ) per year [2], demonstrating that improving energy efficiency in industries could significantly reduce both energy consumption and environmental problems.

Waste heat recovery (WHR) is a crucial strategy for enhancing energy efficiency in the industrial sector. For instance, a 2017 study by Qi Zhang and his team in China found that 20%–50% of the energy used in the iron and steel industry—approximately 1 GW—was wasted in the form of heat [3]. Similarly, N.A. Madlool's 2011 research on cement manufacturing showed that 15.7% of energy consumption is associated with cement production, with developing countries wasting 50%–60% of that energy as heat at temperatures reaching 1500 °C [4]. In the food production sector, energy consumption accounts for 12%–15% of total industrial energy use, with most energy waste being thermal [5]. These examples illustrate that thermal energy represents the largest portion of waste in many industries, making waste heat recovery a feasible and effective strategy for reducing energy losses.

Waste heat recovery is applied across various industries. A study conducted in Europe found that WHR is implemented in iron and steel manufacturing (with heat recovery potential of 11.4%), nonmetallic minerals (11.4%), textiles and leather (11.0%), chemical and petrochemical industries (11.0%), paper and pulp production (10.6%), nonferrous metals (9.6%), and other sectors (20.4%) [6]. The basic principle of WHR involves transferring heat from waste gases to fluids within heat exchangers, allowing the captured energy to be reused in industrial processes. Common WHR technologies include finned pipes, heat pipes, furnace regenerators, passive air preheaters, recuperative and regenerative burners, recuperators, rotary regenerators (heat wheels), shell-and-tube heat exchangers, and waste heat boilers [6].

This structured approach to waste heat recovery provides a practical solution for reducing energy waste and improving overall industrial efficiency.

2 Sources and classification of waste heat in industrial process in industrial processes

To optimize waste heat recovery, it is essential to classify heat sources according to their temperature ranges and apply appropriate recovery methods for each class. Based on research conducted in Germany, heat sources can be broadly categorized into three temperature ranges: low temperature (below 100°C, primarily from steam and exhaust gases), medium temperature (100–400°C, mainly from drying and cooling processes), and high temperature (above 400°C, primarily from furnaces and boilers). Higher temperature sources offer greater potential for efficient heat recovery. However, when the temperature of the source falls below 200°C, it is generally considered insufficiently efficient for reuse, as the energy recovery potential is significantly reduced [7]. This classification is critical for designing and selecting the appropriate waste heat recovery technologies to maximize energy efficiency across different industrial applications.

The heat recovery technologies are mainly separated into two kinds, the direct and indirect. In the direct recovery method, the waste heat sources will be screen out by the heat exchangers or recuperators and then directly output back to the production line themselves or to other heat consuming industries. Besides, the indirect method, the waste heat is transformed to other forms such as electricity by thermoelectric generators or organic Rankine cycle (a more efficient way using waste heat resource).

From various industries, some of them have been defined as high potential industries for WHR. The steel and metal production, which has high heat potential can generate 1.25MW thermal energy in the standard in European countries [8]. At the same time, the waste heat production in the cement industries can reach 2725kw electricity with heat efficiency at 0.218 [9]. Besides, in another high heat potential industry that produce petrol or gas, the efficiency of the heat recovery is varied between single digit to 20%, in other words, a few MW of thermal energy can be reutilized in a manufacture pant consumes 100 MW thermal energy [10].

3 Technology for waste heat recovery and energy reutilization

3.1 Heat exchangers and recuperators

Waste heat treatments are broadly categorized into direct and indirect methods. In direct methods, waste heat is reused either within the industry or supplied to other heat-consuming sectors. Indirect methods involve capturing waste heat and converting it into other forms of energy, such as electricity.

Heat exchangers play a critical role in indirect recovery methods, facilitating the transfer of heat from waste sources to other fluids. Common types of heat exchangers include shell and tube exchangers, and plate exchangers. Shell and tube exchangers consist of a cylindrical shell containing multiple tubes; as hot fluids pass through the tubes; heat is transferred to a cooler fluid circulating outside the tubes. Plate heat exchangers, on the other hand, use thin metal plates with large surface areas to enhance the heat transfer process. The greater surface area allows for more efficient heat capture and transfer, optimizing the recovery process.

3.2 Organic rankine cycle (ORC) systems

The organic Rankine cycles (ORC) which are the improved technic of Rankine cycles that recycle the low-grade heat resources in the industries. With organic fluid with low boiling point in the system, the efficiency of ORC is larger than its prototype the Rankine cycles. In the paper industry, 10% to 15% of the waste heat have been recycled to electricity [11]. In the geothermal energy industries, the efficiencies can reach 12% to 20% [12]. From a review of ORC application in different industries, the efficiencies of the recycle were 8% to 18% [13].

3.3 Thermoelectric generators (TEGs)

Thermoelectric generators (TEGs) represent another advanced technology for converting heat into electricity. These generators typically employ thermoelectric materials such as bismuth telluride for room temperature applications and lead telluride for high-temperature environments. TEGs are constructed using two types of semiconductors, p-type and n-type, connected in parallel. As heat flows from the hot side to the cold side, the temperature difference across the material drives the movement of electrons or holes, generating an electric current [14,15]. Recent advancements have introduced new materials like high-entropy alloys, organic compounds, nanowires, and nanoribbons into TEGs, further enhancing their efficiency. Additionally, improved energy management systems and performance optimization strategies have contributed to increased efficiency and overall effectiveness in energy conversion.

3.4 Heat pump and absorption chillers

The heat pump and absorption chillers are also common in WHR in industries. Heat pumps can transfer the heat from cooler area to a warmer area, in industries, heat pumps are commonly applicated to heating or drying process. In the food and beverage productions, the heat pumps are used to maintain the temperature during the food process and packaging. From the research to the efficiency of heat pumps in the food production, the coefficient of performance (COP), the standard to describe the efficiency of the heat pumps, is in the range of 3.0-5.0, that means by using 1 unit of electric power, the heat pumps can remove 3.0-5.0 units of heat from the environment [16] At the same time, heat pumps also reduce the heat waste in the drying and heating process of textile industry at the COP value 3.0-4.5 [17].

With similar working principle, the absorption chillers are applied to the pharmaceutical industry, with COP value at 0.6 to 0.9 [18].

Besides the normal industries, the heat pumps and absorption chillers are also applied to the space to ensure the crew comfort, recycle the waste heat and control the thermal energy. However, this system should be redesign because of the fluid in the heat pumps and chillers runs motivated by the gravity, however, in the space microgravity and radiated environment, more protection and the motivation of fluid should be reconsidered.

4 Applications of waste heat recovery in industries

4.1 Successful WHR implementations in steel industry

For the WHR in many different industries, several of them are quite successful. The WHR in steel industries for example. In Turkey, the efficiency improvement of WHR system about 70% of the waste energy, and 15% to 30% of the total energy have saved. At the same time, it also reduces 10% to 20% of the CO_2 emission [19]. China, which takes huge amount of world steel production, WHR reaches 40% efficiency of the recovery of waste heat, or 25% of its energy consumption, to electricity or usable heat wand 25% of its CO_2 emission has reduced [20]. For another example, in the UK, the WHR in steel industries has the efficiency around 50% to 60% of its waste heat or 20% to 35% of its total energy consumption. At the same time, 18% of its CO_2 pollution has reduced [21].

4.2 WHR in chemical and petrochemical industries

In the chemical and petrochemical industries, the WHR has also improved the efficiency of energy usage. The researchers in Canada conclude that the WHR system can save around 15% to 30% of the total energy consumption with the efficiency around 60%. From the statistic, the costs on energy saving constructions can be payback in 2 to 5 years. At the same time, in Brazil, the WHR system has the efficiency at around 55% and saves 20% to 25% of its energy consumption [22]. In the economic analyzation, the costs can be reduced significantly, with the average return on investment (ROI) of around 3 to 4 years [23]. Meanwhile, the WHR in Chinese chemical and petrochemical industries has the largest efficiency that up to 70%, recovers 25% to 30% of the energy, and the cost can be paid back in 2 to 3 years [24].

4.3 WHR in cement and glass manufacturing

In the cement and glass manufacturing, the WHR system is more complex than in the other industries mentioned previously. In Brazil, the plants use heat exchangers to capture the heat resources in the kiln exhausted gas and convert it into usable thermal energy or electricity. The efficiency of the recovery is up to 60%, and around 25% of the energy consumption can be reduced [25]. In South Korea, the WHR system in glass manufacturing is consist with heat exchanger and recovery boiler to convert heat in flue gas into available energy for preheating the materials and generate electricity. With efficiency up to 55%, 30% of the energy has recovered [26]. In Vietnam, the recovery system uses heat exchangers and regeneration burners to capture the heat source from kiln gases, improves the energy efficiency, reduces the reliance on external heat source. From the research, 20% to 35% of the energy in both industries has reduced and the efficiency is around 40% to 50% [27].

4.4 Cross-industry comparisons and best practice

In successfully installing the WHR system to the various industries, some key factors are included. From the research around the world, a good WHR system has good system integration, which means the appropriate adoption to different heat resources and industry structure, financial incentives, which means the governments or nations give subsidies or crucial encouragement to the WHR technologies, technology availability, which means that the WHR system should include stable, advanced and effective technologies [28,29].

Following the factors, the cement industry in India achieve 45% in the heat recovery by using pre-heaters and recovery boilers. Germany glass production has the system with average efficiency at 50% reduce 30% of the energy usage [30]. At the same time, the efficiency of steel industries in Japan, Petrochemical sector in Saudi Arabia, aluminum smelting industry in Australia and paper industry in Sweden, achieves 50% or more [31,32].

5 Future trends and research directions in waste heat recovery

With technological advancements, more efficient materials and solutions have emerged for WHR systems. In TEGs, new semiconductor materials with higher conversion rates are increasing efficiency. Additionally, hybrid WHR systems, which combine different recovery methods, can optimize performance by utilizing the remaining heat from other recovery solutions.

WHR systems are also proving beneficial in renewable energy generation. For example, using waste heat to preheat fluids or directly supply power can enhance the efficiency of solar power plants and biomass systems. WHR systems can further stabilize intermittent renewable sources like wind and solar energy, ensuring a more reliable energy output.

Furthermore, as smart grids become more established, WHR systems can be integrated into the grid to supplement electricity during peak demand hours. Excess electricity generated during low-demand periods can be stored and released as needed, improving grid efficiency.

Future improvements in WHR systems should focus on enhancing both efficiency and scalability. The development of new materials and innovative system designs will be crucial in pushing the boundaries of waste heat recovery technology.

6 Conclusion

In conclusion, WHR represents a significant opportunity for industries to enhance energy efficiency, reduce environmental impact, and contribute to global sustainability efforts. As demonstrated by various sectors, including steel, cement, chemical, and petrochemical industries, WHR systems can recover substantial amounts of energy, reduce operational costs, and lower greenhouse gas emissions. The successful implementation of WHR technologies, such as heat exchangers, organic Rankine cycles, and thermoelectric generators, has proven effective in converting waste heat into usable energy or electricity.

Advancements in materials, hybrid recovery systems, and smart grid integration further underscore the potential of WHR in stabilizing renewable energy generation and improving overall energy utilization. However, to fully realize these benefits, ongoing research must focus on improving the efficiency and scalability of WHR systems, optimizing material properties, and exploring innovative designs. By continuing to refine and expand WHR technologies, industries can play a pivotal role in addressing global energy challenges, reducing their environmental footprint, and moving toward a more sustainable future.

References

1. G. P. Beretta, World energy consumption and resources: an outlook for the rest of the century. International journal of environmental technology and management, **7**, 99-112 (2007)
2. E. Efficiency, Tracking industrial energy efficiency and CO2 emissions. International Energy Agency, **34**, 1-12 (2007)

3. Q. Zhang, X. Zhao, H. Lu, T. Ni, & Y. Li. Waste energy recovery and energy efficiency improvement in China's iron and steel industry. Applied Energy, **191**, 502–520 (2017)
4. N. A. Madlool, R. Saidur, M. S. Hossain, & N.A. Rahim. A critical review on energy use and savings in the cement industries. Renewable and sustainable energy reviews, **15**, 2042-2060 (2011)
5. J. M.Clairand, M. Briceno-Leon, G. Escriva-Escriva & A. M. Pantaleo. Review of energy efficiency technologies in the food industry: trends, barriers, and opportunities. IEEE Access, **8**, 48015-48029 (2020)
6. P. Christodoulides, R. Agathokleous, L. Aresti, S. A. Kalogirou, S. A. Tassou, & G. A. Florides. Waste heat recovery technologies revisited with emphasis on new solutions, including heat pipes, and case studies. Energies, **15**, 384 (2022)
7. S. Brückner, S. Liu, L. Miró, M. Radspieler, L. F. Cabeza, & E. Lävemann. Industrial waste heat recovery technologies: An economic analysis of heat transformation technologies. Applied Energy, **151**, 157-167 (2015)
8. M. Biondi, A. Giovannelli, G. Di Lorenzo, & C. Salvini. Techno-economic analysis of as CO2 power plant for waste heat recovery in steel industry. Energy Reports, **6**, 298-304 (2020)
9. M. D. P. Arrieta, & F. R. P. Arrieta, KCS34 evaluation for WHR in cement industry. International Journal of Advanced Engineering Research and Science, **5**, 268245 (2018)
10. R. Agahi, Waste Heat to Power System in Oil and Gas Industry Improves Plant Power Efficiency. In Abu Dhabi International Petroleum Exhibition and Conference (p. D032S213R001). SPE (2019)
11. M. J. Tchanche, L. Y. A. Papadakis, & J. R. J. van der Meer. Low-Temperature Geothermal and Waste Heat Recovery for Power Generation Using Organic Rankine Cycles: Case Study in the Paper Industry. Energy, **102**, 490-500 (2016)
12. Frick, S., & Huenges, E. Geothermal Power Plants: Principles, Application, Case Studies and Environmental Impact. Energy Technology, **1**, 365-366 (2013).
13. L. L. O. Silva, M. J. Alvarado, & D. J. Smith, Organic Rankine Cycle for Low-Temperature Waste Heat Recovery: A Review of Recent Advances. Journal of Energy Resources Technology, **140**, 052301-052315 (2018)
14. Rowe, D. M. (Ed.). Thermoelectrics handbook: macro to nano (CRC press, 2018)
15. Goldsmid, H. J. Introduction to thermoelectricity (Berlin, Springer, 2010)
16. Y. Ma, & W. Liu, Application of heat pumps in the food and beverage industry: A review. Journal of Cleaner Production, **276**, 124046 (2020)
17. X. Li, & J. Xu, Heat pump drying in the textile industry: Performance and applications. Energy Procedia, **152**, 254-259 (2018)
18. A. Khaliq, & A. M. Application of absorption chillers in pharmaceutical manufacturing: Energy and cost savings. Pharmaceutical Engineering, **39**, 20-29 (2019).
19. F. A. Ozturk, M. M. Yavuz, & S. Y. Altintas, Waste heat recovery in steel industry: A review. Journal of Cleaner Production, **242**, 118451 (2020)
20. X. Zhao, C. Liu, & H. Wang, Application of heat recovery in steel manufacturing: Analysis and case studies. Energy Reports, **5**, 224-232 (2019)

21. K. J. Smith, A. H. Brown, & M. P. Johnson, Enhanced waste heat recovery in steel plants: Optimization and performance evaluation. Applied Energy, **222**, 712-724. (2018)
22. M. A. Rosen, B. T. Kar, & S. L. Behie. Waste heat recovery in chemical and petrochemical industries: A review. Energy, **233**, 121076 (2021)
23. J. P. S. Costa, M. A. Lima, & R. J. S. Silva. Economic and environmental benefits of waste heat recovery in the petrochemical industry. Chemical Engineering (2019)
24. K. J. Liu, H. M. Zhang, & Y. Q. Wu, Optimization of waste heat recovery systems in chemical processes: A case study. Applied Thermal Engineering, **129**, 727-735. (2018)
25. R. G. Z. Silva, A. J. D. V. Alves & P. A. S. Ferreira, Analysis and optimization of waste heat recovery systems in cement plants. Energy, **141**, 270-281 (2017)
26. J. H. M. Kim, S. W. Lee, & K. J. Nam, Waste heat recovery from glass manufacturing: System design and economic feasibility. Journal of Cleaner Production, **212**, 1034-1043 (2019)
27. N. T. Pham, L. H. Nguyen, & T. A. Bui, A comparative study of waste heat recovery in cement and glass industries. Energy Reports, **6**, 145-155 (2020)
28. S. M. H. Al-Kayiem, M. M. Abdul-Rahman, & A. N. A. Ibrahim, Key success factors and best practices for waste heat recovery in industrial applications. Energy, **197**, 117199 (2020)
29. L. T. Bui, N. A. Vu, & M. D. Le, A comparative study of waste heat recovery systems across different industries. Renewable and Sustainable Energy Reviews, **101**, 490-505 (2019)
30. M. B. Patel, R. R. Shah, & P. M. Kumar, Global best practices in waste heat recovery: Case studies and analysis. Energy Reports, **7**, 178-188 (2021)
31. S. J. Lee, C. K. Yang, & H. R. Kim, Successful applications of waste heat recovery technologies in industrial processes. Journal of Cleaner Production, **272**, 122900 (2020)
32. J. T. Nguyen, F. R. Smith, & E. W. Johnson, Innovative waste heat recovery solutions: Lessons from leading global projects. Applied Energy, **241**, 64-74 (2019)

Enhancing Energy Efficiency in Photovoltaic Systems through Smart Technology Integration: Innovations and Future Perspectives

Caiwei Lin[1,*]

[1]Huamei International School, Guangzhou, Guangdong, 510520, China

Abstract. Photovoltaic (PV) technology, which converts solar radiation into electricity, has become a key player in the global transition to clean energy. As demand for renewable energy rises, innovations in smart artificial intelligence (AI), the Internet of Things (IoT), and big data analytics are being utilized to enhance the efficiency and reliability of PV systems. The integration of these technologies into PV systems is explored in this review, focusing on how they enhance fault detection, real-time monitoring, and energy optimization. It discusses the role of AI in improving system design and performance, real-time applications leveraging IoT control and data management, and the application of big data analytics for predictive maintenance and energy forecasting. Additionally, the review addresses the environmental and economic benefits of smart PV systems, highlighting their potential for reducing carbon emissions, stabilize energy grids, and provide long-term financial savings. Despite the technological and economic challenges, smart PV systems are poised to play a critical role in future global energy infrastructures by contributing to sustainable development goals and optimizing renewable energy production.

1 Introduction

PV technology, which converts solar radiation directly into electricity, plays an increasingly important role in the global shift toward clean energy. Global installed capacity continues to grow, with significant contributions from regions like Asia and Europe. Despite its widespread use, the current conversion efficiency of PV systems remains below 23%, indicating room for improvement. Havejiri et al. has demonstrated that power and energy rating methods are commonly employed to estimate PV potential, with energy ratings offering greater accuracy [1]. However, conventional PV systems face numerous challenges related to energy efficiency, particularly under adverse climatic conditions such as temperature fluctuations, dust accumulation, and wind speeds. These issues are exacerbated in desert areas, where rising temperatures significantly reduce power output. Furthermore, traditional PV systems face issues of intermittency and undispatchability, complicating grid integration and affecting overall stability. The introduction of smart PV systems, which integrate AI, IoT, and big data analytics, presents a promising solution. These technologies

*Corresponding author: fudea@ldy.edu.rs

enable real-time data analysis to optimize the tilt and alignment of solar panels, utilize machine vision to track the sun's position, and improve predictive maintenance and power distribution efficiency. This, in turn, significantly enhances the overall performance and reliability of PV systems, thereby facilitating the advancement of renewable energy.

This review seeks to examine how AI, IoT, and big data analytics can revolutionize PV systems. by improving fault detection, enhancing real-time monitoring, and optimizing data processing capabilities. The review will also analyze how these technologies can enhance system design., improved operational management, and increased energy efficiency. AI-driven algorithms can monitor system conditions, predict faults, and optimize power generation by analyzing variables such as temperature and voltage. IoT devices enable continuous real-time monitoring, while big data analytics provide insights into optimizing performance and cost reduction. Finally, this review will assess the future role of smart PV systems in the global energy mix. With the rising demand for clean energy, smart PV systems are poised to become essential in achieving sustainable development goals by optimizing energy production and distribution. Continued research, development, the full potential of smart PV systems will be unlocked through vital technological innovations in the future global energy structure.

2 Overview of photovoltaic systems

In a grid-connected PV system, several key components work together to convert solar radiation into usable electricity. The core component, the photovoltaic module (or solar panel), converts sunlight directly into direct current (DC) electricity via the photovoltaic effect. A DC-DC converter then regulates the output voltage to match the system's requirements, ensuring stable and efficient operation. Maximum Power Point Tracking (MPPT) technology is employed to maximize power generation by adjusting the electrical load to operate at the optimal power point under varying sunlight conditions. The inverter is crucial for converting the DC electricity generated by the panels into alternating current (AC) suitable for grid use. Synchronization and island detection mechanisms are also integrated to maintain grid stability and ensure system safety by preventing back-feeding during grid outages, thus protecting both the system and utility workers [2].

Significant evolution has been seen in PV technology, marked by the transition from silicon-based cells of the first generation to the second-generation thin-film cells, which offer greater flexibility but lower efficiency. The efficiency of monocrystalline silicon cells has reached 24.4%, while polycrystalline silicon cells achieve 22.8%. Although these technologies have demonstrated considerable advancements, challenges remain in the development of newer technologies, such as organic and perovskite solar cells. While organic and perovskite cells have the potential for higher efficiency and reduced manufacturing costs, they face critical challenges related to light management and carrier recombination. Solving these issues will be essential for improving their efficiency, reducing costs, and promoting widespread adoption [2].

3 Energy efficiency challenges in traditional PV systems

Energy inefficiencies occur at various stages, from production to end-use, which collectively contribute to increased costs, emissions, and resource wastage. In energy production, losses arise due to outdated equipment, suboptimal power generation processes, and inefficiencies inherent in various energy conversion technologies. For instance, traditional power plants can lose significant amounts of energy as heat. During distribution, transmission losses are common, particularly over long distances, where energy dissipates as it travels through the grid. In the usage stage, particularly in the residential sector, energy waste is prevalent due to inefficient appliances, poor insulation, and inadequate energy management. Addressing these inefficiencies is crucial, and significant improvements can be made through the adoption of technology upgrades, the implementation of best practices, and the deployment of smart energy management systems. Achieving higher energy efficiency, particularly in PV systems, is fraught with technical and operational challenges. As highlighted in the Grid-connected Photovoltaic Systems - Components and Challenges: A Review, design flexibility is an advantage offered by second-generation thin-film solar cells, but they lag behind in efficiency compared to silicon-based cells. While monocrystalline silicon cells have achieved an efficiency of 24.4%, and polycrystalline silicon cells reach 22.8%, both technologies are still costly, especially compared to thin-film alternatives The central challenge in PV technology is increasing the efficiency of thin-film cells without significantly raising costs. Operationally, integrating more efficient materials and improving manufacturing techniques is key to overcoming these barriers, along with continued research into optimizing balance-of-system components like inverters and MPPT controllers to further reduce losses throughout the energy conversion process [2].

4 Environmental and economic impacts of smart photovoltaic systems

4.1 Environmental benefits

Optimizing energy production and minimizing waste through advanced technologies in PV systems has the potential to significantly reduce the carbon footprint. Literature [3] explores the integration of multiple renewable energy sources with the application of system-wide optimization using the whole-station target method, which enhances efficiency and reduces emissions. Similarly, Literature [4] applies genetic algorithms to optimize a hybrid energy system comprising photovoltaic, wind, diesel generators, and energy storage. This approach results in reduced power costs and lower carbon emissions, significantly improving the sustainability of the energy system. By intelligently managing multiple renewable sources, these optimized systems contribute to lower greenhouse gas emissions, directly addressing the environmental impact of traditional energy production methods. Smart PV systems equipped with AI and big data analytics offer significant advantages in enhancing grid stability and increasing the incorporation of renewable energy sources. Literature [5] examines how AI and big data are utilized to predict energy demand and production, enabling intelligent planning and optimization of energy storage. By leveraging machine learning algorithms to forecast weather patterns and solar radiation, smart PV systems can efficiently allocate resources, reducing dependency on non-renewable sources and improving the overall stability of the grid. Furthermore, other renewable energy sources, such as wind and hydropower, can be seamlessly integrated with these systems, enhancing the resilience and reliability of modern energy grids, especially as the proportion of renewable energy in the grid keeps increasing.

4.2 Economic viability

Deploying smart technologies such as AI, IoT, and real-time monitoring in photovoltaic (PV) systems significantly optimizes energy output, reduces downtime, and improves system reliability. Unlike traditional systems, which may suffer from inefficiencies in maintenance and energy management, smart PV systems enable predictive maintenance, allowing for remote control and immediate troubleshooting. While the initial investment in smart technologies is higher, these systems result in long-term cost savings by minimizing energy losses, extending the lifespan of equipment, and reducing operational interruptions. Moreover, smart PV systems often qualify for government subsidies, further reducing the overall financial burden and enhancing their economic viability. The incorporation of AI, IoT, and big data analytics into PV systems provides a significant return on investment (ROI) by optimizing energy production and increasing system efficiency. AI algorithms can predict maintenance needs, preventing system failures and minimizing downtime. IoT enables real-time monitoring of system performance, ensuring PV systems operate at their peak efficiency levels. Big data analytics helps in analyzing energy consumption patterns, enabling better decision-making for energy distribution. These technologies collectively reduce operational costs, enhance energy generation, and maximize system reliability, leading to a quicker ROI as compared to traditional PV systems. For residential, commercial, and industrial applications, the long-term financial benefits of smart PV systems are substantial. These systems reduce operational costs by minimizing energy waste and improving energy management. In residential settings, smart PV systems allow homeowners to decrease electricity bills and earn credits through energy surplus fed back into the grid. In commercial and industrial applications, improved service quality, increased resource efficiency, and higher renewable energy shares result in greater cost savings and lower emissions. Additionally, the accurate forecasting and real-time management provided by smart grid technology reduce carbon emissions, contributing to environmental goals while enhancing financial performance across various sectors.

5 Role of smart technologies in photovoltaic systems

5.1 Artificial intelligence (AI)

A crucial role in enhancing the design and performance of PV systems is played by AI. Literature [6] introduces an AI-based method that integrates an improved PV array model with differential evolution multi-objective optimization to reduce power loss probability and life cycle costs in standalone PV systems. This approach allows for the optimization of module parameters and battery models using the DEAM (Differential Evolution with Adaptive Mutation) algorithm, which enhances the overall system performance. By optimizing these components, AI helps improve the efficiency and longevity of PV systems, ensuring cost-effective and reliable energy production over time. Machine learning models have shown great promise in predicting energy output and demand for PV systems. Literature [7] discusses an optimization approach that combines Geographic Information System (GIS) and Light Detection and Ranging (LiDAR) data to optimize PV panel layouts. This method maximizes energy capture by identifying optimal roof areas and adjusting panel tilt angles. The incorporation of genetic algorithms and the introduction of "skew angle" parameters further enhance the efficiency of energy collection. Machine learning algorithms are widely employed in PV systems and smart grids to predict solar irradiance, optimize performance, and assist with demand-side management, energy storage, and grid stability. These AI frameworks improve system efficiency and facilitate the integration of advanced technologies like microgrids and virtual power plants, ultimately contributing to more resilient and optimized energy systems.

5.2 Internet of things (IoT)

The incorporation of IoT technology into PV systems has revolutionized real-time monitoring and control, allowing for more efficient energy management. An IoT-based data acquisition system, utilizing open-source software and cloud computing, provides a cost-effective and accessible solution for remote monitoring. The deployment of NODEMCU boards and sensors enables real-time monitoring of nine critical PV system parameters, including voltage, current, temperature, and humidity. Outdoor testing validates the stability and reliability of these sensors, offering an economical tool to optimize PV system performance and improve energy generation efficiency [8]. The IoT plays a pivotal role in enhancing smart grid functionalities by using advanced communication technologies to connect controllers, devices, and sensors. IoT enables more efficient energy management by offering remote monitoring, real-time control, and advanced data analytics to optimize grid operations. This integration allows for better decision-making support for users and improves the overall intelligence of the energy grid. Through IoT, smart grids can respond more dynamically to energy demands, manage decentralized energy sources, and ensure greater reliability and flexibility in energy distribution [9]. Adaptive systems designed for real-time environmental monitoring are essential for optimizing PV system performance. These systems, which often rely on wireless sensor networks, utilize a low-power MPPT circuit along with an adjustable voltage converter. Digital algorithms continuously adjust system parameters in real-time, based on environmental factors such as sunlight intensity and temperature, to enhance solar energy capture and conversion efficiency. This adaptive approach, characterized by low power consumption, high adaptability, and scalability, makes it particularly suited for complex outdoor environments where energy optimization is critical. Research highlights several examples of IoT-based control systems applied to smart PV systems [10]. Literature discusses the use of IoT in intelligent PV monitoring systems, which incorporate a range of sensors, such as temperature, humidity, tilt, voltage, current, and light sensors, as well as MEMS and ultrasonic sensors. These sensors connect to a central control system via IoT, allowing for real-time monitoring and data collection [11]. By leveraging cloud computing and big data analytics, these IoT-enabled systems significantly improve the efficiency, reliability, and adaptability of PV installations, making them more resilient to environmental changes and operational challenges.

5.3 Big data analytics

A crucial role in predictive maintenance and performance optimization of PV systems is played by big data analytics. According to the literature [12], intelligent optimization models such as ANN, SVM, decision trees, random forests, boosting algorithms, and deep learning have been effectively applied to improve fault prediction accuracy and energy yield. By leveraging these models, PV systems can optimize maintenance strategies, enhance system reliability, and maximize operational efficiency. Early detection of potential issues, reducing downtime and maintenance costs, is enabled by the analysis of large datasets.

The analysis of large datasets enables accurate energy production forecasting and the identification of operational trends within PV systems. Intelligent optimization techniques, such as ANN and SVM, have been employed to predict energy output, monitor system conditions, and optimize maintenance schedules. These approaches help improve system performance and extend the lifespan of PV installations. Additionally, data mining techniques are essential in processing vast amounts of data for energy forecasting, allowing operators to make informed decisions based on trends and patterns, further enhancing the operational efficiency of photovoltaic power plants. Data analytics is key to improving decision-making processes in PV system operations [12].

By using advanced algorithms and real-time data analysis, operators can optimize energy production and system maintenance. Literature highlights the use of ANN, SVM, and other machine learning techniques for energy prediction and condition maintenance, which significantly contribute to improving energy output and reducing operational risks [12].

These data-driven insights enable PV operators to make more informed decisions, resulting in increased efficiency and reliability.

6 Challenges and future directions

While AI, IoT, and big data present significant opportunities for optimizing PV systems, several limitations persist. These include challenges related to data processing, forecasting accuracy, cybersecurity, and cost-effectiveness. Current algorithms struggle with accurately predicting energy output owing to the sporadic availability of solar powerand rapidly changing environmental conditions. Additionally, the integration of large datasets from multiple sources increases the complexity of data processing, while the high costs of implementing smart technologies in PV systems raise concerns about economic sustainability. Addressing these limitations through further research and technological advancements is crucial for improving system efficiency and reliability.

As PV systems become more interconnected through smart technologies, the risk of cyberattacks increases. The introduction of IoT sensors and cloud-based platforms exposes PV systems to potential vulnerabilities, making cybersecurity a critical concern. To safeguard against attacks, smart PV networks require robust security measures, including firewalls, encrypted communication protocols, and identity authentication mechanisms. Establishing access control systems and conducting regular security audits can further enhance network protection. Maintaining the stability and resilience of clean energy systems requires ensuring cybersecurity in smart PV networks.

Adopting smart technologies such as advanced inverters, monitoring systems, and energy management tools in PV systems can significantly improve efficiency and reliability. However, these technologies also come with increased costs. The upfront investment in smart infrastructure can be substantial, and ongoing maintenance and operational expenses must also be considered. As a result, a careful cost-benefit analysis is required to ensure that the adoption of smart technologies aligns with budgetary constraints and long-term sustainability goals.

Several gaps in the existing research on smart PV systems require addressing. Article highlights challenges related to data security, environmental adaptability, and cost-effective cloud services in the integration of IoT with PV systems [13]. These gaps suggest a need for further exploration of advanced cybersecurity measures, the development of more resilient technologies suited to diverse environmental conditions, and more efficient cloud services are needed to manage the extensive data generated by smart PV systems. Future research should focus on overcoming these challenges to advance the development of IoT applications in smart energy management.

The future of smart PV technology offers exciting opportunities for innovation. Literature discusses the potential of smart grids in integrating PV, energy storage, and building management systems to improve energy efficiency and reduce emissions [14]. Future trends may involve increased public awareness of energy efficiency, the incorporation of multi-energy systems, and the use of blockchain for secure and efficient energy transactions. Research and development in these areas could lead to more advanced, sustainable smart PV systems, contributing to a more resilient and cleaner global energy infrastructure.

7 Conclusion

The integration of smart technologies into PV systems enhances overall performance by optimizing energy harvesting and improving system reliability. Through real-time monitoring of control panel angles and predictive maintenance, smart PV systems maximize energy efficiency while reducing operational costs. Additionally, automation and AI-driven diagnostics help minimize system downtime, promoting long-term sustainability. These advancements enable PV systems to play a crucial role in the development of clean, renewable energy sources. Through the incorporation of smart technologies like AI, IoT, and big data analytics, PV systems can optimize energy production, storage, and consumption. Real-time monitoring allows for precise adjustments to maximize energy harvesting, while predictive algorithms ensure efficient energy distribution. Automated maintenance helps reduce energy waste, and predictive weather and demand models enhance energy storage planning. These technologies significantly improve energy efficiency and contribute to the overall sustainability of PV systems, making them vital for future energy solutions. The role of smart PV systems in enhancing grid stability will become increasingly significant in the global energy landscape, managing demand fluctuations, and reducing operational costs. As they integrate with other renewable energy sources like wind and water, smart PV systems will contribute to a more resilient and adaptive energy grid. Their ability to efficiently harness and manage solar energy positions them as a key component in achieving global SDGs, particularly in reducing carbon emissions and supporting the transition to clean energy.

Reference

1. S. Dubey, J. N. Sarvaiya, & B. Seshadri. Temperature dependent photovoltaic (PV) efficiency and its effect on PV production in the world–a review. Energy procedia, **33**, 311-321 (2013).
2. M.H. Mohamed Hariri, M.K. Mat Desa, S. Masri, M.A.A. Mohd Zainuri, Grid-connected PV generation system—components and challenges: A Review, Energies, **13**, 5-10 (2020).
3. S. Perry, J. Klemeš, I. Bulatov, integrating waste and renewable energy to reduce the carbon footprint of locally integrated energy sectors, Energy, **33**, 1489 - 1497 (2008).
4. A. S. O. Ogunjuyigbe, T. R. Ayodele, O. A. Akinola, Optimal allocation and sizing of PV/wind/split-Diesel/Battery Hybrid Energy System for minimizing life cycle cost, carbon emission and dump energy of Remote Residential Building, Applied Energy, **171**, 153 - 171 (2016).
5. J. Li, M. S. Herdem, J. Nathwani, J. Z. Wen, Methods and applications for artificial intelligence, Big Data, internet of things, and Blockchain in Smart Energy Management, Energy and AI, **11**, 100208 (2023).
6. T. Khatib, D. H. Muhsen, Optimal sizing of standalone photovoltaic system using improved performance model and optimization algorithm, Sustainability, **12**, 2-18 (2020).
7. Q. Zhong, D. Tong, Spatial layout optimization for solar photovoltaic (PV) panel installation, Renewable Energy, **150**, 1-11 (2020).
8. V. Gupta, M. Sharma, R. K. Pachauri, K. N. Babu,A low-cost real-time IOT enabled data acquisition system for monitoring of PV System, Energy Sources, Part A: Recovery, Utilization, and Environmental Effects, **4**, 1-16 (2020)
9. N. Muleta, A. Q. Badar, Study of Energy Management System and IOT integration in smart grid, ICPEE, 1-5 (2021).
10. C. Alippi, C. Galperti, An adaptive system for optimal solar energy harvesting in wireless sensor network nodes, IEEE Transactions on Circuits and Systems I: Regular Papers, **55**, 1742 - 1750 (2008).

11. S. K. Ramu, G. C. Irudayaraj, R. Elango, An iot - based smart monitoring scheme for solar PV applications, Electrical and Electronic Devices, Circuits, and Materials, 211 - 233 (2021).
12. J. Ferrero Bermejo, J. F. Gómez Fernández, R. Pino, A. Crespo Márquez, A. J. Guillén López, Review and comparison of intelligent optimization modelling techniques for energy forecasting and condition-based maintenance in PV plants, Energies, **12**, 1-18 (2019).
13. C. K. Rao, S. K. Sahoo, F. F. Yanine, A literature review on an IOT-based intelligent smart energy management systems for PV Power Generation, Hybrid Advances, **5**, 1-15 (2024).
14. C. Lamnatou, D. Lamnatou, C. Cristofari, Smart grids and smart technologies in relation to photovoltaics, storage systems, buildings and the environment, Renewable Energy, **185**, 1376 - 1391 (2022).

Wave dissipation performance of breakwater

Yufan Hu[1*]

[1]Department of Sino-British College, Liverpool John Moores University, Shanghai, China

Abstract. This paper conducts a comprehensive review of the wave dissipation performance of various types of breakwaters in marine engineering. Over time, breakwaters have undergone significant evolution, and modern ones are constructed using advanced materials to enhance the efficiency of wave energy dissipation. They can be categorized into vertical, sloped, floating, and permeable types, each possessing distinct wave dissipation characteristics, physical principles, advantages, and disadvantages. Vertical breakwaters primarily reflect wave energy but can potentially cause strong reflections, which may have adverse effects in certain areas. Sloped breakwaters utilize the principles of kinetic energy conversion and wave breaking to dissipate energy, making them suitable for a wide range of coastal conditions. Floating breakwaters offer flexibility and adjustability to different wave conditions, while permeable breakwaters allow for water exchange while still reducing wave energy, which is beneficial for maintaining the ecological balance of the coastal area. The choice of breakwater type depends on factors such as wave conditions, topography, and environmental impact. In the future, breakthroughs in materials, design, and technology are anticipated to improve the performance of breakwaters. The selection and combination of breakwater types should take into account local conditions and requirements to ensure the effective protection of coastlines and marine facilities.

1 Introduction

Breakwaters have played a crucial role in marine engineering throughout history. Their development can be traced back to ancient times when simple structures were built to protect coastal areas from the force of waves. Over the years, with advancements in engineering techniques, the design and materials used for breakwaters have evolved significantly. Modern breakwaters are often constructed using concrete, steel, and other advanced materials, and their structures are designed to be more efficient in dissipating wave energy.

The basic concept of a breakwater is to act as a barrier that reduces the energy of waves, thereby protecting coastlines, harbours and marine facilities from damage. Different breakwaters, such as vertical, sloped, floating, and permeable ones, have distinct wave dissipation performances based on their characteristics and the conditions in which they are used.

For instance, vertical breakwaters are suitable for deep water and areas with good foundation conditions. They are designed to reflect most of the wave energy, reducing the

* Corresponding author: 631501010829@mails.cqjtu.edu.cn

intensity of waves that reach the protected area. However, they can also cause strong wave reflections, which may need to be managed carefully to avoid adverse effects on nearby areas. Sloped breakwaters suit areas with poor foundations and abundant stone sources. They utilize the principles of kinetic energy conversion and wave breaking to dissipate wave energy. When a wave approaches a sloped breakwater, its energy is partially absorbed as the wave climbs up the slope, and the wave may also break on the slope, further reducing its energy. Floating breakwaters are used in deep water with small wave heights and large wave slopes. They are usually made of buoyant materials and are anchored to the seabed. These breakwaters work by absorbing and dissipating the energy of waves through the movement and deformation of the floating structure. Permeable breakwaters allow some wave energy to pass through while still providing some degree of wave dissipation. They can reduce the wave height and energy by causing turbulence and friction within the structure [1].

Disasters, such as the 1993 Hokkaido earthquake and tsunami in Japan and Hurricane Katrina's impact on New Orleans in 2005, highlight the importance of effective breakwaters in protecting coastlines and marine facilities [2]. The performance of breakwaters can be challenged by extreme events, and their design and construction need to consider a range of factors to ensure their effectiveness. In this paper, the wave dissipation performance of these different types of breakwaters and their effects in practical applications will be reviewed. By understanding the characteristics and performance of each type, more informed decisions can be made in the design and construction of breakwaters to better protect the coastlines and marine facilities.

2 Wave dissipation characteristics of different types of breakwaters

Breakwaters can be classified into several types based on their structure and function, including vertical, sloped, floating, and permeable types. Each type has its unique wave dissipation characteristics and physical principles[3].

2.1 Vertical breakwaters

Vertical breakwaters are designed to mainly reflect the wave energy. When a wave approaches a vertical breakwater, part of the wave is reflected, and the remaining energy is transferred to the breakwater. The wave force acting on a vertical breakwater can be analyzed using the following formulas:

• Horizontal wave force:

$$F_H = \frac{1}{2}\rho C_p H^2 L \tag{1}$$

where (FH) is the horizontal wave force, (ρ) is the density of water, (Cp) is the pressure coefficient, (H) is the wave height, and (L) is the wavelength.

• Vertical wave force:

$$F_V = \rho g C_v \frac{H^2 L}{8} \tag{2}$$

Where (Fv) is the vertical wave force, (g) is the acceleration due to gravity, and (Cv) is the vertical force coefficient.

• Wave moment:

$$M = F_H \cdot h_c \tag{3}$$

Where (M) is the wave moment and (hc) is the height of the centre of pressure from the base. However, vertical breakwaters can cause strong wave reflections, which may lead to increased wave energy in some areas. To reduce the reflection, measures such as adding curved surfaces or using porous materials can be taken.

Advantages of vertical breakwaters include their strong ability to reflect waves, providing good protection for the areas behind them. They are suitable for areas where strong wave reflection is acceptable or can be managed, such as in some industrial ports.

However, a major disadvantage is that they can cause strong wave reflections, which may lead to increased wave energy in some areas. This can cause erosion and instability in the surrounding areas. To reduce the reflection, measures such as adding curved surfaces or using porous materials can be taken, but these additions may increase the complexity and cost of the breakwater.

2.2 Sloped Breakwaters

Sloped breakwaters mainly utilize the principles of kinetic energy conversion to gravitational potential energy and wave breaking. When a wave approaches a sloped breakwater, its energy is partially absorbed as the wave climbs up the slope. The wave may also break on the slope, further dissipating its energy. The Goda formula [4], proposed by Yoshimi Goda, is often used to calculate the wave forces on sloped breakwaters.

• The horizontal wave force (Ps) is calculated as:

$$P_s = (\alpha_1 + \alpha_2)\gamma H \qquad (4)$$

where (α_1) and (α_2) are coefficients, (γ) is the specific weight of water, and (H) is the wave height.

• The total wave force (P) is calculated as:

$$P = 0.5(P_s + P_b)d_1 + 0.5(P_s + P_m)\mathrm{h}_\mathrm{c}^* \qquad (5)$$

where (Pb) and (Pm) are related to (Ps), (d1) is the height from the base of the parapet wall to the still water level, and (h_c^*) is the height of the parapet wall above the still water level.

• The energy dissipation rates associated with wave breaking on slopes can be expressed using the following empirical formula (Yoshimi Goda formula)[5]:

$$\gamma = A\{1 - exp[-1.5\frac{\pi d_b}{L_0}(1 + 15\,tan^{4/3}\,\theta)]\}/\frac{d_b}{L_0} \qquad (6)$$

In this formula:

- γ represents the wave energy dissipation coefficient.
- db is the water depth at the breaking point.
- L0 is the deep water wavelength.
- θ is the slope angle.

Advantages of sloped breakwaters include their effective wave dissipation ability, reducing the energy of waves through kinetic energy conversion and wave breaking. The construction of sloped breakwaters may require more materials and space compared to vertical breakwaters. Additionally, the stability of the slope needs to be carefully considered to ensure its performance, as the slope may be subject to erosion and instability over time.

2.3 Floating Breakwaters

Floating breakwaters are usually made of buoyant materials and are anchored to the seabed. They work by absorbing and dissipating the energy of waves through the movement and deformation of the floating structure.

• The wave force acting on a floating breakwater can be analyzed using the following formula:

$$F = C_d \frac{1}{2}\rho U^2 A \qquad (7)$$

Where (F) is the wave force, (Cd) is the drag coefficient, (ρ) is the density of water, (U) is the velocity of the water particles, and (A) is the projected area of the breakwater.

• The movement of the floating breakwater can be described by the equation of motion:

$$m\frac{d^2x}{dt^2} + C\frac{dx}{dt} + kx = F \tag{8}$$

Where (m) is the mass of the breakwater, (x) is the displacement, (C) is the damping coefficient, (k) is the stiffness coefficient, and (F) is the wave force.

Floating breakwaters can be easily installed and removed, making them suitable for areas with specific requirements or temporary wave protection needs.

However, they may require regular maintenance and monitoring to ensure their stability and performance. They may also be affected by extreme weather conditions and require additional measures for protection, which can increase the cost and complexity of their operation.

2.4 Permeable Breakwaters

Permeable breakwaters allow some wave energy to pass through while still providing some degree of wave dissipation. They can reduce the wave height and energy by causing turbulence and friction within the structure. The wave dissipation effect of permeable breakwaters can be analysed using the concept of transmission coefficient, which is defined as the ratio of the transmitted wave height to the incident wave height. The transmission coefficient can be calculated using empirical formulas or numerical models.

3 Advantages and Disadvantages of Different Types of Breakwaters

Compared to other types, vertical breakwaters are suitable for areas where strong wave reflection needs to be managed, such as some industrial ports, but they may not be suitable for areas where wave reflection could cause problems due to increased wave energy. Sloped breakwaters are commonly used in areas where wave dissipation is the main goal, as they are more effective in reducing wave energy through kinetic energy conversion and wave breaking, and they are suitable for a wide range of coastal conditions. Floating breakwaters are more flexible and can be adjusted to different wave conditions, making them suitable for areas with specific or temporary needs. Permeable breakwaters are a good choice in areas where water exchange is desirable while still reducing wave energy, which is beneficial for the ecological balance of the coastal area. The choice of breakwater type depends on factors like local wave conditions, coastal topography, protection purpose, and environmental impact. Engineers need to carefully consider these factors to select the most appropriate type for a specific location to ensure effective wave dissipation and coastal protection.

4 Conclusion

In conclusion, to ensure effective wave dissipation and protection of the coastal area, a rational combination of different types of breakwaters should be adopted based on specific conditions.

For instance, in areas where strong wave reflection is manageable, such as some industrial ports, vertical breakwaters can be utilized. The horizontal wave force acting on a vertical breakwater is calculated by the formula $F_H = \frac{1}{2}\rho C_p H^2 L$. However, their strong wave reflections may cause problems in areas where it is undesirable, so careful consideration is needed.

Sloped breakwaters, which utilize the principles of kinetic energy conversion and wave breaking, are often a good choice in areas where wave dissipation is the primary goal. The

Goda formula for calculating the wave forces on sloped breakwaters, such as $P_s = (\alpha_1 + \alpha_2)\gamma H$ and $P = 0.5(P_s + P_b)d_1 + 0.5(P_s + P_m)h_c^*$, demonstrates their ability to reduce wave energy. They are suitable for a wide range of coastal conditions and are commonly used in many ports and harbours.

Floating breakwaters, with their flexibility and adjustability to different wave conditions, are suitable for areas with specific or temporary wave protection needs. The formula $F = C_d \frac{1}{2}\rho U^2 A$ shows how they work to absorb and dissipate wave energy.

Permeable breakwaters are a suitable option in areas where allowing some water exchange while reducing wave energy is desirable to maintain the ecological balance of the coastal area.

In the future, there may be several potential breakthroughs in the performance of breakwaters. Advances in materials science could lead to the development of more durable and efficient materials for breakwater construction. Their resistance will be enhanced. Further research and innovation in the design and layout of breakwaters, based on the understanding of the physical principles represented by these formulas, could improve their wave dissipation capabilities and reduce the impact of wave reflections. The integration of smart technologies and monitoring systems may enable real-time monitoring of breakwater performance and early detection of potential issues, which can improve maintenance and adjustments.

Overall, the selection and combination of breakwater types should be based on a comprehensive understanding of the local conditions and requirements, considering the characteristics and formulas associated with each type. Continued research and development in this field will contribute to the improvement of their performance and effectiveness in protecting our coastlines and marine facilities.

References

1. Li Yanbao, Wu Yongqiang, Jiang Xuedong, A Review of the Development of Research on Breakwater Failures at Home and Abroad. China Harbour Engineering. **6**,(2004)
2. Lei Songlin, Zheng Yonglai, Discussion on the Structural Type and Application of Breakwater. China Water Transport. **06**,07,(2007).
3. Industry standard of the People's Republic of China Code for design of breakwaters and revetments. People's Communications Publishing House Co., Ltd. (2018).
4. Goda Y, A performance test of nearshore wave height prediction with CLASH datasets. Coastal Engineering. **56**,220-229,(2009).
5. Kamphuis J W, Incipient wave breaking[J]. Coastal Engineering. **15**,185-203,(1991).

Multifunctional integration of breakwaters: taking ecological protection and restoration as an example

Yuyang Xia[*]

Shanghai Ocean University, College of Marine Science and Ecology, China

Abstract. This paper focuses on breakwater design and development in coastal and ecosystem protection contexts. It investigates traditional breakwaters' negative impacts on the marine ecosystem, like disrupting sediment transport and reducing water exchange, causing sediment accumulation and pollution. Current research deficiencies are identified, including a lack of long-term ecological impact studies, insufficient research on sustainable materials, and incomplete impact assessments on local communities and the economy. The review investigates breakwater-ecosystem integration in three aspects. For water exchange, it discusses culverts and permeable breakwaters, considering their effectiveness and factors like location and size. Eco-friendly materials like ecological concrete and its benefits for organisms and the environment are presented, along with other recycled materials' economic advantages. The importance of providing biological habitats to increase biodiversity is also investigated. Case studies illustrate these concepts. However, current breakwater design and research have limitations. Future research should focus on ecological functions, compensation and restoration strategies, sustainable materials, and impact assessments. International cooperation should be enhanced for more effective breakwater design and sustainable coastal development.

1 Introduction

In the past decade, economic development and population migration have led to a significant increase in coastal population density, for example, in China's southeastern coastal cities from 2000 to 2010[1]. Moreover, China's floating population continues to migrate to the eastern coast. At the same time, disasters such as rising sea levels threaten coastal and island developing countries, disrupting the ecological balance and economy. Coastal protection projects are crucial for local development [2]. Breakwaters, as traditional protection structures, are built offshore to reduce the power of waves and protect coastal areas. Ancient coastal residents used stones to resist waves. In Roman times, concrete promoted the construction of breakwaters. Since the 20th century, their development has diversified, and the number of structural combinations has increased. Nowadays, in addition to wave dissipation, breakwater designs also incorporate sustainable development elements. When

[*]Corresponding author: 2229517@st.shou.edu.cn

designing breakwaters, five aspects should be considered: determining wave conditions, selecting the appropriate type, ensuring the source of materials, and verifying and improving the results of model tests.

Under the concept of sustainable development, there are currently deficiencies in breakwater design research, which limits innovation and also poses challenges to marine ecological protection, resource utilization, and regional development. For example, environmentally unfriendly situations are common. In terms of ecological protection, research on the long-term impact of breakwater construction and operation on the ecosystem is not comprehensive, and there is little research on ecological compensation and restoration measures for its negative impacts. In terms of material selection, there has been insufficient research on sustainable and environmentally friendly materials. Most breakwaters still rely on traditional materials, which increases resource consumption and environmental pressure. The impact of breakwater construction on local communities and economies has not been fully assessed. With the continuous progress of engineering technology, the materials of breakwaters have become more diverse, ranging from the initial natural stone to concrete and more. There have also been innovations in the shape and structure of breakwaters to adapt to different marine environments and protection needs. Research has found that in recent years there has been a trend that more and more breakwaters are designed to incorporate wave attenuation and other functions. The materials, shapes, and structures of traditional breakwaters have been modified to minimize these adverse effects by increasing their ecological functions. For example, breakwaters are integrated with wave dissipation dikes and ecosystem protection and restoration, which can help enhance ecosystem stability and biodiversity.

This study explores the coordinated development of coastal protection and ecosystem protection and restoration in the form of a literature review based on a large amount of literature on the multi-functional integration of breakwaters.

2 Types of integration of breakwaters with ecosystem protection and restoration

2.1 Breakwaters to Promote Water Exchange

Traditional breakwaters can impede the transport of sediment by the exchange of water along the coast, thereby affecting the natural sediment transport process and leading to the accumulation of sediment in the water area [3]. In addition, the reduction in water exchange can also cause polluted water nearby to accumulate in the harbour basin and surrounding sea areas under the push of waves. For example, on the coast of the East China Sea with a relatively high sediment content, the encircling port breakwaters are commonly seen in muddy coastal areas. These encircling port breakwaters are composed of traditional gravity breakwaters. Although they provide good stability for moored ships, there are problems such as water pollution in the harbour basin and a decline in water environment capacity [4]. To mitigate the impact of sediment deposition and declining water quality on the marine ecosystem, there are currently two improved designs for traditional breakwaters: breakwaters with culverts and open-cell breakwaters.

Excavating a culvert on the surface of a traditional breakwater is an economical and effective way. It can connect the basin to the open sea to a certain extent and enhance water exchange by utilizing the hydrodynamic force of rising and falling tides. However, due to the differences in hydrodynamic conditions of each water area, the location and size of the culvert will affect the stability of ship mooring and water exchange in the port [5]. For example, Chaofan Lv and Xizeng Zhao found through experimental calculations that an increase in relative length leads to a reduction in the maximum oscillatory flow and thus weakens the water exchange capacity [6]. In coastal areas with small tidal ranges, setting the culvert location at sea level has a better water exchange effect than being located below sea level [7].

Weigel proposed in 1961 that a permeable breakwater composed of multiple rows of rigid piles spaced closely can effectively improve the water exchange capacity [8]. The lower support structure of the permeable breakwater is different from the traditional breakwater surface. There is a certain gap between the structures below the water surface, and water can pass freely below. A wave-blocking structure can be set at the water surface to effectively reduce wave energy. The gap can cause the incoming waves to form vortices inside, thereby further dissipating wave energy. There are four structures of permeable breakwaters, namely perforated baffle type, pile foundation type, curtain wall type, and caisson pier type [9]. In recent years, three new types of permeable breakwaters, namely comb type, buttress type, and perforated caisson type, have been proposed [10]. The semi-permeable breakwater is made of block materials with pores (such as porous concrete blocks or irregular gravel). Another type of semi-permeable breakwater is a combination of a rigid breakwater surface and an internal partially permeable structure.

2.2 Breakwaters made of eco-friendly materials

Rock and concrete from quarries are the two common construction materials used in traditional breakwaters. However, quarrying causes serious damage to local vegetation, water bodies and soil, leading to soil erosion and a decrease in biomass [11].

Ecological concrete is a concrete material specially designed for environmental protection and ecological restoration. Unlike commonly used concrete, the aggregate of ecological concrete can be recycled from construction waste, industrial waste and domestic waste that meets certain standards. Some natural aggregates can also be used. This is in line with the concept of sustainable development and resource recycling [12]. In addition, ecological concrete has a uniformly and continuously distributed pore structure and a certain degree of water permeability. It can also provide a habitat for marine animals and plants, forming a biological community. This makes ecological concrete breakwaters part of the ecosystem, compensating for the traditional breakwaters' inability to serve the ecosystem. As a result, ecological concrete breakwaters combine the dual functions of being environmentally friendly and providing a habitat for organisms.

Some recycled waste materials can exhibit great potential if properly processed. They can be reused in the construction of eco-friendly breakwaters, thereby achieving resource recycling and reducing negative environmental impacts. From the perspective of resource consumption, the production of traditional breakwater construction materials usually requires a large amount of mining of natural resources such as ore and gravel. The use of recycled waste materials for breakwater construction can effectively reduce the demand for new resources and the environmental pressure caused by resource mining. Furthermore, the use of recycled waste materials as construction materials has significant advantages in terms of economic costs. Recycled materials are generally low-cost, as the acquisition costs of waste such as used tyres and waste plastics are relatively low, often lower than the cost of purchasing new building materials. In some areas, these wastes can even be obtained for free, and builders or construction companies only need to pay the transportation and processing costs to turn these wastes into usable building materials.

2.3 Breakwaters with Biological Habitat Functions

The smooth and repetitive surface of traditional breakwaters is not conducive to the habitat and reproduction of organisms. Instead, it attracts invasive species that are highly tolerant of various environments, changing the competition pattern. This results in the diversity of marine animal and plant communities on the surface of infrastructure being generally lower than that in natural communities [11]. Moisés A. Aguilera et al. compared the biodiversity of a granite breakwater with that of a natural rocky shore. They found that the species diversity and abundance of the natural habitat were higher. The lack of microhabitats on the breakwater results in a decrease in biodiversity. In addition, Moisés A. Aguilera et al. studied the thermal patterns of artificial breakwaters and their impact on biodiversity. They found that the rock temperatures of artificial breakwaters were higher and less variable spatially, resulting in lower biodiversity and reduced species richness and species density [13].

3 Case studies on the integration of breakwaters and ecosystem protection and restoration

3.1 Research and case studies on breakwaters that promote water exchange

V.K. Tsoukala and C.I. Moutzouris proposed that in coastal areas with low tidal ranges such as those along the Mediterranean coast, for example, in Greece. To be more in line with the main mechanism of tidal hydrodynamics, the flushing culvert formed by aligning the longitudinal axis of the culvert on the breakwater with sea level is a common layout. The water exchange through the culvert creates a flushing effect in the port and helps reduce the pollution levels of chemicals, organisms and floating solids. The flushing culvert can reduce the pollutant concentration in the port basin by between 70% and 90% within 24 hours. The effect of wave penetration, that is, the effect of water exchange, can be quantified by the wave transmission coefficient. The higher the wave transmission coefficient, the better the effect of water exchange. From the perspective of culvert geometric characteristics, studies have found that increasing the width, height and submergence depth of the flushing culvert and reducing the length of the culvert can increase the transmission coefficient, thereby enhancing water circulation and exchange, helping to reduce pollutant concentration and improve water quality. From the perspective of wave characteristics, studies have found that the wave transmission coefficient increases with the increase of wave period and decreases with the increase of incident wave height. For vertically incident waves, the wave transmission coefficient is relatively higher [14,15].

Chaofan Lv, Xizeng Zhao and Mingchang Li proposed a new type of ecologically protective breakwater, the inclined culvert vertical breakwater, based on the original horizontal culvert vertical breakwater to further promote water exchange and reduce sediment deposition. The characteristic of this breakwater is that inclined culverts are set on the vertical breakwater. When the inlet of the culvert located in the open sea is higher than the outlet of the culvert located in the port, the high-turbidity water at the bottom of the port will flow into the sea through oscillatory flow, and the low-turbidity water on the sea surface will flush the port area. The lower foundation is a permeable gravel pile, which can effectively dissipate wave energy, reduce the reflection coefficient and wave force on the inclined culvert vertical breakwater, and maintain the stability of the structure. The inclined culvert vertical breakwater is more conducive to engineering applications than the horizontal culvert vertical breakwater. It can effectively improve the stability of ship berthing and structure, further promote water exchange, and improve the port environment.

Charles K Sollitt and Ralph H Cross pointed out the relationship between the wave transmission coefficient and the geometric characteristics of the permeable breakwater. The wave transmission coefficient decreases with the decrease of the porosity and permeability of the permeable breakwater, decreases with the increase of the width of the breakwater, and decreases with the increase of the steepness of the breakwater [16, 17].

The Huangqi fishing port is in the Huangqi Bay on the south side of the Huangqi Peninsula in Lianjiang County, Fujian Province. Huaidong Xie pointed out that the lower permeable structure of the new permeable breakwater of the Huangqi fishing port is a pile foundation. Several inclined piles in different directions form a group. Different pile foundation structure units are staggered in two rows back and forth and connected by longitudinal and transverse beam systems. The superstructure is a spatial frame structure, which is composed of upper and lower pile caps and trapezoidal rigid columns connecting them and is integrated with the pile foundation. The vertical piles are staggered back and forth. Multiple rows of permeable baffle plates are set between the piles. Multiple rows of cantilever wave-blocking structures are set under the lower pile cap, and the bottom end extends below the checked low water level. A proper amount of riprap is laid under the breakwater and together with multiple rows of inclined pile groups in different directions, it forms a wave dissipation system at medium and low water levels [18].

The new port area of Yeosu in South Korea plans to expand the northern port area on the original basis. A new type of water exchange breakwater structure, namely the perforated caisson permeable breakwater, is adopted in a local section of about 33m at the root of the breakwater on the outside of the northern port area. Yusheng Shen and others pointed out that the outside of this breakwater is a sloping breakwater, and the inside is a caisson structure. The caisson includes four compartments, and each compartment is perforated. The innermost perforation position is relatively low. This new type of water exchange breakwater has a certain water exchange capacity [19].

3.2 Research and cases of breakwaters using eco-friendly materials

Ecological concrete has a pore structure that is evenly and continuously distributed and has a certain degree of water permeability. It can also provide a habitat for marine animals and plants, forming a biological community, so that the breakwaters built with ecological concrete become an integral part of the ecosystem. Bas W. Borsje and others proposed that the introduction of ecological concrete in the construction of ecologically enhanced breakwaters can effectively alleviate the problem of declining biodiversity. The holes and gaps in the breakwater provide shelter and hiding places for small benthic organisms. These organisms usually live in caves and crevices. The holes and gaps in the breakwater also provide a habitat for the larval stages of non-benthic organisms. In the study, it was found that the proportion of invasive species to native species on the ecologically enhanced breakwater was significantly lower than that on the standard breakwater. Although 12 invasive invertebrates were also found on the ecologically enhanced breakwaters, there were also 21 native species. In contrast, there were only 10 invasive species and 5 native species on the standard breakwater. This suggests that the design of the ecologically enhanced breakwater helps to attract and support communities of native species, which are better able to compete with invasive species for space [20].

Steven B. Scyphers et al. conducted experiments at two sites along the rapidly eroding coastline of Alabama. At each site, two subtidal breakwaters made of loose oyster shells were constructed, and an unaltered plot was set up as a control, using a randomized paired design. The fish and mobile invertebrate communities near the breakwaters were more diverse than those in the control areas. The total abundance of species captured in 10-cm gill nets was higher near the breakwaters. For benthic fish, the 10 cm gill net catch was richer near the breakwater. From the data captured by the purse seine, there was no difference in the total abundance of benthic fish, but the total abundance of decapod crustaceans such as crabs was higher near the breakwater. The abundance of some economically important species such as blue crabs, red drums, spotted trout and flatfish increased significantly near the breakwater [21].

Lea T. Mamo et al. found that the rock elements that make up the breakwater have a significant impact on benthic organisms. In an ecological engineering modification of an upgraded breakwater in Coffs Harbour, Australia, the impact of different rock elements on benthic organisms was assessed through research. The assemblage of primary grey rocks was more diverse than the assemblage of granites at one site, and the diversity of endangered algae on the rock bottom was higher than on the top at this site. The results suggest that incorporating different sizes of primary rocks into the blocks and layers that make up the breakwater can improve the biodiversity of coastal protection infrastructure [22].

4 Conclusion

Breakwaters play a vital role in coastal protection and the protection of port facilities, and their effectiveness should not be overlooked. However, the materials traditionally used for breakwaters have several negative impacts on the marine ecosystem. With the advancement of engineering technology, environmentally friendly materials such as ecological concrete have begun to be used in the construction of breakwaters. The shape and structure of breakwaters are also constantly being innovated and evolved. For example, new structures such as breakwaters with culverts and permeable breakwaters with open cells can effectively promote water exchange, reduce sedimentation, and improve water quality. In addition, breakwater designs with biological habitats are beneficial for increasing biodiversity and maintaining ecosystem balance.

However, there are still some limitations in the current design and research of breakwaters. For example, the long-term impact of breakwater construction and operation on the ecosystem has not been fully studied, and there is a lack of quantitative research on the effect of permeable breakwaters on water quality improvement. There has been relatively little research on ecological compensation and restoration measures for their negative impacts. In terms of material selection, research and application of sustainable and environmentally friendly materials still need to be strengthened. At the same time, the impact of breakwater construction on local communities and the economy has not been fully assessed. Future research should focus on the following areas:

Moreover, further research on the ecological functions of breakwaters should be conducted, a comprehensive exploration of the impact mechanisms on the marine ecosystem should be made, and more scientific and reasonable ecological compensation and restoration strategies should be developed. For example, research and development efforts on sustainable and environmentally friendly materials should be intensified, newer environmentally friendly materials should be actively explored, and their wide application in breakwater construction should be promoted, to effectively reduce resource consumption and environmental pressure. Furthermore, a comprehensive assessment of the impact of breakwater construction on local communities and the economy should be carried out, and more comprehensive and holistic planning and decision-making should be formulated to achieve coordinated and sustainable development of the economy, society and environment.

Finally, international cooperation and exchanges should be vigorously strengthened to widely share the experience and technology of breakwater construction and jointly respond to global challenges such as sea level rise. In summary, through continuous research and innovation, breakwaters are expected to achieve more significant ecological benefits while improving their protective effectiveness, contributing to the sustainable development of coastal areas.

References

1. Guojun Zhang, Wanling Huang, Chunshan Zhou,Yongwang Cao,Spatio-temporal characteristics of demographic distribution in China from the perspective of urban agglomeration. Acta Geographica Sinica. **73**, (2018)
2. Hernández-Delgado, Edwin A, Coastal Restoration Challenges and Strategies for Small Island Developing States in the Face of Sea Level Rise and Climate Change.Coasts.**4**, (2024)
3. Byrnes, Mark R, Hiland, Matteson W, Large-scale sediment transport patterns on the continental shelf and influence on shoreline response: St. Andrew Sound, Georgia to Nassau Sound, Florida, USA.Marine Geology.**126**, (1995)
4. Haichen Pan, Research on the ecological transformation of the ring harbour. (2021)
5. Chaofan Lv, Development of boundary element water-wave model and its application in environment-friendly breakwaters. (2023)
6. Lv C, Zhao X. Environmentally friendly vertical wall breakwater with a culvert for encircled harbor basin. Journal of Engineering Mechanics. **147**, (2021)
7. TSOUKALA V, MOUTZOURIS C. Wave transmission in harbors through flushing culverts. Ocean Engineering.**36**, (2009)
8. WIEGEL R L. Closely spaced piles as a breakwater, (Hydraulic Engineering Laboratory, Wave Research, (1961)
9. Weiyi Chen, Yinan Lin. A review of the research on the characteristics of permeable breakwaters.China Water Transport.**16**, (2016)
10. Jun Zhao. Research on the wave-dissipating performance of new permeable breakwaters.China Water Transport.**17**, (2017)
11. Bo Huang. Analysis of the impact of open-pit quarries on the ecological environment and prevention.Friends of Science.**2**, (2012)
12. Yiren Wang, Dongmin Wang. Research progress on the influence of aggregate type and quality on the performance of permeable concrete. Materials Guide.**31**, (2017)
13. Aguilera M A, Broitman B R, Thiel M. Spatial variability in community composition on a granite breakwater versus natural rocky shores: lack of microhabitats suppresses intertidal biodiversity. Marine pollution bulletin.**87**, (2014)
14. Tsoukala, VK,Moutzouris, CI.Wave transmission in harbors through flushing culverts.Ocean Engineering. Ocean Engineering.**36**, (2009)
15. Tsoukala, VK,Moutzouris, CI.A 3-D experimental study on the beneficial role of flushing culverts in sea water quality.Coastal Engineering.**5**, (2008)
16. Li, Mingchang,Chaofan,Lv, Xizeng,Zhao.Hydrodynamic performance of a new-type vertical wall breakwater with inclined culvert on a permeable rubble mound foundation.Applied Ocean Research.**126**, (2022)
17. Sollitt C K, Cross R H. Wave transmission through permeable breakwaters.Coastal Engineering. (1972)
18. Huaidong Xie.Design study of a new type of permeable breakwater in Huangqi Fishing Harbour.Journal of Hohai University.**4**, (1997)
19. Shen Yusheng, Sun Zhongbin, Zhou Zijun, Jin Zhentian. Experimental study on the

characteristics of water exchange breakwater, in Proceedings of the Twenty-seventh National Symposium on Hydrodynamics, Nanjing, Jiangsu Province, November 6-11(2015)

20. Borsje B W, van Wesenbeeck B K, Dekker F, et al. How ecological engineering can serve in coastal protection. Ecological Engineering. **37**, (2011)
21. Scyphers S B, Powers S P, Heck Jr K L, et al. Oyster reefs as natural breakwaters mitigate shoreline loss and facilitate fisheries. PloS one. **6**, (2011)
22. Mamo L T, Coleman M A, Kelaher B P. Ecological enhancement of breakwater upgrades: size and type of rocks used influence benthic communities. Marine Ecology Progress Series. **661**, (2021)

E3S Web of Conferences **606**, 05007 (2025) https://doi.org/10.1051/e3sconf/202560605007
ICNAOE 2024

A Review of Biomass Energy: Comparison of Utilization Methods and Future Prospects

Kai Zhang*

Cornell University, Civil and Environmental Engineering Department, 14850 Ithaca, USA

Abstract. Biomass refers to the biodegradable portion of products and residues from agriculture, forestry, and related industries, including municipal and industrial waste. It encompasses molecular and macromolecular compounds derived from plants and other organic sources. Biomass energy, considered sustainable and eco-friendly, can reduce greenhouse gas emissions and enhance energy security, contributing significantly to net-zero emission goals. Various methods for biomass utilization include thermochemical and biochemical conversions, as well as co-production. Thermochemical conversion methods like pyrolysis, gasification, and torrefaction transform biomass into fuels and chemicals, offering solutions for waste management and renewable energy. Biochemical methods, such as anaerobic digestion and fermentation, use biological processes to produce biogas and biofuels. Co-production, combining biomass with other energy sources (e.g., geothermal or coal), optimizes resource use and increases energy efficiency. While challenges remain, such as cost and technical barriers, biomass energy provides an effective, carbon-neutral path for reducing fossil fuel dependence. The paper highlights biomass energy's diverse technologies and its importance in the sustainable development and future energy landscape.

1 Introduction

Biomass is defined as the biodegradable portion of products and agricultural residues (including plant and livestock waste), forestry and related industries (including fisheries and aquaculture) residues, as well as other wastes such as industrial and municipal waste. Biomass includes all molecular and macromolecular compounds derived from vegetables, agricultural products, forestry products, and any remaining materials. Biomass energy is a sustainable and eco-friendly energy source that has garnered widespread attention [1]. It offers a promising solution to address environmental issues and the urgent need for a transition to green energy, as it can reduce greenhouse gas emissions while supporting national energy security. Integrating biomass energy with the pursuit of net-zero emission goals appears to be a symbiotic strategy with substantial potential to mitigate global climate issues [2]. Utilizing biomass energy is an effective means of reducing greenhouse gas emissions due to its renewable and carbon-neutral characteristics [3]. This is because biomass sources are carbon-neutral; the carbon dioxide emitted during energy production is offset by

* Corresponding author: kz368@cornell.edu

the carbon absorbed during the biomass growth cycle. As nations embark on the challenging journey toward net-zero emissions, leveraging the substantial potential of biomass energy seems to be a well-considered approach [3].

The types of biomass energy supply globally can be categorized into five major groups: solid biomass, liquid biofuels, municipal waste, industrial waste, and biogas, as shown in fig.1. In 2020, the global supply of biomass energy reached 57.5 exajoules (EJ). Of this, 86% came from solid biomass sources, including wood chips, wood pellets, and traditional biomass sources. Liquid biofuels accounted for 7%, municipal and industrial waste represented 2% to 3%, and biogas contributed 2%.

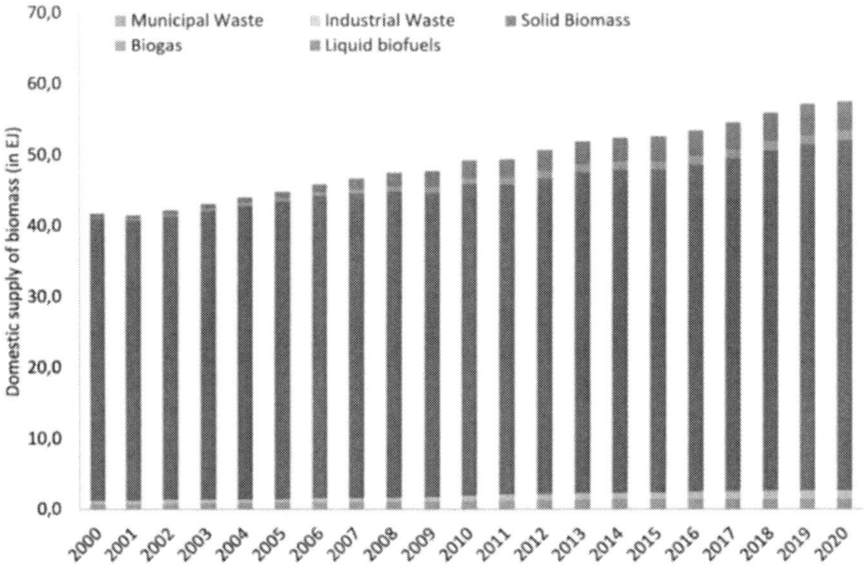

Fig. 1. Biomass Supply in 2020 [4].

There are currently several methods for utilizing biomass energy, including thermochemical conversion, biochemical conversion, biomass liquefaction, and co-production. Thermochemical conversion refers to the use of biomass through thermochemical processes, including pyrolysis, gasification, and torrefaction. Biochemical conversion involves the transformation of biomass through biological processes, including anaerobic digestion, fermentation, and enzymatic hydrolysis. Biomass liquefaction technology converts biomass into liquid fuels, including methods such as hydro-liquefaction and biodiesel production. Co-production combines biomass energy with other forms of energy, such as solar or geothermal, to enhance energy production. This article aims to introduce various biomass energy utilization methods and their environmental benefits.

2 Thermochemical Conversion Technology

Thermochemical conversion technology involves transforming biomass or other solid fuels into usable energy or chemicals through high-temperature heating. This technology efficiently utilizes waste or biomass resources, reduces environmental pollution, and provides sustainable energy.

2.1 Pyrolysis

Pyrolysis is the decomposition of biomass through heating in the absence of oxygen, producing charcoal, liquid, and gaseous fuels. The pyrolysis process can be classified into three types: conventional pyrolysis, fast pyrolysis, and flash pyrolysis. These categories are determined by variables such as temperature, heating rate, and vapor residence time [5]. Hemicellulose decomposes first at temperatures between 470 and 530 K, followed by cellulose in the range of 510 to 620 K, with lignin as the final component to undergo pyrolysis between 550 and 770 K. A process with low temperatures, high heating rates, and short gas residence times favors liquid fuel production. To increase charcoal yield, a process with low temperature and low heating rate is preferred. For maximizing gaseous fuel production, high temperature, low heating rate, and extended gas residence time should be considered.
Pyrolysis is commonly used for converting agricultural and forestry biomass, serving as a practical method to address both renewable energy security and waste management. Several agricultural and forestry residues, such as camelina meal, mustard meal, flax straw, hemp straw, and spruce wood, can be processed through pyrolysis [6].

2.2 Gasification

Gasification converts biomass into syngas (primarily composed of CO and H_2) under high temperatures and partial oxidation conditions. Syngas can be further used for power generation, hydrogen production, or the synthesis of liquid fuels. Gasification has the highest energy conversion efficiency, as it transforms biomass into fuel gas (known as syngas) through a series of chemical reactions. Syngas is a gas mixture composed of CO, CH_4, and H_2. The syngas generated from the gasification process can serve as a future alternative to fossil fuels. Using syngas produced by biomass gasification as a dual fuel in diesel power plants can reduce biodiesel consumption by 60% [7]. When using syngas with a calorific value of 4 MJ/kg produced from municipal solid waste (MSW) gasification as a dual-fuel material, diesel consumption can be reduced by 44% [8]. However, syngas is currently not widely used as a fuel for generators due to power loss issues. According to Molino et al. [9], the production of syngas depends on biomass type, gasification reactor, and operating parameters used during gasification, such as gasifying agent and temperature (T). To produce high-quality syngas, the gasification process requires $T > 750$ °C [9]. Air and O_2 are commonly used as gasifying agents, with air being the most widely used due to its accessibility and ease of use.

2.3 Torrefaction

Torrefaction, also known as biomass roasting, is referred to industrially as roasting, slow and mild pyrolysis, wood cooking, or high-temperature drying. This process involves low-temperature heating to remove moisture and volatile components from biomass, increasing its energy density and improving combustion performance. Torrefied biomass resembles low-rank coal and is suitable for power generation and heating. Torrefaction is often applied to the treatment of lignocellulosic biomass. The basic principle of this treatment method is the preferential and selective degradation of extractable compounds and polysaccharides (constituting the low-calorific-value portion) within the biomass [10]. The less thermally stable components of lignocellulosic biomass degrade first as the material is gradually heated. Compared with cellulose and lignin, hemicellulose in biomass is an organic compound with low calorific value and a high O/C atomic ratio [10]. Selectively removing hemicellulose increases the proportion of carbon-rich compounds and compounds with higher energy density in torrefied biomass, thereby reducing their reactivity and water absorption capacity while enhancing thermal stability. Common biomass sources for torrefaction include maritime pine, chestnut, oak, eucalyptus, Caribbean pine, birch, pine, bagasse, bamboo, wood blocks, willow, beech, sessile oak, meranti, and oil palm waste.

3 Biochemical Conversion Technology

Biochemical conversion technology refers to the process of converting biomass or organic waste into useful products through biochemical reactions. This process relies on microorganisms, enzymes, or other biocatalysts to decompose organic matter, producing energy or other chemicals.

3.1 Anaerobic Digestion

Anaerobic digestion uses microorganisms to decompose organic matter under anaerobic conditions, producing biogas (primarily composed of CH_4, CO_2, hydrogen sulfide (H_2S), water vapor, etc.) and bio-fertilizer, which is rich in nitrogen, phosphorus, potassium, and trace nutrients that plants derive from fundamental resources such as soil and water. Biogas can be used for power generation, heating, or as a substitute for natural gas [11]. One of the advantages of biogas is its flexibility in applications, which range from heat and electricity production to vehicle fuel, provided it undergoes upgrading (removal of CO_2) or purification (removal of H_2S) [12]. Anaerobic digestion is considered one of the most sustainable and eco-friendly processes and plays a vital role in renewable energy production. The anaerobic nitrification process involves a series of four biochemical reaction stages: hydrolysis, acidogenesis, acetogenesis, and methanogenesis [13]. Process performance is highly dependent on operational strategies and process parameters, such as reactor type, inoculum and substrate used, hydraulic retention time, feed rate, and operating temperature. Additionally, parameters such as pH, alkalinity, and volatile fatty acid concentration determine microbial metabolism. Recent advancements in anaerobic digestion have opened up new possibilities for energy independence in urban, rural, and remote areas. Recent studies have highlighted the potential of anaerobic nitrification to reduce greenhouse gas emissions at national or regional levels [14].

3.2 Fermentation

Fermentation is a process that uses microbial fermentation of biomass to produce clean fuels, such as biohydrogen. This section focuses on bio-based hydrogen production, one of the currently practical clean fuels. Bio-based hydrogen can be produced through two main methods: dark fermentation and photofermentation. Dark fermentation has garnered considerable attention due to its high yield, short reaction time, and ability to use various organic waste materials as feedstock. In dark fermentation, anaerobic bacteria (e.g., Enterobacteriaceae, Clostridium, and Bacillus species) utilize organic waste and water to produce hydrogen. The amount of biohydrogen produced depends on the type of microorganism and is influenced by low substrate decomposition rates and the consumption of substrates by methanogens and acetogens [15]. Pure cultures in dark fermentation yield higher hydrogen output, but mixed cultures offer broader substrate options and easier process control [16]. Most hydrogen production processes are based on pure sugars (e.g., glucose, xylose, sucrose), making the process costly [17].

Another major challenge in the fermentation process is the recalcitrance of biomass due to its complex structure and lignin content, which hinders enzyme access to hydrolysable sugars. Biomass requires pretreatment prior to fermentation. Pretreatment methods are classified into two categories: conventional methods (e.g., acid, alkali) and advanced methods (e.g., ionic liquids, supercritical fluids). The yield of biohydrogen and other by-products (such as acetate, butyrate, lactate, ethanol, and CH_4) from lignocellulosic biomass in dark fermentation depends on the type of microorganisms used and the operational conditions [18].

4 Co-Production (Polygeneration)

Biomass co-production is a method that uses biomass resources to simultaneously produce multiple products and forms of energy. By generating fuels, electricity, chemicals, and other products in a single process, it enhances resource utilization and economic efficiency. Key characteristics of biomass co-production include efficient resource use to maximize biomass utilization and minimize waste, improved economic benefits through diversified revenue streams, and environmental friendliness by reducing dependence on fossil fuels and lowering pollutant emissions. Common co-production models include biorefining (converting biomass into fuels and chemicals), combined heat and power (CHP) (simultaneously generating electricity and heat), and waste-to-energy (converting organic waste into energy and fertilizers). Despite challenges such as technical complexity, high costs, and supply chain management difficulties, biomass co-production contributes to sustainable development and is expected to play a larger role in energy and environmental sectors as technology advances and policy support increases.

4.1 Biomass and Geothermal Co-Production

Moret et al. explored the potential of combining woody biomass power plants with deep geothermal energy to meet urban energy demands [19]. Geothermal energy, a well-known green renewable energy source, is stored within the Earth. One of its main advantages is independence from weather conditions. This energy can reach the Earth's surface through volcanic eruptions, hot springs, or heat-bearing underground water and appears widely in various forms on the surface. Many countries use geothermal energy for various purposes, such as direct heating, power generation, cooling, and drying. In Italy, geothermal energy is the third-largest renewable thermal energy resource. By 2050, geothermal energy is projected to contribute up to 8.3% of global electricity generation, serving 17% of the world's population. However, factors such as technology limitations hinder geothermal energy development. For example, the emission of non-condensable gases in geothermal production poses a significant environmental issue, and the low power generation efficiency from low-temperature resources is another challenge. These obstacles have led to much lower geothermal system installations compared to wind or solar, despite geothermal energy's advantages over these renewables. One solution to this issue is to supplement geothermal energy with biomass. The combination of these two renewable energy sources allows for more efficient utilization of renewable energy potential.

4.2 Biomass and Coal Co-Production

Biomass is considered an alternative clean energy source. Partially replacing coal with biomass in power generation can reduce CO emissions, and in combination with carbon capture technology, it can even achieve negative CO emissions. Studies have also shown that co-firing coal and biomass can reduce emissions of CO, NO , SO , and PM2.5 [20]. Beyond theoretical validation through numerical simulations and experimental research, the fact that co-firing coal with biomass can reduce certain pollutant emissions has been proven in practice. Currently, most power plants are designed for coal combustion, with very few boilers designed specifically for biomass. To fully utilize existing coal-fired power plants with minimal modification, co-firing coal with biomass waste is the most feasible option. Compared with burning biomass alone for power generation, coal and biomass co-firing technology offers higher power generation efficiency, lower equipment and operational costs, and greater flexibility with different types of biomass. Although co-firing coal and biomass has many advantages, the addition of biomass in coal-fired boilers, which are specifically designed for coal, inevitably leads to certain issues different from those encountered when burning coal alone. This is one of the current challenges facing this technology.

5 Conclusion

This paper provides a comprehensive review of biomass energy, focusing on various utilization methods and their future prospects. The article is divided into four sections. The first section introduces the importance of biomass as a renewable energy source and its potential in sustainable development. The second section analyzes thermochemical conversion technologies, such as pyrolysis, gasification, and combustion, comparing the efficiency and applications of these technologies. The third section discusses biochemical conversion technologies, including anaerobic digestion and fermentation, highlighting their role in biofuel production. The final section introduces the concept of co-production (polygeneration), emphasizing its potential in optimizing resource use and enhancing energy efficiency. The article concludes with a discussion on the future outlook of biomass energy, analyzing its role in the global energy transition.

References

1. Popp J, Kovács S, Oláh J, et al. Bioeconomy: Biomass and biomass-based energy supply and demand[J]. New Biotechnology, **60**: 76-84 (2021) https://doi.org/10.1016/j.nbt.2020.10.004
2. Tan X, Tan T. Biofuels from biomass toward a net-zero carbon and sustainable world[J]. Joule, **6(7)**: 1396-1399 (2022) https://doi.org/10.1016/j.joule.2022.06.005
3. Balakrishnan D, Sharma P, Bora B J, et al. Harnessing biomass energy: Advancements through machine learning and AI applications for sustainability and efficiency[J]. Process Safety and Environmental Protection, **191**:193-205 (2024) https://doi.org/10.1016/j.psep.2024.08.084
4. World Bioenergy Association, GLOBAL BIOENERGY STATISTICS 2020 (10/18/2024)
5. Adelawon B O, Latinwo G K, Eboibi B E, et al. Comparison of the slow, fast, and flash pyrolysis of recycled maize-cob biomass waste, box-benhken process optimization and characterization studies for the thermal fast pyrolysis production of bio-energy. Chemical Engineering Communications, **209,** 1246-1276 (2022) https://doi.org/10.1016/j.renene.2017.04.060
6. Jha S, Pattnaik F, Nanda S, et al. Investigations of thermal effects during pyrolysis of agro-forestry biomass and physicochemical characterizations of biofuel products[J]. Biocatalysis and Agricultural Biotechnology, **61,** 103379 (2024) https://doi.org/10.1016/j.bcab.2024.103379
7. Sudarmanta B. Dual fuel engine performance using biodiesel and syn-gas from rice husk downdraft gasification fo power generation. (2015) http://dx.doi.org/10.13140/RG.2.1.2503.4968

8. Singh J, Singh S, Mohapatra S K. Production of syngas from agricultural residue as a renewable fuel and its sustainable use in dual-fuel compression ignition engine to investigate performance, emission, and noise characteristics. Energy Sources, Part A: Recovery, Utilization, and Environmental Effects, **42,** 41-55 (2020) https://doi.org/10.1080/15567036.2019.1587053
9. Molino A, Larocca V, Chianese S, et al. Biofuels Production by Biomass Gasification: A Review. Energies (2018) https://doi.org/10.3390/en11040811
10. Shen D K, Gu S, Bridgwater A V. The thermal performance of the polysaccharides extracted from hardwood: Cellulose and hemicellulose[J]. Carbohydrate Polymers, **82,** 39-45 (2010) https://doi.org/10.1016/j.carbpol.2010.04.018
11. Isha A, Kumar S, Jha B, et al. Development of stabilization methods using a pilot scale anaerobic digester for seasonal variations in kitchen wastes for improved methane production with zero breakdowns. Cleaner Engineering and Technology, **1**, 100015 (2020) https://doi.org/10.1016/j.clet.2020.100015
12. Kapoor R, Subbarao P M V, Vijay V K, et al. Factors affecting methane loss from a water scrubbing based biogas upgrading system[J]. Applied Energy, **208**, 1379-1388 (2017) https://doi.org/10.1016/j.apenergy.2017.09.017
13. Bhujbal S K, Kumar M, Vijay V K, et al. Potential of termite gut microbiota for biomethanation of lignocellulosic wastes: current status and future perspectives. Reviews in Environmental Science and Bio/Technology, **20,** 419-438 (2021) https://doi.org/10.1007/s11157-021-09576-y
14. Chaudhary V P, Chandra R, Denis D M, et al. Agri-biomass-based bio-energy supply model: An inclusive sustainable and circular economy approach for a self-resilient rural India[J]. Biofuels, Bioproducts and Biorefining, **16,** 1284-1296 (2022) https://doi.org/10.1002/bbb.2373
15. Yang J, Liu X, Liu X, et al. Enhanced dark fermentative hydrogen production from waste activated sludge by combining potassium ferrate with alkaline pretreatment[J]. Science of The Total Environment, **707**, 136105 (2020) https://doi.org/10.1016/j.scitotenv.2019.136105
16. Mahata C, Ray S, Das D. Optimization of dark fermentative hydrogen production from organic wastes using acidogenic mixed consortia[J]. Energy Conversion and Management, **219**, 113047 (2020) https://doi.org/10.1016/j.enconman.2020.113047
17. Moradian J M, Xu Z-A, Shi Y-T, et al. Efficient biohydrogen and bioelectricity production from xylose by microbial fuel cell with newly isolated yeast of Cystobasidium slooffiae. International Journal of Energy Research, **44,** 325-333(2020) https://doi.org/10.1002/er.4922
18. Anna S, Anna D, Aleksandra C, et al.: Anaerobic Digestion: I. A Common Process Ensuring Energy Flow and the Circulation of Matter in Ecosystems. II. A Tool for the Production of Gaseous Biofuels, Angela Faustino J, editor, Fermentation Processes, Rijeka: IntechOpen, **Ch. 14** (2017) http://dx.doi.org/10.5772/64645
19. Moret S, Peduzzi E, Gerber L, et al. Integration of deep geothermal energy and woody biomass conversion pathways in urban systems. Energy Conversion and Management, **129**, 305-318(2016) https://doi.org/10.1016/j.enconman.2016.09.079

20. Jiang Y, Mori T, Naganuma H, et al. Effect of the optimal combination of bituminous coal with high biomass content on particulate matter (PM) emissions during co-firing. Fuel, **316**, 123244(2022) https://doi.org/10.1016/j.fuel.2022.123244

AUTHOR INDEX